STRUCTURE OF LIQUID CRYSTAL PHASES

World Scientific Lecture Notes in Physics — Vol. 23

STRUCTURE OF
LIQUID CRYSTAL PHASES

P S PERSHAN

Division of Applied Science and
The Physics Department
Harvard University

World Scientific

Singapore • New Jersey • Hong Kong

Published by

World Scientific Publishing Co. Pte. Ltd.
P. O. Box 128, Farrer Road, Singapore 9128

U. S. A. office: World Scientific Publishing Co., Inc.
687 Hartwell Street, Teaneck NJ 07666, USA

Library of Congress Cataloging-in-Publication data is available.

STRUCTURE OF LIQUID CRYSTAL PHASES

ISBN 9971-50-668-8
     9971-50-705-6 pbk

Printed in Singapore by Kim Hup Lee Printing Co. Pte. Ltd.

DEDICATED TO MY WIFE

Patricia Barbara Pershan

Preface:

The text of this book was originally prepared to be a chapter in Volume B of the forthcoming edition of the INTERNATIONAL TABLES FOR CRYSTALLOGRAPHY and it will eventually appear as a chapter entitled: "Scattering From Mesomorphic Structures." I am grateful to Prof. Uri Shmueli of Tel Aviv University for inviting me to participate in this project on behalf of the INTERNATIONAL UNION OF CRYSTALLOGRAPHY. I would also like to express my gratitude to Prof. Carl Garland of MIT for reading a rough draft and suggesting that I have it published separately in a form that might result in a wider distribution. Others that I would like to thank for reading it and offering helpful comments include S. Chandrasekhar, J. Doucet, C. C. Huang, A. J. Leadbetter, A. M. Levelut, B. Ocko, R. Pindak, and E. Sirota.

I would also like to thank my several students and collaborators for their participation in the research program in our laboratory. Without their dedication I would never have become sufficiently knowledgeable to prepare this book. I am grateful to Ms. Lori Gevelinger for her help in typing, library research, and other matters involved in the preparation of this manuscript.

I owe a special debt of gratitude to Professors David Litster and Bob Birgeneau for their hospitality in allowing me to learn modern X-ray techniques at MIT during my sabbatical in the 1978-1979 academic year.

I would also like to express my appreciation to the National Science Foundation, and Dr. Fred Stafford in particular, for the support for my research program: Grant NSF-DMR-85-13523. The Harvard Materials Research Laboratory supplied the basic funding for the construction and continuing operation of the X-ray laboratory that was constructed following the MIT sabbatical with the help of L. Sorensen: current Grant NSF-DMR-86-14003. In the initial phase of this research, partial support was also provided by the Department of Defense through the Joint Services Electronics Program: current contract N0014-84-K-0465.

I am especially grateful to my wife for her support during the several years of travel to one or another synchrotron and for the many other days when I might as well have been far from home.

TABLE OF CONTENTS

Preface   v

PART ONE

1. Introduction   3

2. Nematic Phase   14

3. Smectic-A and Smectic-C Phases   25
   3.1. Homogeneous Smectic-A and Smectic-C Phases   25
   3.2. Modulated Smectic-A and Smectic-C Phases   35
   3.3. Surface Effects   36

4. Phases with In-Plane Order   42
   4.1. Hexatic Phases in Two Dimensions   47
   4.2. Hexatic Phases in Three Dimensions   52
      4.2.1. Hexatic-B   53
      4.2.2. Smectic-F, Smectic-I   54
   4.3. Crystalline Phases with Molecular Rotation   62
      4.3.1. Crystal-B   62
      4.3.2. Crystal-G, Crystal-J   73
   4.4. Crystalline Phases with Herringbone Packing   74
      4.4.1. Crystal-E   74
      4.4.2. Crystal-H, Crystal-K   77

5. Discotic Phases   78

6. Other Phases   82

7. References   86

PART TWO — SELECTED REPRINTS

1. Nematic to Smectic-A

   1.1. De Gennes, P. G. (1972). An analogy between superconductors
        and smectics A. *Solid State Commun.* **10**, 753–756   115

   1.2. McMillan, W. L. (1973). Measurement of smectic-A-phase order-
        parameter fluctuations near a second-order smectic-A-nematic-phase
        transition. *Phys. Rev.* **A7**, 1419–1422   119

1.3. Ocko, B. M., Birgeneau, R. J., Litster, J. D. and Neubert, M. E. (1984). Critical and tricritical behavior at the nematic to smectic-A transition. *Phys. Rev. Lett.* **52**, 208–211.   123

1.4. Chan, K. K., Deutsch, M., Ocko, B. M., Pershan, P. S. and Sorensen, L. B. (1985). Integrated x-ray scattering intensity measurement of the order parameter at the nematic to smectic-A phase transition. *Phys. Rev. Lett.* **54**, 920–923   127

2. Landau-Peierls Instability

2.1. Caillé, A. (1972). Remarques sur la diffusion des rayons X dans les smectiques A. *C. R. Acad. Sc. Paris* **274B**, 891–893*   133

2.2. Als-Nielsen, J., Litster, J. D., Birgeneau, R. J., Kaplan, M., Safinya, C. R., Lindegaard-Andersen, A. and Mathiesen, S. (1980). Observation of algebraic decay of positional order in a smectic liquid crystal. *Phys. Rev.* **B22**, 312–320   136

2.3. Safinya, C. R., Roux, D., Smith, G. S., Sinha, S. K., Dimon, P., Clark, N. A. and Bellocq, A. M. (1986). Steric interactions in a model multimembrane system: a synchrotron x-ray study. *Phys. Rev. Lett.* **57**, 2718–2721   145

3. Nematic, Smectic-A, Smectic C Multi-Critical Point

3.1. Chen, J. H. and Lubensky, T. C. (1976). Landau-Ginzberg mean-field theory for the nematic to smectic-C and nematic to smectic-A phase transitions. *Phys. Rev.* **A14**, 1202–1207   151

3.2. Martinez-Miranda, L. J., Kortan, A. R. and Birgeneau, R. J. (1986). X-ray study of fluctuations near the nematic-smectic-A-smectic-C multicritical point. *Phys. Rev. Lett.* **56**, 2264–2267   157

4. Multiple Smectic-A Phases

4.1. Levelut, A. M., Tarento, R. J., Hardouin, F., Achard, M. F. and Sigaud, G. (1981). Number of SA phases. *Phys. Rev.* **A24**, 2180–2186   163

4.2. Prost, J. (1984). The Smectic State. *Adv. Phys.* **33**, 1–46   170

* The exponent in Eq. 9b should read $4 - 2X$ rather than $2 - 2X$.

4.3. Wang, J. & Lubensky, T. C. (1984). Theory of the S_{A_1}-S_{A_2} phase transition in liquid crystal. *Phys. Rev.* **A29**, 2210–2217 — 216

4.4. Ratna, B. R., Shashidhar, R. and Raja, V. N. (1985). Smectic-A phase with two collinear incommensurate density modulations. *Phys. Rev. Lett.* **55**, 1476–1478 — 224

4.5. Chan, K. K., Pershan, P. S., Sorensen, L. B. and Hardouin, F. (1985). X-ray scattering study of the smectic-A_1 to smectic-A_2 transition. *Phys. Rev. Lett.* **54**, 1694–1697 — 227

4.6. Chan, K. K., Pershan, P. S., Sorensen, L. B. and Hardouin, F. (1986). X-ray studies of transitions between nematic, smectic-A_1, -A_2, and -A_d phases. *Phys. Rev.* **A34**, 1420–1433 — 231

5. In Plane Order-Hexatic Phases-Theory

5.1. Halperin, B. I. and Nelson, D. R. (1978). Theory of two-dimensional melting. *Phys. Rev. Lett.* **41**, 121–124. [Erratum: (1978) *Phys. Rev. Lett.* **41**, 519] — 247

5.2. Birgeneau, R. J. and Litster, J. D. (1978). Bond orientational order model for smectic B liquid crystals. *J. Phys. (Paris) Lett.* **39**, L399–L402 — 251

5.3. Nelson, D. R. and Halperin, B. I. (1980). Solid and fluid phases in smectic layers with tilted molecules. *Phys. Rev.* **B21**, 5312–5329 — 255

6. In Plane Order-Experiment

6.1. Levelut, A. M. (1976). Étude de l' ordre local lié a la rotation des molécules dans la phase smectique B. *J. Phys. (Paris)* **37**, C3-51-C3-54 — 275

6.2. Leadbetter, A. J., Frost, J. C. and Mazid, M. A. (1979). Interlayer correlations in smectic B phases. *J. Phys. (Paris) Lett.* **40**, L325–L329 — 279

6.3. Moncton, D. E. and Pindak, R. (1979). Long-range order in two- and three-dimensional smectic-B liquid-crystal films. *Phys. Rev. Lett.* **43**, 701–704 — 284

6.4. Aeppli, G., Litster, J. D., Birgeneau, R. J. and Pershan, P. S. (1981). High resolution x-ray study of the smectic A-smectic B phase transition and the smectic B phase in butyloxybenzylidene octylaniline. *Mol. Cryst. Liq. Cryst.* **67**, 205–214 — 288

6.5. Pindak, R., Moncton, D. E., Davey, S. C. and Goodby, J. W. (1981). X-ray observation of a stacked hexatic liquid-crystal B phase. *Phys. Rev. Lett.* **46** 1135–1138 — 298

6.6. Collett, J., Sorensen, L. B., Pershan, P. S., Litster, J., Birgeneau, R. J. and Als-Nielsen, J. (1982). Synchrotron x-ray study of novel crystalline-B phases in heptyloxybenzylidene-heptylanaline (70.7). *Phys. Rev. Lett.* **49**, 553–556 — 302

6.7. Hirth, J. P., Pershan, P. S., Collett, J., Sirota, E. and Sorensen, L. B. (1984). Dislocation model for restacking phase transitions in crystalline-B liquid crystals. *Phys. Rev. Lett.* **53**, 473–476 — 306

6.8. Collett, J., Sorensen, L. B., Pershan, P. S. and Als-Nielsen, J. (1985). X-ray scattering study of restacking transitions in the crystalline-B phases of heptyloxybenzylidene-heptylanaline (70.7). *Phys. Rev.* **A32**, 1036–1043 — 310

6.9. Sirota, E. B., Pershan, P. S., Sorensen, L. B. and Collett, J. (1987). X-ray and optical studies of the thickness dependence of the phase diagram of liquid crystal films. *Phys. Rev.* **A36**, 2890–2901 — 318

7. Tilted Hexatic Phases

7.1. Benattar, J. J., Doucet, J., Lambert, M. and Levelut, A. M. (1979). Nature of the smectic F phase. *Phys Rev.* **A20**, 2505–2509 — 333

7.2. Collett, J., Pershan, P. S., Sirota, E. B. & Sorensen, L. B. (1984). Synchrotron x-ray study of the thickness dependence of the phase diagram of thin liquid-crystal films. *Phys. Rev. Lett.* **52**, 356–359 — 338

7.3. Sirota, E. B., Pershan, P. S., Sorensen, L. B. and Collett, J. (1985). X-ray studies of tilted hexatic phases in thin liquid-crystal films. *Phys. Rev. Lett.* **55**, 2039–2042 — 342

7.4. Aharony, A., Birgeneau, R. J., Brock, J. D. and Litster, J. D. (1986). Multicriticality in hexatic liquid crystals. *Phys. Rev. Lett.* **57**, 1012–1015 — 346

7.5. Brock, J. D., Aharony, A., Birgeneau, R. J., Evans-Lutterodt, K. W., Litster, J. D., Horn, P. M., Stephenson, G. B. and Tajbakhsh, A. R. (1986). Orientational and positional order in a tilted hexatic liquid-crystal phase. *Phys. Rev. Lett.* **57**, 98–101 — 350

8. Smectic-D

8.1. Diele, S., Brand, P. and Sackmann, H. (1972). X-ray diffraction and polymorphism of smectic liquid crystals. II. D and E modifications. *Mol. Cryst. Liq. Cryst.* **17**, 163–169   357

8.2. Tardieu, A. and Billard, J. (1976). On the structure of the "smectic D modification". *J. Phys. (Paris) Colloq.* **37**, C3-79-C3-81   364

8.3. Etherington, G., Leadbetter, A. J., Wang, X. J., Gray, G. W. and Tajbakhsh, A. (1986). Structure of the smectic D phase. *Liquid Crystals*, **1**, 209–214   367

9. Structure of Surfaces

9.1. Pershan, P. S. and Als-Nielsen, J. (1984). X-ray reflectivity from the surface of a liquid crystal: surface structure and absolute value of critical fluctuations. *Phys. Rev. Lett.* **52**, 759–762   375

9.2. Ocko, B. M., Braslau, A., Pershan, P. S., Als-Nielsen, J. and Deutsch, M. (1986). Quantized layer growth at liquid-crystal surfaces. *Phys. Rev. Lett.* **57**, 94–97   379

9.3. Pershan, P. S., Braslau, A., Weiss, A. H. and Als-Nielsen, J. (1987). Smectic layering at the free surface of liquid crystals in the nematic phase: X-ray reflectivity. *Phys. Rev.* **A35**, 4800–4813   383

10. Discotic Phases

10.1. Chandrasekhar, S., Sadashiva, B. K. and Suresh, K. A. (1977). Liquid crystals of disk like molecules. *Pramana* **9**, 471–480   399

10.2. Levelut, A. M. (1983). Structures des phases mesomorphes formées de molecules discoïdes. *J. Chim. Phys.* **80**, 149–161   409

10.3. Safinya, C. R., Liang, K. S., Varady, W. A., Clark, N. A. and Andersson, G. (1984). Synchrotron x-ray study of the orientational ordering D2-D1 structural phase transition of freely suspended discotic strands in triphenylene hexa-n-dodecanoate. *Phys. Rev. Lett.* **53**, 1172–1175   422

PART ONE

1. Introduction:

The term mesomorphic is derived from the prefix meso-
that is defined in the dictionary as "a word element meaning
middle;" and the term "-morphic" that is defined as "an ad-
jective termination corresponding to morph or form. Thus
mesomorphic order implies some "form," or order, that is "in
the middle," or intermediate between that of liquids and
crystals. The name liquid crystalline was coined by re-
searchers who found it to be more descriptive, and the two
are used synonymously. It follows that a mesomorphic, or
liquid crystalline phase, must have more symmetry than any
one of the 230 space groups that characterize crystals.

A major source of confusion in the early liquid crystal
literature had to do with the fact that many of the molecules
that form liquid crystals also form true three dimensional
crystals with diffraction patterns that are only subtly
different from those of other liquid crystalline phases.
Since most of the original mesomorphic phase identifications
were done using a "miscibility" procedure, that depends on
optically observed changes in textures accompanying variation
in the samples chemical composition, it is not surprising
that some three dimensional crystalline phase were mistakenly
identified as mesomorphic. Phases were identified as being
either the same as, or different than, phases that were
previously observed (Liebert, 1978; Gray & Goodby, 1984),

and although many of the workers were very clever in deducing
the microscopic structure responsible for the microscopic
textures, the phases were labeled in the order of discovery
as "smectic-A, smectic-B, etc." without any attempt to
develop a systematic nomenclature that would reflect the
underlying order. Although different groups did not always
assign the same letters to the same phases, the problem is
now resolved and the assignments used in this article are
commonly accepted (Gray & Goodby, 1984).

Figure 1 illustrates the way in which increasing order
can be assigned to the series of mesomorphic phases in three
dimension listed in Table 1a. Although the phases in this
series are the most thoroughly documented mesomorphic phases,
there are others not included in the table that we will
discuss below.

The progression from the completely symmetric isotropic
liquid through the mesomorphic phases into the crystalline
phases can be described in terms of three separate types of
order. The first, or the molecular orientational order,
describes the fact that the molecules have some preferential
orientation analogous to the spin orientational order of
ferromagnetic materials. In the present case the molecular
quantity that is oriented is a symmetric second rank tensor,
like the moment of inertia or the electric polarizability,
rather than a magnetic moment. This is the only type of long
range order in the nematic phase, and as a consequence its

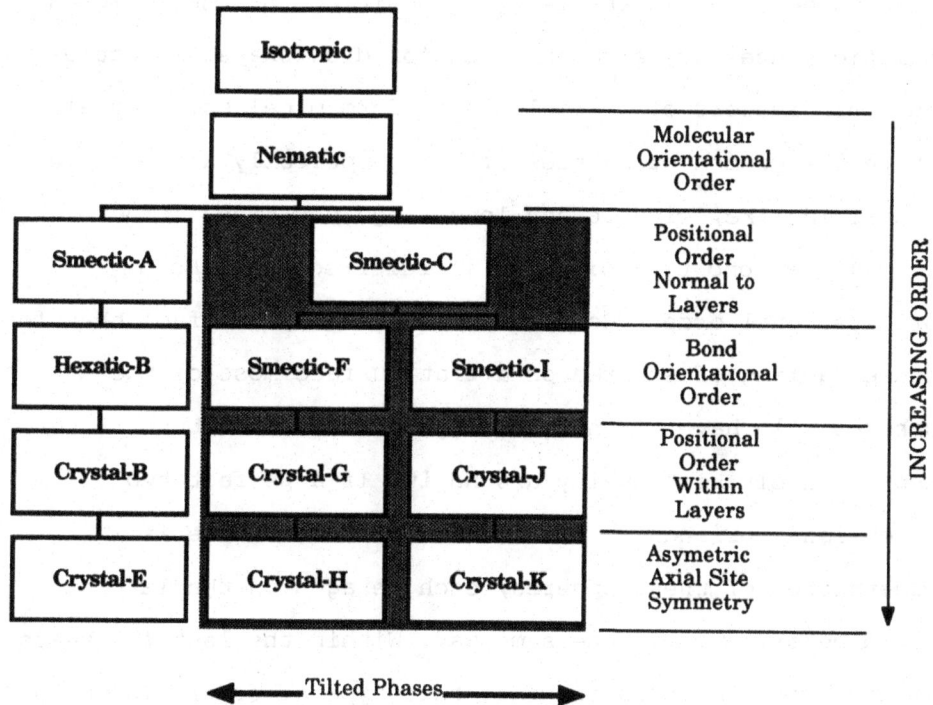

Figure 1: Illustration of the progression of order throughout the sequence of mesomorphic phases that are based on "rod like" molecules. The cross-hatched section indicates phases in which the molecules are tilted with respect to the smectic layers.

physical properties are those of an anisotropic fluid and this is the origin of the name liquid crystal. Figure 2a is a schematic illustration of the nematic order assuming the molecules can be represented by oblong ellipses. The average orientation of the ellipses is aligned; however there is no long range order in the relative positions of the ellipses. Nematic phases are also observed for disk shaped molecules and for clusters of molecules that form micelles. They all share the common properties of being optically anisotropic and fluid-like, without any long range positional order.

The second type of order is referred to as bond orientational order. Consider, for example, the fact that for dense packing of spheres on a flat surface most of the spheres will have six neighboring spheres distributed approximately hexagonally around it. If a perfect two dimensional triangular lattice of indefinite size is constructed of these spheres, each hexagon on the lattice would be oriented in the same way. Within the last few years we have come to recognize that this type of order, in which the hexagons are everywhere parallel to one another, is possible even when there is not a lattice. This type of order is referred to as bond orientational order, and bond orientational order in the absence of a lattice is the essential property defining the hexatic phases (Halperin & Nelson, 1978; Nelson & Halperin, 1979; Young, 1979; Birgeneau & Litster, 1978).

The third type of order is the positional order of an indefinite lattice of the type that defines the 230 space groups of conventional crystals. In view of the fact that some of the mesomorphic phases have a layered structure, it

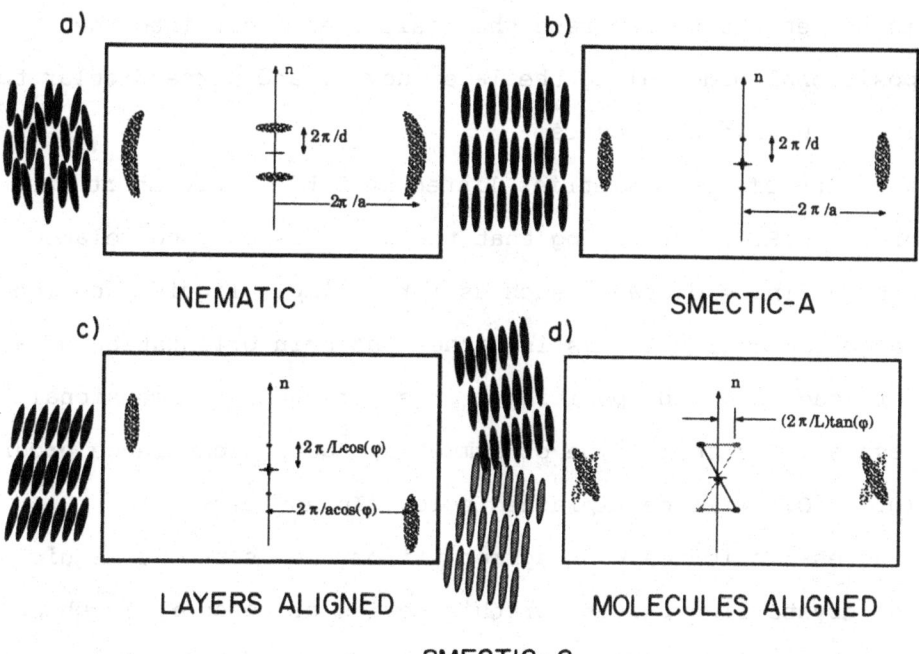

Figure 2: Schematic illustration of the real space molecular order and the scattering cross sections in reciprocal space for the: (a) nematic; (b) smectic-A; and (c, d) smectic-C phases. The scattering cross sections are enclosed in the boxes. Figure (c) indicates the smectic-C phase for an oriented monodomain and (d) indicates for a polydomain smectic-C structure in which the molecular axes are aligned.

is convenient to separate the positional order into the positional order along the layer normal and perpendicular to it, or within the layers.

Two of the symmetries listed in Table 1 are short range order (S.R.O.), implying that the order is only correlated over a finite distance such as for a simple liquid, and long range order (L.R.O.) as in either the spin orientation of a ferromagnet or the positional order of a three dimensional crystal. The third type of symmetry, "quasi long range order" (Q.L.R.O.) will be explained below. In any case, the progressive increase in symmetry from the isotropic liquid to the crystalline phases for this series of mesomorphic phases is illustrated in Figure 1. One objective of this chapter is to describe the reciprocal space structure of the phases listed in the table and the phase transitions between them.

Table 1a: Some of the symmetry properties of the series of
three dimensional phases described in Figure 1. The terms
L.R.O. and S.R.O. imply long range or short range order
respectively and Q.L.R.O. refers to "quasi long range order" as
explained in the text.

Phase	Molecular Orientation Order Within Layer	Bond Orientation Order	Positional Order Normal Within Layer	
Smectic-A (SmA)	S.R.O.	S.R.O.	S.R.O.	S.R.O.
Smectic-C (SmC)	L.R.O.	L.R.O.*	S.R.O.	S.R.O.
hexatic-B	L.R.O.*	L.R.O.	Q.L.R.O.	S.R.O.
Smectic-F (SmF)	L.R.O.	L.R.O.	Q.L.R.O.	S.R.O.
Smectic-I (SmI)	L.R.O.	L.R.O.	Q.L.R.O.	S.R.O.
Crystalline-B (CrB)	L.R.O.	L.R.O.	L.R.O.	L.R.O.
Crystalline-G (CrG)	L.R.O.	L.R.O.	L.R.O.	L.R.O.
Crystalline-J (CrJ)	L.R.O.	L.R.O.	L.R.O.	L.R.O.
Crystalline-E (CrE)	L.R.O.	L.R.O.	L.R.O.	L.R.O.
Crystalline-H (CrH)	L.R.O.	L.R.O.	L.R.O.	L.R.O.
Crystalline-K (CrK)	L.R.O.	L.R.O.	L.R.O.	L.R.O.

* Theoretically the existence of L.R.O. in the molecular
orientation, or tilt, implies that there must be some L.R.O.
in the bond orientation and visa versa.

Table 1b: The symmetry properties of the **two dimensional** hexatic and crystalline phases.

Phase	Molecular Orientation Order Within Layer	Bond Orientation Order	Positional Order Within Layer
Smectic-A (SmA)	S.R.O.	S.R.O.	S.R.O.
Smectic-C (SmC)	Q.L.R.O.	Q.L.R.O.	S.R.O.
hexatic-B	Q.L.R.O.	Q.L.R.O.	S.R.O.
Smectic-F (SmF)	Q.L.R.O.	Q.L.R.O.	S.R.O.
Smectic-I (SmI)	Q.L.R.O.	Q.L.R.O.	S.R.O.
Crystalline-B (CrB)	L.R.O.	L.R.O.	Q.L.R.O.
Crystalline-G (CrG)	L.R.O.	L.R.O.	Q.L.R.O.
Crystalline-J (CrJ)	L.R.O.	L.R.O.	Q.L.R.O.
Crystalline-E (CrE)	L.R.O.	L.R.O.	Q.L.R.O.
Crystalline-H (CrH)	L.R.O.	L.R.O.	Q.L.R.O.
Crystalline-K (CrK)	L.R.O.	L.R.O.	Q.L.R.O.

Finally, in most of the crystalline phases that we want to discuss, the molecules have considerable amounts of rotational disorder. For example, one series of molecules that form mesomorphic phases consists of long thin molecules that might be described as "blade shaped." Although the cross section of these molecules is quite anisotropic the site symmetry of the molecule is often symmetric, as though the molecule is rotating freely about its long axis. On cooling, many of the mesomorphic systems undergo transitions to the phases, listed at the bottom of Figure 1, for which the site symmetry is anisotropic as though some of the rotational motions about the molecular axis have been frozen out. A similar type of transition, in which rotational motions are frozen out, occurs on cooling systems such as succinonitrile ($NCCH_2CH_2CN$) that form optically isotropic "plastic crystals" (Springer, 1977).

There are two broad classes of liquid crystalline systems, the thermotropic and the lyotropic, and since the former are much better understood, this chapter will emphasize results on thermotropic systems (Liebert, 1978). The historical difference between these two, and also the origin of their names, is that the lyotropic are always mixtures, or solutions, of unlike molecules in which one is a normal, or nonmesogenic liquid. Solutions of soap and water are prototypical examples of lyotropics, and their mesomorphic phases appear as a function of either concentration or temperature. In contrast the thermotropic systems are usually formed from a single chemical component, and the mesomorphic phases appear primarily as a function of temperature changes. The molecular distinction between the two is that one of the molecules in the lyotropic solution always has a hydrophilic part, often called the "head group," and one or more hydrophobic alkane chains called "tails." These molecules will often form mesomorphic phases as single component or neat systems; however, the general belief is that in solution with either water or oil most of the phases are the result of competition between the hydrophilic and hydrophobic interactions, as well as other things such as packing and steric constraints (Pershan, 1979; Safran & Clark, 1987). To the extent that molecules that form thermotropic liquid crystalline phases have hydrophilic and hydrophobic parts, the disparity in the affinity of these parts for either water or oil is much less and most of these molecules are relatively insoluble in water. These molecules are called thermotropic because their phase transformations are primarily studied as a function of only temperature. This is not to say that there are not numerous examples of interesting studies of the concentration dependence of phase diagrams involving mixtures of thermotropic liquid crystals.

Figure 3 displays some common examples of molecules that form lyotropic and thermotropic phases. In spite of the above remarks it is interesting to observe that different parts of typical thermotropic molecules do have some of the same features as the lyotropic molecules. For example, although the rod-like thermotropic molecules always have an alkane chain at one or both ends of a more rigid section, the chain lengths are rarely as long as those of the lyotropic molecules, and although the solubility of the parts of the thermotropic molecules, when separated, are not as disparate as those of the lyotropic molecules, they are definitely different. We suspect that this may account for the subtler features of the phase transformations between the mesomorphic phases to be discussed below. On the other hand the

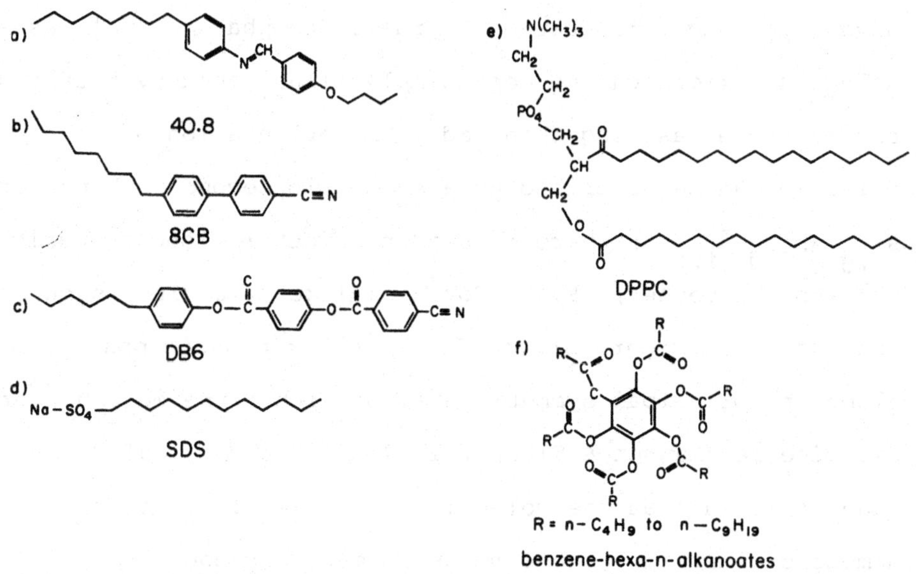

Figure 3: Chemical formulas for some of the molecules that form thermotropic liquid crystals; (a) n-(4-(n-butyl)oxybenzylidene)-4'-(n-octyl)aniline(40.8); (b) 4-cyano-4'-n-octylbiphenol (8CB); (c) hexylphenyl cyanobenzyloxy benzoate (DB$_6$): lyotropic liquid crystals; (d) sodium dodecyl sulfate; (e) 1,2-dipalmitoyl-L-phosphatidylcholine (DPPC); and a discotic liquid crystal (f) benzene-hexa-n-alkanoate.

inhomogeneity of the molecule is probably not important for the nematic phase.

2. Nematic Phase:

The nematic phase is a fluid for which the molecules have long range orientational order. The phase as well as its molecular origin can be most simply illustrated by treating the molecules as long thin rods. The orientation of each molecule can be described by a symmetric second rank tensor $s_{i,j} = (n_i n_j - \delta_{i,j}/3)$ where n is a unit vector along the axis of the rod (De Gennes, 1974). For disk-like molecules, such as that shown in Figure 3f, or for micellar nematic phases, n is along the principal symmetry axis of either the molecule or the micelle (Lawson & Flautt, 1967). Since physical quantities such as the molecular polarizability, or the moment of inertia, transform as symmetric second rank tensors, either one of these could be used as specific representations of the molecular orientational order. The macroscopic order, however, is given by the statistical average $S_{i,j} = <s_{i,j}> = S(<n_i><n_j> - \delta_{i,j}/3)$, where $<n>$ is a unit vector along the macroscopic symmetry axis and S is the order parameter of the nematic phase.

The microscopic origin of the phase can be understood in terms of steric constraints that occur on filling space with highly asymmetric objects such as long rods or flat disks. Maximizing the density requires some degree of short range orientational order, and theoretical arguments can be invoked to demonstrate long range order. Onsager presented quantitative arguments of this type to explain the nematic order observed in concentrated solutions of the long thin rods of tobacco mosaic viruses (Onsager, 1949; Lee & Meyer, 1986), and qualitatively similar ideas explain the nematic order for the shorter thermotropic molecules (Maier & Saupe, 1958; Maier & Saupe, 1959).

The existence of nematic order can also be understood in terms of a phenomenological mean field theory (De Gennes, 1969b; De Gennes, 1971; Fan & Stephen, 1970). If the free energy difference ΔF between the isotropic and nematic phases can be expressed as an analytic function of the nematic order parameter $S_{i,j}$ one can expand $\Delta F(S_{i,j})$ as a power series in which the successive terms all transform as the identity representation of the point group of the isotropic phase, i.e. as scalers. The most general form is given by:

Eq. 1

$$\Delta F(S_{i,j}) = \frac{A}{2}\sum_{ij}S_{ij}S_{ji} + \frac{B}{3}\sum_{ijk}S_{ij}S_{jk}S_{ki} + \frac{D}{4}\left|\sum_{ij}S_{ij}S_{ji}\right|^2 + \frac{D'}{4}\sum_{ijkl}S_{ij}S_{jk}S_{kl}S_{li}$$

The usual mean field treatment assumes that the coefficient of the leading term is of the form $A = a(T-T^*)$ where T is the absolute temperature and T^* is the temperature at which A=0. Taking a, D and D'>0 one can show that for either positive or negative values of B, but for sufficiently large T, the minimum value of $\Delta F=0$ occurs for $S_{i,j}=0$ corresponding to the isotropic phase. For $T<T^*$, ΔF can be minimized, at some negative value, for a non-zero $S_{i,j}$ corresponding to nematic order. The details of how this is worked out for a tensorial order parameter can be found in the literature (De Gennes, 1974); however the basic idea can be understood by treating $S_{i,j}$ as a scaler. If we write

Eq. 2

$$\Delta F = \frac{1}{2}AS^2 + \frac{1}{3}BS^3 + \frac{1}{4}DS^4 = \left\{\frac{A}{2} - \frac{B^2}{9D}\right\} S^2 + \frac{D}{4}\left(S + \frac{2B}{3D}\right)^2 S^2$$

and if T_{NI} is defined by the condition $A = a \cdot (T_{NI} - T^*) = 2B^2/9D$, $\Delta F = 0$ for both $S = 0$ and $S = -2B/3D$. This value for T_{NI} marks the transition temperature from the isotropic phase, when $A > 2B^2/9D$ and the only minimum is at $S = 0$ with $\Delta F = 0$, to the nematic case when $A < 2B^2/9D$ and the absolute minimum with $\Delta F < 0$ is slightly shifted from $S = -2B/3D$. The symmetry properties of second rank tensors imply that there will usually be a nonvanishing value for B, and this implies that the transition from the isotropic to nematic transition will be first order with a discontinuous jump in the nematic order parameter $S_{i,j}$. Although most nematic systems are uniaxial, biaxial nematic order is theoretically possible (Freiser, 1971; Alben, 1973; Lubensky, 1987) and it has been observed in certain lyotropic nematic liquid crystals (Neto, Galerne, Levelut & Liebert, 1985; Hendrikx, Charvolin & Rawiso, 1986; Yu & Saupe, 1980) and in one thermotropic system (Malthête, Liébert, Levelut & Galerne, 1986).

The X-ray scattering cross section of an oriented monodomain sample of the nematic phase with rod-like molecules usually exhibits a diffuse spot like that illustrated in Figure 2a where the maximum of the cross section is along the average molecular axis <**n**> at a value of $|\mathbf{q}| \approx 2\pi/d$, where $d \approx 20.0$Å to 40.0Å is of the order of the molecular length L. This a precursor to the smectic-A order that develops at lower temperatures for many materials. In addition there is a diffuse ring along the directions normal to <**n**> at $|\mathbf{q}| \approx 2\pi/a$, where $a \approx 4.0$Å is comparable to the average radius of the molecule. In some nematic systems the near neighbor correlations favor antiparallel alignment and molecular centers tend to form pairs such that the peak of the scattering cross section can actually have values anywhere in the range from $2\pi/L$ to $2\pi/2L$. There are also other cases where there are two diffuse peaks, corresponding to both $|\mathbf{q}_1| \approx 2\pi/L$ and $|\mathbf{q}_2| \approx |\mathbf{q}_1|/2$ that are precursors of a richer smectic-A morphology (Prost & Barois, 1983; Prost, 1984; Sigaud, Hardouin, Achard & Gasparoux, 1979; Wang & Lubensky, 1984; Hardouin, Levelut, Achard & Sigaud, 1983; Chan, Pershan, Sorensen & Hardouin, 1985b). In some cases $|\mathbf{q}_2| \neq \frac{1}{2}|\mathbf{q}_1|$ and competition between the order parameters at incommensurate wavevectors gives rise to modulated phases. For the moment we will restrict the discussion to those systems for which the order parameter is characterized by a single wavevector.

On cooling many nematic systems undergo a 2nd order phase transition to a smectic-A phase and as the temperature approaches the nematic to smectic-A transition the widths of these diffuse peaks become infinitesimally small. In 1972 De Gennes demonstrated that this phenomena could be understood in analogy with either the transitions from normal fluidity to superfluidity in liquid helium, or normal conductivity to superconductivity in metals (De Gennes, 1972). Since the electron density of the smectic-A phase is (quasi-)periodic in one dimension he represented it by the form:

$$\rho(\mathbf{r}) = \langle\rho\rangle + \text{Re}\left\{ \Psi \cdot e^{i(2\pi/d)z} \right\}$$

where d is the thickness of the smectic layers lying in the x-y plane. The complex quantity $\psi = |\psi|\exp(i\phi)$ is similar to the superfluid wave function except that in this analogy the amplitude $|\psi|$ describes the electron density variations normal to the smectic layers, and the phase ϕ describes the position of the layers along the z-axis. De Gennes proposed a mean field theory for the transition in which the free energy difference between the nematic and smectic-A phase $\Delta F(\Psi)$ was represented by

Eq. 3

$$\Delta F(\psi) = \frac{A}{2}|\psi|^2 + \frac{D}{4}|\psi|^4 + \frac{E}{2}\left[\left|\left\{\frac{\partial}{\partial z} - i\left(\frac{2\pi}{d}\right)\right\}\psi\right|^2 + \left(\frac{\partial\psi}{\partial x}\right)^2 + \left(\frac{\partial\psi}{\partial y}\right)^2 \right]$$

This mean field theory differs from the one for the isotropic to nematic transition in that the symmetry for the latter allowed a term that was cubic in the order parameter, while no such term is allowed for the nematic to smectic-A transition. In both cases, however, the coefficient of the leading term is taken to have the form $a(T-T^*)$. If $D>0$, without the cubic term the free energy has only one minimum when $T>T*$ at $|\psi|=0$, and two equivalent minima at $|\psi|=\{a(T^*-T)/D\}^{0.5}$ for $T<T^*$. On the basis of this free energy the nematic to smectic-A transition can be second order with a transition temperature $T_{NA}=T^*$ and an order parameter that varies as the square root of $(T_{NA}-T)$. There are conditions that we will not discuss in detail when D can be negative. In that case the nematic to smectic-A transition will be first order (McMillan, 1972; McMillan, 1973a; McMillan, 1973b; McMillan 1973c). McMillan pointed out by allowing coupling between the smectic and nematic order parameters a more general free energy can be developed in which D is negative. McMillan's prediction that for systems in which the difference $T_{IN}-T_{NA}$ is small the nematic to smectic-A transition will be first order is supported by experiment (Ocko, Birgeneau & Litster, 1986; Ocko, Birgeneau, Litster & Neubert, 1984; Thoen, Marynissen & Van Dael, 1984). Although the mean field theory is not quantitatively accurate, it does explain the principal qualitative features of the nematic to smectic-A transition.

The differential scattering cross section for X-rays can be expressed in terms of the Fourier transform of the density-density correlation function $\langle\rho(\mathbf{r})\rho(0)\rangle$. The expectation value is calculated from the thermal average of the order parameter that is obtained from the free energy density $\Delta F(\psi)$. On taking the transform

Eq. 4

$$\Psi(Q) \equiv \frac{1}{(2\pi)^3} \int d^3 r e^{i(Q \cdot r)} \rho(r)$$

the free energy density in reciprocal space has the form

Eq. 5

$$\Delta F(\psi) = \frac{A}{2}|\psi|^2 + \frac{D}{4}|\psi|^4 + \frac{E}{2}\left[\{Q_z - (2\pi/d)\}^2 + Q_x^2 + Q_y^2 \right] |\psi|^2$$

and one can show that for $T > T_{NA}$ the cross section obtained from the above form for the free energy is

Eq. 6

$$\frac{d\sigma}{d\Omega} \approx \frac{\sigma_0}{A + E[(Q_z - (2\pi/d))^2 + Q_x^2 + Q_y^2]}$$

where the term in $|\psi|^4$ has been neglected. The mean field theory predicts that the peak intensity should vary as $\sigma_0/A \approx 1/(T - T_{NA})$ and that the half width of the peak in any direction should vary as $(A/E)^{1/2} \approx (T - T_{NA})^{1/2}$. The physical interpretation of the half width is that the smectic fluctuations in the nematic phase are correlated over lengths $\xi = \sqrt{(E/A)} \sim (T - T_{NA})^{-1/2}$.

One of the major shortcomings of all mean field theories is that they do not take into account the difference between the average value of the order parameter $\langle\psi\rangle$ and the instantaneous value $\psi \equiv \langle\psi\rangle + \delta\psi$, where $\delta\psi$ represents the thermal fluctuations (Ma, 1976). The usual effect expected from theories for this type of critical phenomena is a "renormalization" of the various terms in the free energy such that the temperature dependence of correlation length has the form $\xi(t) \propto t^{-\nu}$, where $t \equiv (T-T^*)/T^*$, $T^* \equiv T_{NA}$ is the 2nd order transition temperature and ν is expected to have some universal value that is generally not equal to 0.5. One of the major unsolved problems of the nematic to smectic-A phase transition is that the width along the scattering vector $\mathbf{q}$ varies as $1/\xi_{//} \propto t^{\nu_{//}}$ with a different temperature dependence than the width perpendicular to $\mathbf{q}$, $1/\xi_{\perp} \propto t^{\nu_{\perp}}$ and that neither $\nu_{//}$ nor $\nu_{\perp}$ have the expected universal values (Lubensky, 1983; Nelson & Toner, 1981).

The correlation lengths are measured by fitting the differential scattering cross sections to the empirical form:

Eq. 7

$$\frac{d\sigma}{d\Omega} = \frac{\sigma}{1+(Q_z-|\mathbf{q}|)^2\xi_{//}^2 + Q_\perp^2\xi_\perp^2 + c(Q_\perp^2\xi_\perp^2)^2}$$

The amplitude $\sigma \propto t^{-\gamma}$ where the measured values of γ are empirically found to be very close to the measured values for the sum $\nu_{//}+\nu_{\perp}$. Most of the systems that have been measured to date have values for $\nu_{//}>0.66>\nu_{\perp}$ and $\nu_{//} - \nu_{\perp} \approx 0.1$ to 0.2. Table 2 lists sources of the observed values for γ, $\nu_{//}$ & $\nu_{\perp}$. The theoretical and experimental studies of this pretransition effect account for a sizable fraction of all of the liquid crystal research in the last ten or fifteen years, and as of this writing the explanation for these two different temperature dependences remains one of the major unresolved theoretical questions in equilibrium statistical physics.

Table 2: Summary of critical exponents from
X-ray scattering studies of the nematic to
smectic-A phase transition.

Molecule	γ	$\nu_{//}$	$\nu_\perp$	ref
40.7	1.46	0.78	0.65	a
$\bar{8}$S5	1.53	0.83	0.68	b, g
CBOOA	1.30	0.70	0.62	c, d
40.8	1.31	0.70	0.57	e
80CB	1.32	0.71	0.58	d, f
$\bar{9}$S5	1.31	0.71	0.57	b, g
8CB	1.26	0.67	0.51	h, i
1$\bar{0}$S5	1.10	0.61	0.51	b, g
9CB	1.10	0.57	0.39	g, j

References:
a. Garland, Meichle, Ocko, Kortan, Safinya, Yu,
Litster & Birgeneau, 1983
b. Brisbin, De Hoff, Lockhart & Johnson, 1979
c. Djurek, Baturic-Rubcic & Franulovic, 1974
d. Litster, Als-Nielsen, Birgeneau, Dana, Davidov,
Garcia-Golding, Kaplan, Safinya & Schaetzing, 1979
e. Birgeneau, Garland, Kasting & Ocko, 1981
f. Kasting, Lushington & Garland, 1980
g. Ocko, Birgeneau, Litster & Neubert, 1984
h. Thoen, Marynissen & Van Dael, 1982
i. Davidov, Safinya, Kaplan, Dana, Schaetzing,
Birgeneau & Litster, 1979
j. Thoen, Marynissen & Van Dael, 1984

It is very likely that the origin of the problem is the
Q.L.R.O. in the position of the smectic layers. Lubensky
attempted to deal with this by introducing a gauge
transformation in such a way that the thermal fluctuations of
the transformed order parameter did not have the logarithmic
divergence. While this approach has been informative, it has
not yet yielded an agreed upon understanding. Experimentally,
the effect of the phase can be studied in systems where there
are two competing order parameters with wavevectors that are
at q_2 and $q_1 \approx 2 \, q_2$ (Sigaud, et al., 1979; Hardouin, et al.,
1983; Prost & Barois, 1983; Prost, 1984; Wang & Lubensky,
1984; Chan, et al., 1985b). On cooling, mixtures of
hexylphenyl cyanobenzyloxy benzoate (DB_6) and terephthal-bis-
butylaniline (TBBA) first undergo a second order transition
from the nematic to a phase that is designated as smectic-A_1.
The various smectic-A and smectic-C morphologies will be
described in more detail in the following section; however
the smectic-A_1 phase is characterized by a single peak at $q_1 = 2\pi/d$ due to a one dimensional density wave with wavelength d
of the order of the molecular length L. In addition, however,
there are thermal fluctuations of a second order parameter
with a period of 2L that give rise to a diffuse peak at $q_2 = \pi/L$. On further cooling this system undergoes a second 2nd
order transition to a smectic-A_2 phase with Q.L.R.O. at $q_2 \approx \pi/L$, with a second harmonic that is exactly at $q = 2q_2 \approx 2 \, \pi/L$.
The critical scattering on approaching this transition is

similar to that of the nematic to smectic-A_1, except that the preexisting density wave at $q_1 = 2\pi/L$ quenches the phase fluctuations of the order parameter at the subharmonic $q_2 = \pi/L$. The measured values of $\nu_{//} = \nu_{\perp} \approx 0.74$ (Chan, et al., 1985b) agree with those expected from the appropriate theory (Huse, 1985). A mean field theory that describes this effect is discussed in section 3.2 below.

It is interesting to note that even those systems for which the nematic to smectic-A transition is first order show some pretransitional lengthening of the correlation lengths $\xi_{//}$ and $\xi_{\perp}$. In these cases the apparent T^* at which the correlation lengths would diverge is lower than T_{NA} and the divergence is truncated by the first order transition (Ocko, et al., 1984).

3. Smectic-A and Smectic-C Phases:

3.1. Homogeneous Smectic-A and Smectic-C Phases

In the smectic-A and smectic-C phases the molecules organize themselves into layers, and from a naive point of view one might describe them as forming a one dimensional periodic lattice in which the individual layers are two dimensional liquids. In the smectic-A phase the average molecular axis <n> is normal to the smectic layers while for the smectic-C it makes a finite angle. It follows from this that the smectic-C phase has lower symmetry than the smectic-A, and the phase transition from the smectic-A to smectic-C can be considered as the ordering of a two component order parameter, i.e. the two components of the projection of the molecular axis on the smectic layers (De Gennes, 1973). Alternatively, Chen and Lubensky have developed a mean field theory in which the transition is described by a free energy density of the Lifshitz form (Chen & Lubensky, 1976). This will be described in more detail below, however it corresponds to replacing Eq. 5 for the free energy $\Delta F(\psi)$ by an expression for which the minimum is obtained when the wavevector $\mathbf{q}$, of the order parameter $\psi \propto \exp[i\mathbf{q} \cdot \mathbf{r}]$, tilts away from the molecular axis.

The X-ray cross section for the prototypical aligned monodomain smectic-A sample is shown in Figure 2b. It consists of a single sharp spot along the molecular axis at

|**q**| somewhere between $2\pi/2L$ and $2\pi/L$ that reflects the Q.L.R.O. along the layer normal, and a diffuse ring in the perpendicular direction at |**q**|$\approx 2\pi/a$ that reflects the S.R.O. within the layer. The scattering cross section for an aligned smectic-C phase is similar to that of the smectic-A except that the molecular tilt alters the intensity distribution of the diffuse ring. This is illustrated in Figure 2c for a monodomain sample. Figure 2d illustrates the scattering pattern for a polydomain smectic-C sample in which the molecular axis remained fixed, but where the smectic layers are randomly distributed azymuthally around the molecular axis.

The naivety of describing these as periodic stacks of 2-dimensional liquids derives from the fact that the sharp spot along the molecular axis has a distinct temperature dependent shape indicative of Q.L.R.O. that distinguishes it from the Bragg peaks due to true L.R.O. in conventional three dimensional crystals. Landau and Peierls discussed this effect for the case of two dimensional crystals (Landau, 1965; Peierls, 1934) and Caillé extended the argument to the mesomorphic systems (Caillé, 1972).

The usual treatment of thermal vibrations in three dimensional crystals estimates the Debye-Waller factor by integrating the thermal expectation value for the mean square amplitude over reciprocal space (Kittel, 1963):

Eq. 8

$$W \sim \frac{k_B T}{c^3} \int_0^{k_D} \frac{k^{(d-1)}}{k^2} dk$$

where c is the sound velocity, $\omega_D \equiv ck_D$ is the Debye frequency and d=3 for three dimensional crystals. In this case the integral converges and the only effect is to reduce the integrated intensity of the Bragg peak by a factor proportional to exp(-2W). For two dimensional crystals d=1, and the integral, of the form of dk/k, obtains a logarithmic divergence at the lower limit (Fleming, Moncton, McWhan & DiSalvo, 1980). A more precise treatment of thermal vibrations, necessitated by this divergence, is to calculate the relative phase of X-rays scattered from two points in the sample a distance $|r|$ apart. The appropriate integral that replaces the Debye-Waller integral is

Eq. 9

$$< (u(r)-u(0))^2 > \sim \frac{k_B T}{c^3} \int \sin^2 (k \cdot r) \frac{dk}{k} d\{\cos(k \cdot r)\}$$

and the divergence due to the lower limit is cut off by the fact that $\sin^2 (k \cdot r)$ vanishes as $k \to 0$. More complete analysis obtains $< (u(r)-u(0))^2 > \sim (k_B T/c^2) \ln(|r|/a)$ where $a \approx$ the atomic size. If this is exponentiated, as for the Debye-Waller factor, the density-density correlation function can be shown to have the form $< \rho(r) \rho(0) > \sim |r/a|^{-\eta}$ where $\eta \sim |q|^2 (k_B T/c^2)$ and $|q| \sim 2\pi/a$. In place of the usual periodic density-density correlation function of three dimensional crystals, the periodic correlations of two dimensional crystals decay away as some power of the distance. This type of positional order, in which the correlations decay as some power of the distance, is the quasi long range order (Q.L.R.O.) that appears in Table 1. It is distinguished from true long range order (L.R.O) where the correlations continue indefinitely, and short range order (S.R.O.) where the positional correlations decay exponentially as in either a simple fluid or a nematic liquid crystal.

The usual prediction of Bragg scattering for three dimensional crystals is obtained from the Fourier transform of the three dimensional density-density correlation function. Since the correlation function is made up of periodic and random parts, it follows that the scattering cross section is made up of a delta-function at the Bragg condition superposed on a background of thermal diffuse scattering from the random part. In principal these two types of scattering can be separated empirically by using a high resolution spectrometer that integrates all of the δ-function Bragg peak, but only a small part of the thermal diffuse scattering. Since the two dimensional lattice is not strictly periodic there is no formal way to separate the periodic and random parts, and the Fourier transform for the algebraic correlation function obtains a cross section that is described by an algebraic singularity of the form $|Q-q|^{\eta-2}$ (Gunther, Imry & Lajzerowicz, 1980). The following argument that the X-ray scattering line shape for the one dimensional periodicity of the smectic-A system in three dimensions is similar to that of two dimensional crystal as given by Caillé in 1972 (Caillé, 1972).

In three dimensional crystals both the longitudinal and shear sound waves satisfy linear dispersion relations of the form $\omega=ck$. In simple liquids, and also for nematic liquid crystals, only the longitudinal sound wave has such a linear dispersion relation. Shear soundwaves are overdamped and the decay rate $1/\tau$ is given by the imaginary part of a dispersion relation of the form $\omega=i(\eta/\rho)k^2$ where η is a viscosity coefficient and ρ is the liquid density. The intermediate order of the smectic-A mesomorphic phase, between the three dimensional crystal and the nematic, results in one of the modes for shear soundwaves having the curious dispersion relation $\omega^2=c^2k_\perp^2k_z^2/(k_\perp^2+k_z^2)$ where $k_\perp$ and k_z are the magnitudes of the components of the acoustic wavevector perpendicular and parallel to <n> respectively (De Gennes, 1969a; Martin, Parodi & Pershan, 1972). More detailed analysis, including terms of higher order in $k_\perp^2$, obtains the equivalent of the Debye-Waller factor for the smectic-A as

Eq. 10

$$W \sim k_BT \int_0^{k_D} \frac{k_\perp dk_\perp dk_z}{Bk_z^2+Kk_\perp^4}$$

where B and K are smectic elastic constants, $k_\perp^2=k_x^2+k_y^2$ and k_D is the Debye wavevector. On substitution of $u^2=(K/B)k_\perp^2+k_z^2$ the integral can be manipulated into the form

∫du/u which diverges logarithmically at the lower limit in exactly the same way as the integral for the Debye-Waller factor of the two dimensional crystal. The result is that the smectic-A phase has a sharp peak, described by an algebraic cusp, at the place in reciprocal space where one would expect a true δ-function Bragg cross section from a truly periodic one dimensional lattice. In fact the lattice is not truly periodic and the smectic-A system has only Q.L.R.O. order along the direction <**n**>.

X-ray scattering experiments to test this idea were carried out on one thermotropic smectic-A system, but the results, while consistent with the theory, were not adequate to provide an unambiguous proof of the algebraic cusp (Als-Nielsen, Litster, Birgeneau, Kaplan, Safinya, Lindegaard & Mathiesen, 1980). One of the principal difficulties was due to the fact that when thermotropic samples are oriented in an external magnetic field in the higher temperature nematic phase and then gradually cooled through the nematic to smectic-A phase transition, the smectic-A samples usually have mosaic spreads of the order of a fraction of a degree and this is not sufficient for detailed line shape studies near to the peak. A second difficulty is that in most of the thermotropic smectic-A phases that have been studied to date, only the lowest order peak is observed. It is not clear whether this is due to a large Debye-Waller type effect or whether the form factor for the smectic-A layer falls off this rapidly. Nevertheless, since the factor η in the exponent of the cusp $|Q-q|^{\eta-2}$ depends quadratically on the magnitude of the reciprocal vector $|q|$, the shape of the cusp for the different orders would constitute a severe test of the theory.

Fortunately, it is common to observe multiple orders for lyotropic smectic-A systems and such an experiment, recently carried out on the lyotropic smectic-A system formed from a quaternary mixture of sodium dodecyl sulfate, pentanol, water and dodecane, confirmed the theoretical predictions for the Landau-Peierls effect in the smectic-A phase (Safinya, Roux, Smith, Sinha, Dimon, Clark & Bellocq, 1986a). The problem of sample mosaic was resolved by using a three dimensional powder. Although the conditions on the analysis are delicate, Safinya et al. demonstrated that for a perfect powder, for which the microcrystals are sufficiently large, the powder line shape does allow unambiguous determination of all of the parameters of the anisotropic line shape.

The only other X-ray study of a critical property on the smectic-A side of the transition has been a measurement of the temperature dependence of the integrated intensity of the peak. For three dimensional crystals the integrated intensity of a Bragg peak can be measured for samples with poor mosaic distributions, and because the differences between Q.L.R.O and true L.R.O. are only manifest at long distances in real space, or at small wavevectors in reciprocal space, the same is true for the "quasi-Bragg peak" of the smectic-A phase. Chan et al. measured the temperature dependence of the integrated intensity of the smectic-A peak across the nematic to smectic-A phase transition for a number of liquid crystals with varying exponents $\nu_{//}$ and $\nu_{\perp}$ (Chan, Deutsch, Ocko, Pershan & Sorensen, 1985a). For the Landau-De Gennes free energy density (Eq. 5), the theoretical prediction is that the critical part of the integrated intensity should vary as $|t|^{x}$ where $x = 1 - \alpha$ when the critical part of the heat capacity diverges according to the power law $|t|^{-\alpha}$. Six samples were measured with values of α varying from 0 to 0.5. Although for samples with $\alpha \approx 0.5$, the critical intensity did vary as $x \approx 0.5$, there were systematic deviations for smaller values of α, and for $\alpha \approx 0$ the measured values of x were in the range of 0.7 to 0.76. The origin of this discrepancy is not presently understood.

Similar integrated intensity measurements in the vicinity of the first order nematic to smectic-C transition cannot easily be done in smectic-C samples since the magnetic field aligns the molecular axis <n>, and when the layers form at some angle φ to <n> the layer normals are distributed along the full 2π of azimuthal directions around <n> as shown in Figure 3d. The X-ray scattering pattern for such a sample is a partial powder with a peak intensity distribution that forms a ring of radius $|q|\sin(\varphi)$. The opening of the single spot along the average molecular axis <n> into a ring can be used to study either the nematic to smectic-C or the smectic-A to smectic-C transition (Martinez-Miranda, Kortan & Birgeneau, 1986).

The statistical physics in the region of the phase diagram surrounding the triple point, where the nematic, smectic-A and smectic-C phases meet, has been the subject of considerable theoretical speculation (Chen & Lubensky, 1976; Chu & McMillan, 1977; Benguigui, 1979; Huang & Lien, 1981; Grinstein & Toner, 1983). The best representation of the observed X-ray scattering structure near the nematic to smectic-A, the nematic to smectic-C, and the nematic, smectic-A, smectic-C multicritical point is obtained from the mean field theory of Chen and Lubensky, the essence of which is expressed in terms of an energy density of the form

Eq. 11

$$\Delta F(\psi) = \frac{A}{2} |\psi|^2 + \frac{D}{4} |\psi|^4 +$$

$$\frac{1}{2}\left[E_{//}(Q_{//}^2 - Q_0^2)^2 + E_{\perp}Q_{\perp}^2 + E_{\perp\perp}Q_{\perp}^4 + E_{//\perp}Q_{\perp}^2 (Q_{//}^2 - Q_0^2) \right] |\psi(Q)|^2$$

where $\psi = \psi(Q)$ is the Fourier component of the electron density:

Eq. 12

$$\psi(Q) \equiv \frac{1}{(2\pi)^3} \int d^3r \; e^{i(Q \cdot r)} \rho(r).$$

The quantities $E_{//}$, $E_{\perp\perp}$, and $E_{//\perp}$ are all positive definite, however the sign of A and $E_{\perp}$ depends on temperature. For A>0 and $E_{\perp}>0$ the free energy, including the higher order terms, is minimized by $\psi(Q)=0$ and the nematic is the stable phase.

For A<0 and $E_\perp$ >0 the minimum in the free energy occurs for a nonvanishing value for $\psi(\mathbf{Q})$ in the vicinity of $Q_{//} \approx Q_0$, corresponding to the uniaxial smectic-A phase; however for $E_\perp$ <0 the free energy minimum occurs for a nonvanishing $\psi(\mathbf{Q})$ with a finite value of $Q_\perp$, corresponding to smectic-C order. The special point in the phase diagram where two terms in the free energy vanish simultaneously is known as a "Lifshitz point" (Hornreich, Luban & Shtrikman, 1975). In the present problem this occurs at the triple point where the nematic, smectic-A and smectic-C phases coexist. Although there have been other theoretical models for this transition, the best agreement between the observed and theoretical line shapes for the X-ray scattering cross sections are based on the Chen-Lubensky model. Most of the results from light scattering experiments in the vicinity of the NAC triple point also agree with the main features predicted by the Chen-Lubensky model; however, there are some discrepancies that are not explained (Solomon & Litster, 1986).

The nematic to smectic-C transition in the vicinity of this point is particularly interesting in that on approaching the nematic to smectic-C transition temperature from the nematic phase, the X-ray scattering line shapes first appear to be identical to the shapes usually observed on approaching the nematic to smectic-A phase transition; however within approximately 0.1°C of the transition they change to shapes that clearly indicate smectic-C type fluctuations. Details of

this crossover are among the strongest evidence supporting
the Lifshitz idea behind the Chen-Lubensky model.

3.2.   Modulated Smectic-A and Smectic-C Phases

Previously we mentioned that although the reciprocal
lattice spacing $|q|$ for many smectic-A phases corresponds to
$2\pi/L$, where L is the molecular length, there are a number of
others for which $|q|$ is between π/L and $2\pi/L$ (Leadbetter et
al, 1979c; Leadbetter, Durrant & Rugman, 1977). This suggests
the possibility of different types of smectic-A phases in
which the bare molecular length is not the sole determining
factor of the period d. In 1979 workers at Bordeaux optically
observed some sort of phase transition between two phases
that both appeared to be of the smectic-A type (Sigaud,
et al., 1979). Subsequent X-ray studies indicated that in the
nematic phase these materials simultaneously displayed
critical fluctuations with two separate periods (Levelut,
Tarento, Hardouin, Achard & Sigaud, 1981; Hardouin, et al.,
1980, 1983; Ratna, Shashidar & Raja, 1985; Ratna,
Nagabhushana, Raja, Shashidhar & Chandrasekhar, 1986; Chan,
et al., 1985b, 1986; Safinya, Varady, Chiang & Dimon, 1986b;
Fontes, Heiney, Haseltine & Smith, 1986) and confirmed phase
transitions between phases that have been designated
smectic-A_1 with period $d \approx L$, -A_2, with period $d \approx 2L$, and -A_d with
period $L < d < 2L$. Stimulated by the experimental results Prost
and co-workers generalized the De Gennes mean field theory by
writing

$$\rho(\mathbf{r}) = <\rho> + \mathrm{Re}\left(\Psi_1 . e^{i\mathbf{q}_1 \cdot \mathbf{r}} + \Psi_2 . e^{i\mathbf{q}_2 \cdot \mathbf{r}}\right)$$

where 1 and 2 refer to two different density waves (Prost, 1979; Prost & Barois, 1983; Barois, Prost & Lubensky, 1985). In the special case that $\mathbf{q}_1 \approx 2\mathbf{q}_2$ the free energy represented by Eq. 3 must be generalized to include terms like

$$(\Psi_2{}^*)^2 \Psi_1 e^{[i(\mathbf{q}_1 - 2\mathbf{q}_2) \cdot \mathbf{r}]} + c.c.$$

that couple the two order parameters. Suitable choices for the relative values of the phenomenological parameters of the free energy then result in minima that correspond to any one of these three smectic-A phases. Much more interesting, however, was the observation that even if $|\mathbf{q}_1| < 2|\mathbf{q}_2|$ the two order parameters could still be coupled together if $\mathbf{q}_1$ and $\mathbf{q}_2$ were not co-linear, as illustrated in Figure 4a, such that $2\mathbf{q}_1 \cdot \mathbf{q}_2 = |\mathbf{q}_1|^2$. Prost et al. predicted the existence of phases that are modulated in the direction perpendicular to the average layer normal with a period $4\pi/[|\mathbf{q}_2|\sin(\varphi)] = 2\pi/|\mathbf{q}_m|$. Such a modulated phase has been observed and is designated as the smectic-$\tilde{A}$ (Hardouin, Sigaud, Tinh & Achard, 1981). Similar considerations apply to the smectic-C phases and the modulated phase is designated smectic-$\tilde{C}$ (Hardouin, Tinh, Achard & Levelut, 1982; Huang, Lien, Dumrongrattana & Chiang, 1984; Safinya, et al., 1986b).

3.3.   Surface Effects

The effects of surfaces in inducing macroscopic alignment of mesomorphic phases have been important for both

a)

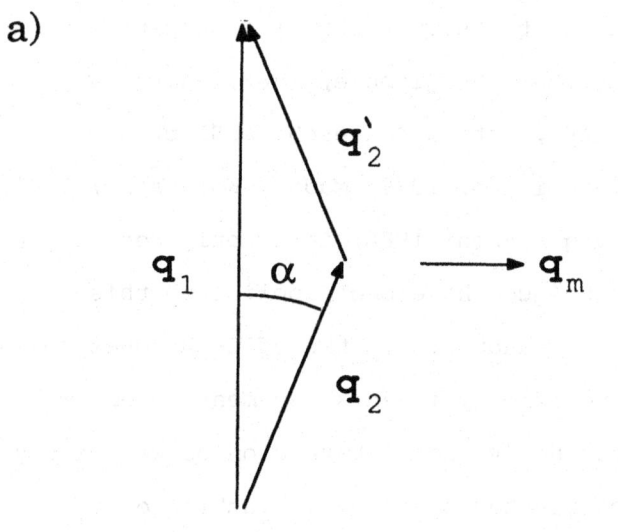

b)

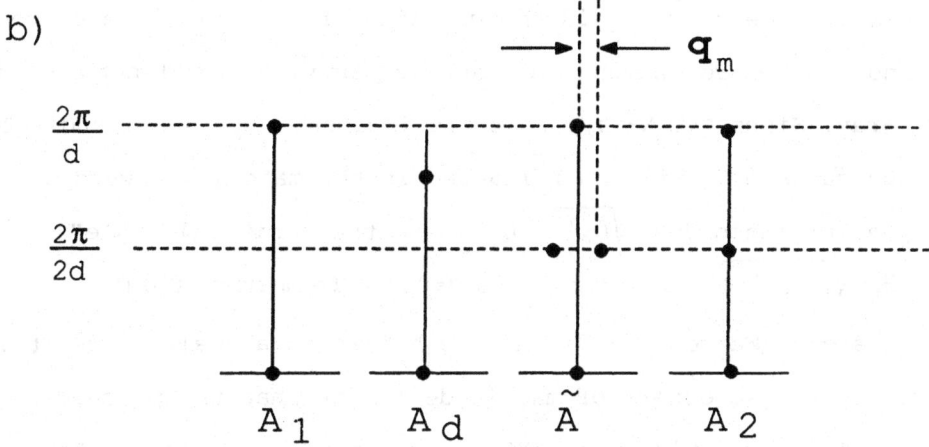

Figure 4: (a) Schematic illustration of the necessary condition for coupling between order parameters when $|\mathbf{h}_2|<2|\mathbf{h}_1|$; $|\mathbf{q}|=\sqrt{|\mathbf{h}_2|^2-|\mathbf{h}_1|^2}=|\mathbf{h}_1|\sin(\alpha)$. (b) Positions of the principal peaks for the indicated smectic-A phases.

technological applications and for basic research
(Sprokel,1980; Gray & Goodby, 1984). Although there are a
variety of experimental techniques that are sensitive to
mesomorphic surface order (Beaglehole, 1982; Faetti &
Palleschi, 1984; Faetti, Gatti, Palleschi & Sluckin, 1985;
Gannon & Faber, 1978; Miyano, 1979; Mada & Kobayashi, 1981;
Guyot-Sionnest, Hsiung & Shen, 1986) it is only recently that
X-ray scattering techniques have been applied to this
problem. In one form or another all the of techniques for
obtaining surface specificity in an X-ray measurement make
use of the fact that the average interaction between X-rays
and materials can be treated by the introduction of a
dielectric constant $\varepsilon \approx 1 - (4\pi\rho e^2/m\omega^2) = 1 - \rho r_e \lambda^2/\pi$ where ρ is the
electron density, r_e is the classical radius of the electron,
and ω and λ are the angular frequency and the wavelength of the
X-ray. Since $\varepsilon < 1$ X-rays that are incident at a small angle to
the surface θ_o will be refracted in the material toward a
smaller angle $\theta_T \approx \sqrt{\theta_o^2 - \theta_c^2}$ where the "critical angle"
$\theta_c \approx \sqrt{\rho r_e \lambda^2/\pi} \approx 0.003$ radians ($\approx 0.2°$) form most liquid
crystals (Warren, 1968). Although this is a small angle it is
at least two orders of magnitude larger than the practical
angular resolution available in modern X-ray spectrometers
(Als-Nielsen, Christensen & Pershan, 1982; Pershan & Als-
Nielsen, 1984; Pershan, Braslau, Weiss & Als-Nielsen, 1987).
One can demonstrate that for many conditions the specular
reflection $R(\theta_o)$ is given by:

$$R(\theta_0) \approx R_F(\theta_0) \, |\rho^{-1}\!\int\! dz \, \exp(-iQz) \langle \partial\rho/\partial z\rangle|^2$$

where $Q \equiv (4\pi/\lambda)\sin(\theta_0)$, $\langle \partial\rho/\partial z\rangle$ is the normal derivative of the electron density averaged over a region in the surface that is defined by the coherence area of the incident X-ray, and

$$R_F(\theta_0) \approx \left(\frac{\theta_0 - \sqrt{\theta_0{}^2 - \theta_c{}^2}}{\theta_0 + \sqrt{\theta_0{}^2 - \theta_c{}^2}}\right)^2$$

is the Fresnel reflection law that is calculated from classical optics for a flat interface between the vacuum and a material of dielectric constraint ε. Since the condition for specular reflection, that the incident and scattered angles are equal and in the same plane, requires that the scattering vector $Q = \hat{z}(4\pi/\lambda)\sin(\theta_0)$ be parallel to the surface normal, it is quite practical to obtain, for flat surfaces, an unambiguous separation of the specular reflection signal from all other scattering events.

Figure 5a illustrates the specular reflectivity from the free nematic-air interface for the liquid crystal octyloxycyanobiphenyl (8OCB) 0.050°C above the nematic to smectic-A phase transition temperature (Pershan & Als-Nielsen, 1984). The dashed line is the Fresnel reflection $R_F(\theta_0)$ in units of $\sin(\theta_0)/\sin(\theta_c)$ where the peak at $\theta_c = 1.39°$ corresponds to surface induced smectic order in the nematic phase: i.e. the selection rule for specular reflection has been used to separate the specular reflection from the critical scattering from the bulk. Since the full width at half maxima is exactly equal to the reciprocal of the

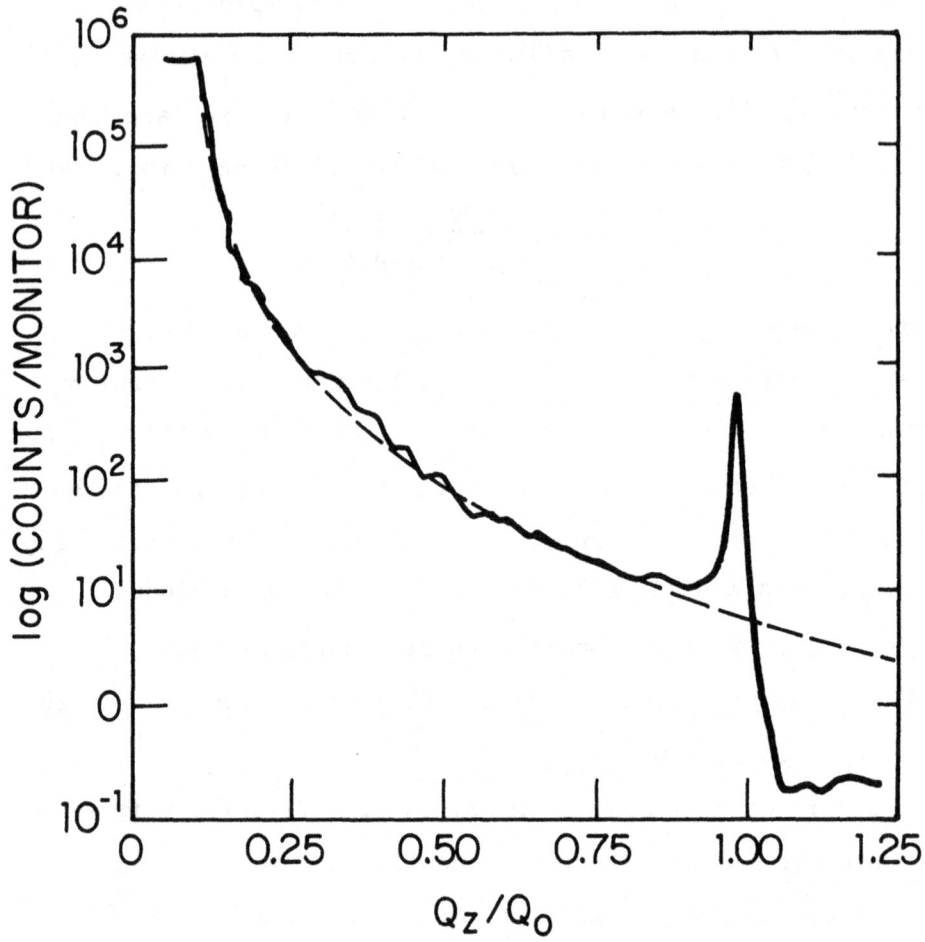

Figure 5: Specular reflectivity of ≈8kev X-rays from the air-liquid interface of the nematic liquid crystal 8OCB 0.05°C above the nematic to smectic-A transition temperature. The dashed line is the Fresnel reflection law as described in the text.

correlation length for critical fluctuations in the bulk, $2/\xi_{//}$ at all temperatures from $T-T_{NA}\approx0.006°C$ up to values near to the nematic to isotropic transition, $T-T_{NA}\approx3.0°C$, it is clear this is an example where the gravitationally induced long range order in the surface position has induced mesomorphic order that has long range correlations parallel to the surface. Along the surface normal the correlations have only the same finite range as the bulk critical fluctuations. Studies on a number of other nematic (Gransbergen, De Jeu & Als-Nielsen, 1986; Ocko, Pershan, Safinya & Chiang, 1987) and isotropic surfaces (Ocko, Braslau, Pershan, Als-Nielsen & Deutsch, 1986) indicate features that are specific to local structure of the surface.

4. Phases With In-Plane Order

Although the combination of optical microscopy and X-ray scattering studies on unoriented samples identified most of the mesomorphic phases, there remain a number of subtle features that were only discovered by spectra from well oriented samples (see the extensive references contained in Gray & Goodby, 1984). Nematic phases are sufficiently fluid that they are easily oriented by either external electric or magnetic fields, or surface boundary conditions, but similar alignment techniques are not generally successful for the more ordered phases because the combination of strains, induced by thermal expansion and the enhanced elasticity that accompanies the order, create defects that do not easily anneal. Other defects that might have been formed during initial growth of the phase also become trapped and it is difficult to obtain well oriented samples by cooling from a higher temperature aligned phase. Nevertheless, in some cases it has been possible to obtain crystalline-B samples with mosaic spreads of the order of a fraction of a degree by slowing cooling samples that were aligned in a the nematic phase. In other cases, mesomorphic phases were obtained by heating up and melting single crystals that were grown from solution (Benattar, Doucet, Lambert & Levelut, 1979; Leadbetter, Mazid & Malik, 1980a).

Moncton and Pindak (Moncton & Pindak, 1979) were the first to realize that X-ray scattering studies could be carried out on the freely suspended films that Freidel described in his classical treatise on liquid crystals (Freidel, 1922). These samples, formed across a plane aperture (i.e. approximately 1 cm in diameter) in the same manner as soap bubbles, have mosaic spreads that are an order of magnitude smaller. The geometry is illustrated in Figure 6a. The substrate in which the aperture is cut can be either glass (i.e. microscope cover slips), steel or copper sheets, etc. A small amount of the material, usually in the high temperature region of the smectic-A phase, is spread around the outside of an aperture that is maintained at the necessary temperature, and a wiper is used to drag some of the material across the aperture. If a stable film is successfully drawn, it is detected optically by its finite reflectivity. In particular, against a dark background and with the proper illumination it is quite easy to detect the thinnest free films.

In contrast to conventional soap films that are stabilized by electrostatic effects, smectic films are stabilized by their own layer structure. Films as thin as two molecular layers can be drawn and studied for weeks (Young, Pindak, Clark & Meyer, 1978). Thicker films of the order of thousands of layers can also be made and with some experience in depositing the raw material around the aperture and the speed of drawing, it is possible to draw films of almost any desired thickness(Moncton, Pindak Davey & Brown, 1982). For films thinner than approximately 20 to 30 molecular layers (i.e, 600 Å to 1000Å) the thickness is determined from the reflected intensity of a small helium neon laser. Since the reflected intensities for films of 2,3,4,5,... layers are in the ratio of 4,9,16,25,..., the measurement can be calibrated by drawing and measuring a reasonable number of thin films. The most straightforward for thick films is to measure the ellipticity of the polarization induced in laser light transmitted through the film at an oblique angle (Collett, 1983; Collett, Sorensen, Pershan & Als-Nielsen, 1985); however a subtler method that makes use of the colors of white light reflected from the films is also practical (Sirota, Pershan, Sorensen & Collett, 1987). In certain circumstances the thickness can also be measured using the X-ray scattering intensity in combination with one of the other methods.

Figure 6b illustrates the scattering geometry used with these films. Although recent unpublished work has demonstrated the possibility of a reflection geometry (Sorensen, 1987-private communication), all of the X-ray scattering studies to be described here were done in transmission. Since the in-plane molecular spacings are typically between 4Å and 5Å, while the layer spacing is closer to 30Å, it is difficult to study the $(0,0,L)$ peaks in this geometry.

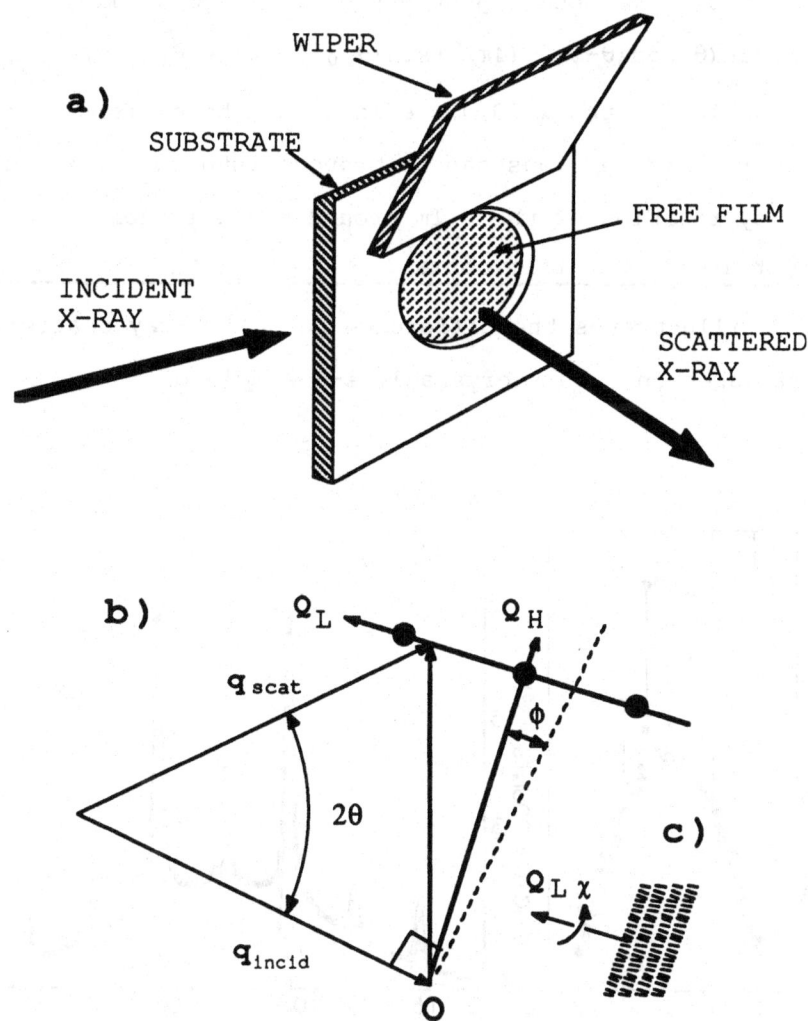

Figure 6: (a) Schematic illustration of the geometry and (b) kinematics of x-ray scattering from a freely suspended smectic film. The insert (c) illustrates the orientation of the film in real space corresponding to the reciprocal space kinematics in (b). If the angle $\phi = \theta$ the film is oriented such that the scattering vector is parallel to the surface of the film, i.e., parallel to the smectic layers. A "Q_L-scan" is

taken by simultaneous adjustment of ϕ and 2θ to keep

$(4\pi/\lambda)\sin(\theta)\cos(\theta-\phi) = (4\pi/\lambda)\sin(\theta_{1,0,0})$ where $\theta_{1,0,0}$ is the

Bragg angle for the $(1,0,0)$ reflection. The different in-

plane Bragg reflections can be brought into the scattering

plane by rotation of the film around the film normal by the

angle χ around the film normal.

Figure 7 illustrates the difference between X-ray scattering

spectra taken on a bulk crystalline-B sample of

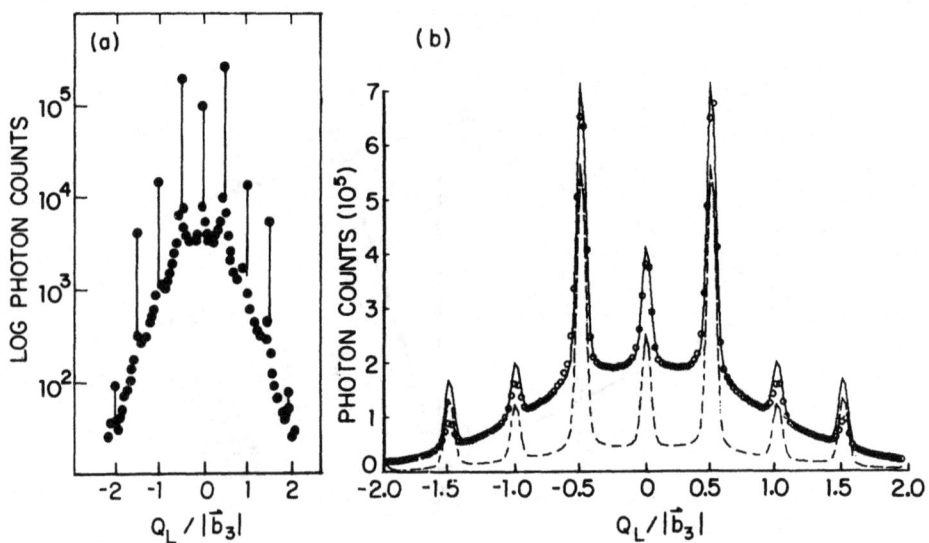

Figure 7: Typical Q_L-scans from the crystalline-B phases

of a free film of 70.7, displayed on a logarithmic scale to

illustrate the reduced level of the diffuse scattering

relative to the Bragg reflection (a) and a bulk sample of

40.8 oriented by a magnetic field (b).

n-(4-(n-butyl)oxybenzylidene)-4'-(n-octyl)aniline(40.8) that was oriented in an external magnetic field while in the nematic phase and then cooled through the smectic-A phase into the crystalline-B phase (Aeppli, Litster, Birgeneau & Pershan, 1981), and one taken on a thick freely suspended film of n-(4-(n-heptyl)oxybenzylidene)-4'-(n-heptyl)aniline(70.7) (Collett, Sorensen, Pershan, Litster, Birgeneau, & Als-Nielsen, 1982; Collett, et al., 1985). Note that the data for 70.7 is plotted on a semi-logarithmic scale in order to simultaneously display both the Bragg peak and the thermal diffuse background. The scans are along the Q_L direction, at the appropriate value of Q_H to intersect the peaks associated with the intralayer periodicity. In both cases the widths of the Bragg peaks are essentially determined by the sample mosaicity and as a result of the better alignment the ratio of the thermal diffuse background to the Bragg peak is nearly an order of magnitude smaller for the free film sample.

4.1.   Hexatic Phases in Two Dimensions

The hexatic phase of matter was first proposed independently by Halperin and Nelson (Halperin & Nelson 1978, Nelson & Halperin 1979) and Young (Young, 1979) on the basis of theoretical studies into the melting process in two dimensions. Following work by Kosterlitz and Thouless (Kosterlitz & Thouless, 1973), they observed that since the interaction energy between pairs of dislocations in two dimensions decreases logarithmically with their separation, the enthalpy and the entropy terms in the free energy have the same functional dependence on the density of dislocations. It follows that the free energy difference between the crystalline and hexatic phase has the form $\Delta F = \Delta H - T\Delta S \approx T_c S(\rho) - T S(\rho) = S(\rho)(T_c - T)$, where $S(\rho) \approx \rho \log(\rho)$ is the entropy as a function of the density of dislocations ρ, and T_c is defined such that $T_c S(\rho)$ is the enthalpy. Since the prefactor of the enthalpy term is independent of temperature while that of the entropy term is linear, there will be a critical temperature, T_c, at which the sign of the free energy changes from positive to negative. For temperatures greater than T_c the entropy term will dominate and the system will be unstable against the spontaneous generation of dislocations. When this happens the two dimensional crystal, with positional Q.L.R.O., but true long range order in the orientation of neighboring atoms, can melt into a new phase in which the positional order is short range, but for which there is Q.L.R.O. in the orientation of the six neighbors surrounding any atom. The reciprocal space structures for the two dimensional crystal and hexatic phases are illustrated in Figs. 8b and 8c respectively. That of the two dimensional solid consists of a hexagonal lattice of sharp rods (i.e. algebraic line shapes in the plane of the crystal). For a finite size sample, the reciprocal space structure of the two dimensional hexatic phase is a hexagonal lattice of diffuse

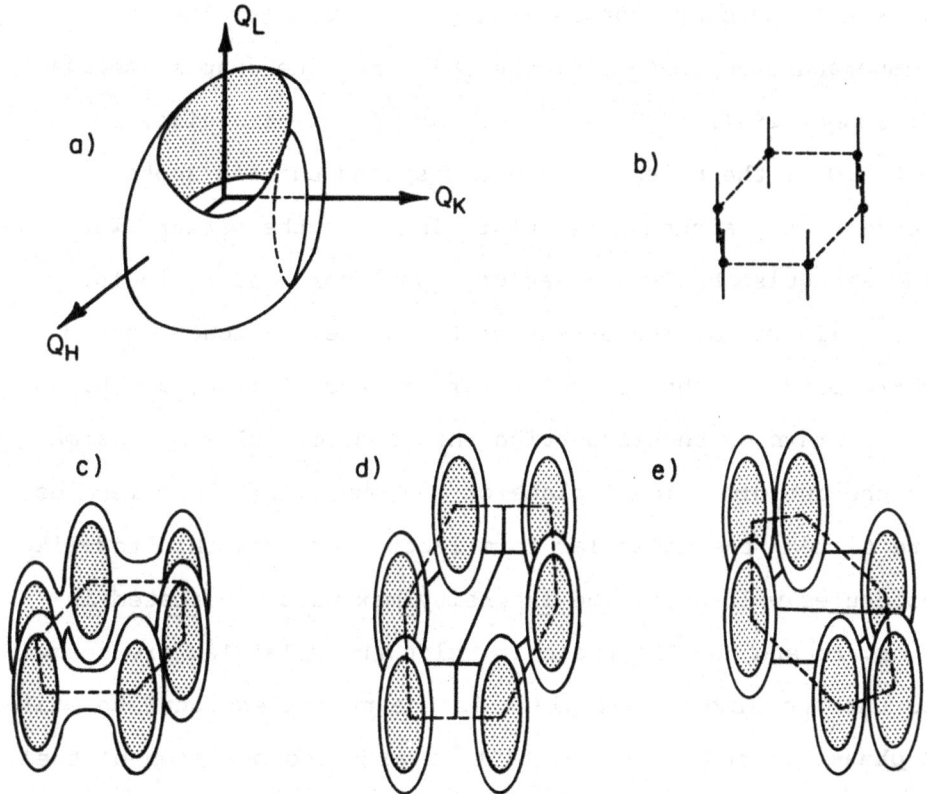

Figure 8: Scattering intensities in reciprocal space
from two dimensional (a) liquid; (b) crystal; (c) normal
hexatic; and tilted hexatics in which the tilt is (d)
towards the nearest neighbors as for the smectic-I or (e)
between the nearest neighbors as for the smectic-F. The thin
rods of scattering in (b) indicate the singular cusp for
peaks with algebraic line shapes in the H-K plane.

rods and there are theoretical predictions for the temperature dependence of the in-plane line shapes (Aeppli & Bruinsma, 1984). If the sample was of infinite size, the Q.L.R.O of the orientation would spread the six spots continuously around a circular ring, and the pattern would be indistinguishable from that of a well correlated liquid, i.e., Figure 8a. The extent of the patterns along the rod corresponds to the molecular form factor. Figures 8a, b, and c are drawn on the assumption that the molecules are normal to the two dimensional plane of the phase. If the molecules are tilted, the molecular form factor for long thin rod-like molecules will shift the intensity maxima as indicated in Figs. 8d and 8e. The phase in which the molecules are normal to the two dimensional plane is the two dimensional **hexatic-B phase**. If the molecules tilt towards the position of their nearest neighbors (in real space), or in the direction that is between the lowest order peaks in reciprocal space, the phase is the two dimensional **smectic-I**, Figure 8d. The other tilted phase, for which the tilt direction is between the nearest neighbors in real space or in the direction of the lowest order peaks in reciprocal space, is the **smectic-F**, Figure 8e.

Although theory (Halperin & Nelson, 1978; Nelson & Halperin, 1979; Young, 1979) predicts that the two dimensional crystal can melt into a hexatic phase, it does not say that it must happen, and the crystal can melt directly into a two dimensional liquid phase. Obviously, the hexatic phases will also melt into a two dimensional liquid phase. Figure 8a illustrates the reciprocal space structure for the two dimensional liquid in which the molecules are normal to the two dimensional surface. Since the longitudinal (i.e. radial) width of the hexatic spot could be similar to the width that might be expected in a well correlated fluid, the direct X-ray proof of the transition from the hexatic-B to the normal liquid requires a hexatic sample in which the domains are sufficiently large that the sample is not a two dimensional powder. On the other hand, the elastic constants must be sufficiently large that the Q.L.R.O. does not smear the six spots into a circle. The radial line shape of the the powder pattern of the hexatic-B phase can also be subtly different from that of the liquid and this is another possible way that X-ray scattering can detect melting of the hexatic-B phase (Aeppli & Bruinsma, 1984).

Changes that occur on the melting of the tilted hexatics, i.e. smectic-F and smectic-I, are usually easier to detect and this will be discussed in more detail below. On the other hand, there is a fundamental theoretical problem concerning the way of understanding the melting of the titled hexatics. These phases actually have the same symmetry as the two dimensional tilted fluid phase, i.e. the smectic-C. In two dimensions they all have Q.L.R.O. in the tilt orientation, and since the simplest phenomenological argument says that there is a linear coupling between the tilt order and the near neighbor positional order (Nelson & Halperin, 1980; Bruinsma & Nelson, 1981), it follows that the Q.L.R.O. of the smectic-C tilt should induce Q.L.R.O. in the near neighbor positional order. Thus, by the usual arguments, if there is to be a phase transition between the smectic-C and one of the tilted hexatic phases, the transition must be a first order transition (Landau & Lifshitz, 1958). This is analogous to the three dimensional liquid to vapor transition which is first order up to a critical point, and beyond the critical point there is no real phase transition.

4.2.   Hexatic Phases in Three Dimensions:

Based on both this theory and the various X-ray scattering patterns that had been reported in the literature (Gray & Goodby, 1984), Litster and Birgeneau (Birgeneau & Litster, 1978) suggested that some of the three dimensional systems that were previously identified as mesomorphic were actually three dimensional hexatic systems. They observed that it is not theoretically consistent to propose that the smectic phases are layers of two dimensional crystals randomly displaced with respect to each other since, in thermal equilibrium, the interactions between layers of two dimensional crystals must necessarily cause the layers to lock together to form a three dimensional crystal.[1] On the other hand, if the layers were two dimensional hexatics, then the interactions would have the effect of changing the Q.L.R.O. of the hexagonal distribution of neighbors into the true long range order orientational distribution of the three dimensional hexatic. In addition interactions between layers in the three dimensional hexatics can also result in interlayer correlations that would sharpen the width of the diffuse peaks in the reciprocal space direction along the layer normal.

4.2.1. Hexatic-B

Although Leadbetter et al. (Leadbetter, Frost & Mazid, 1979b) had remarked on the different types of X-ray structures that were observed in materials identified as "smectic-B" the first proof for the existence of the hexatic-B phase of matter was the experiment by Pindak and Moncton (Pindak, Moncton, Davey & Goodby, 1981) on thick freely suspended films of the liquid crystal n-hexyl-4'-pentyloxybiphenyl-4-carboxylate (65OBC). A second study on free films of the liquid crystal n-butyl 4'-n-hexyloxybiphenyl-4-carboxylate (46OBC) demonstrated that as the hexatic-B melts into the smectic-A phase, the position

---

[1]Prior to the paper by Litster and Birgeneau, it was commonly believed that some of the smectic phases consisted of uncorrelated stacks of two dimensional crystals.

and the in-plane width of the X-ray scattering peaks varied continuously. In particular, in-plane correlation length evolved continuously from 160 Å, nearly 10°C below the hexatic to smectic-A transition, to only 17 Å, a few degrees above. Similar behavior was also observed in a film only 2 layers thick (Davey, Budai, Goodby, Pindak & Moncton, 1984). Since the observed width of the peak along the layer normal corresponded to the molecular form factor, these systems have negligible interlayer correlations.

4.2.2.   Smectic-F, Smectic-I

In contrast to the hexatic-B phase, the principal reciprocal space features of the smectic-F phase were clearly determined before the theoretical work that proposed hexatic phase. Demus and co-workers (Demus, Diele, Klapperstück, Link & Zaschke, 1971) identified a new phase in one material in 1971, and subsequent X-ray studies by Leadbetter and colleagues (Leadbetter, Mazid & Richardson, 1980b; Leadbetter, Gaughan, Kelley, Gray & Goodby, 1979d; Gane & Leadbetter, 1981) and by Benattar and colleagues (Benattar, Levelut & Strzelecki, 1978; Benattar, Moussa & Lambert, 1983; Benattar, Moussa & Lambert, 1980; Guillon, Skoulios & Benattar, 1986) showed it to have the reciprocal space structure illustrated in Figure 9b. There are interlayer correlations in the three dimensional smectic-F phases, and as a consequence the reciprocal space structure has maxima along the diffuse rods. Benattar et al. obtained monodomain

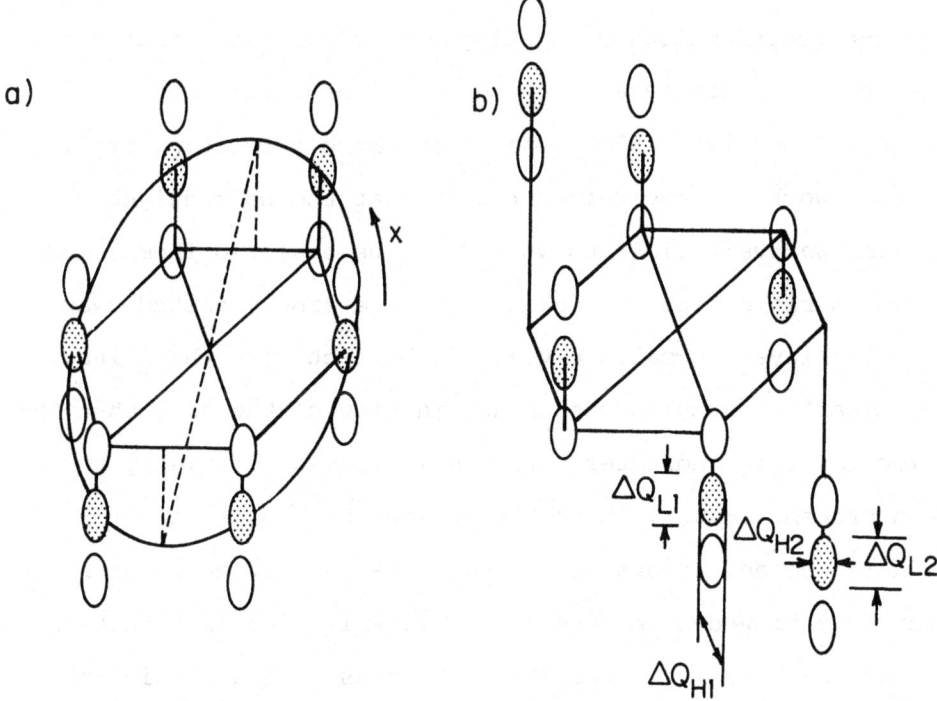

Figure 9: Scattering intensities in reciprocal space from three dimensional tilted hexatic phases: (a) the smectic-I and (b) the smectic-F. The variation of the intensity along the Q_L-direction indicates interlayer correlations that are absent in Figures 6(d) and 6(e). The peak widths $\Delta Q_{L1,2}$ and $\Delta Q_{H1,2}$ correspond to the four inequivalent widths in the smectic-F phase. Similar inequivalent widths exist for the smectic-I phase. The dark circle in (a) indicates the reciprocal space scan that directly measures the hexatic order. A similar scan in the smectic-C phase would have intensity independent of χ.

smectic-F samples of the liquid crystal terephthalylidene-bis-4-n-pentylaniline by melting a single crystal that was previously precipitated from solution (Benattar, Doucet, Lambert & Levelut, 1979). One of the more surprising results of this work was the demonstration that the near neighbor packing was very close to what would be expected from a model in which rigid closely packed rods were simply tilted away from the layer normal. In view of the fact that the molecules are clearly not cylindrical, and in view of the fact that the molecular tilt indicates that the macroscopic symmetry has been broken, it would have been reasonable to expect significant deviations from local hexagonal symmetry when the system is viewed along the molecular axis. The fact that this is not the case indicates that this phase has a considerable amount of rotational disorder around the long axis of the molecules.

Other important features of the smectic-F phase are, firstly, that the local molecular packing is identical to that of the tilted crystalline-G phase (Benattar, et al., 1979; Sirota, Pershan, Sorensen & Collett, 1985; Guillon, Skoulios & Benattar, 1986). Secondly, there is considerable temperature dependence to the widths of the various diffuse peaks. Figure 9b indicates the four inequivalent line widths that Sirota et al. measured in freely suspended films of the liquid crystal 4-n-heptyloxybenzylidene-4-n-heptylaniline (70.7). Parenthetically, bulk samples of this material do not have a smectic-F phase; however the smectic-F is observed in freely suspended films as thick as ~ 200 layers. Figure 10 illustrates the thickness-temperature phase diagram of 70.7 between 52°C and 69°C (Sirota, et al., 1985; Sirota, Pershan & Deutsch, 1987a). Bulk samples and thick films have a first order transition from the crystalline-B to the smectic-C at 69°C. Thinner films indicate a surface phase above 69°C that will be discussed below. Furthermore, although there is a strong temperature dependence to the widths of the diffuse scattering peaks, the widths are independent of film thickness. This demonstrates that although the free film boundary conditions have stabilized the smectic-F phase, the properties of the phase are not affected by the boundaries. Finally, the fact that the widths $\Delta \mathbf{Q}_{L1}$ and $\Delta \mathbf{Q}_{L2}$ along the $\mathbf{L}$ direction and $\Delta \mathbf{Q}_{H1}$ and $\Delta \mathbf{Q}_{H2}$ along the in-plane directions are not equal indicates that the correlations are very anisotropic (Brock, Aharony, Birgeneau, Evans-Lutterodt, Litster, Horn, Stephenson & Tajbakhsh, 1986; Sirota, et al., 1985). We will discuss one possible model for these properties after presenting other data on thick films of 70.7. From the fact that the positions of the intensity maxima for the diffuse spots of the smectic-F phase of 70.7 correspond exactly to the positions of the Bragg peaks in the crystalline-G phase, we learn that the local molecular packing must be identical in the two phases. The major difference between the crystalline-G and the tilted hexatic

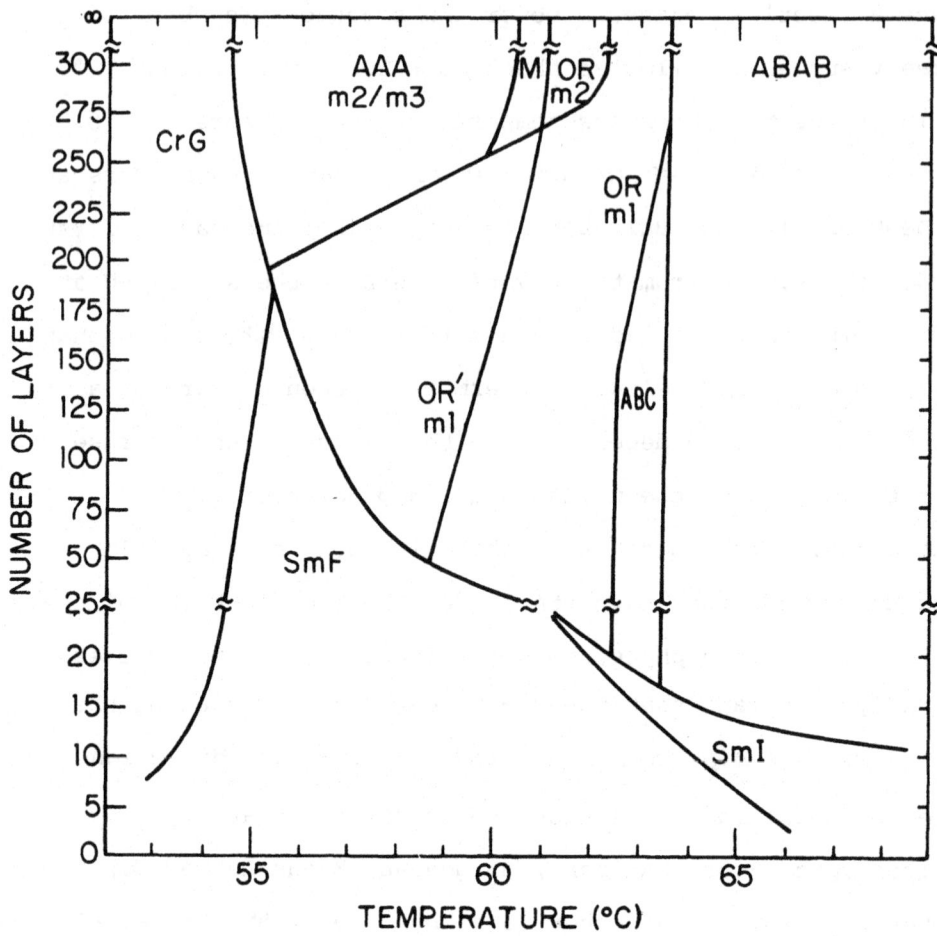

Figure 10: The phase diagram for free films of 70.7 as a function of thickness and temperature. The phases ABAB, AAA, OR_{m1}, OR_{m2}, OR'_{m1}, M, and ABAB are all crystalline-B with varying interlayer stacking, or long wavelength modulations; CrG, SmF and SmI are crystalline-G, smectic-F, and smectic-I (Sirota, et al., 1985; Sirota, et al., 1987a; Sirota, et al., 1987b).

smectic-F phase is that in the latter, defects destroy the long range positional order of the former (Benattar, et al., 1979; Sirota, et al., 1985). Although this is consistent with the existing theoretical model that attributes hexatic order to a proliferation of unbounded dislocations, it is not obvious that the proliferation is attributable to the same Kosterlitz-Thouless mechanism that Halperin and Nelson, and Young discussed for the transition from the two dimensional crystal to the hexatic phase. We will say more on this point below.

The only identified difference between the two tilted hexatic phases, the smectic-F and the smectic-I, is the direction of the molecular tilt relative to the near neighbor positions. For the smectic-I the molecules tilt towards one of the near neighbors, while for the smectic-F they tilt between the neighbors (Gane & Leadbetter, 1983). There are a number of systems that have both smectic-I and smectic-F phases, and in all cases that we are aware of the smectic-I is the higher temperature phase (Gray & Goodby, 1984; Sirota, et al., 1985; Sirota, et al., 1987b).

Optical studies of freely suspended films of materials in the n0.m series indicated tilted surface phases at temperatures for which the bulk had uniaxial phases (Farber, 1985). As mentioned above, X-ray scattering studies of 70.7 demonstrated that the smectic-F phase set in for a narrow temperature range in films as thick as 180 layers, and that the temperature range increases with decreasing layer number. For films of the order of 25 layers thick the smectic-I phase is observed at approximately 61°C, and with decreasing thickness the temperature range for this phase also increases. Below approximately 10 to 15 layers the smectic-I phase extends up to ~69°C where bulk samples undergo a first order transition from the crystalline-B to the smectic-C phase. Synchrotron X-ray scattering experiments show that in thin films (say 5 layers for example) the homogeneous smectic-I film undergoes a first order transition to one in which the two surface layers are smectic-I and the three interior layers are smectic-C (Sirota, et al., 1985; Sirota, et al., 1987b). The fact that two phases with the same symmetry can coexist in this manner tells us that in this material there is some important microscopic difference between them. This is reaffirmed by the fact that the phase transition from the surface smectic-I to the homogeneous smectic-C phase has been observed to be first order (Sorensen, Amador, Sirota, Stragler & Pershan, 1987).

In contrast to 70.7, Birgeneau and colleagues found that in racemic 4-(2'-methylbutyl)phenyl 4'-octyloxylbiphenyl-4-carboxylate (8OSI) (Brock, et al., 1986), the X-ray structure of the smectic-I phase evolves continuously into that of the smectic-C. By applying a magnetic field to a thick freely suspended sample, Birgeneau et al. were able to obtain a large monodomain sample. They measured the X-ray scattering intensity around the circle in the reciprocal space plane shown in Figure 9a that passes through the peaks. For higher temperatures, when the sample is in the smectic-C phase, the intensity is essentially constant around the circle; however on cooling, it gradually condenses into six peaks, separated by 60°. The data was analyzed by expressing the intensity as a Fourier series of the form

$$S(\chi) = I_0 \left[\frac{1}{2} + \sum_{n=1}^{\infty} C_{6n} \cos 6n (90° - \chi) \right] + I_B$$

where I_0 fixes the absolute intensity and I_B fixes the background. The temperature variation of the coefficients scaled according to the relation $C_{6n} = C_6^{\sigma n}$ where the empirical relation $\sigma_n = 2.6(n-1)$ is in good agreement with a theoretical form predicted by Aharony, et al. (Aharony, Birgeneau, Brock & Litster, 1986). The only other system in which this type of measurement has been done was the smectic-C phase of 70.7 (Collett, 1983). In that case the intensity around the circle was constant, indicating the absence of any tilt induced bond orientational order (Aharony, et al., 1986).

It would appear that the near neighbor molecular packing of the smectic-I and the crystalline-J phase are the same, in just the same way as for the packing of the smectic-F and the crystalline-G phases. The four smectic-I widths analogous to those illustrated in Figure 9a are, like that of the smectic-F, both anisotropic and temperature dependent (Sirota, et al., 1985; Sirota, et al., 1987b; Brock, et al., 1986; Benattar, et al., 1979).

4.3. Crystalline Phases with Molecular Rotation

4.3.1. Crystal-B

Recognition of the distinction between the hexatic-B and crystalline-B phases provided one of the more important keys to understanding the ordered mesomorphic phases. There are a number of distinct phases called crystalline-B that are all true three dimensional crystals, with resolution limited Bragg peaks (Moncton & Pindak, 1979; Pershan, et al., 1981). The feature common to them all is that the average molecular orientation is normal to the layers, and within each layer the molecules are distributed on a triangular lattice. In view of the "blade-like" shape of the molecule, the hexagonal site symmetry implies that the molecules must be rotating rapidly (Levelut & Lambert, 1971; Levelut, 1976; Richardson, Leadbetter & Frost, 1978). We have previously remarked that this apparent rotational motion characterizes all of the phases listed in Table 1 except for the crystalline-E, -H, and -K. In the most common crystalline-B phase, adjacent layers have ABAB type stacking (Leadbetter, et al., 1979d; Leadbetter, Mazid & Kelly, 1979). High resolution studies on well oriented samples show that in addition to the Bragg peaks the crystalline-B phases have rods of relatively intense diffuse scattering distributed along the $(1,0,L)$

Bragg peaks (Moncton & Pindak, 1979; Pershan, et al., 1981).
The widths of these rods in the reciprocal space direction,
parallel to the layers, are very sharp, and without a high
resolution spectrometer their widths would appear to be
resolution limited. In contrast along the reciprocal space
direction normal to the layers, their structure corresponds
to the molecular form factor.

If the intensity of the diffuse scattering can be
represented as proportional to $<\mathbf{Q}\cdot\mathbf{u}>^2$, where $\mathbf{u}$ describes the
molecular displacement, the fact that there is not a rod of
diffuse scattering through the $(0,0,L)$ peaks indicates that
the rods through the $(1,0,L)$ peaks originate from random
disorder in "sliding" displacements of adjacent layers. It is
likely that these displacements are thermally excited phonon
vibrations; however, we cannot rule out some sort of non-
thermal static defect structure. In any event assuming this
diffuse scattering originates in a thermal vibration for
which adjacent layers slide over one another with some
amplitude $<\mathbf{u}^2>^{1/2}$, and assuming strong coupling between this
shearing motion and the molecular tilt, we can define an
angle $\phi = \tan^{-1}(<\mathbf{u}^2>^{1/2}/d)$, where d is the layer thickness. The
observed diffuse intensity corresponds to angles ϕ between $3°$
and $6°$ (Pershan, et al., 1981).

Leadbetter and colleagues demonstrated that in the n0.m
series various molecules undergo a series of restacking
transitions and that crystalline-B phases exist with ABC and

AAA stacking as well as the more common ABAB (Leadbetter, Mazid & Richardson, 1980b; Leadbetter, et al., 1979e). Subsequent high resolution studies on thick freely suspended films revealed that the restacking transitions were actually subtler, and in 70.7, for example, on cooling the hexagonal ABAB phase one observes an orthorhombic and then a monoclinic phase before the hexagonal AAA (Collett, et al., 1982; Collett, et al., 1985). Furthermore, the first transition from the hexagonal ABAB to the monoclinic phase is accompanied by the appearance of a relatively long wavelength modulation within the plane of the layers. The polarization of this modulation is along the layer normal, or orthogonal to the polarization of the displacements that gave rise to the rods of thermal diffuse scattering (Gane & Leadbetter, 1983).

It is also interesting to note that the AAA simple hexagonal structure does not seem to have been observed outside of liquid crystalline materials and were it not for the fact that the crystalline-B hexagonal AAA is always accompanied by long wavelength modulations it would be the only case that we are aware of.

Figures 11a and 11b illustrate the reciprocal space positions of the Bragg peaks (dark dots) and modulation induced side bands (open circles) for the unmodulated hexagonal ABAB and the modulated orthorhombic phase (Collett, Pershan, Sirota & Sorensen, 1984). For convenience we only

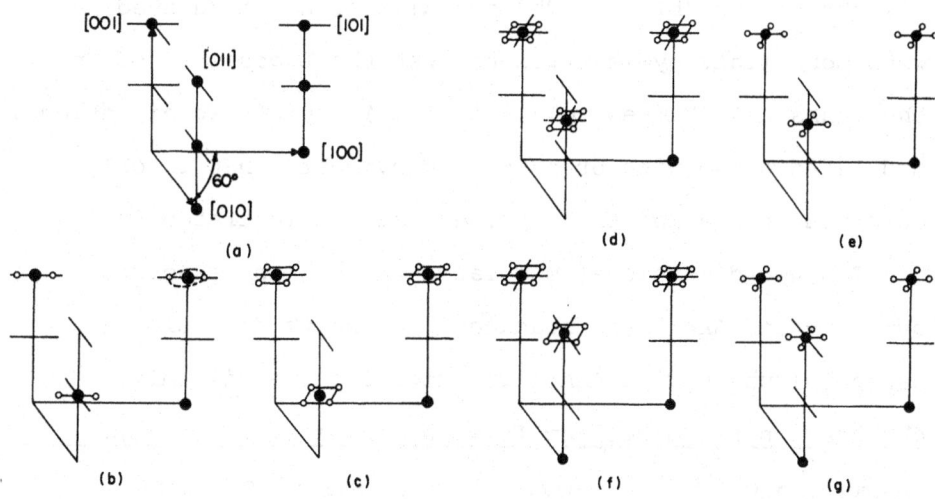

Figure 11: Location of the Bragg peaks in one 60°
section of reciprocal space for the three dimensional
crystalline-B phases observed in thick films of 70.7: (a)
the normal hexagonal crystalline-B phase with ABAB stacking.
(b) The one dimensional modulated phase with orthorhombic
symmetry. The closed circles are the principal Bragg peaks
and the open circles indicate side bands associated with the
long wavelength modulation. (c) The two dimensional
modulated phase with orthorhombic symmetry. Only the lowest
order side bands are shown. They are situated on the corners
of squares surrounding the Bragg peak. The squares are
oriented as shown and the amplitude of the square diagonal
is equal to the distance between the two side bands

illustrated in (b). (d) The two dimensional modulated phase with monoclinic symmetry. Note that the L-position of one of the peaks has shifted relative to (c). (e) A two dimensional modulated phase with orthorhombic symmetry that is only observed on heating the quenched phase illustrated in (g). (f) The two dimensional modulated phase with hexagonal symmetry and AAA layer stacking. (g) A two dimensional hexagonal phase with AAA layer stacking that is only observed on rapid cooling from the phase shown in (c).

display one 60° sector. Hirth et al. (Hirth, Pershan, Collett, Sirota & Sorensen, 1984) explained how both the reciprocal space structure and the modulation of the orthorhombic phase could result from an ordered array of partial dislocations. They were not, however, able to provide a specific model for the microscopic driving force for the transition. Sirota et al. (Sirota, et al., 1987a) proposed a variation of the Hirth model in which the dislocations pair up to form a wall of dislocation dipoles such that within the wall the local molecular packing is essentially identical to the packing in the crystalline-G phase that appears at temperatures just below the crystalline-B phase. This model explains: 1) the macroscopic symmetry of the phase, 2) the period of the modulation, 3) the polarization of the modulation, and 4) the size of the observed deviations of the reciprocal space structure from the hexagonal symmetry of the

ABAB phase and suggests a microscopic driving mechanism that we will discuss below.

On further cooling there is a first order transition in which the one dimensional modulation that appeared at the transition to orthorhombic symmetry is replaced by a two dimensional modulation as shown in Figure 11c. On further cooling there is another first order transition in which the positions of the principal Bragg spots change from having orthorhombic to monoclinic symmetry as illustrated in Figure 11d. On further cooling the Bragg peaks shift continually until there is one more first order transition to a phase with hexagonal AAA positions as illustrated in Figure 11f. On further cooling the AAA symmetry remains unchanged, and the modulation period is only slightly dependent on temperature, but the modulation amplitude increases dramatically. Eventually, as indicated in the phase diagram shown in Figure 10, the system undergoes another first order transition to the tilted crystalline-G phase. The patterns in Figures 11e and 11g are observed by rapid quenching from the temperatures at which the patterns in Figure 11b are observed.

Although there is not yet an established theoretical explanation for the origin of the "restacking-modulation" effects, there are a number of experimental facts that we can summarize, and which indicate a probable direction for future research. Firstly, if one ignores the long wavelength modulation, the hexagonal ABAB phase is the only phase in the diagram for 70.7 for which there are two molecules per unit cell. There must be some basic molecular effect that determines this particular coupling between every other layer. In addition it is particularly interesting that it only manifests itself for a small temperature range and then vanishes as the sample is cooled. Secondly, any explanation for the driving force of the restacking transition must also explain the modulations that accompany it. In particular, unless one cools rapidly, the same modulation structures with the same amplitudes always appear at the same temperature, regardless of the sample history, i.e., whether heating or cooling. No significant hysteresis is observed and Sirota argued that the structures are in thermal equilibrium.

There are a number of physical systems for which the development of long wavelength modulations is understood, and in each case they are the result of two or more competing interaction energies that cannot be simultaneously minimized (Blinc & Levanyuk, 1986; Safinya, Varady, Chiang & Dimon, 1986b; Lubensky & Ingersent, 1986; Winkor & Clarke, 1986; Moncton, Stephens, Birgeneau, Horn & Brown, 1981; Fleming, Moncton, McWhan & DiSalvo, 1980; Villain, 1980; Frank & van der Merwe, 1949; Bak, Mukamel, Villain & Wentowska, 1979; Pokrovsky & Talapov, 1979). The easiest one to visualize is epitaxial growth of one crystalline phase on the surface of another when the two lattice vectors are slightly incommensurate. The first atomic row of adsorbate molecules can be positioned to minimize the attractive interactions with the substrate. This is slightly more difficult for the second row, since the distance that minimizes the interaction energy between the first and second rows of adsorbate molecules is not necessarily the same as the distance that would minimize the interaction energy between the first row and the substrate. As more and more rows are added, the energy price of this incommensurability builds up, and one possible configuration that minimizes the global energy is a modulated structure.

In all known cases the very existence of modulated structures implies that there must be competing interactions, and the only real question about the modulated structures in the crystalline-B phases is the identification of the competing interactions. It appears that one of the more likely possibilities is the difficulty in packing the 70.7 molecules within a triangular lattice while simultaneously optimizing the area per molecule of the alkane tails and the conjugated rings in the core (Carlson & Sethna, 1987; Sadoc & Charvolin, 1986). Typically the mean cross sectional area for a straight alkane in the all trans configuration is between 18 $Å^2$ and 19 $Å^2$, while the mean area per molecule in the crystalline-B phase is closer to 24 $Å^2$. While these two could be reconciled by assuming that the alkanes are tilted with respect to the conjugated core, there is no reason why the angle that reconciles the two should also be the same angle that minimizes the internal energy of the molecule. Even if it were the correct angle at some temperature by accident, the average area per chain is certainly temperature dependent. Even without attempting to include the rotational dynamics that are necessary to understanding the axial site symmetry it is obvious that there can be a conflict in the packing requirements of the two different parts of the molecule.

A possible explanation of these various structures might be as follows: at high temperatures both the alkane chain, as well as the other degrees of freedom, have considerable thermal motions that make it possible for the conflicting packing requirements to be simultaneously reconciled by one or another compromise. On the other hand with decreasing temperature some of the thermal motions become frozen out, and the energy cost of the reconciliation that was possible at higher temperatures becomes too costly. At this point the system must find another solution, and the various modulated phases represent the different compromises. Finally, all of the compromises involving inhomogeneities, like the modulations or grain boundaries, become impossible and the system transforms into a homogeneous crystalline-G phase.

If this type of argument could be made more specific, it would also provide a possible explanation for the molecular origin of the three dimensional hexatic phases. The original suggestion for the existence of hexatic phases in two dimensions was based on the fact that the interaction energy between dislocations in two dimensions was logarithmic, such that the entropy and the enthalpy had the same functional dependence on the density of dislocations. This gave rise to the observation that above a certain temperature two dimensional crystals would be unstable against thermally generated dislocations. Although Litster and Birgeneau's suggestion that some of the observed smectic phases might be stacks of two dimensional hexatics is certainly correct, it is not necessary that the observed three dimensional hexatics originate from entropy driven thermally excited dislocations. For example, the temperature-layer number phase diagram for 70.7 that is shown in Figure 10 has the interesting property that the temperature region over which the tilted hexatic phases exists in thin films is almost the same as the temperature region for which the modulated phases exist in thick films and in bulk samples.

From the fact that molecules in the n0.m series that only differ by one or two $-CH_2-$ groups have different sequences of mesomorphic phases, we learn that within any one molecule the difference in chemical potentials between the different mesomorphic phases must be very small (Leadbetter, et al., 1979e; Doucet & Levelut, 1977; Leadbetter, et al., 1979b; Leadbetter, et al., 1980b; Smith, Garland & Curtis, 1973; Smith & Garland, 1973). For example, although in 70.7 the smectic-F phase is only observed in finite thickness films, both 50.6 and 90.4 have smectic-F phase in bulk. Thus in bulk 70.7 the chemical potential for the smectic-F phase must be only slightly larger than that of the modulated crystalline-B phases, and the effect of the surfaces must be sufficient to reverse the order in samples of finite thickness.

As far as the appearance of the smectic-F phase in 70.7 is concerned, it is well known that the interaction energy between dislocation pairs is very different near a free surface than it is in the bulk (Pershan, 1974; Pershan & Prost, 1975). The origin of this is that the elastic properties of the surface will usually cause the stress field of a dislocation near to the surface to either vanish or to be considerably smaller than it would in the bulk. Since the interaction energy between dislocations depends on this stress field, the surface significantly modifies the dislocation-dislocation interaction. This is a long range effect, and it would not be surprising if the interactions that stabilized the dislocation arrays to produce the long wavelength modulations in the thick samples would be sufficiently weaker that in the samples of finite thickness, the dislocation arrays are disordered. Alternatively, there is evidence that specific surface interactions favor a finite molecular tilt at temperatures where the bulk phases are uniaxial (Farber, 1985). Incommensurability between the period of the tilted surface molecules and the crystalline-B phases below the surface would increase the density of dislocations, and this would also modify the dislocation-dislocation interactions in the bulk.

Sirota et al. (Sirota, et al., 1985; Sirota, et al., 1987b) demonstrated, that while the correlation lengths of the smectic-F phase have a significant temperature dependence, the lengths are independent of film thickness, and this supports the argument that although the effects of the surface are important in stabilizing the smectic-F phase in 70.7, once the phase is established it is essentially no different than the smectic-F phases observed in bulk samples of other materials. Brock et al. observed anisotropies in the correlation lengths of thick samples of 8OSI that are similar to those observed by Sirota (Brock, et al., 1986).

These observations motivate the hypothesis that the dislocation density in the smectic-F phases are determined by the same incommensurability that gives rise to the modulated crystalline-B structures. Although all of the experimental evidence supporting this hypothesis was obtained from the smectic-F tilted hexatic phase, there is no reason why this speculation could not apply to both the tilted smectic-I and the untilted hexatic-B phase.

4.3.2.   Crystal-G, Crystal-J

The crystalline-G and crystalline-J phases are the ordered versions of the smectic-F and smectic-I phases respectively. The positions of the principal peaks illustrated in Figure 9 for the smectic-F(I) are identical to the positions in the smectic-G(J) phase if small thermal shifts are discounted. In both the hexatic and crystalline phases the molecules are tilted with respect to the layer normals by approximately 25° to 30° with nearly hexagonal packing around the tilted axis (Doucet & Levelut, 1977; Levelut, Doucet & Lambert, 1974; Levelut, 1976; Leadbetter, et al., 1979e; Sirota, et al., 1987b). The interlayer molecular packing appears to be end to end, in an AAA type of stacking (Benattar, et al., 1983; Benattar, Moussa, Lambert & Germian, 1981; Levelut, 1976; Gane, et al., 1983). There is only one molecule per unit cell and there is no evidence for the long wavelength modulations that are so prevalent in the crystalline-B phase that is the next higher temperature phase, above the crystalline-G in 70.7.

4.4.   Crystalline Phases with Herringbone Packing

4.4.1.   Crystal-E

Figure 12 illustrates the intralayer molecular packing proposed for the crystalline-E phase (Levelut, 1976; Doucet, 1979; Levelut, et al., 1974; Doucet, Levelut, Lambert, Lievert & Strzelecki, 1975; Leadbetter, Richardson & Carlile, 1976; Richardson, Leadbetter & Frost, 1978; Leadbetter, Frost, Gaughan & Mazid, 1979a; Leadbetter, et al., 1979b). The molecules are, on average, normal to the layers; however from the optical birefringence it is apparent that the site symmetry is not uniaxial. X-ray diffraction studies on single crystals by Doucet and coworkers demonstrated that the biaxiality was not attributable to molecular tilt and subsequent work by a number of others resulted in the arrangement shown in Figure 12a. The most important distinguishing reciprocal space feature associated with the intralayer "herringbone" packing is the appearance of Bragg peaks at $\sin(\theta)$ equal to $\sqrt{7}/2$ times the value for the lowest order in-plane Bragg peak for the triangular lattice (Pindak, et al., 1981). These are illustrated by the open circles in Figure 12b. The shaded circles correspond to peaks that are missing because of the glide plane that relates the two molecules in the rectangular cell.

Leadbetter et al. (Leadbetter, Mazid & Malik, 1980a) carried out detailed studies on both the crystalline-E phase of isobutyl-4-(4'-phenylbenzylideneamino) cinnamate (IBPBAC) and on the crystalline phase immediately below the crystalline-E phase. Partially ordered samples of the crystalline-E phase were obtained by melting the lower temperature crystalline phase. Although the data for the crystalline-E phase left some ambiguity they argued that the phase they were studying might well have had molecular tilts of the order of 5° or 6° degrees. This is an important distinction, since the crystalline-H and crystalline-J phases are essentially tilted versions of the crystalline-E. Thus,

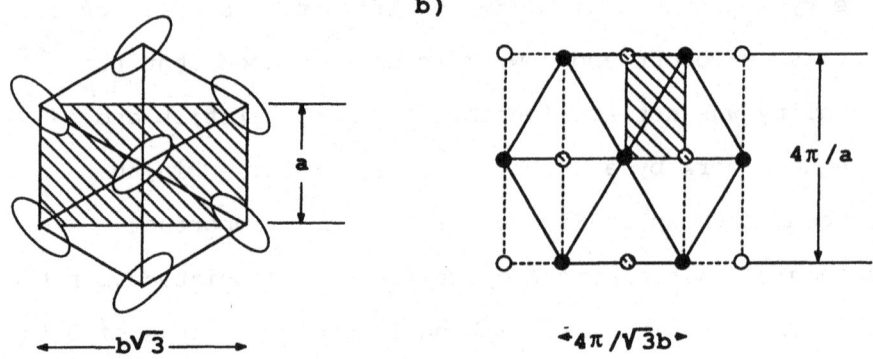

Figure 12: (a) The "herringbone" stacking suggested for the crystalline-E phase in which molecular rotation is partially restricted. The primitive rectangular unit cell containing two molecules is illustrated by the shaded region. The lattice has rectangular symmetry and a ≠ b. (b) The position of the Bragg peaks in the plane in reciprocal space that is parallel to the layers. The dark circles indicate the principal Bragg peaks that would be the only ones present if all molecules were equivalent. The open circles indicate additional peaks that are observed for the model illustrated in (a). The cross hatched circles indicate peaks that are missing because of the glide plane in (a).

one important symmetry difference that might distinguish the crystalline-E from the others is the presence of a mirror plane parallel to the layers. In view of the low symmetry of the individual molecules, the existence of such a mirror plane would imply residual molecular motions. In fact using neutron diffraction Leadbetter et al. (Leadbetter, et al., 1976) demonstrated for a different liquid crystal that even though the site symmetry is not axially symmetric, there is considerable residual rotational motion in the crystalline-E phase about the long axis of the molecules. Since the in-plane spacing is too small for neighboring molecules to be rotating independently of each other, they proposed what might be interpreted as large, partially hindered rotations.

4.4.2. Crystal-H, Crystal-K

The crystalline-H and crystalline-K phases are tilted versions of the crystalline-E. The crystalline-H is tilted in the direction between the near neighbors, with the convenient mnemonic that on cooling the sequence of phases with the same relative orientation of tilt to near neighbor position is F→G→H. Similarly the tilt direction for the crystalline-K phase is similar to that of the smectic-I and crystalline-J so that the expected phase sequence on cooling might be I→J→K. In fact both of these sequences are only intended to indicate the progression in lower symmetry; the actual transitions vary from material to material.

5. Discotic Phases

In contrast to the long thin rod-like molecules that formed most of the other phases discussed in this chapter, the discotic phases are formed by molecules that are more disk-like. See Figure 3e for example. There was evidence that mesomorphic phases were formed from disk-like molecules as far back as 1960 (Brooks & Taylor, 1968); however the first identification of a discotic phase was by Chandrasekhar et al. in 1977 with benzene-hexa-n-alkanoates compounds (Chandrasekhar, Sadashiva & Suresh, 1977). Disk-like molecules can form either a fluid nematic phase in which the disk normals are aligned, without any particular long range order in the molecular center of mass; or more ordered "columnar" (Helfrich, 1979) or "discotic" (Billard, et al., 1981) phases in which the molecular positions are correlated such that the disks stack on top of one another to form columns. Some of the literature designates this nematic phase as N_D to distinguish it from the phase formed by "rod-like" molecules (Destrade, Gasparoux, Foucher, Tinh, Malthête & Jacques, 1983). In the same way that the appearance of layers characterizes order in smectic phases, the order for the discotic phases is characterized by the appearance of columns. Chandrasekhar (Chandrasekhar, 1982, Chandrasekhar, 1983) and Destrade (Destrade, et al., 1983) have reviewed this area and have summarized the several notations for various phases that appear in the literature. Levelut (Levelut, 1983) has also written a review and presented a table listing the space groups for columnar phases formed by eighteen different molecules. Unfortunately, it is not absolutely clear which of these are mesomorphic phases and which are crystals with true long range positional order.

Figure 13 illustrates the molecular packing in two of the well identified discotic phases that are designated as D_1 and D_2 (Chandrasekhar, 1982). The phase D_2 consists of a hexagonal array of columns for which there is not any intracolumnar order. The system is uniaxial and, as originally proposed, the molecular normals were supposed to be along the column axis. However, recent X-ray scattering studies on oriented free standing fibers of the D_2 phase of triphenylene hexa-n-dodecanoate indicate that the molecules are tilted with respect to the layer normal (Safinya, Clark, Liang, Varady & Chiang, 1985; Safinya, Liang, Varady, Clark & Andersson, 1984). The D_1 phase is definitely a tilted phase, and consequently the columns are packed in a rectangular cell. According to Safinya et al., the D_1 to D_2 transition corresponds to an order-disorder transition in which the molecular tilt orientation is ordered about the column axis in the D_1 phase and disordered in the D_2 phase. The reciprocal space structure of the D_1 phase is similar to that of the crystalline-E phase shown in Figure 12b.

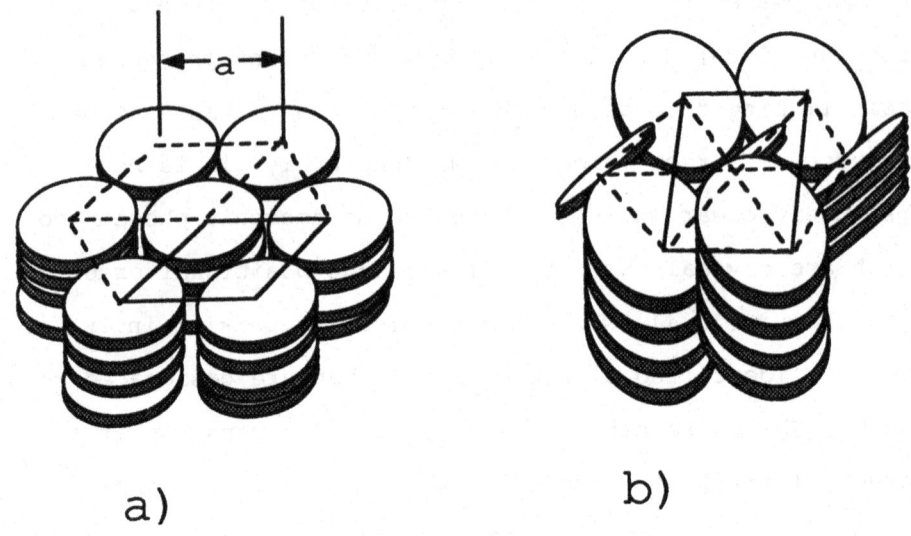

a)   b)

Figure 13: Schematic illustration of the molecular stacking for the discotic (a) D2 and (b) D1 phases. In neither of these two phases is there any indication of long range positional order along the columns. The hexagonal symmetry of the D1 phase is broken by "herringbone-like" correlations in the molecular tilt from column to column.

Other discotic phases that have been proposed would have the molecules arranged periodically along the column, butdisordered between columns. This does not seem physically realistic since it is well known that thermal fluctuations rule out the possibility of a one dimensional periodic structure even more strongly than for the two dimensional lattice that was discussed above (Landau, 1965; Peierls,

1934). On the other hand, in the absence of either more high resolution studies on oriented fibers, or further theoretical studies, we prefer not to speculate on the variety of possible true discotic or discotic like crystalline phases that might exist. This is a subject for future research.

6. Other Phases

We have deliberately chosen not to discuss the
properties of the cholesteric phase in this book because the
length scales that characterize the long range order are of
the order of microns and are more easily studied by optical
scattering than by X-rays (De Gennes, 1974; De Vries, 1951).
Nematic phases formed from chiral molecules develop long
range order in which the orientation of the director <n>
varies in a plane wave-like manner that can be described as
$\mathbf{x}\cos(2\pi z/\lambda) + \mathbf{y}\sin(2\pi z/\lambda)$, where $\mathbf{x}$ and $\mathbf{y}$ are unit vectors and
$\lambda/2$ is the cholesteric "pitch" that can be anywhere from 0.1
to 10 microns depending on the particular molecule. Even more
interesting is that for many cholesteric systems there is a
small temperature range, of the order of 1°C between the
cholesteric and isotropic phases for which there is a phase
known as the "blue phase" (Coates & Gray, 1975; Stegemeyer &
Bergmann, 1981; Meiboom, Sethna, Anderson & Brinkman, 1981;
Bensimon, Domany & Shtrikman, 1983; Hornreich & Shtrikman,
1983; Crooker, 1983). In fact there is more than one "blue
phase" but they all have the property that the cholesteric
twist forms a three dimensional lattice twisted network
rather than the plane wave-like twist of the cholesteric
phase. Three dimensional Bragg scattering from blue phases
using laser light indicates cubic lattices; however, since
the optical cholesteric interactions are much stronger than

the usual interactions between x-rays and atoms, interpretation of the results is subtler.

Gray and Goodby discuss a "smectic-D" phase that is otherwise omitted from this chapter (Gray & Goodby, 1984). Gray and co-workers first observed this phase in the homologous series of 4'-n-alkoxy-3'-nitrobiphenyl-4-carboxylic acids (Gray, Jones & Marson, 1957). In the hexadecyloxy compound this phase exists for a region of about $16°C$ between the smectic-C and smectic-A phases: smectic-C ($171.0°C$) smectic-D ($197.2°C$) smectic-A. It is optically isotropic and X-ray studies by Diele, Brand and Sackmann (Diele, Brand & Sackmann, 1972) and by Tardieu and Billard (Tardieu & Billard, 1976) indicate a number of similarities to the "cubic-isotropic" phase observed in lyotropic systems (Luzzati & Reiss-Husson, 1966; Tardieu & Luzzati, 1970). More recently, Etherington (Etherington, Leadbetter, Wang, Gray & Tajbakhsh, 1986) studied the "smectic-D" phase of 4'-n-octadecyloxy-3'-cyanobinphenyl-4-carboxylic acid. Since this material appears to be more stable than some of the others that were previously studied they were able to perform sufficient measurements to determine that space group is either the cubic P23 or Pm3 with a lattice parameter of 86Å. Etherington et al. suggested that the "smectic-D" phase that they studied is a true three dimensional cubic crystal of micelles and noted that the designation of "smectic-D" is not accurate.Guillon and Skoulios have recently proposed a

molecular model for this, and related phases (Guillon &
Skoulios, 1987).

Fontell has reviewed the literature on the X-ray
diffraction studies of lyotropic mesomorphic systems and the
reader is referred there for more extensive information on
those cubic systems (Fontell, 1974). The mesomorphic
structures of lyotropic systems are much richer than that of
the thermotropic and, in addition to all structures mentioned
here, there are lyotropic systems in which the smectic-A
lamella seem to break up into cylindrical rods which seem to
have the same macroscopic symmetry as some of the discotic
phases. On the other hand, it is also much more difficult to
prepare a review for the lyotropic systems in the same type
of detail as for the thermotropic. The extra complexity
associated with the need to control water concentration as
well as temperature has made both theoretical and
experimental progress more difficult, and since there has not
been very much experimental work on well oriented samples,
detailed knowledge of many of these phases is also limited.
Aside from the simpler lamella systems, which seem to have
the same symmetry as the thermotropic smectic-A phase, it is
not at all clear which of the other phases are three
dimensional crystals and which are true mesomorphic
structures. For example, dipalmitoyl-phosphatidylcholine has
a L_β-phase that appears for temperatures and (or) water
content that is lower than that of the smectic-A L_α-phase

(Shipley, Hitchcock, Mason & Thomas, 1974; Small, 1967;
Chapman, Williams & Ladbrooke, 1967). The diffraction pattern
for this phase contains sharp large angle reflections that
may well correspond to a phase that is like one of the
crystalline phases listed in Table 1, Figure 1. On the other
hand this phase could also be hexatic and we do not have
sufficient information to decide. The interested reader is
referred to the referenced articles for further detailed
information.

*reprinted in this volume

7. References

AEPPLI, G. & BRUINSMA, R. (1984). *Hexatic order and liquid crystal density fluctuations. Phys. Rev. Lett.* **53**, 2133-2136.

*AEPPLI, G., LITSTER, J.D., BIRGENEAU, R.J. & PERSHAN, P.S. (1981). *High resolution x-ray study of the smectic A-smectic B phase transition and the smectic B phase in butyloxybenzylidene octylaniline. Mol. Cryst. Liq. Cryst.* **67**, 205-214.

*AHARONY, A., BIRGENEAU, R.J., BROCK, J.D. & LITSTER, J.D. (1986). *Multicriticality in hexatic liquid crystals. Phys. Rev. Lett.* **57**, 1012-1015.

ALBEN, R. (1973). *Phase transitions in a fluid of biaxial particles. Phys. Rev. Lett.* **30**, 778-781.

ALS-NIELSEN, J., BIRGENEAU, R.J., KAPLAN, M., LITSTER, J.D. & SAFINYA, C.R. (1977). *High resolution x-ray study of a second-order nematic-smectic-A phase transition. Phys. Rev. Lett.* **39**, 352-355.

ALS-NIELSEN, J., CHRISTENSEN, F. & PERSHAN, P.S. (1982). *Smectic-A order at the surface of a nematic liquid crystal: synchrotron x-ray diffraction. Phys. Rev. Lett.* **48**, 1107-1110.

*ALS-NIELSEN, J., LITSTER, J.D., BIRGENEAU, R.J., KAPLAN, M., SAFINYA, C.R., LINDEGAARD, A. & MATHIESEN, S. (1980). *Observation of algebraic decay of positional order in a smectic liquid crystal. Phys. Rev. B* **22**, 312-320.

BAK, P., MUKAMEL, D., VILLAIN, J. & WENTOWSKA, K. (1979). *Commensurate-incommensurate transitions in rare-gas monolayers adsorbed on graphite and in layered charge-density-wave systems. Phys. Rev. B* **19**, 1610-1613.

BAROIS, P., PROST, J. & LUBENSKY, T.C. (1985). *New critical points in frustrated smectics. J. Phys. (Paris)* **46**, 391-399.

BEAGLEHOLE, D. (1982). *Pretransition order on the surface of a nematic liquid crystal. Mol. Cryst. Liq. Cryst.* **89**, 319-325.

[*]BENATTAR, J.J., DOUCET, J., LAMBERT, M. & LEVELUT, A.M. (1979). *Nature of the smectic F phase. Phys. Rev. A* **20**, 2505-2509.

BENATTAR, J.J., LEVELUT, A.M. & STRZELECKI, L. (1978). *Etude de l'influence de la longueur moleculaire les characteristiques des phases smectique ordonnees. J. Phys. (Paris)* **39**, 1233-1240.

BENATTAR, J.J., MOUSSA, F. & LAMBERT, M. (1980). *Two-dimensional order in the smectic F phase. J. Phys. (Paris)* **41**, 1371-1374.

BENATTAR, J.J., MOUSSA, F. & LAMBERT, M. (1983). *Two dimensional ordering in liquid crystals: the SmF and Smi phases. J. Chim. Phys.* **80**, 99-107.

BENATTAR, J.J., MOUSSA, F., LAMBERT, M. & GERMIAN, C. (1981). *Two kinds of two-dimensional order: the SmF and SmI phases. J. Phys. (Paris) Lett.* **42**, L67-L70.

BENGUIGUI, L. (1979). *A Landau theory of the NAC point. J. Phys. (Paris) Colloq.* **40**, C3-419-C3-421.

BENSIMON, D., DOMANY, E. & SHTRIKMAN, S. (1983). *Optical activity of cholesteric liquid crystals in the pretransitional regime and in the blue phase. Phys. Rev. A* **28**, 427-433.

BILLARD, J., DUBOIS, J.C., VAUCHER, C. & LEVELUT, A.M. (1981). *Structures of the two discophases of rufigallol hexa-n-octanoate. Mol. Cryst. Liq. Cryst.* **66**, 115-122.

*BIRGENEAU, R.J. & LITSTER, J.D. (1978). *Bond orientational order model for smectic B liquid crystals. J. Phys. (Paris) Lett.* **39**, L399-L402.

BIRGENEAU, R.J., GARLAND, C.W., KASTING, G.B. & OCKO, B.M. (1981). *Critical behavior near the nematic-smectic-A transition in butyloxybenzilidene octylaniline (40.8). Phys. Rev. A* **24**, 2624-2634.

BLINC, R. & LEVANYUK, A.P., Editors (1986). *Modern Problems in Condensed Matter Sciences, Incommensurate Phases in Dielectrics, 2. Materials.* Amsterdam: North-Holland.

BRISBIN, D., DE HOFF, R., LOCKHART, T.E. & JOHNSON, D.L. (1979). *Specific heat near the nematic-smectic-A tricritical point. Phys. Rev. Lett.* **43**, 1171-1174.

*BROCK, J.D., AHARONY, A., BIRGENEAU, R.J., EVANS-LUTTERODT, K.W., LITSTER, J.D., HORN, P.M., STEPHENSON, G.B. & TAJBAKHSH, A.R. (1986). *Orientational and positional* *order in a tilted hexatic liquid crystal phase. Phys. Rev. Lett.* **57**, 98-101.

BROOKS, J.D. & TAYLOR, G.H. (1968). In *Chemistry and Physics of Carbon*, *Vol. 3*, edited by Phillip L. Walker, Jr., pp. 243-286. New York: Marcel Dekker.

BROWNSEY, G.L. & LEADBETTER, A.J. (1981). *Novel liquid crystal structures in cyano bicyclohexanes. J. Phys. (Paris) Lett.* **42**, L135-L139.

BRUINSMA, R. & NELSON, D.R. (1981). *Bond orientational order in smectic liquid crystals. Phys. Rev. B* **23**, 402-410.

BUDAI, J., PINDAK, R., DAVEY, S.C. & GOODBY, J.W. (1984). *A structural investigation of the liquid crystal phase of 4-(2'-methylbytyl)phenyl 4'-n-octylbiphenyl-4-carboxylate. J. Phys. (Paris) Lett.* **45**, L1053-L1062.

*CAILLÉ, A. (1972). *Remarques sur la diffusion des rayons X dans les smectiques A. C. R. Adad. Sc. Paris* **274B**, 891-893.

CARLSON, J.M. & SETHNA, J.P. (1987). *Theory of the ripple phase in hydrated phospholipid bilayers. Preprint.*

*CHAN, K.K., DEUTSCH, M., OCKO, B.M., PERSHAN, P.S. & SORENSEN, L.B. (1985a). *Integrated x-ray scattering intensity measurement of the order parameter at the nematic to smectic-A phase transition. Phys. Rev. Lett.* **54**, 920-923.

CHAN, K.K., PERSHAN, P.S., SORENSEN, L.B. & HARDOUIN, F. (1985b). *X-ray scattering study of the smectic-A1 to smectic-A2 transition. Phys. Rev. Lett.* **54**, 1694-1697.

CHAN, K.K., PERSHAN, P.S., SORENSEN, L.B. & HARDOUIN, F. (1986). *X-ray studies of transitions between nematic, smectic-A1, -A2, and -Ad phases. Phys. Rev. A* **34**, 1420-1433.

CHANDRASEKHAR, S. (1982). In *Advances in Liquid Crystals, Vol. 5,* edited by G.H. Brown, pp. 47-78. London and New York: Academic Press.

CHANDRASEKHAR, S. (1983). *Liquid crystals of disk-like molecules. Phil. Trans. R. Soc. Lond. A* **309**, 93-103.

CHANDRASEKHAR, S., SADASHIVA, B.K. & SURESH, K.A. (1977). *Liquid crystals of disk like molecules. Pramana* **9**, 471-480.

CHAPMAN, D., WILLIAMS, R.M. & LADBROOKE, B.D. (1967). *Physical studies of phospholipids. VI. Thermotropic and lyotropic mesomorphism of some 1,2-diacyl-phosphatidylcholines (lecithins). Chem. Phys. Lipids* **1**, 445-475.

CHEN, J.H. & LUBENSKY, T.C. (1976). *Landau-Ginzberg mean-field theory for the nematic to smectic-C and nematic to smectic-A phase transitions. Phys. Rev. A* **14**, 1202-1207.

CHU, K.C. & MCMILLAN, W.L. (1977). *Unified Landau theory for the nematic, smectic A and smectic C phases of liquid crystals. Phys. Rev. A* **15**, 1181-1187.

COATES, D. & GRAY, G.W. (1975). *A correlation of optical features of amorphous liquid-cholesteric liquid crystal transitions. Phys. Lett. A* **51**, 335-336.

COLLETT, J. (1983). *Ph.D. Thesis. Harvard University. Unpublished.*

[*]COLLETT, J., PERSHAN, P.S., SIROTA, E.B. & SORENSEN, L.B. (1984). *Synchrotron x-ray study of the thickness dependence of the phase diagram of thin liquid-crystal films. Phys. Rev. Lett.* **52**, 356-359.

[*]COLLETT, J., SORENSEN, L.B., PERSHAN, P.S. & ALS-NIELSEN, J. (1985). *X-ray scattering study of restacking transitions in the crystalline-B phases of heptyloxybenzylidene heptylaniline 70.7. Phys. Rev. A* **32**, 1036-1043.

[*]COLLETT, J., SORENSEN, L.B., PERSHAN, P.S., LITSTER, J., BIRGENEAU, R.J. & ALS-NIELSEN, J. (1982). *Synchrotron x-ray study of novel crystalline-B phases in 70.7. Phys. Rev. Lett.* **49**, 553-556.

CROOKER, P.P. (1983). *The cholesteric blue phase: a progress report. Mol. Cryst. Liq. Cryst.* **98**, 31-45.

DAVEY, S.C., BUDAI, J., GOODBY, J.W., PINDAK, R. & MONCTON, D.E. (1984). *X-ray study of the hexatic-B to smectic-A phase transition in liquid crystal films. Phys. Rev. Lett.* **53**, 2129-2132.

DAVIDOV, D., SAFINYA, C.R., KAPLAN, M., DANA, S.S., SCHAETZING, R., BIRGENEAU, R.J. & LITSTER, J.D. (1979). *High-resolution x-ray and light-scattering study of* critical behavior associated with the nematic-smectic-A transition in 4-cyano-4'-octylbiphenyl. *Phys. Rev. B* **19**, 1657-1663.

DE GENNES, P.G. (1969a). *Conjectures sur l'état smectique. J. Phys. (Paris) Colloq.* **30**, C4-65-C4-71.

DE GENNES, P.G. (1969b). *Phenomenology of short-range-order effects in the isotropic phase of nematic materials. Phys. Lett. A* **30**, 454-455.

DE GENNES, P.G. (1971). *Short range order effects in the isotropic phase of nematics and cholesterics. Mol. Cryst. Liq. Cryst.* **12**, 193-214.

[*]DE GENNES, P.G. (1972). *An analogy between super conductors and smectics A. Solid State Commun.* **10**, 753-756.

DE GENNES, P.G. (1973). *Some remarks on the polymorphism of smectics. Mol. Cryst. Liq. Cryst.* **21**, 49-76.

DE GENNES, P.G. (1974). *The Physics of Liquid Crystals.* Oxford: Clarendon Press.

DEMUS, D., DIELE, S., KLAPPERSTÜCK, M., LINK, V. & ZASCHKE, H. (1971). *Investigation of a smectic tetramorphous substance. Mol. Cryst. Liq. Cryst.* **15**, 161-174.

DESTRADE, C., GASPAROUX, H., FOUCHER, P., TINH, N.H., MALTHETE, J. & JACQUES, J. (1983). *Molecules discoides et polymorphisme mesomorphe. J. Chim. Phys.* **80**, 137-148.

DE VRIES, H.L. (1951). *Rotary power and other optical properties of liquid crystals. Acta Cryst.* **4**, 219-226.

[*]DIELE, S., BRAND, P. & SACKMANN, H. (1972). *X-ray diffraction and polymorphism of smectic liquid crystals. II D and E modifications. Mol. Cryst. Liq. Cryst.* **17**, 163-169.

DJUREK, D., BATURIC-RUBCIC, J. & FRANULOVIC, K. (1974). *Specific-heat critical exponents near the nematic-smectic-A phase transition. Phys. Rev. Lett.* **33**, 1126-1129.

DOUCET, J. & LEVELUT, A.M. (1977). *X-ray study of the ordered smectic phases in some benzylidenianilines. J. Phys. (Paris)* **38**, 1163-1170.

DOUCET, J. (1979). In *Molecular Physics of Liquid Crystals*, edited by G.W. Gray, G.R. Luckhurst, pp. 317-341. London and New York: Academic Press.

DOUCET, J., LEVELUT, A.M., LAMBERT, M., LIEVERT, L. & STRZELECKI, L. (1975). *Nature de la phase smectique. J. Phys. (Paris) Colloq.* **36**, C1-13-C1-19.

[*]ETHERINGTON, G., LEADBETTER, A.J., WANG, X.J., GRAY, G.W. & TAJBAKHSH, A. (1986). *Structure of the smectic D phase. Liquid Crystals* **1**, 209-214.

FAETTI, S. & PALLESCHI, V. (1984). *Nematic isotropic interface of some members of the homologous series of the 4-cyano-4'-(n alkyl)biphenyl liquid crystals. Phys. Rev. A* **30**, 3241-3251.

FAETTI, S., GATTI, M., PALLESCHI, V. & SLUCKIN, T.J. (1985). *Almost critical behavior of the anchoring energy at the interface between a nematic liquid crystal and a substrate.* Phys. Rev. Lett. **55**, 1681–1684.

FAN, C.P. & STEPHEN, M.J. (1970). *Isotropic-nematic phase transition in liquid crystals.* Phys. Rev. Lett. **25**, 500–503.

FARBER, A.S. (1985). *Ph.D. Thesis. Brandeis University (Unpublished).*

FLEMING, R.M., MONCTON, D.E., MCWHAN, D.B. & DISALVO, F.J. (1980). In *Ordering in Two Dimensions, Proceedings of an International Conference held at Lake Geneva, Wisconsin,* edited by Sunil K. Sinha, pp. 131–134. New York: North Holland.

FONTELL, K. (1974). In *Liquid Crystals and Plastic Crystals, Vol. II,* edited by G.W. Gray, P.A. Winsor, pp. 80–109. Chichester, U.K.: Ellis Horwood Publishers.

FONTES, E., HEINEY, P.A., HASELTINE, J.H. & SMITH, A.B. III (1986). *High resolution x-ray scattering study of the multiply reentrant polar mesogen DB90NO$_2$.* J. Phys. *(Paris)* **47**, 1533–1539.

FRANK, F.C. & VAN DER MERWE, J.H. (1949). *One-dimensional dislocations. I. Static theory.* Pro. Roy. Soc., London **A198**, 205–216.

FREISER, M.J. (1971). *Successive transitions in a nematic liquid.* Mol. Cryst. Liq. Cryst. **14**, 165–182.

FRIEDEL, G. (1922). *Les etats mesormorphes de la matiere.* Ann. Phys. *(Paris)* **18**, 273–474.

GANE, P.A.C. & LEADBETTER, A.J. (1981). *The crystal and molecular structure of N-(4-n octyloxybenzylidene)-4'-butylaniline (80.4) and the crystal-smectic G transition. Mol. Cryst. Liq. Cryst.* **78**, 183-200.

GANE, P.A.C. & LEADBETTER, A.J. (1983). *Modulated crystal B phases and the B- to G- phase transition in two types of liquid crystalline compounds. J. Phys. C* **16**, 2059-2067.

GANE, P.A.C., LEADBETTER, A.J., WRIGHTON, P.G., GOODBY, J.W., GRAY, G.W. & TAJBAKHSH, A.R. (1983). *The phase behavior of bis-(4'-n-heptyloxybenzylidene)-1,4-phenylenediamine(HEPTOBPD), crystal J and K phases. Mol. Cryst. Liq. Cryst.* **100**, 67-74.

GANNON, M.G.J. & FABER, T.E. (1978). *The surface tension of nematic liquid crystals. Phil. Mag. A* **37**, 117-135.

GARLAND, C.W., MEICHLE, M., OCKO, B.M., KORTAN, A.R., SAFINYA, C.R., YU, L.J., LITSTER, J.D. & BIRGENEAU, R.J. (1983). *Critical behavior at the nematic-smectic-A transition in butyloxybenzylidene heptylaniline (40.7). Phys. Rev. A* **27**, 3234-3240.

GRANSBERGEN, E.F., DE JEU, W.H. & ALS-NIELSEN, J. (1986). *Antiferroelectric surface layers in a liquid crystal as observed by synchrotron x-ray scattering. J. Phys. (Paris)* **47**, 711-718.

GRAY, G.W. & GOODBY, J.W. (1984). *Smectic Liquid Crystals: Textures and Structures.* Glasgow: Leonard Hill.

GRAY, G.W., JONES, B. & MARSON, F. (1957). *Mesomorphism and chemical constitution. Part VIII. The effect of 3'-substituents on the mesomorphism of the 4'-n-alkoxydiphenyl-4-carboxylic acids and their alkyl esters.* J. Chem. Soc. **1**, 393-401.

GRINSTEIN, G. & TONER, J. (1983). *Dislocation-loop theory of the nematic-smectic A-smectic C multicritical point.* Phys. Rev. Lett. **51**, 2386-2389.

GUILLON, D. & SKOULIOS, A. (1987). *Molecular model for the R smectic DS mesophase.* Europhys. Lett. **3**, 79-85.

GUILLON, D., SKOULIOS, A. & BENATTAR, J.J. (1986). *Volume and x-ray diffraction study of terephthal-bis-4,n-decylaniline (TBDA).* J. Phys. (Paris) **47**, 133-138.

GUNTHER, L., IMRY, Y. & LAJZEROWICZ, J. (1980). *X-ray scattering in smectic-A liquid crystals.* Phys. Rev. A **22**, 1733-1740.

GUYOT-SIONNEST, P., HSIUNG, H. & SHEN, Y.R. (1986). *Surface polar ordering in a liquid crystal observed by optical second-harmonic generation.* Phys. Rev. Lett. **57**, 2963-2966.

[*]HALPERIN, B. I. & NELSON, D. R. (1978), *Theory of two-dimensional melting.* Phys. Rev. Lett. **41**, 121-124. **41**, 519(E).

HARDOUIN, F., LEVELUT, A.M., ACHARD, M.F. & SIGAUD, S. (1983). *Polymorphisme des substances mesogenes a* molecules polaires. I. Physico-chimie et structure. J. Chim. Phys. **80**, 53-64.

HARDOUIN, F., LEVELUT, A.M., BENATTAR, J.J. & SIGAUD, G. (1980). X-rays investigations of the smectic A1-smectic A2 transition. Solid State Comm. **33**, 337-340.

HARDOUIN, F., SIGAUD, G., TINH, N.H. & ACHARD, M.F. (1981). A fluid smectic A antiphase in a pure nitro rod-like compound. J. Phys. (Paris) Lett. **42**, L63-L66.

HARDOUIN, F., TINH, N.H., ACHARD, M.F. & LEVELUT, A.M. (1982). A new thermotropic smectic phase made of ribbons. J. Phys. (Paris) Lett. **43**, L327-L331.

HELFRICH, W. (1980). Structure of liquid crystals especially order in two dimensions. J. Phys. (Paris) Colloq. **40**, C3-105-C3-114.

HENDRIKX, Y., CHARVOLIN, J. & RAWISO, M. (1986). Uniaxial-biaxial transition in lyotropic nematic solutions: local biaxiality in the uniaxial phase. Phys. Rev. B **33**, 3534-3537.

[*]HIRTH, J.P., PERSHAN, P.S., COLLETT, J., SIROTA, E. & SORENSEN, L.B. (1984). Dislocation model for restacking phase transitions in crystalline-B liquid crystals. Phys. Rev. Lett. **53**, 473-476.

HORNREICH, R.M. & SHTRIKMAN, S. (1983). Theory of light scattering in cholesteric blue phases. Phys. Rev. A **28**, 1791-1807.

HORNREICH, R.M., LUBAN, M. & SHTRIKMAN, S. (1975). *Critical behavior at the onset of k-space instability on the lamada line. Phys. Rev. Lett.* **35**, 1678-1681.

HUANG, C.C. & LIEN, S.C. (1981). *Nature of a nematic-smectic-A-smectic-C multicritical point. Phys. Rev. Lett.* **47**, 1917-1920.

HUANG, C.C., LIEN, S.C., DUMRONGRATTANA, S. & CHIANG, L.Y. (1984). *Calorimetric studies near the smectic-A1-smectic-Ã phase transition of a liquid crystal compound. Phys. Rev. A* **30**, 965-967.

HUSE, D.A. (1985). *Fisher renormalization at the smectic-A1 to smectic-A2 transition in a mixture. Phys. Rev. Lett.* **55**, 2228-2228.

KASTING, G.B., LUSHINGTON, K.J. & GARLAND, C.W. (1980). *Critical heat capacity near the nematic-smectic-A transition in octyloxycyanobiphenyl in the range 1-2000 bar. Phys. Rev. B* **22**, 321-331.

KITTEL, C. (1963). *Quantum Theory of Solids.* New York: Wiley.

KOSTERLITZ, J.M. & THOULESS, D.G. (1973). *Ordering, metastability and phase transitions in two-dimensions. J. Phys. C* **6**, 1181-1203.

LANDAU, L.D. & LIFSHITZ, E.M. (1958). *Statistical Physics.* London: Pergamon Press.

LANDAU, L.D. (1965). In *Collected Papers of L.D. Landau*, edited by D. ter haar, pp. 193-216. New York: Gordon Breach.

LAWSON, K.D. & FLAUTT, T.J. (1967). *Magnetically oriented lyotropic liquid crystalline phases. J. Am. Chem. Soc.* **89**, 5489-5491.

LEADBETTER, A.J., DURRANT, J.L.A. & RUGMAN, M. (1977). *The density of 4-n-octyl-4'-cyano-biphenyl (8CB). Mol. Cryst. Liq. Cryst.* **34**, 231-235.

*LEADBETTER, A.J., FROST, J.C. & MAZID, M.A. (1979b). *Interlayer correlations in smectic B phases. J. Phys. (Paris) Lett.* **40**, L325-L329.

LEADBETTER, A.J., FROST, J.C., GAUGHAN, J.P. & MAZID, M.A. (1979a). *The structure of the crystal, smectic E and smectic B forms of IBPAC. J. Phys. (Paris) Colloq.* **40**, C3-185-C3-192.

LEADBETTER, A.J., FROST, J.C., GAUGHAN, J.P., GRAY, G.W. & MOSLEY, A. (1979c). *The structure of smectic A phases of compounds with cyano end groups. J. Phys. (Paris) Colloq.* **40**, C3-375-C3-380.

LEADBETTER, A.J., GAUGHAN, J.P., KELLEY, B., GRAY, G.W. & GOODBY, J.J. (1979d). *Characterisation and structure of some new smectic F phases. J. Phys. (Paris) Colloq.* **40**, C3-178-C3-184.

LEADBETTER, A.J., MAZID, M.A. & KELLY, B.A. (1979e). *Structure of the smectic-B phase and the nature of the smectic-B to H transition in the N-(4-n alkoxybenzylidene)-4' alkylanilines. Phys. Rev. Lett.* **43**, 630-633.

LEADBETTER, A.J., MAZID, M.A. & MALIK, K.M.A. (1980a). *The crystal and molecular structure of isobutyl 4-(4'-phenylbenzylidene-amino)cinnamate(IBPBAC)-and the crystal smectic E transition. Mol. Cryst. Liq. Cryst.* **61**, 39-60.

LEADBETTER, A.J., MAZID, M.A. & RICHARDSON, R.M. (1980b). In *Liquid Crystals*, edited by S. Chandrasekhar, pp. 65-79. London: Heyden and Sons.

LEADBETTER, A.J., RICHARDSON, R.M. & CARLILE, C.J. (1976). *The nature of the smectic E phase. J. Phys. (Paris) Colloq.* **37**, C3-65-C3-68.

LEE, S.D. & MEYER, R.B. (1986). *Computations of the phase equilibrium, elastic constants, and viscosities of a hard-rod nematic liquid crystal. J. Chem. Phys.* **84**, 3443-3448.

LEVELUT, A.M. & LAMBERT, M. (1971). *Structure des cristaux liquides smectic B. C.R. Acad Sci.* **272B**, 1018-1021.

[*]LEVELUT, A.M. (1976). *Étude de l'ordre local lié a la rotation des molécules dans la phase smectique B. J. Phys. (Paris) Colloq.* **37**, C3-51-C3-54.

[*]LEVELUT, A.M. (1983). *Structures des phases mesomorphes formee de molecules discoides. J. Chim. Phys.* **80**, 149-161.

LEVELUT, A.M., DOUCET, J. & LAMBERT, M. (1974). *Etude par diffusion de rayons X de la nature des phases smectiques B et de la transition de phase solide-smectique B. J. Phys. (Paris)* **35**, 773-779.

[*]LEVELUT, A.M., TARENTO, R.J., HARDOUIN, F., ACHARD, M.F. & SIGAUD, G. (1981). *Number of SA phases. Phys. Rev. A* **24**, 2180-2186.

LIEBERT, L.E., Editor (1978). *"Liquid Crystals" In Solid State Physics: Advances in Research and Applications*, edited by H. Ehrenreich, F. Seitz, D. Turnbull, Supp. 14. New York: Academic Press.

LITSTER, J.D., ALS-NIELSEN, J., BIRGENEAU, R.J., DANA, S.S., DAVIDOV, D., GARCIA-GOLDING, F., KAPLAN, M., SAFINYA, C.R. & SCHAETZING, R. (1979). *High resolution x-ray and light scattering studies of bilayer smectic A compounds. J. Phys. (Paris) Colloq.* **40**, C3-339-C3-344.

LUBENSKY, T.C. & INGERSENT, K. (1986). *Patterns in systems with competing incommensurate lengths. Preprint.*

LUBENSKY, T.C. (1983). *The nematic to smectic A transition: a theoretical overview. J. Chim. Phys.* **80**, 31-43.

LUBENSKY, T.C. (1987). *Mean field theory for the biaxial nematic phase and the NNUAC critical point. Mol. Cryst. Liq. Cryst.* **146**, 55-69.

LUZZATI, V. & REISS-HUSSON, F. (1966). *Structure of the cubic phase of lipid-water systems. Nature* **210**, 1351-1352.

MA, S.K. (1976). *Modern Theory of Critical Phenomena*. Reading, Massachusetts: W. A. Benjamin, Inc.

MADA, H. & KOBAYASHI, (1981). *Surface order parameter of 4-n-heptyl-4'-cyanobiphenyl. Mol. Cryst. Liq. Cryst.* **66**, 57-60.

MAIER, W. & SAUPE, A. (1958). *Eine einfache molekulare Theorie des nematischen kristallinglüssigen Zustandes. Z. Naturf* **13A**, 564-566.

MAIER, W. & SAUPE, A. (1959). *Eine einfache molekulare-statistiche Theorie der nematischen kristallinglüssigen phase. Teil I.. Z. Naturf* **14A**, 882-889.

MALTHÉTE, J., LIÉBERT, L., LEVELUT, A.M. & GALERNE, Y. (1986). *Némmatic biaxe thermotrope. C.R. Adac. Sc. Paris* **303**, 1073-1076.

MARTIN, P.C., PARODI, O. & PERSHAN, P.S. (1972). *Unified hydrodynamic theory for crystals, liquid crystals, and normal fluids. Phys. Rev. A* **6**, 2401-2420.

*MARTINEZ-MIRANDA, L.J., KORTAN, A.R. & BIRGENEAU, R.J. (1986). *X-ray study of fluctuations near the nematic-smectic-A-smectic-C multicritical point. Phys. Rev. Lett.* **56**, 2264-2267.

MCMILLAN, W.L. (1972). *X-ray scattering from liquid crystals. I. Cholesteryl nonanoate and myristate. Phys. Rev. A* **6**, 936-947.

*MCMILLAN, W.L. (1973a). *Measurement of smectic-A-phase order-parameter fluctuations near a second-order smectic-A-nematic-phase transition. Phys. Rev. A* **7**, 1419-1422.

MCMILLAN, W.L. (1973b). *Measurement of smectic-A-phase order-parameter fluctuations in the nematic phase of p-n-octyloxybenzylidene-p'-toluidine. Phys. Rev. A* **7**, 1673-1678.

MCMILLAN, W.L. (1973c). *Measurement of smectic-phase order-parameter fluctuations in the nematic phase of heptyloxybenzene. Phys. Rev. A* **8**, 328-331.

MEIBOOM, S., SETHNA, J.P., ANDERSON, P.W. & BRINKMAN, W.F. (1981). *Theory of the blue phase of cholesteric liquid crystals. Phys. Rev. Lett.* **46**, 1216-1219.

MIYANO, K. (1979). *Wall-induced pretransitional birefringence: a new tool to study boundary aligning forces in liquid crystals. Phys. Rev. Lett.* **43**, 51-54.

[*]MONCTON, D.E. & PINDAK, R. (1979). *Long-range order in two- and three-dimensional smectic-B liquid crystal films. Phys. Rev. Lett.* **43**, 701-704.

MONCTON, D.E., PINDAK, R., DAVEY, S.C. & BROWN, G.S. (1982). *Melting of variable thickness liquid crystal thin films: a synchrotron x-ray study. Phys. Rev. Lett.* **49**, 1865-1868.

MONCTON, D.E., STEPHENS, P.W., BIRGENEAU, R.J., HORN, P.M. & BROWN, G.S. (1981). *Synchrotron x-ray study of the commensurate-incommensurate transition of monolayer krypton on graphite. Phys. Rev. Lett.* **46**, 1533-1536.

MOUSSA, F., BENATTAR, J.J. & WILLIAMS, C. (1983). *Positional order and bond orientation order in the liquid crystal smectic F phase. Mol. Cryst. Liq. Cryst.* **99**, 145-154.

NELSON, D.R. & HALPERIN, B.I. (1979). *Dislocation-mediated melting in two dimensions. Phys. Rev. B* **19**, 2457-2484.

*NELSON, D.R. & HALPERIN, B.I. (1980). *Solid and fluid phases in smectic layers with tilted molecules. Phys. Rev. B* **21**, 5312-5329.

NELSON, D.R. & TONER, J. (1981). *Bond-orientational order, dislocation loops, and melting of solids and smectic-A liquid crystals. Phys. Rev. B* **24**, 363-387.

NETO, A.M.F., GALERNE, Y., LEVELUT, A.M. & LIEBERT, L. (1985). *Pseudo lamellar ordering in uniaxial and biaxial lyotropic nematics: a synchrotron x-ray diffraction experiment. J. Phys. (Paris) Lett.* **46**, L499-L505.

OCKO, B.M., BIRGENEAU, R.J. & LITSTER, J.D. (1986). *Crossover to tricritical behavior at the nematic to smectic A transition: an x-ray scattering study. Z. Physik B* **62**, 487-497.

*OCKO, B.M., BIRGENEAU, R.J., LITSTER, J.D. & NEUBERT, M.E. (1984). *Critical and tricritical behavior at the nematic to smectic-A transition. Phys. Rev. Lett.* **52**, 208-211.

*OCKO, B.M., BRASLAU, A., PERSHAN, P.S., ALS-NIELSEN, J. & DEUTSCH, M. (1986). *Quantized layer growth at liquid crystal surfaces. Phys. Rev. Lett.* **57**, 94-97.

OCKO, B.M., PERSHAN, P.S., SAFINYA, C.R. & CHIANG, L.Y. (1987). *Incommensurate smectic order at the free surface in the nematic phase of DB7NO$_2$. Phys. Rev. A* **35**, 1868-1872.

ONSAGER, L. (1949). *The effects of shapes on the interaction of colloidal particles. Ann. N.Y. Acad. Sci.* **51**, 627-659.

PEIERLS, R.E. (1934). *Transformation temperatures. Helv. Phys. Acta Suppl.* **2**, 81-83.

[*]PERSHAN, P.S. & ALS-NIELSEN, J. (1984). *X-ray reflectivity from the surface of a liquid crystal: surface structure and absolute value of critical fluctuations. Phys. Rev. Lett.* **52**, 759-762.

[*]PERSHAN, P.S., BRASLAU, A., WEISS, A.H. & ALS-NIELSEN, J. (1987). *Smectic layering at the free surface of liquid crystals in the nematic phase: X-ray reflectivity. Phys. Rev. A* **35**, 4800-4813.

PERSHAN, P.S. & PROST, J. (1975). *Dislocation and impurity effects in smectic-A liquid crystals. J. Appl. Phys.* **46**, 2343-2353.

PERSHAN, P.S. (1974). *Dislocation effects in smectc-A liquid crystals. J. Appl. Phys.* **45**, 1590-1604.

PERSHAN, P.S. (1979). *Amphiphilic molecules and liquid crystals. J. Phys. (Paris) Colloq.* **40**, C3-423-C3-432.

PERSHAN, P.S. (1982). *Lyotropic liquid crystals. Physics Today* **35**, 34-39.

[*]PINDAK, R., MONCTON, D.E., DAVEY, S.C. & GOODBY, J.W. (1981). *X-ray observation of a stacked hexatic liquid-crystal B phase. Phys. Rev. Lett.* **46**, 1135-1138.

POKROVSKY, V.L. & TALAPOV, A.L. (1979). *Ground state, spectrum, and phase diagram of two-dimensional incommensurate crystals. Phys. Rev. Lett.* **42**, 65-67.

[*]PROST, J. (1984). *The Smectic State. Adv. Phys.* **33**, 1-46.

PROST, J. & BAROIS, P. (1983). *Polymorphism in polar mesogens. II. - Theoretical aspects. J. Chim. Phys.* **80**, 65-81.

PROST, J. (1979). *Smectic A to smectic A phase transition. J. Phys. (Paris)* **40**, 581-587.

RATNA, B.R., NAGABHUSHANA, C., RAJA, V.N., SHASHIDHAR, R. & CHANDRASEKHAR, S. (1986). *Density, dielectric and x-ray studes of smectic A-smectic A transitions. Mol. Cryst. Liq. Cryst.* **138**, 245-257.

[*]RATNA, B.R., SHASHIDHAR, R. & RAJA, V.N. (1985). *Smectic-A phase with two collinear incommensurate density modulations. Phys. Rev. Lett.* **55**, 1476-1478.

RICHARDSON, R.M., LEADBETTER, A.J. & FROST, J.C. (1978). *The structure and dynamics of the smectic B phase. Ann. Phys.* **3**, 177-186.

SADOC, J.F. & CHARVOLIN, J. (1986). *Frustration in bilayers and topologies of liquid crystals of amphilic molecules. J. Phys. (Paris)* **47**, 683-691.

SAFINYA, C.R., CLARK, N.A., LIANG, K.S., VARADY, W.A. & CHIANG, L.Y. (1985). *Synchrotron x-ray scattering study of freely suspended discotic strands. Mol. Cryst. Liq. Cryst.* **123**, 205-216.

[*]SAFINYA, C.R., LIANG, K.S., VARADY, W.A., CLARK, N.A. & ANDERSSON, G. (1984). *Synchrotron x-ray study of the orientational ordering D2-D1 structural phase transition*

of freely suspended discotic strands in triphenylene hexa-n-dodecanoate. *Phys. Rev. Lett.* **53**, 1172-1175.

[*]SAFINYA, C.R., ROUX, D., SMITH, G.S., SINHA, S.K., DIMON, P., CLARK, N.A. & BELLOCQ, A.M. (1986a). *Steric interactions in a model multimembrane system: a synchrotron x-ray study. Phys. Rev. Lett.* **57**, 2718-2721.

SAFINYA, C.R., VARADY, W.A., CHIANG, L.Y. & DIMON, P. (1986b). *X-ray study of the nematic phase and smectic-A1 to smectic-Ã phase transition in heptylphenyl nitrobenzoloxybenzoate (DB7NO2). Phys. Rev. Lett.* **57**, 432-435.

SAFRAN, S.A. & CLARK, N.A., Editors. (1987). *Physics of Complex and Supermoleclar Fluids.* New York: Wiley.

SHIPLEY, C.G., HITCHCOCK, P.B., MASON, R. & THOMAS, K.M. (1974). *Structural chemistry of 1,2 diauroyl-DL-phosphatidylethanolamine: Molecular conformation and intermolecular packing of phospholipids. Proc. Nat. Acad. Sci. (USA)* **71**, 3036-3040.

SIGAUD, G., HARDOUIN, F., ACHARD, M.F. & GASPAROUX, H. (1979). *Anomalous transitional behaviour in mixtures of liquid crystals: a new transition of SA-SA type?. J. Phys. (Paris) Colloq.* **40**, C3-356-C3-359.

SIROTA, E.B., PERSHAN, P.S. & DEUTSCH, M. (1987a). *Modulated crystalline-B phases in liquid crystals. Phys. Rev. A* **36**, 2902-2913.

[*]SIROTA, E.B., PERSHAN, P.S., SORENSEN, L.B. & COLLETT, J. (1985). *X-ray studies of tilted hexatic phases in thin liquid-crystal films.* Phys. Rev. Lett. **55**, 2039-2042.

[*]SIROTA, E.B., PERSHAN, P.S., SORENSEN, L.B. & COLLETT, J. (1987b). *X-ray and optical studies of the thickness dependence of the phase diagram of liquid crystal films.* Phys. Rev. A **36**, 2890-2901.

SMALL, D. (1967). *Phase equilibria and structure of dry and hydrated egg lecithin.* J. Lipid Res. **8**, 551-557.

SMITH, G.W. & GARLAND, Z.G. (1973). *Liquid crystalline phases in a doubly homologous series of benzylideneanilines-textures and scanning calorimetry.* J. Chem. Phys. **59**, 3214-3228.

SMITH, G.W., GARLAND, Z.G. & CURTIS, R.J. (1973). *Phase transitions in mesomorphic benzylideneanilines.* Mol. Cryst. Liq. Cryst. **19**, 327-330.

SOLOMON, L. & LITSTER, J.D. (1986). *Light scattering measurements in the $\overline{7}S5$-8OCB nematic-smectic-A-smectic-C liquid-crystal system.* Phys. Rev. Lett. **56**, 2268-2271.

SORENSEN, L.B., AMADOR, S., SIROTA, E.B., STRAGLER, H. & PERSHAN, P.S. (1987) *(unpublished)*.

SPRINGER, T. (1977). In *Current Physics, Vol.3, Dynamics of Solids and Liquids by Neutron Scattering*, edited by S.W. Lovesey and T. Springer, Berlin: Springer Verlag. pp 255-300: see page 281.

SPROKEL, G. E. (1980). *The physics and chemistry of liquid crystal devices*. New York: Plenum Press.

STEGEMEYER, H. & Bergmann, K. (1981). In *Liquid Crystals of One- and Two-Dimensional Order, Springer Series in Chemical Physics 11*, edited by W. Helfrich and G. Heppke, pp. 161-175. Berlin, Heidelberg and New York: Springer-Verlag.

[*]TARDIEU, A. & BILLARD, J. (1976). *On the structure of the «smectic D modification»*. J. Phys. (Paris) Colloq. **37**, C3-79-C3-81.

TARDIEU, A. & LUZZATI, V. (1970). *Polymorphism of lipids: A novel cubic phase - a cage-like network of rods with enclosed spherical micelles*. Biochim. biophys. Acta **219**, 11-17.

THOEN, J., MARYNISSEN, H. & VAN DAEL, W. (1982). *Temperature dependence of the enthalpy and the heat capacity of the liquid-crystal octylcyanobiphenyl (8CB)*. Phys. Rev. A **26**, 2886-2905.

THOEN, J., MARYNISSEN, H. & VAN DAEL, W. (1984). *Nematic-smectic-A tricritical point in alkylcyanobiphenyl liquid crystals*. Phys. Rev. Lett. **52**, 204-207.

VILLAIN, J. (1980). In *Order in Strongly Fluctuating Condensed Matter System*, edited by T. Riste, pp. 221-260. New York: Plenum Press.

WADATI, M. & ISIHARA, A. (1972). *Theory of liquid crystals*. Mol. Cryst. Liq. Cryst. **17**, 95-108.

[*]WANG, J. & LUBENSKY, T.C. (1984). *Theory of the SA1-SA2 phase transition in liquid crystal*. Phys. Rev. A **29**, 2210-2217.

WARREN, B.E. (1968) *X-ray diffraction*. Reading, Massachusetts, Addison-Wesley.

WINKOR, M.J. & CLARKE, R. (1986). *Long-period stacking transitions in intercalated graphite*. Phys. Rev. Lett. **56**, 2072-2075.

YOUNG, A.P. (1979). *Melting and the vector Coulomb gas in two dimensions*. Phys. Rev. B **19**, 1855-1866.

YOUNG, C.Y., PINDAK, R., CLARK, N.A. & MEYER, R.B. (1978). *Light-scattering study of two-dimensional molecular-orientation fluctuations in a freely suspended ferroelectric liquid crystal film*. Phys. Rev. Lett. **40**, 773-776.

YU, L.J. & SAUPE, A. (1980). *Observation of a biaxial nematic phase in potassium laurate-1-decanol-water mixtures*. Phys. Rev. Lett. **45**, 1000-1003.

PART TWO
SELECTED REPRINTS

1. NEMATIC TO SMECTIC-A

1.1. De Gennes, P. G. (1972). An analogy between superconductors and smectics A. *Solid State Commun.* **10**, 753–756 — 115

1.2. McMillan, W. L. (1973). Measurement of smectic-A-phase order-parameter fluctuations near a second-order smectic-A-nematic-phase transition. *Phys. Rev.* **A7**, 1419–1422 — 119

1.3. Ocko, B. M., Birgeneau, R. J., Litster, J. D. and Neubert, M. E. (1984). Critical and tricritical behavior at the nematic to smectic-A transition. *Phys. Rev. Lett.* **52**, 208–211. — 123

1.4. Chan, K. K., Deutsch, M., Ocko, B. M., Pershan, P. S. and Sorensen, L. B. (1985). Integrated x-ray scattering intensity measurement of the order parameter at the nematic to smectic-A phase transition. *Phys. Rev. Lett.* **54**, 920–923 — 127

Solid State Communications, Vol. 10, pp. 753–756, 1972. Pergamon Press.

AN ANALOGY BETWEEN SUPERCONDUCTORS AND SMECTICS A

P.G. de Gennes

Laboratoire de Physique des Solides, Bât. 510, Faculté des Sciences Orsay 91

(Received 27 January 1972 by P.G. de Gennes)

The conformation of a smectic A can be described by a phase function $\phi(R)$, the n-th layer corresponding to $\phi(R) = 2\pi n$. The role of the phase in smectics A and in superfluids is similar. This analogy leads to the following predictions for a *second order* smectic A⟷nematic transition: (1) the transition temperature is lowered if twist, or bend distortions are imposed: these distortions correspond to a magnetic field in superconductors. (2) the Frank coefficients K_2 and K_3 of the nematic phase must show pretransitional anomalies.

1. PRINCIPLES

SMECTICS A are layered systems :[1] the layers may be planar, or deformed into focal conic textures. But the interlayer distance d is essentially fixed. In terms of the director n (a unit vector parallel to the local optical axis) this implies that the contour integral

$$\frac{1}{d} \oint n.dl$$

on a closed circuit, must be equal to 0 in a dislocation-free case, and to an integer ν in more general situations. This is the analog of flux quantization in a superfluid,[2] n playing the role of the magnetic vector potential.

This remark becomes particularly useful if we have a second order smectic A⟷nematic transition point T_o. (That such a transition may be of second order was noticed first by McMillan,[3] using a specific interaction model). Let us consider such a transition and start from a nematic phase aligned along the z axis ($n = n_0$). In terms of the Fourier components ρ_k of the density, the free energy may be expanded as :

$$F = F_0 + \tfrac{1}{2} \sum_k A(k,T) |\rho_k|^2 \qquad (1)$$

The X-ray intensity for a scattering wave vector k is then proportional to $T/A(k,T)$. The function $A(k)$ is a minimum at $k = \pm 2\pi/d.n_0$.

Near one of these points :

$$A\left(k \to \frac{2\pi}{d} n_0\right) = A(T) + \frac{1}{2M_v} (k_z - q_0)^2 + \frac{1}{2M_T} (k_x^2 + k_y^2) \qquad (2)$$

At the transition point T_o, $A(T_o) = 0$. In the vicinity of T_o, we may write the density $\rho(R)$ as :

$$\rho(R) = 2^{-1/2} e^{2\pi i z/d} \psi(R) + cc \qquad (3)$$

where $\psi(R)$ is a slowly varying function of R, and is complex. The phase of ψ indicates the position of the layers.[4] In terms of ψ, including terms up to order 4, the free energy[1] becomes :

$$F - F_0 = A(T) |\psi|^2 + \frac{1}{2} B |\psi|^4 + \nabla\psi : \frac{1}{2M} : \nabla\psi \qquad (1')$$

where $1/2M$ is the tensor defined in equation (2). Let us now include the fluctuations of the director $n = n_0 + \delta n$. F must be invariant if we rotate simultaneously n and the layers. This imposes :

$$F - F_0 = A|\psi|^2 + \frac{1}{2} B|\psi|^4 +$$

$$\left(\nabla + i \frac{2\pi}{d} \delta n\right)\psi^* : \frac{1}{2M} : \left(\nabla - i \frac{2\pi}{d} \delta n\right)\psi. \quad (4)$$

This has the Landau–Ginsburg form,[2] familiar for a charged superfluid: many effects which occur in these systems should have their counterpart in smectics A. We list a few of them below. Note that, for practical applications, equation (4) must be supplemented by the Frank elastic terms of the unperturbed nematic:[5]

$$F_{el} = \frac{1}{2} K_1 (\text{div } n)^2 + \frac{1}{2} K_2 (n \cdot \text{curl } n)^2 +$$

$$\frac{1}{2} K_3 (n \cdot \text{curl } n)^2 \quad (5)$$

2. APPLICATIONS BELOW T_c

Neglecting for the moment the difference between M_v and M_T, we find two characteristic lengths:

(a) *a coherence length* $\xi(T) = (2MA)^{-1/2}$ giving the distance over which a local perturbation affects the amplitude of ψ: for instance, $\xi(T)$ is the core radius for dislocations in the smectic phase.

(b) *a penetration depth* $\lambda(T) = d/2\pi$ $(MKB/A)^{1/2}$ describing the following gedanke experiment: at the free surface of the smectic, a *weak* twist or bend deformation is imposed (Fig. 1): the deformation penetrates only in a thickness λ (K is the appropriate elastic constant, K_2 for twist, K_3 for bend).

The ratio $\kappa = \lambda/\xi$ is temperature independent. κ controls the phase diagram under imposed deformations (Fig. 2)

(1) a splay deformation does not change T_c

(2) when $h = 2\pi/d \text{ curl } \delta n$ is non zero (i.e. under twist or bend), T_c is decreased. If $\kappa > 2^{-1/2}$, we expect a 'Shubnikov phase' (see reference 2) where the deformation is relaxed by a regular array of dislocations (Fig. 3).

The (h, T) limiting curve is defined by:

$$A(T) \cong A'(T_c - T) = h/2M \quad (6)$$

where $M = M_T$ for twist (h parallel to n_0) and

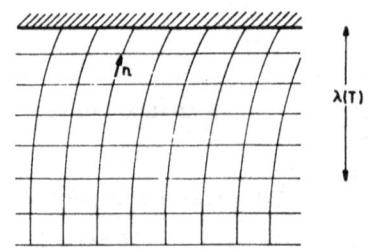

FIG. 1. Analog of the penetration depth for a smectic A. The layers are parallel to the free surface S. A weak bend deformation is imposed from S. The director n (initially normal to S) is perturbed in a thickness $\lambda(T)$. A similar penetration depth may also be defined for twist deformations.

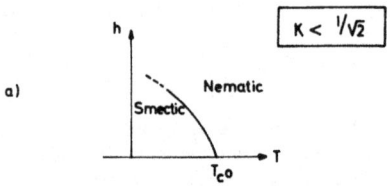

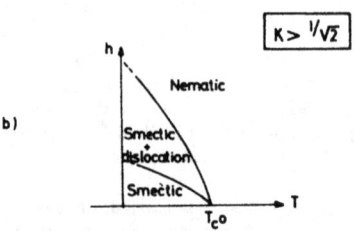

FIG. 2. Phase diagram near a second order smectic A $\longleftrightarrow$ nematic transition point T_{c0}. h measures the amplitude of an imposed twist (or bend) deformation. Depending on the numerical value of the Landau–Ginsburg parameter κ one may have a 'type I' material [Fig. 2(a)] or a 'type II' material [Fig. 2(b)].

$M = (M_T M_v)^{1/2}$ for bend (h normal to n_0). Qualitatively, if the distortion described by h

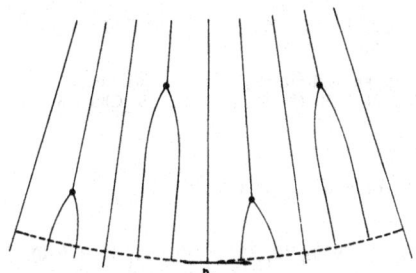

FIG. 3. The 'Shubnikov phase' of a smectic A with $K > 1/\sqrt{2}$ under an imposed bending deformation. The planes of the layers tend to remain equidistant. This is achieved by a regular network of edge dislocations. (The specific model chosen for the cores need not be correct). A similar network (with screw dislocations) is expected under twist.

takes place over a length L ($h \sim 1/dL$), equation (6) gives $T_c - T/T_c \sim d/L$. This may be easily measurable; instead of using a mechanical agent, twist may be imposed by starting from a cholectric phase.

If $\kappa < 2^{-1/2}$, we expect a first order transition under a finite h; the threshold curve corresponding to:

$$\frac{A^2}{2B} = \frac{1}{2} K_i (\text{curl } \underline{n})^2 \qquad (7)$$

(where $K_i = K_2$ for twist, and K_3 for bend). For distortions weaker than the threshold,[7] and in suitable geometric conditions, it may be possible to obtain an 'intermediate state'[2] with coexisting nematic and smectic regions.

3. PRETRANSITIONAL EFFECTS IN THE NEMATIC PHASE

Above T_c, the order parameter ψ vanishes on the average, but it has significant fluctuations, leading to the 'cybotactic clusters' of de Vries:[6] These clusters have a size $\xi(T)$.

They do not accept twist and bend, and thus give rise to an increase δK_i in the twist and bend elastic constants. A rise δK_3 has indeed been observed by Gruler[7] in certain nematics just above the smectic domain. The fluctuations of ψ, when incorporated in equation (4) give extra terms proportional to h^2 in the free energy: this is the analog of the fluctuation diamagnetism in superconductors above T_c, which has been calculated most simply by Schmid[8] in the limit of weak perturbations ($h\xi^2 \ll 1$). For the present problem, the new h^2 contributions represent a correction to the elastic constants. A transposition of the results of reference 7 gives:

$$K_2 = \frac{\pi k_B T}{6 M_T d^2} \left(\frac{M_v}{A(T)} \right)^{\frac{1}{2}} \qquad (8)$$

$$K_3 = \frac{\pi k_B T}{6 d^2} (M_v A(T))^{-1/2}$$

Thus, in the mean field approximation, δK_i should be proportional to $(T - T_c)^{-1/2}$. However, the mean field results are probably not very accurate for the smectic–nematic transition (while they are quite sufficient for superconductors). In fact, using the Wilson calculation of critical exponents,[9] applied to a complex order parameter ψ, one expects $A(T) \sim (T - T_c)^{\gamma}$ with $\gamma = 1.30$, and $\xi(T) \sim (T - T_c)^{-\gamma/2 - \eta}$ with $\eta \sim 0.04$. Neglecting η and using simple scaling arguments, one is led to $\delta K \sim \xi(T) \sim (T - T_c)^{-0.66}$

Similarly, below T_c, the critical 'distortion fields' h defined in equations (6) and (7) will probably not be exactly linear in $T_c - T$, and will contain weak logarithmic operation.

Acknowledgements – This work was done at P.U.C (Rio de Janeiro) during the 4th Brazilian Symposium on theoretical physics; it is a pleasure to thank Professors E. Ferreira and R. Lobo for their hospitality on this occasion. The author is also indebted to P. Martin for various related discussions.

REFERENCES

1. General references on smectics A: FRIEDEL G., *Ann. Phys.* **18**, 273 (1922); DE GENNES P. G., *J. Phys.* **30**, Colloque C 4 (suppt to No. 11–12) P. C4–65 (1969).

2. General references on superconductors: LYNTON E. A., *Superconductivity*, 2nd ed., Methuen, London (1964); *Superconductivity*, ed. R. D. Parks, M. Dekker, N.Y. (1969). In particular Chap. 6 and 8.

3. McMILLAN W. L., *Phys. Rev.* **A4**, 1238 (1971).

4. The phase of ψ differs from the phase defined in the abstract by a term $2\pi z/d$.

5. FRANK F. C., *Disc. Faraday Soc.* **25**, 19 (1958).

6. DE VRIES A., *Molecular Crystals and Liquid Crystals*, **10**, 31 (1970); **10**, 219 (1970); **11**, 361 (1970).

7. GRULER H., Oral presentation at the Pont-à-Mousson meeting, June (1971).

8. SCHMID A., *Phys. Rev.* **180**, 527 (1969).

9. WILSON K. G., *Phys. Rev. Lett.* (to be published).

La conformation d'un spécimen de smectique A peut être décrite par une 'fonction phase' $\phi(R)$, telle que la $n^{\text{ème}}$ couche à $\phi(R) = 2n\pi$. Le rôle de la phase dans les smectiques A et dans les superfluides chargés est essentiellement le même. Cette analogie conduit aux prédictions suivantes, pour le voisinage d'une transition smectique A $\leftrightarrow$ nématique du $2^{\text{ème}}$ ordre: (1) la température de transition biasse si une distorsion de torsion ou de flexion est imposée: ces distortions sont l'analogue d'un champ magnétique pour un supraconducteur. (2) les coefficients de Frank K_2 et K_3 dans la phase nématique doivent montrer des anomalies prétransitionnelles.

PHYSICAL REVIEW A   VOLUME 7, NUMBER 4   APRIL 1973

Measurement of Smectic-A-Phase Order-Parameter Fluctuations near a Second-Order Smectic-A–Nematic-Phase Transition

W. L. McMillan*

Bell Laboratories, Murray Hill, New Jersey 07974

(Received 13 November 1972)

The anisotropic liquid-structure factor of p-cyanobenzylidene-amino-p-n-octyloxybenzene has been measured in the nematic phase using Cu $K\alpha$ x rays. This material exhibits a second-order phase transition to the smectic A phase at $T_c = 82.8$ °C. The liquid-structure factor shows a non-Lorentzian peak at a wave number of $q_0 = 0.179$ Å^{-1} (equivalent d spacing 35.0 Å) of the following form: $S_q = 1 + \{3.5\epsilon^\gamma + 10 \ [(q_\| - q_0)/q_0]^2 + 0.6(q_\perp/q_0)^\eta\}^{-1}$, where $\epsilon = (T - T_c)/T_c$, $\gamma = 1.49 \pm 0.1$, $\eta = 2.5 \pm 0.2$, and $\hbar q_\| \ (\hbar q_\perp)$ are the momentum transfers parallel (perpendicular) to the orienting field.

I. INTRODUCTION

The author[1] has recently published a measurement of the anisotropic-liquid-structure factor in the nematic phase of p-n-octyloxybenzylidene-p-toluidine (OBT), a material which has a first-order nematic-smectic-A-phase transition. The liquid-structure factor is peaked in field direction and at a scattering angle equal to the smectic-A Bragg angle. The peak shape is Lorentzian and the peak height grows as one approaches the transition to the smectic-A phase. This pretransition phenomenon is physically due to small regions of the nematic fluctuating into a smectic-A-like configuration. Mathematically one describes the phenomenon using a Landau theory of the phase transition and calculating the scattering due to order-parameter fluctuations. The Landau theory is due to the author[2] and to deGennes.[3] The theory predicts a Lorentzian peak in the liquid-structure factor with the peak height varying as $(T - T^*)^{-1}$, where T^* is a critical temperature somewhat below the first-order transition temperature. The measurements on OBT agreed well with the peak shape and temperature dependence predicted by the Landau theory. The correlation length is 84 Å, 0.3 °C above the phase transition.

In the present paper we present a measurement of the anisotropic-liquid-structure factor in the nematic phase of p-cyanobenzylidene-amino-p-n-octyloxybenzene (CBAOB). In this material the smectic-A-nematic-phase transition is second order. The peak height varies as $(T - T_c)^{-\gamma}$ with $\gamma = 1.49 \pm 0.1$ and the peak shape is no longer Lo-

rentzian but falls off faster in the transverse direction. The longitudinal correlation length is very long, ~ 2500 Å, 0.2 °C above the phase transition.

The x-ray apparatus has been described previously. The present sample showed stronger scattering near the phase transition and it was, therefore, possible to work at higher resolution; collimators of 0.3×3 mm were used in addition to the 1×3-mm collimators used previously.

The sample of CBAOB obtained from Eastman (No. 923247) was relatively pure and was recrystallized once from ethanol. The transition temperatures were measured with a polarizing microscope equipped with a Mettler FP5 hot stage and the transition entropies were measured on a Perkin–Elmer DSC-1B differential scanning calorimeter; these results are reported in Table I. The smectic-A-nematic transition was unobservable on the calorimeter, which can detect a transition entropy of about 0.02R_0; this transition is presumably second order.

The experimental results are presented in Sec. II and analyzed in Sec. III.

TABLE I. Transition temperatures and transition entropies of CBAOB. The smectic-A-nematic transition is unobservable on the scanning calorimeter.

Transition	Temperature (°C)	Entropy
Crystal → smectic A	73.2	9.1R_0
Smectic A → nematic	82.8	< 0.02R_0
Nematic-isotropic	107.5	0.26R_0

W. L. McMILLAN

II. EXPERIMENTAL RESULTS

The nematic phase is uniaxial and the liquid-structure factor is a function of momentum transfer parallel to the external field $\hbar q_{\parallel}$ and transverse to the external field $\hbar q_{\perp}$. The sample is aligned in a field of 10 kG. In the nematic phase the liquid structure exhibits a peak in the field direction ($q_{\perp} = 0$) for $q_{\parallel} = 0.179$ Å^{-1}; the equivalent d spacing is 35.0 Å which is equal to the interplanar spacing in the smectic-A phase. The intensities measured at three temperatures are presented in Fig. 1 for the longitudinal section (I vs $q_{\parallel}$ for $q_{\perp} = 0$) and in Fig. 2 for the transverse section (I vs $q_{\perp}$ for $q_{\parallel} = 0.179$ Å^{-1}). The peak intensity (for $q_{\parallel} = 0.179$ Å^{-1}, $q_{\perp} = 0$) versus temperature was also measured from 83 to 91 °C. The data close to the peak were taken with the high-resolution 0.3×3-mm collimators and the data in the wings (open circles) were taken with 1×3-mm collimators which gave a factor of 43 greater intensity. The vertical scale in these figures is the number of counts per 4-min counting period for the 0.3×3-mm collimators. The 1×3-mm data approach a constant 70 counts/counting period (after subtracting a background of 30 counts)

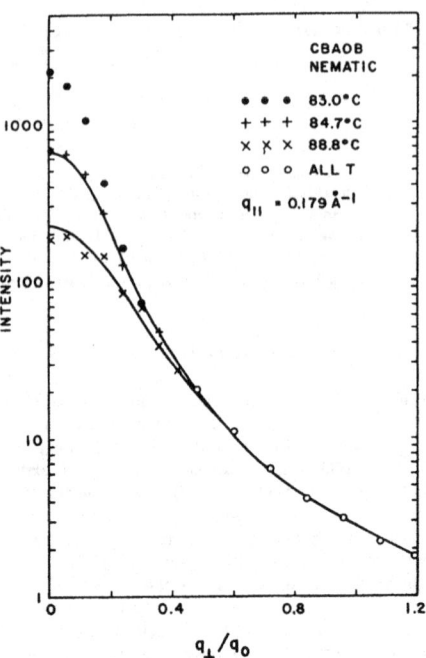

FIG. 2. Intensity versus $q_{\perp}$ for $q_{\parallel} = q_0 = 0.179$ Å^{-1} for CBAOB in the nematic phase. The data are labeled as in Fig. 1.

for large q in both the longitudinal and transverse directions. This is the random-gas contribution to the liquid-structure factor and permits one to determine an absolute normalization for the data. This constant term has been subtracted from the data in Figs. 1 and 2.

With the 0.3×3-mm collimators the instrumental resolution in both longitudinal and transverse directions is adequate at the two higher temperatures; however at the lowest temperature the peak height is resolution limited. An attempt to fit these data with a two-dimensional Lorentzian failed; the data in the longitudinal direction are accurately Lorentzian but in the transverse direction the intensity falls off more rapidly than $q_{\perp}^{-2}$. In order to probe this behavior, the collimators were rotated 90° (3×0.3 mm) to provide high resolution in the transverse direction, and the final collimator was opened up to 8×0.3 mm. In this configuration one is measuring the integral of the liquid-structure factor over $q_{\parallel}$ but with high resolution in the transverse direction. These data are shown at the lowest temperature in Fig. 3. The peak intensity versus temperature was also measured.

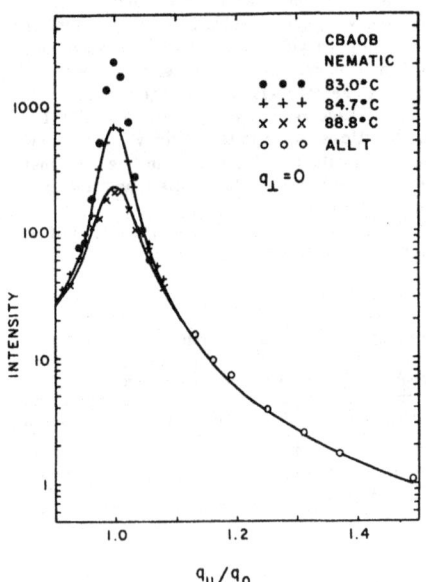

FIG. 1. Intensity versus $q_{\parallel}$ for $q_{\perp} = 0$ for CBAOB in the nematic phase. The open circles are data taken with 1×3-mm collimators at all three temperatures; these data are independent of temperature. The solid lines are from Eq. (3) convoluted with the instrumental resolution function.

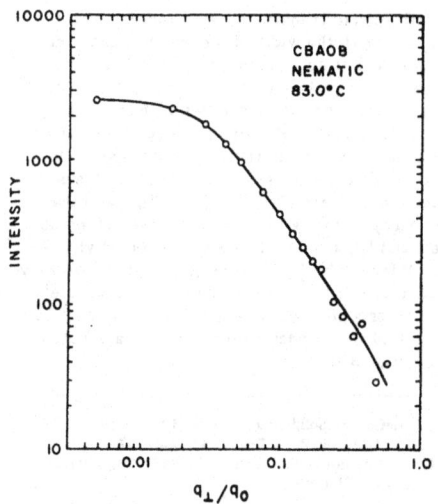

FIG. 3. Intensity versus $q_\perp$ for the 3×0.3-mm collimators for CBAOB in the nematic phase. The solid line is from Eq. (3) integrated over $q_\parallel$. Finite slit-height corrections are included here.

III. DATA ANALYSIS

The Landau theory of the smectic-A phase predicts a Lorentzian peak in the liquid-structure factor:

$$S_q = 1 + \left[\beta(T) + \alpha_\parallel \left(\frac{q_\parallel - q_0}{q_0} \right)^2 + \alpha_\perp \left(\frac{q_\perp}{q_0} \right)^2 \right]^{-1}, \quad (1)$$

with β varying linearly with temperature,

$$\beta = \beta_0 (T - T_c) . \quad (2)$$

In order to fit the data on CBAOB we must modify this expression in two ways. The exponent of $q_\perp$ must be increased and the exponent of $T - T_c$ must be increased. The data of Figs. 1–3 are fitted nicely by choosing the exponent of $q_\perp$ to be 2.5 rather than 2. The solid lines in these figures show the fit obtained with the parameters given in Eq. (3); the instrumental resolution has been folded in. Once the other parameters have been fixed one can determine β at any temperature from a measurement of peak height at that temperature. Values of $\beta/\alpha_\parallel$ determined from the peak height versus temperature for both the 0.3×3-mm and 3×0.3-mm collimator runs are shown in Fig. 4. The straight line is drawn with an exponent of 1.49. Thus, we find experimentally that the liquid-structure factor of CBAOB is given by

$$S_q = 1 + \left[\beta_0 \epsilon^\gamma + \alpha_\parallel \left(\frac{q_\parallel - q_0}{q_0} \right)^2 + \alpha_\perp \left(\frac{q_\perp}{q_0} \right)^\eta \right]^{-1},$$

with $\epsilon \equiv (T - T_c)/T_c$, $\gamma = 1.49 \pm 0.1$, $\eta = 2.5 \pm 0.2$, $\beta_0 = 3.5$, $\alpha_\parallel = 10$, and $\alpha_\perp = 0.6$. The absolute normalization is accurate only within $\pm 20\%$ and $\alpha_\parallel$ and $\alpha_\perp$ are independent of temperature within $\pm 5\%$. The solid lines in Figs. 1–3 were computed from this expression with instrumental resolution folded in.

IV. CONCLUSIONS

We have found a material with a second-order smectic-A-nematic-phase transition and have measured the liquid-structure factor in the nematic phase. The order-parameter fluctuations are in a critical regime with peak height $\alpha \epsilon^{-1.49}$ and longitudinal and transverse coherence length proportional to $\epsilon^{-0.75}$ and $\epsilon^{-0.6}$, respectively. The value of $\gamma = 1.49$ is somewhat larger than that found for other phase transitions and the anisotropy of the exponents for the coherence lengths was unexpected. The hot stage used here was not designed for critical-point work, and it was not possible to work closer than 0.2 °C to the phase transition. The longitudinal coherence length is 2500 Å, 0.2 °C above the transition and the longitudinal resolution is ~ 300 Å. It is desirable to have this experiment repeated with an improved hot stage and with the

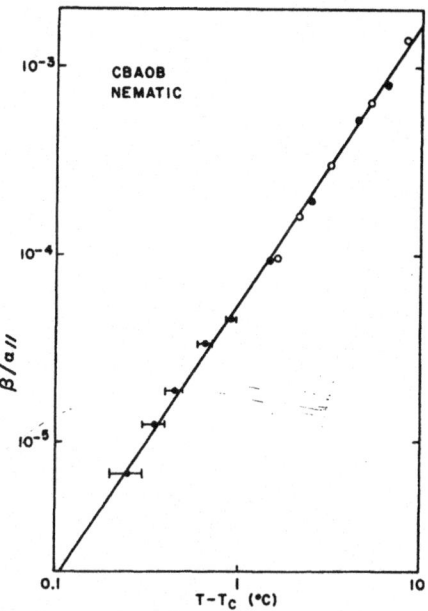

FIG. 4. Inverse peak height versus temperature from the 0.3×3-mm collimator data (open circles) and the 3×0.3-mm collimator data (filled circles). The data have been corrected for instrumental resolution. The solid line is drawn with a slope of 1.49.

full resolution of single-crystal techniques. Because of the resolution problem it was not possible to separate the fluctuation and Bragg-scattering contributions in the smectic-A phase. It is desirable to have measurements on this material of other properties (e. g., heat capacity and elastic constants) which are expected to exhibit critical behavior.

Some feeling for the domain of validity of the "classical" Landau[2,3] and microscopic[4,5] theories is beginning to emerge from this series of measurements. The phase diagram predicted by the microscopic theory appears to be qualitatively correct. In a given homologous series the shorter members exhibit a wide-temperature range of the nematic phase and a second-order smectic-A-nematic-phase transition. With increasing molecular length the width of the nematic phase decreases and at some point the smectic-A-nematic-phase transition becomes first order; the entropy of this transition then increases with increasing molecular length. Finally the smectic-A-nematic and nematic-isotropic transitions coalesce. In the region where the smectic-A-nematic transition entropy is moderately large ($> 0.5R_0$), both the microscopic theory (tested on cholesteryl myristate[2]) and the Landau theory (tested on octyloxy-benzylidene-toluidene[1]) work quite well. For very small transition entropy (cholesteryl nonanoate[2]) the microscopic theory does not work, and we have shown that the Landau theory fails for a second-order transition.

*Present address: Department of Physics, University of Illinois, Urbana, Ill.

[1]W. L. McMillan, Phys. Rev. (to be published).
[2]W. L. McMillan, Phys. Rev. A **6**, 936 (1972).
[3]P. G. deGennes, Solid State Commun. 10, 753 (1972).
[4]W. L. McMillan, Phys. Rev. A **4**, 1238 (1971).
[5]K. K. Kobayashi, Phys. Lett. 31A, 125 (1970); J. Phys. Soc. Jap. 29, 101 (1970).

Critical and Tricritical Behavior at the Nematic to Smectic-A Transition

B. M. Ocko, R. J. Birgeneau, and J. D. Litster

Department of Physics and Center for Materials Science and Engineering,
Massachusetts Institute of Technology, Cambridge, Massachusetts 02139

and

M. E. Neubert

Liquid Crystal Institute, Kent State University, Kent, Ohio 44242
(Received 24 October 1983)

High-resolution x-ray scattering measurements of the nematic to smectic-A transitions in two different homologous series are reported; it is found that the correlation-length exponents $\nu_\parallel$ and $\nu_\perp$ and the susceptibility exponent γ decrease continuously with decreasing nematic range; the transitions ultimately become first order. The exponents at the crossover point are close, but not identical, to mean-field tricritical values. For both series and all samples $\nu_\parallel \neq \nu_\perp$ while $\alpha + \nu_\parallel + 2\nu_\perp - 2 = 0$ to within the errors.

PACS numbers: 64.60.Kw, 61.30.-v, 64.70.Ew

Tricritical behavior,[1] in which a phase transition crosses over from being second to first order, has now been studied extensively in a variety of systems including He^3-He^4 mixtures[2] and metamagnets such as $FeCl_2$ and $Dy_3Al_5O_{12}$.[3] In general it is found that the tricritical region is well described by mean-field theory as expected on the basis of Ginzburg-criterion and renormalization-group arguments.[4] Alben[5] has predicted that a He^3-He^4-like tricritical point may occur in binary liquid crystal mixtures for the nematic to smectic-A (N-S_A) transition. So far very little information is available on N-S_A tricritical behavior except from some heat-capacity measurements by Brisbin et al.[6] in the homologous series pentylbenzenethioalkoxybenzoate ($\overline{n}$S5) and by Thoen and co-workers[7] in the series alkoxycyanobiphenyl (nCB). In this letter we report detailed studies using high-resolution x-ray scattering techniques of the phase-transition behavior in these two homologous series with emphasis on the tricritical region.

Studies of N-S_A tricritical behavior are important both because of intrinsic interest in tricritical phenomena and because of the possible light they can shed on the general N-S_A problem. Alben's analogy with He^3-He^4 mixtures was based on the McMillan–Kobayashi–de Gennes (M-K-deG) model[8] in which the N-S_A transition is in the universality class of a superconductor ($d = 3$, $n = 2$) albeit with an anisotropic gauge. The tricritical behavior which occurs with decreasing nematic range is driven by the coupling between the nematic and smectic order parameters. There is an important difference between the binary liquid crystal mixture and He^3-He^4 cases in that in the former the concentration represents

the nonordering field whereas in the latter it represents the nonordering density. Unfortunately, the elegant M-K-deG analogy to superconductivity does not properly describe the empirical behavior. Specifically, the longitudinal and transverse correlation lengths diverge at different rates.[9] This may be due to the crossover effect discussed by Lubensky and co-workers[10] in which $\nu_\perp$ should evolve from $\nu_\parallel$ to $\nu_\parallel/2$ as T_{NA}, the N-S_A transition temperature, is approached. In addition, the power laws for all measured quantities exhibit nonuniversal behavior with exponents depending on the McMillan ratio[8] T_{NA}/T_{NI}, where T_{NI} is the N-isotropic transition temperature.[9] Further, for no value of T_{NA}/T_{NI} do all of the measured exponents agree with the $d = 3$, $n = 2$ values. Recently Brisbin et al.[6] and Garland et al.[9] have postulated that these continuously varying exponents represent a smooth crossover from critical to tricritical behavior, although the tricritical region itself has not yet been properly explored. In order to test these ideas and to elucidate the N-S_A problem as a whole we have studied in detail this critical-tricritical crossover.

As noted above, experiments were carried out in the two homologous series $\overline{n}$S5 and nCB. By mixing neighboring homologs one can effectively vary n continuously. Heat-capacity experiments[6,7] suggest that the tricritical point occurs for $n \simeq 10$ for $\overline{n}$S5 and $n \simeq 9$ for nCB. The nCB material was purchased from British Drug House and used as received while the $\overline{n}$S5 compounds were synthesized by one of us (M.E.N.).[11] The x-ray experiments involved standard techniques in our laboratory.[12] A longitudinal in-plane resolution of $\sim 5 \times 10^{-4}$ Å^{-1} or $\sim 1 \times 10^{-4}$ Å^{-1} half width at half maximum (HWHM) was achieved by employing

124

VOLUME 52, NUMBER 3   PHYSICAL REVIEW LETTERS   16 JANUARY 1984

perfect Ge(111) or Si(111) crystals as monochromator and analyzer using Cu $K\alpha$ radiation. The transverse resolution in the vertical direction was ~ 0.02 Å^{-1} HWHM while the transverse in-plane resolution was essentially perfect.

The samples were contained in beryllium-window cells placed in a two-stage oven. The director was aligned by an 8-kG magnetic field. In these measurements we defined T_{NA} as the temperature at which the longitudinal scan first became resolution limited; using this criterion T_{NA} could be determined to within 2×10^{-3} °C.

In the nematic phase, the critical scattering is centered about the points $(0, 0, \pm q_\parallel{}^0)$. It has been found previously[12] that the scattering is well represented by the form

$$\sigma(\vec{q}) = \frac{\sigma_0}{1 + \xi_\parallel{}^2 (q_\parallel - q_\parallel{}^0)^2 + \xi_\perp{}^2 q_\perp{}^2 (1 + C\xi_\perp{}^2 q_\perp{}^2)} \quad (1)$$

convoluted with the instrumental resolution function. Thus from scans through $(0, 0, q_\parallel{}^0)$ along q_z and q_x it is possible to measure σ_0, $\xi_\parallel$, and $\xi_\perp$.

We show in Fig. 1 values for $\xi_\parallel$ obtained in this fashion for a series of $\overline{n}$S5 samples. For concentrations exhibiting a second order N-S_A transition each of $\xi_\parallel$, $\xi_\perp$, and σ_0 exhibit single power-law behavior over the reduced temperature range ~ 10^{-2} to ~ 5×10^{-5}. However, if the transition is

first order then T_{NA}, defined by the appearance of a resolution-limited smectic peak, should be too high and accordingly the data should saturate at small reduced temperatures. Such an effect is indeed observed for $\overline{10.2}$S5 and $\overline{10.4}$S5, suggesting that these samples exhibit first-order N-S_A transitions. Thus to within our accuracy we find that the N-S_A tricritical point for the $\overline{n}$S5 series occurs for $n = 10 \pm 0.2$. This is consistent with the heat-capacity measurements of Brisbin et al.[6] We find quite similar results in the nCB series; in that system we find the tricritical point for $n = 9.0 \pm 0.1$; this agrees to within the errors with the heat-capacity study by Thoen and co-workers.[7] The corresponding McMillan ratios for $\overline{n}$S5 and nCB are 0.984 and 0.994, respectively.

We now discuss the power laws observed for each series. It is evident from Fig. 1 that both the critical exponents and absolute lengths decrease dramatically as the tricritical concentration is approached. Nevertheless, single power laws describe the data quite well over the complete temperature range. Similar results hold for $\xi_\perp$ and σ_0 in both the $\overline{n}$S5 and nCB series. The results at the tricritical concentrations are shown in Fig. 2. Again single power laws work

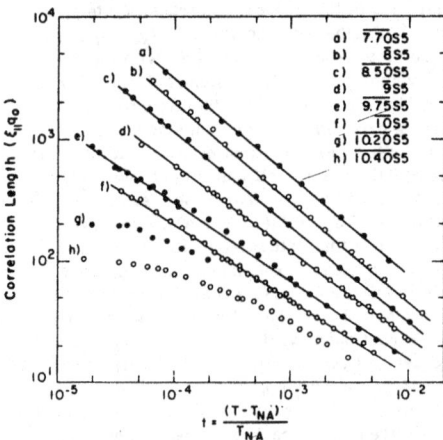

FIG. 1. Longitudinal correlation length vs reduced temperature for $\overline{n}$S5. The solid lines are single power-law fits. Deviations from a single power law for $\overline{10.2}$S5 and $\overline{10.4}$S5 are characteristic of first-order behavior. The data for $\overline{7.7}$S5 and $\overline{8}$.S5 were taken by C. R. Safinya (unpublished).

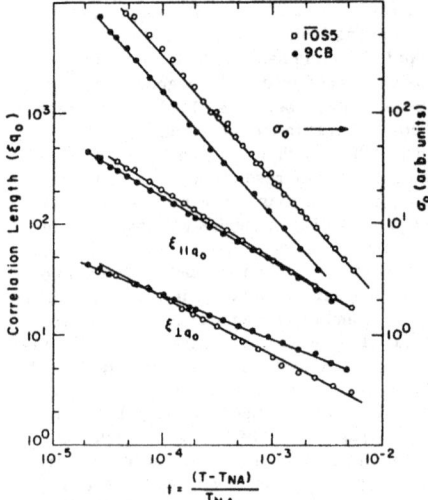

FIG. 2. Susceptibility and the longitudinal and transverse correlation lengths vs reduced temperature for 9CB and $\overline{10}$S5. The solid lines are single power fits. The susceptibility amplitude difference between the two samples is for display purposes only.

well over the reduced temperature range $\sim 2 \times 10^{-5}$ to 8×10^{-3} except possibly for $\xi_\perp$ in 9CB. One of the most surprising features is that the absolute lengths for $\overline{10}$S5 and 9CB are closely similar, suggesting universal behavior. For $\overline{10}$S5 we find $\gamma = 1.10 \pm 0.05$, $\nu_\parallel = 0.61 \pm 0.03$, and $\nu_\perp = 0.51 \pm 0.05$ while in 9CB $\gamma = 1.10 \pm 0.05$, $\nu_\parallel = 0.57 \pm 0.03$, and $\nu_\perp = 0.37^{+0.07}_{-0.03}$. These may be compared with the mean-field tricritical values[4] $\gamma = 1$, and $\nu_\parallel = \nu_\perp = 0.5$. The measured exponents are close to these values although the discrepancies are significant. These results and those for several other compounds in the series are given in Table I. It has been found previously that the anisotropic hyperscaling relation

$$\alpha + \nu_\parallel + 2\nu_\perp = 2 \qquad (2)$$

holds quite well for the N-S_A transitions away from the tricritical point. As shown in Table I, for $\overline{10}$S5, $\alpha + \nu_\parallel + 2\nu_\perp = 2.08 \pm 0.18$, while for 9CB, $\alpha + \nu_\parallel + 2\nu_\perp = 1.84^{+0.22}_{-0.14}$. Thus Eq. (2) is satisfied to within the errors at the N-S_A tricritical point as well. Finally, in previous studies of N-S_A transitions[9] it has been found that $\nu_\parallel - \nu_\perp = 0.13 \pm 0.03$. For $\overline{10}$S5 and 9CB we find $\nu_\parallel - \nu_\perp = 0.10$ and 0.20, respectively, consistent with previous results.

These experiments thus have dramatically confirmed the suggestion of Brisbin et al.[6] and Garland et al.[9] that there is a continuous evolution of the measured exponents for N-S_A transitions from the saturated, that is, the long nematic range, limit to the tricritical region. Somewhat ironically, all of the systems studied in the first-generation N-S_A experiments were in the middle of this critical to tricritical crossover region.[9] We now discuss the exponents themselves. As discussed

earlier, in the M-K-deG model[8] the exponents should cross over from those characteristic of the $d = 3$, $n = 2$ universality class, $\gamma = 1.32$, $\nu_\parallel = \nu_\perp = 0.67$, $\alpha = -0.03$, to the tricritical mean-field values $\gamma = 1$, $\nu_\parallel = \nu_\perp = 0.5$, $\alpha = 0.5$. However, the model neglects the divergent phase fluctuations characterizing the smectic-A phase. Lubensky and co-workers[10] have treated the phase fluctuations by making a gauge transformation to an order parameter with nondivergent fluctuations and then, after making an appropriate decoupling approximation, transforming back to the laboratory gauge. This model predicts that the measured $\xi_\parallel$ should scale like $\tau^{-\nu_\parallel}$ over the accessible temperature range whereas $\xi_\perp$ should exhibit a crossover from $\tau^{-\nu_\parallel}$ to $\tau^{-\nu_\parallel/2}$. For the $d = 3$, $n = 2$ critical behavior this also necessitates that the appropriate hyperscaling expression is $\alpha + 3\nu_\parallel = 2$, not Eq. (2). Similar predictions have recently been given by Grinstein and Toner[13] using a different approach.

In the context of butyloxybenzlideneheptylaniline (4O.7), it has been shown[9] that the crossover form is capable of accounting for the measured $\xi_\perp$; a single power law seems to work slightly better for $\xi_\parallel$ but the differences between the crossover and single power-law forms are within the overall uncertainties. We do not have enough information about the elastic constants to test the crossover form at the tricritical point. Nevertheless, by analogy with 4O.7 we expect for both $\overline{10}$S5 and $\overline{9}$CB the anisotropy can be ascribed to this mechanism. Indeed, γ and $\nu_\parallel$ are identical in both systems whereas $\nu_\perp$ differs considerably (see Fig. 1). This difference may simply reflect the nonuniversal nature of the crossover behavior. We should also emphasize that Eq. (1) is

TABLE I. Evolution of critical exponents in mixtures approaching the tricritical point.

Material	T_{NA}/T_{NI}	$\alpha = \alpha'$	$\nu_\perp$	$\nu_\parallel$	γ	$\alpha + \nu_\parallel + 2\nu_\perp$
XY		-0.026	0.67 to 0.33	0.67	1.32	2
$\overline{8}$S5	0.936	-0	0.68	0.83	1.53	2.19 ± 0.16
$\overline{9}$S5	0.967	0.22	0.57	0.71	1.31	2.01 ± 0.18
$\overline{8}$CB	0.977	0.31	0.51	0.67	1.26	2.00 ± 0.13
$\overline{10}$S5	0.984	0.45	0.51	0.61	1.10	2.08 ± 0.18
$\overline{9}$CB	0.994	0.53	0.37	0.57	1.10	$1.84^{+0.22}_{-0.14}$
Tricritical	0.87	0.5	0.5 to 0.25	0.5 0.25	1	2

strictly empirical. In 9CB especially we find large deviations from Lorentzian behavior over the whole temperature range; typically $c = 0.1$ to 0.25. The transverse scans can be equally well fitted by a Lorentzian raised to the power 1.6. A theoretical prediction for the form of $S(q_\perp)$ in the crossover region is needed.

It has been shown previously[9] that for systems with long nematic ranges both γ and $\nu_\parallel$ are respectively about 0.15 and 0.11 larger than the $d = 3$, $n = 2$ values. At the tricritical points for $\overline{10}S5$ and 9CB we find $\gamma(\text{meas}) - \gamma(\text{theor}) = 0.10 \pm 0.05$ and $\nu_\parallel(\text{meas}) - \nu_\parallel(\text{theor}) = 0.09 \pm 0.04$ over the reduced temperature range $\sim 2 \times 10^{-5}$ to $\sim 10^{-2}$. These discrepancies remain unexplained. The anisotropic hyperscaling relation Eq. (2) holds for all materials measured to date including our tricritical systems. However, averaging over all systems in the literature (excluding $\overline{10}S5$ and 9CB), we find $\alpha + 3\nu_\parallel = 2.36 \pm 0.1$. For $\overline{10}S5$ and 9CB we find as an average $\alpha + 3\nu_\parallel = 2.26 \pm 0.14$. Thus, even at the tricritical point anisotropic rather than isotropic scaling seems to hold. This discrepancy with the theory remains unexplained. Finally, we note that the presently known elastic behavior below T_c is inconsistent with the helium model.

In summary, with this work we have empirically characterized the mass density fluctuations near the N-S_A transition from the critical to tricritical limits. The general trends are consistent with current theory. However, there are persistent quantitative discrepancies which remain unexplained. Finally, only limited elastic-constant data above and below T_{NA} are available, especially in the tricritical region. Such measurements are strongly needed in order to complete the empirical picture.

We would like to thank Geoff Grinstein, David

Johnson, Tom Lubensky, and John Toner for invaluable discussions of these experiments. We are indebted to R. F. Griffith, R. E. Cline, and T. A. Santora for their help in preparing the $\overline{n}S5$ samples.

This work was supported by the National Science Foundation, Materials Research Laboratories Program, under Contract No. DMR-81-19295, and by the Joint Services Electronics Program under Grant No. DAAG-29-83-K-0003.

[1]R. B. Griffiths, Phys. Rev. Lett. 24, 715 (1970).
[2]See, for example, E. K. Riedel, H. Meyer, and R. Behringer, J. Low Temp. Phys. 22, 369 (1976).
[3]For a recent review see W. P. Wolf, to be published.
[4]R. Bausch, Z. Phys. 254, 81 (1972); E. K. Riedel and F. J. Wegner, Phys. Rev. Lett. 29, 349 (1972).
[5]R. Alben, Solid State Commun. 13, 1783 (1972).
[6]D. Brisbin, R. DeHoff, T. E. Lockhart, and D. L. Johnson, Phys. Rev. Lett. 43, 1171 (1979).
[7]J. Thoen, H. Marynissen, and W. Van Dael, preceding Letter [Phys. Rev. Lett. 52, 204 (1984)].
[8]W. L. McMillan, Phys. Rev. A 4, 1238 (1971); K. Kobayashi, J. Phys. Soc. Jpn. 29, 101 (1970); P. G. de Gennes, Solid State Commun. 10, 753 (1972), and Mol. Cryst. Liq. Cryst. 21, 49 (1973).
[9]For a summary of most available high-resolution data see C. W. Garland, M. Meichle, B. M. Ocko, A. R. Kortan, C. R. Safinya, L. J. Yu, J. D. Litster, and R. J. Birgeneau, Phys. Rev. A 27, 3234 (1983).
[10]For a review see T. C. Lubensky, J. Chim. Phys. Phys.-Chim. Biol. 80, 31 (1983).
[11]M. E. Neubert, R. E. Cline, M. J. Zawaski, P. J. Wildman, and Arun Ehachai, Mol. Cryst. Liq. Cryst. 76, 43 (1981).
[12]J. Als-Nielsen, R. J. Birgeneau, M. Kaplan, J. D. Litster, and C. R. Safinya, Phys. Rev. Lett. 39, 352 (1977).
[13]G. Grinstein and J. Toner (unpublished work).

Integrated X-Ray-Scattering Intensity Measurement of the Order Parameter at the Nematic–to–Smectic-A Phase Transition

Kelby K. Chan, Moshe Deutsch,[a] B. M. Ocko, P. S. Pershan, and L. B. Sorensen[b]

Division of Applied Sciences, Harvard University, Cambridge, Massachusetts 02138

(Received 22 October 1984)

The temperature dependence of the square of the smectic order parameter, $|\psi|^2$, was determined from the integrated x-ray-scattering intensity. The data are described very well by the form $I(t)/I(0) = 1 \mp A^{(\pm)}|t|^x$, where $t = (T - T_{NA})/T_{NA}$. The determined values for x do not agree with the theoretically expected value $x = 1 - \alpha$, where α is the heat-capacity exponent. This disagreement raises questions concerning the quantitative validity of the Landau–de Gennes free energy.

PACS numbers: 61.30.−v, 61.10.−i, 64.60.Fr

In spite of considerable research activity over the past ten years the nematic–to–smectic-A (N-A) phase transition remains one of the principal unsolved problems in equilibrium statistical physics.[1] Although the Landau free-energy density proposed by de Gennes in 1972 does explain most of the qualitative features of the transition,[2] there are significant quantitative discrepancies between the current experimental results and the predictions based on this free energy. Recent experimental evidence of tricritical behavior in different materials may explain the apparent nonuniversality of critical exponents,[3–5] however, there is still no consensus about either the universality class of the transition, or the effects of the inherent anisotropy of the critical correlations. Experiments that might improve our understanding have been hampered by the technical difficulty of maintaining well-oriented monodomain samples on the smectic side of the transition, and, except for heat-capacity[6] and light scattering[7] measurements, most of the experimental results are for the critical effects on the nematic side of the transition.

In this Letter we describe measurements of the integrated x-ray-scattering intensity which sum over the scattering from differently oriented domains. This is the classical crystallographers' solution[8] to the problem of sample mosaic and it allows a quantitative measure of the mean square amplitude of the smectic order parameter, $\langle |\psi|^2 \rangle$, on both the nematic and the smectic sides of the transition. One very general consequence of the Landau–de Gennes free energy is that $\langle |\psi|^2 \rangle = L \mp M^{\pm} |t|^x$, where x is related to the heat-capacity exponent by $x = 1 - \alpha$ and $t = (T - T_{NA})/T_{NA}$ (T_{NA} is the N-A transition temperature). We show from the data that x is not generally equal to $1 - \alpha$ and therefore propose that the Landau–de Gennes free energy is not adequate for quantitative predictions regarding the N-A transition.

The x-ray-scattering spectrometer was configured to integrate over a large region of reciprocal space.[9] A germanium monochromator crystal collimates the Cu $K\alpha$ x rays in the scattering plane. A beam monitor

between the monochromator and the sample is used to measure drifts in the incident beam intensity, and the data are normalized to a constant monitor level. The sample is held in a beryllium holder that is placed inside a two-stage oven capable of ± 0.5 mK stability over ~ 8 h. Permanent cobalt-samarium magnets inside the oven provide a 4.3 kG field that aligns the nematic director in the scattering plane parallel to the z axis in the sample reference frame (see Fig. 1). A receiving slit 4 mm wide by 8 mm high, located 270 mm from the sample, is placed at $2\theta = 2 \sin^{-1}(\lambda/2d)$, where d is the smectic layer spacing and λ ($= 1.542$ Å) is the Cu $K\alpha$ wavelength. The vector $Q_0 = (2\pi/d)\hat{z}$. For $\theta = 1.5°$, the height of the spectrometer resolution function (dotted line in Fig. 1) corresponds to $\Delta Q_z/Q_0 = 0.15$. The crosshatched region indicates the (x-z) cross-sectional area in the sample reciprocal space that is swept out by the resolution function when

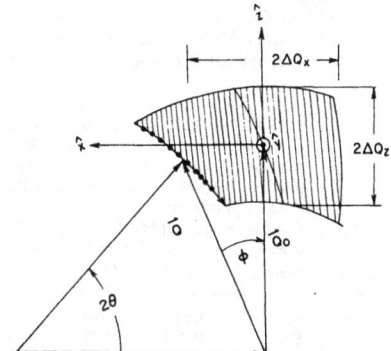

FIG. 1. Scattering geometry for the measurement of integrated intensity. The shaded region is the (x-z) cross section of the reciprocal-space volume swept out by the resolution function as the sample reference frame is rotated from $-\phi$ to ϕ.

128

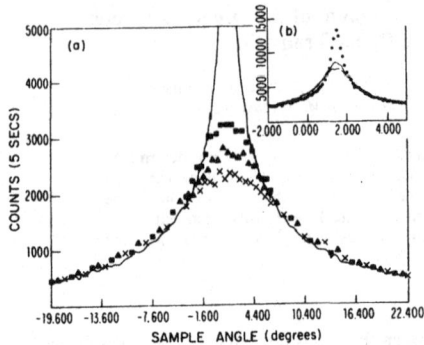

FIG. 2. Intensity vs sample rotation ϕ for 8OCB; $T_{NA} = 66.205\,°C$. (a) Solid line, $T = 66.214\,°C$; squares, $66.560\,°C$; triangles, $66.832\,°C$, and crosses, $67.234\,°C$. (b) Dots, $T = 66.203\,°C$; solid lines, $66.208\,°C$ and $66.211\,°C$. (a) Displays only every sixth data point.

the sample is rotated by ϕ. The thickness of the spectrometer resolution function (normal to the scattering plane) is $\Delta Q_y / Q_0 = 0.30$.

Figure 2(a) displays the measured scattering intensity versus sample orientation ϕ for octyloxycyanobiphenyl (8OCB) at different temperatures. The sample oven limits $|\phi|$ to $\leq 20°$. The intensity obtained by integrating the ϕ scan has the effect of sweeping the spectrometer resolution over a region $\Delta Q_x / Q_0 = \sin(20°) = 0.34$. Therefore, the integrated intensity $I(t)$ is proportional to $\int_{\Delta^3 Q} S(\mathbf{Q}) d^3 Q$, where $S(\mathbf{Q})$ is the static structure factor and, for small θ, the resolution volume $\Delta^3 Q$ is approximately a rectangular parallelepiped $(2\Delta Q_x)(2\Delta Q_y)(2\Delta Q_z)$.

Figure 3 displays $I(t)/I(0)$ for six materials. With the assumption that the error in each point is deter-

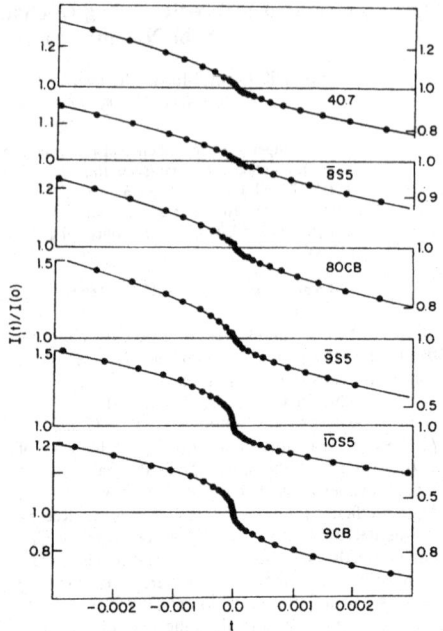

FIG. 3. Integrated intensity $I(t)/I(0)$. The solid lines are the results of nonlinear least-squares fits described in Table I.

mined by counting statistics, one standard deviation corresponds to $\sim 0.2\%$ uncertainty in $I(t)$. Table I displays the results of fitting the data by the form $I(t)/I(0) = 1 \mp A^{\pm}|t|^x$.[10] Except for the case of $\overline{10}S5$, independent fits of the data for $T < T_{NA}$ result

TABLE I. Results of nonlinear least-squares fits for the data in Fig. 3. Values of α, $\nu_{\parallel,\perp}$, and γ are obtained from Refs. 4, 5, and 17. The last column is the empirically motivated scaling relation $x_{sc} = (\nu_{\parallel} + \nu_{\perp})/2$. 4O.7, 4-$n$-butyloxybenzylidene-4'-$n$-heptylaniline; $\overline{8}S5$, 4-n-pentylphenylthiol-4'-n-octyloxybenzoate; $\overline{9}S5$, 4-n-pentylphenylthiol-4'-n-nonyloxybenzoate; $\overline{10}S5$, 4-n-pentylphenylthiol-4'-n-decyloxybenzoate; 9CB, 4-cyano-4'-n-nonylbiphenyl.

Material	T_{NA}/T_{NI}	A^+/A^-	x	α	$\nu_{\perp}$	$\nu_{\parallel}$	γ	x_{sc}
4O.7	0.926	0.69	0.70 ± 0.05	-0.03	0.65	0.78	1.46	0.72
$\overline{8}S5$	0.936	0.90	0.76 ± 0.05	0.0	0.68	0.83	1.53	0.76
8OCB	0.963	0.87	0.68 ± 0.05	0.2	0.58	0.71	1.32	0.65
$\overline{9}S5$	0.967	0.73	0.63 ± 0.05	0.22	0.57	0.71	1.31	0.64
$\overline{10}S5$	0.984	$\cdots$	$0.50/0.43^a$	0.45	0.51	0.61	1.10	0.56
9CB	0.994	0.99	0.47 ± 0.05	0.53	0.37	0.57	1.10	0.47

$^a x_+/x_-$.

921

in $x^+ = x^-$ within experimental error. We discuss 10S5 below. The solid lines in Fig. 3 display the results of the fits for $|T - T_{NA}| < 1$ K. The transition temperature, T_{NA}, can be identified with a precision of $\sim \pm 3$ mK without any numerical analysis. For example, Fig. 2(b) displays the relevant scans for 8OCB at 66.211, 66.208, and 66.203 °C. The first two scans are nematic and the wings superpose perfectly. The last scan is smectic and is easily recognized by the changes in the wings and a sudden decrease in the full width at half maximum. In addition, there is often a slight shift in the peak position at the transition. If T_{NA} is allowed to vary over a reasonable range of values (i.e., ± 3 mK) the exponent, x, varies by less than approximately ± 0.01.

To understand the significance of these results, consider the definition of the structure factor

$$S(\mathbf{Q}) = (2\pi)^{-3} \int \langle \delta\rho(\mathbf{r}) \, \delta\rho(0) \rangle \exp(-i\mathbf{Q}\cdot\mathbf{r}) d^3 r,$$

where $G(\mathbf{r}) = \langle \rho(\mathbf{r}) \, \delta\rho(0) \rangle$ is the correlation function for density fluctuations. Let $H(\mathbf{Q}) = 1$ if $\mathbf{Q}$ is inside the shaded region in Fig. 1 and zero elsewhere. Then the integrated intensity

$$I(t) = \int S(\mathbf{Q}) H(\mathbf{Q}) d^3 Q$$
$$= \int h(\mathbf{r}) \langle \delta\rho(\mathbf{r}) \, \delta\rho(0) \rangle d^3 r,$$

where $h(\mathbf{r}) = (2\pi)^{-3} \int H(\mathbf{Q}) \exp(-i\mathbf{Q}\cdot\mathbf{r}) d^3 Q$ is localized to a region in real space of size $(2\pi)^3/\Delta^3 Q$ centered around the origin. If we substitute $\delta\rho(\mathbf{r}) = \mathrm{Re}[\psi \exp(+iQ_0 z)]$, and assume[11]

$$G(\mathbf{r}) = \mathrm{Re}[\langle \psi(r)\psi(0)^* \rangle]$$
$$= C_0(\mathbf{r}) \mp |t|^x C_\pm(r_\perp/\xi_\perp, z/\xi_\parallel),$$

it follows that since $(\xi_\parallel \Delta Q_z)^{-1}$ and $(\xi_\perp \Delta Q_{xy})^{-1}$ vanish as $t \to 0$,

$$I(t) = L \mp M^\pm |t|^x,$$

where

$$L = \int C_0(\mathbf{r}) h(\mathbf{r}) d^3 r$$

and

$$M^\pm = \int C_\pm(r_\perp/\xi_\perp, z/\xi_\parallel) h(\mathbf{r}) d^3 r$$
$$= C_\pm(0,0) \int h(\mathbf{r}) d^3 r.$$

For each data set the stability of the results were tested by range shrinking in which points furthest from the transition temperature were successively discarded and the data refitted. The absence of any clear trend due to range shrinking together with the fact that the bulk of the temperature dependence of $I(t)$ results from the central portion of the scan (i.e., $Q_x < 0.1\Delta Q_x$) indicates that the resolution volume $\Delta^3 Q$ is large enough to obtain all of the critical intensity.

The uncertainty in the critical exponent $x = 0.68 \pm 0.05$ includes statistical errors, variations accompanying the range-shrinking process, and the uncertainty in T_{NA}.

If we take the asymptotic form for $G(\mathbf{r})_{t=0} = C_0(\mathbf{r}) = (r_\perp)^{-(1+\eta_\perp)}$ or $(z)^{-(1+\eta_\parallel)}$ and note that for $t > 0$ and for large r the correlations must fall off faster than $(r)^{-(1+\eta)}$, then the leading terms in $C_\pm(r_\perp/\xi_\perp, z/\xi_\parallel)$ must cancel the algebraic decay. This requires forms for $C_+ \sim (\xi_\perp/r_\perp)^{1+\eta_\perp}$ or $(\xi_\parallel/z)^{1+\eta_\parallel}$. To effect cancellation over a range of t, both $t^x \xi_\parallel^{(1+\eta_\parallel)}$ and $t^x \xi_\perp^{(1+\eta_\perp)}$ must be temperature independent. Since $\xi \sim t^{-\nu}$, this requires that $x = \nu_\parallel (1+\eta_\parallel) = \nu_\perp (1+\eta_\perp)$. For the observed anisotropic exponents ($\nu_\parallel \neq \nu_\perp$) this result is not consistent with the previous result of anisotropic scaling: $\gamma = \nu_\parallel (2 - \eta_\parallel) = \nu_\perp (2 - \eta_\perp)$.[1] Furthermore, since $\nu_\perp < x \approx \gamma/2 < \nu_\parallel$ the pair of relations involving either $\nu_\parallel, \eta_\parallel$ or $\nu_\perp, \eta_\perp$ are not individually satisfied. However, it is interesting to observe that within experimental error, the following empirical generalizations of isotropic scaling (with $\eta = 0$), $x_{sc} = (\nu_\parallel + \nu_\perp)/2$, and $\gamma = (\nu_\parallel + \nu_\perp)$, are very well satisfied.[12]

There is a general argument[13,14] that predicts $x = 1 - \alpha$ for the Landau–de Gennes free energy. The simplest form of the argument starts with the free energy $U = \int [a't|\psi|^2 + \delta U(\psi)] d^3 r$, and calculates the partition function $Z = \int \exp(-U/k_B T) d\psi$. Differentiation with respect to temperature then yields the excess entropy $\Delta S \sim \langle |\psi|^2 \rangle$. Since $\langle |\psi|^2 \rangle \sim |t|^x$ the critical part of the heat capacity $C \sim t^{-\alpha} \sim t^{x-1}$, so that $x = 1 - \alpha$. The data in Table I show unambiguously that this reduction is not universally satisfied.

In view of this direct relation between the Landau free-energy density and the relation $x = 1 - \alpha$, the discrepancy suggests that the Landau–de Gennes form is missing some essential physics. One possibility is the existence of a nearby tricritical point.[3-5] Unfortunately, the data in Table I show that the discrepancy is most serious for materials with the smallest McMillan ratio,[15] T_{NA}/T_{NI}, and these are the materials which should be furthest from the tricritical point. In fact the best agreement is for 9CB which is believed to be right at a tricritical point.

A second possibility concerns the data for $T < T_{NA}$. Even though the smectic-A phase has only quasi-long-range order (QLRO), the integrated intensity measured in this low-resolution experiment might behave similarly to systems with true long-range order (TLRO). However, in either case (QLRO or TRLO) $\int S(\mathbf{Q}) d^3 Q \sim \langle |\psi|^2 \rangle \sim |t|^{1-\alpha}$. If the system had TLRO, $S(\mathbf{Q})$ would have one term proportional to $|\langle \psi \rangle|^2 \delta(\mathbf{Q} - \mathbf{Q}_0)$ and a second diffuse term, with a different temperature dependence related to $\langle (\delta\psi)^2 \rangle$. In the case of TLRO scaling theory defines β such that $|\langle \psi \rangle|^2 \sim |t|^{-2\beta}$ with $2\beta = 2 - \gamma - \alpha$.[16] Since for the

VOLUME 54, NUMBER 9   PHYSICAL REVIEW LETTERS   4 MARCH 1985

data in Table I, $\gamma > 1$, if the system were behaving like one with TRLO it would have $2\beta < 1 - \alpha$. Thus if the measurement did not include all of the fluctuations, so that the measured value of $\langle (\delta\psi)^2 \rangle$ was less than the true value, the data would yield an effective exponent $2\beta < x^{\text{eff}} < 1 - \alpha$. We discount effects of this kind since independent fits of the data for $T \gtrsim T_{NA}$ obtain equal values for $x^{\pm}$ and the extracted values of x^- do not vary with the range-shrinking process.

A third possibility is that $C_{\pm}(0,0) = 0$. Since M must remain finite for all t, the only possibility is that $M \sim |t|^{\delta x}$, where $\delta x > 0$. In this case the measured exponent $x^m = x + \delta x$ would be greater than $1 - \alpha$. In fact, the measured values are les than or equal to $1 - \alpha$. A final possibility is that because of chemical impurities, surface effects, or some other mechanism, the transitions are slightly first order. In this case, the data for $I(t)$ would be asymmetric about $t = 0$ with a steeper slope on the smectic side. Again, since $x^+ = x^-$, we discount this for every material, except $\overline{10}\text{S5}$. For $\overline{10}\text{S5}$, however, other measurements[4,5] indicate that it is near to a tricritical point and we believe that trace impurities have induced a first-order transition in the present $\overline{10}\text{S5}$ sample.

If we ignore the possibility that x is only equal to $1 - \alpha$ in a critical regime that is restricted to a range of $|T - T_{NA}|$ smaller than probed here, our present understanding of the critical properties predicted by the Landau–de Gennes free energy suggests that this free energy is not completely adequate. This is particularly distressing since the deficiencies are most serious for the materials, such as 4O.7 and $\overline{8}\text{S5}$, which were expected to be described best. These have the smallest McMillan ratios, heat-capacity exponents $\alpha = 0$, and are furthest from a possible tricritical point.

We gratefully acknowledge discussions with B. Halperin and D. Nelson. This work was supported by the National Science Foundation under Grants No. DMR-82-12189 and No. DMR-80-20247.

(a)Permanent address: Physics Department, Bar-Ilan University, Ramat-Gan, Israel.

(b)Permanent address: Department of Physics, FM-15, University of Washington, Seattle, Wash. 98105.

[1]For a recent review see T. C. Lubensky, J. Chim. Phys. 80, 31 (1983).

[2]P. G. de Gennes, Solid State Commun. 10, 753 (1972).

[3]D. Brisbin, R. De Hoff, T. E. Lockhart, and D. L. Johnson, Phys. Rev. Lett. 43, 1171 (1979).

[4]J. Thoen, H. Marynissen, and W. Van Dael, Phys. Rev. Lett. 52, 204 (1984).

[5]B. M. Ocko, R. J. Birgeneau, J. D. Litster, and M. E. Neubert, Phys. Rev. Lett. 52, 208 (1984).

[6]See Ref. 1 for a description of relevant experimental studies.

[7]M. R. Fisch, P. S. Pershan, and L. B. Sorensen, Phys. Rev. A 29, 2741 (1984), and Phys. Rev. Lett. 48, 943 (1982).

[8]See, for example, B. E. Warren, X-Ray Diffraction (Addision-Wesley, Reading, Mass., 1969), Chap. 4.

[9]J. Als-Nielsen, R. J. Birgeneau, M. Kaplan, J. D. Litster, and C. R. Safinya, Phys. Rev. Lett. 39, 352 (1977).

[10]Addition of a linear term, Bt, to this form reduces the resultant values of x by less than 0.02 with no significant improvement in the quality of the fit.

[11]The assumption that $G(\mathbf{r}) = C_0(r) \mp |t|^x C_{\pm}(r/\xi)$ is a general consequence of isotropic scaling [see M. E. Fischer and A. Aharony, Phys. Rev. Lett. 31, 1238 (1973)]. The anisotropic form used here is a natural consequence of the anisotropic scaling hypothesis. The only feature of this form that is essential to the present discussion is the assumption that the product of the coefficient in the leading term in the temperature expansion of $G(\mathbf{r}) - C_0(\mathbf{r})$ and the resolution function $h(\mathbf{r})$ are integrable.

[12]P. S. Pershan and J. Als-Nielsen, Phys. Rev. Lett. 52, 759 (1984).

[13]The most explicit statement of this result appears in Fischer and Aharony, Ref. 11, Eq. (4); however, general arguments supporting it are given by their several references.

[14]A similar form was introduced by Daves et al. for the hexatic order parameter near the hexatic-to-smectic-A transition. S. C. Daves, J. Budai, J. W. Goodby, and R. Pindak, "X-Ray Study of the Hexatic-B to Smectic-A Phase transition in Liquid Crystal Films" (to be published).

[15]W. McMillan, Phys. Rev. A 7, 1419 (1973).

[16]H. Eugene Stanley, Introduction to Phase Transitions Critical Phenomena (Oxford Univ. Press, New York, 1971), p. 185.

[17]C. W. Garland, M. Meichle, B. M. Ocko, A. R. Kortan, C. R. Safinya, L. J. Yu, J. D. Litster, and R. J. Birgeneau, Phys. Rev. A 27, 3234 (1983).

2. LANDAU-PEIERLS INSTABILITY

2.1. Caillé, A. (1972). Remarques sur la diffusion des rayons X dans les smectiques. A. *C. R. Acad. Sc. Paris* **274B**, 891–893* 133

2.2. Als-Nielsen, J., Litster, J. D., Birgeneau, R. J., Kaplan, M., Safinya, C. R., Lindegaard-Anderson, A. and Mathiesen, S. (1980). Observation of algebraic decay of positional order in a smectic liquid crystal. *Phys. Rev.* **B22**, 312–320 136

2.3. Safinya, C. R., Roux, D., Smith, G. S., Sinha, S. K., Dimon, P., Clark, N. A. and Bellocq, A. M. (1986). Steric interactions in a model multimembrane system: a synchrotron x-ray study. *Phys. Rev. Lett.* **57**, 2718–2721 145

* The exponent in Eq. 9b should read 4 −2X rather than 2 − 2X.

C. R. Acad. Sc. Paris, t. 274, p. 891-893 (5 avril 1972)   Série B

PHYSIQUE CRISTALLINE. — *Remarques sur la diffusion des rayons X dans les smectiques* A. Note (*) de M. **Alain Caille**, transmise par M. André Guinier.

L'intensité $I\left(\vec{q}\right)$ diffusée par un monocristal smectique A est analysée pour $\vec{q}$ dans le voisinage du vecteur d'onde de Bragg : $|\vec{q_z}| = 2\pi/d$. Les lois différent pour des directions différentes du vecteur d'onde $\vec{q}$. Les exposants caractéristiques sont fonction de la température.

Les cristaux liquides, dans la phase smectique, possèdent à l'équilibre une configuration en couches équidistantes, d'épaisseur d, perpendiculaires à un axe (OZ). L'intensité diffusée des rayons X possède des maximums correspondant à la condition de Bragg : $|\vec{q_z}| = 2\pi/d$, où $\vec{q} = \vec{k_r} - \vec{k_i}$; $\vec{k_r}$ et $\vec{k_i}$ sont respectivement les vecteurs d'onde réfléchi et incident. Nous calculons l'intensité diffusée par les fluctuations thermiques du déplacement $u\left(\vec{r}\right)$ des couches dans la direction OZ, couplé à la dilatation de volume θ. Nous supposons θ et $\partial u/\partial z$ petits (approximation harmonique); pour ce cas, les moyennes thermodynamiques des composantes de Fourier $U_{\vec{q}}$, ont été calculées dans (¹) :

$$(1) \qquad \left\langle \left| U_{\vec{q}} \right|^2 \right\rangle = \frac{k_B T}{\bar{B} q_z^2 + K_1 q_\perp^4},$$

où

$$U_{\vec{q}} = \int e^{i\vec{q} \cdot \vec{r}} u\left(\vec{r}\right) d^3\vec{r} \qquad \text{et} \qquad \bar{B} = B - \frac{C^2}{A} > 0.$$

A, B et C représentent des rigidités isothermes et K_1 une constante de Frank. Nous supposons que la largeur du maximum de Bragg, due aux défauts statiques d'une structure lamellaire parfaitement ordonnée, est faible par rapport à l'élargissement dû aux fluctuations thermiques. Nous considérons que les rayons X sont diffusés par un monodomaine infini de surfaces définies par l'équation

$$(2) \qquad z_n = nd + u_n\left(\vec{\rho}\right),$$

n est un entier et $u_n\left(\vec{\rho}\right)$ est le déplacement de la couche n dans la direction OZ en un point $\vec{\rho}$ (coordonnée polaire dans le plan des couches). L'intensité des rayons X diffusée à l'équilibre thermique par ce système (exprimée en unité électronique) est

$$(3) \qquad I = |f_0|^2 \sum_n \int d\vec{\rho} \, e^{iq_z \cdot nd} \, e^{i\vec{q}_\perp \cdot \vec{\rho}} \left\langle e^{iq_z \cdot (u_n(\vec{\rho}) - u_0(0))} \right\rangle,$$

(2)

où $\vec{q}_\perp$ est la composante vectorielle de $\vec{q}$ perpendiculaire à l'axe OZ et $\langle\ \rangle$ représente une moyenne thermique. f_0 est le facteur atomique diffusant pour le système défini en 2. Dans l'approximation harmonique, toute combinaison linéaire des $u_n(\vec{\rho})$ est une variable aléatoire de distribution gaussienne. Donc

(4) $$\left\langle e^{iq_z(u_n(\vec{\rho})-u_0(0))}\right\rangle = \exp\left(-\frac{1}{2}q_z^2\left\langle\,|\,u_n(\vec{\rho})-u_0(0)\,|^2\,\right\rangle\right).$$

Pour n grand, nous remplaçons la sommation sur n par une intégrale sur z; la valeur quadratique moyenne devient alors

(5) $$\left\langle\,|\,u_n(\vec{\rho})-u_0(0)\,|^2\,\right\rangle = \frac{2}{(2\,\pi)^3}\int d^3\vec{q}\,(1-e^{-i\vec{q}.\vec{r}})\left\langle\,|\,U_{\vec{q}}\,|^2\,\right\rangle.$$

Utilisant l'expression (1) dans (5), nous obtenons pour $z \gg a^2/\alpha$:

(6) $$G(\vec{r}) = \left\langle e^{iq_z(u_n(\vec{\rho})-u_0(0))}\right\rangle = \exp(-2\,M)\,\exp\left(-X E_1\left(\frac{\rho^2}{4\,\alpha\,z}\right)\right)\left(\frac{2\,a}{\rho}\right)^{2\,X},$$

où

$$x = \left(\frac{K}{B}\right)^{\frac{1}{2}} \quad \text{et} \quad X = \frac{\pi}{2}\frac{k_B T}{K\,a^2}\,x.$$

M joue le rôle du facteur de Debye-Waller. Il est linéaire en T et quadratique en fonction du nombre d'onde $|\vec{q}|$:

(7) $$M = -\,q^2\,\frac{k_B T\,\gamma}{(2\,\pi)\,B\,\alpha},$$

γ est la constante d'Euler ($\gamma = 0{,}577\ldots$). $E_1(x)$ est la fonction intégrale exponentielle

$$E_1(x) = \int_x^\infty \frac{e^{-t}}{t}\,dt.$$

La fonction de corrélation $G(\vec{r})$ s'écrit

(8 a) $$G(\vec{r}) = 2\,e^{-M}\left(\frac{a^2}{\alpha\,z}\right)^X, \qquad \rho \ll z,$$

(8 b) $$G(\vec{r}) = 2\,e^{-2\,M}\left(\frac{2\,a}{\rho}\right)^{2\,X}, \qquad \rho \gg z.$$

L'intensité diffusée est obtenue à partir de l'équation (6) : pour les deux directions principales (z et x) :

(9 a) $$I \sim \frac{1}{\left(q_z - \dfrac{2\,\pi}{d}\right)^{2-X}}, \qquad q_\perp = 0,$$

(9 b) $$I \sim \frac{1}{q_\perp^{2-2X}}, \qquad q_z = 0.$$

(3)

Les lois sont très différentes de celles que l'on trouve pour la diffusion par un solide (^{2}) où X pour la diffusion à un phonon serait nul. Pour le cristal liquide (smectique A), X est de l'ordre de l'unité. Il faut aussi noter que l'exposant de l'expression (9) est différent si nous nous déplaçons suivant q_z ou $q_\perp$. Le déplacement u des couches est plus fortement corrélé dans la direction z que dans le plan des couches. Cette anisotropie se traduit par un maximum d'intensité allongé dans la direction $q_\perp$. Ce phénomène est observé sur des diagrammes photographiques (^{3}) de rayons X pour un smectique A orienté dans un champ magnétique. Le paramètre X dépend linéairement de α. En s'approchant de la température de transition smectique-nématique T_{SN} et si la transition à T_{SN} est du deuxième ordre, on attend $\overline{B} \to 0$ et $X \to \infty$. Toutefois, pour X grand, il est probable que l'approximation harmonique n'est plus acceptable. Il serait donc intéressant :

a. de vérifier la loi (9) pour T nettement inférieur à T_{SN}, par des mesures sur des spécimens monodomaines : les expériences existantes (^{4}) (Mc Millan) ne permettent pas cette vérification. Une étude plus fine est en cours (^{5});

b. de mettre en évidence des déviations aux lois harmoniques pour $T \to T_{SN}$.

P. G. de Gennes et J. Jouffroy ont attiré mon attention sur ces problèmes.

(*) Séance du 27 mars 1972.
(¹) P. G. DE GENNES, *J. Phys.*, C, 4, 1969, p. 65-71.
(²) A. A. MARADUDIN, E. W. MONTROLL et G. H. WEISS, *Solid State Physics*, Suppl. 3, 1963.
(³) S. DIELE, P. BRAND et H. SACKMANN, *Molecular Crystals and Liquid Crystals*, 16, 1972, p. 105-116.
(⁴) W. L. MC MILLAN, *X-Ray scattering from liquid crystals* (à paraître).
(⁵) J. JOUFFROY, Communication privée.

Laboratoire de Physique des Solides,
Faculté des Sciences d'Orsay,
Bât. 510,
91-Orsay, Essonne.

PHYSICAL REVIEW B   VOLUME 22, NUMBER 1   1 JULY 1980

Observation of algebraic decay of positional order in a smectic liquid crystal

J. Als-Nielsen and J. D. Litster[*]

Risø National Laboratory, DK-4000 Roskilde, Denmark

R. J. Birgeneau, M. Kaplan, and C. R. Safinya

*Department of Physics, Massachusetts Institute of Technology,
Cambridge, Massachusetts 02139*

A. Lindegaard-Andersen and S. Mathiesen

*Technical Physics Laboratory III, Technical University of Denmark,
DK-2800 Lyngby, Denmark*

(Received 17 September 1979)

A smectic-A liquid crystal in three dimensions has been predicted to exhibit algebraic decay of the layer correlations rather than true long-range order. As a consequence, the smectic Bragg peaks are expected to be power-law singularities of the form $q_{\parallel}^{-2+\eta}$ and $q_{\perp}^{-4+2\eta}$ where $\parallel (\perp)$ is along (perpendicular to) the smectic density wave vector direction, rather than δ function peaks. Observation of these phenomena requires very high instrumental resolution together with a resolution function with wings which drop off much more rapidly than $q_{\parallel}^{-2} (q_{\perp}^{-4})$. We show that these requirements may be met by using a three crystal x-ray spectrometer with multiple-reflection channel cut crystals as monochromator and analyzer. We find that the smectic-A Bragg peaks observed in the liquid-crystal octyloxy-cyanobiphenyl are indeed consistent with the predicted power-law singularity form. Furthermore, the explicit values of η required to describe the measured profiles are in accordance with calculations of η using the harmonic approximation with empirically determined splay and layer compressibility elastic constants.

I. INTRODUCTION

It has been known theoretically for some time that translational order as it occurs in a solid cannot exist in two dimensions because it is destroyed by thermally excited fluctuations.[1,2] Thus for crystalline solids d^0, the lower marginal dimensionality, is two. This is the spatial dimension at which thermal fluctuations prevent the establishment of the long-range order which the interactions between the atoms would favor. For many systems d^0 is two, and this is one of the reasons for the intense recent experimental and theoretical interest in two-dimensional materials. For these systems it is predicted that a transition occurs to a state of quasi-long-range order in which the positional correlation functions do not extend to infinity, but decay algebraically as some power of the distance.[3] A similar algebraic decay of correlation functions occurs at an ordinary critical point and in both cases the power-law decay of correlations is accompanied by an infinite susceptibility. For various experimental reasons, this predicted behavior has not yet been observed directly in any two-dimensional system.[4]

A closely related phenomenon is predicted to occur in smectic-A and -C liquid crystals.[5,6] These are phases of liquid crystals which possess both orientational long-range order (LRO) of the anisotropic molecules and translational-order intermediate between that of a liquid and a solid.[6] Explicitly, in the A and C phases one has a one-dimensional mass-density wave in a three-dimensional liquid. The density wave may be either along (A phase) or at an angle ϕ (C phase) to the nematic director. As we shall discuss below, it may be readily shown that the positional fluctuations diverge logarithmically at large distances at all temperatures in exact analogy with 2D crystals. Thus the smectic-A-to-nematic phase transition is not only a relatively simple example of melting, but such liquid crystals also provide a three-dimensional system in which the effects of divergent long-wavelength acoustic fluctuations can be studied experimentally. Ultimately one may hope that study of smectic phases will lead to better understanding of melting, a problem which still eludes condensed-matter physicists.

In this paper we report the results of a high-resolution x-ray-diffraction study of the density wave in the smectic-A phase of octyloxy-cyanobiphenyl (8OCB).

The traditional picture[6] of the smectic-A phase (SmA) is shown in the left part of Fig. 1. The molecular centers are arranged in layers with a well-

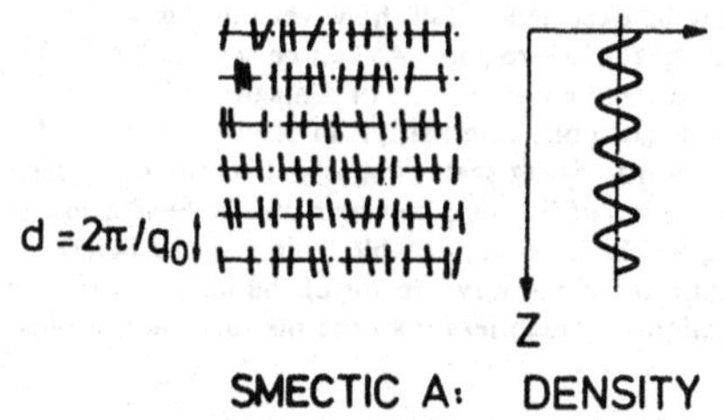

$$d = 2\pi/q_0$$

SMECTIC A: DENSITY

FIG. 1. The smectic-A phase consists of a one-dimensional density wave along the average direction of the molecular axis. The sinusodial shape of the density wave is reflected in the diffraction pattern by the absence of higher-order reflections.

defined layer spacing d but with a liquidlike order within each layer. The average orientation, $\bar{n}$, of the molecules is perpendicular to the layers in the SmA phase. If this picture were literally correct, Bragg scattering of x-rays should occur whenever the difference between incident and scattered wave vectors equals a multiple of $q_0 = 2\pi/d$, that is $\bar{k}_i - \bar{k}_f = p(2\pi/d)\bar{n}$ with $p = 1, 2 \ldots$. In an earlier study[7] we found that very near T_c the higher-order Bragg peaks are absent, or precisely, their intensity is at least a factor of 10^4 less than the fundamental reflection. A more correct description of the SmA phase is therefore that the density forms a sinusoidal ripple as shown in the right-hand part of Fig. 1, and the dot-dashed lines through the molecules should not be interpreted as lattice planes but rather as planes of a certain phase of the density wave. The question is now whether this ripple has true long-range order.

Let us briefly recall the arguments of Landau[2] and Lifshitz[5] and by Peierls[1] on this point. The long-wavelength acoustic modes shown pictorially in Fig. 2 arising from the SmA ordering involve a displacement u of the smectic layers in the z direction, or more properly gradients in the phase $\phi = q_0 u$ of the density wave. These modes have unusually anisotropic elastic behavior. With the wave vector $\bar{q}$ along z

($\bar{q} = q_\| \hat{z}$, see upper part of Fig. 2) the displacement u is longitudinal, and the energy is of the usual elastic form $\frac{1}{2}B_{q_\|} u^2(\bar{q})$ with B a compressibility for the smectic layers. However with $\bar{q}$ normal to $\hat{z}$, the displacement is transverse and the layer separation is unchanged as is evident from the lower part of Fig. 2; thus, to second order in $q_\perp$ the displacement requires no elastic energy. The restoring force for $q_\| = 0$ then arises because the director field $\bar{n}$ remains normal to the layers and thus a director splay distortion results with an elastic energy density $\frac{1}{2}K(\bar{\nabla}\cdot\bar{n})^2$, where K is the nematic phase splay elastic constant. Since $\bar{n}$ is normal to the layers, one has $\delta\bar{n} = -\bar{\nabla}_\perp u$ and an elastic energy $\frac{1}{2}Kq_\perp^4 u^2(\bar{q})$. Experimentally one finds that K has the same value in both the nematic and smectic phases near the SmA-N transition.[8] Since a magnetic field exerts a torque on the molecules, the complete expression for the elastic energy density of the SmA elastic modes of wave vector $\bar{q}$ is

$$F_{\bar{q}} = \frac{1}{2}(Bq_\|^2 + \chi_a H^2 q_\perp^2 + Kq_\perp^4)u^2(q) , \qquad (1)$$

where χ_a is the volume diamagnetic susceptibility anisotropy that results from nematic ordering. We note that B and $\chi_a H^2$ have the dimensions of energy per unit length. The characteristic lengths of the smectic phase are thus the sample dimension L, the density wavelength $2\pi/q_0$, and two additional lengths $\lambda = (K/B)^{1/2}$ and $\xi_M = (K/\chi_a H^2)^{1/2}$. Both λ (the penetration depth) and ξ_M (the magnetic coherence length) are relaxation lengths; λ determines the decay of an undulation distortion (pure splay director distortion) imposed at the surface of a uniform smectic. A surface-imposed director alignment in a nematic phase will relax to that favored by a magnetic field with a characteristic distance ξ_M; in the smectic-A phase ξ_M determines the long-wavelength cutoff of the diverging fluctuations in u.

The mean-squared fluctuations $\langle u^2 \rangle$ can be calculated by applying the equipartition theorem to Eq. (1) and summing over all wave vectors $< q_0$. One readily finds

$$\langle u^2 \rangle = (2\pi)^{-3}kT \int_{-q_0}^{q_0} dq_\| \int_{q_{min}}^{q_{max}} (Bq_\|^2 + \chi_a H^2 q_\perp^2 + Kq_\perp^4)^{-1} 2\pi q_\perp dq_\perp$$

$$= (4\pi)^{-1}kT(BK)^{-1/2} \int_{q_{min}}^{q_{max}} (q_\perp^2 + \xi_M^{-2})^{-1/2} dq_\perp$$

$$= (4\pi)^{-1}(BK)^{-1/2}kT \ln(q_{max}L/2\pi) \quad \text{for} \quad \xi_M = 0 , \qquad (2a)$$

$$\simeq (4\pi)^{-1}(BK)^{-1/2}kT \ln(2\xi_M q_{max}) \quad \text{for} \quad 0 < \xi_M << L . \qquad (2b)$$

The mean-squared displacement thus diverges logarithmically with the smaller of L or ξ_M and, in this model, an infinite three-dimensional smectic-A liquid crystal in zero field does not have true long-range positional order at any finite temperature. A similar divergence due to long-wavelength modes makes two-dimensional crystals

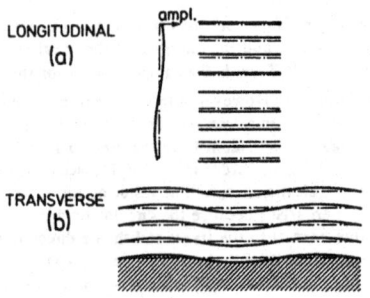

FIG. 2. (a), (b) The two fundamental long-wavelength modes of the smetic-*A* phase. The energy density is proportional to $q_{\parallel}^2$ for the longitudinal mode but to $q_{\perp}^4$ for the transverse mode.

unstable. We should note, however, that there are a number of important differences in the predicted phase transition behavior of these two systems. For example, the 2D crystal should exhibit no observable specific-heat singularity on melting whereas the 3D smectic-*A* exhibits a dramatic λ anomaly in its specific heat at the N-A transition.[9] There is not yet a fully satisfactory theory for the nematic–smetic-*A* transition. However, in the approach presented by Halperin and Lubensky[10] one may define a nonlocal order parameter which has true LRO in three dimensions in spite of the fact that the positional correlations are expected to decay algebraically with distance in the smectic-*A* phase.

The consequences of the divergent real-space positional fluctuations for x-ray scattering may be calculated in the harmonic approximation.[11] The details of this calculation are given in Appendix A. In this case the instantaneous correlations exhibit an anisotropic power-law dependence on molecular separation. On Fourier transforming for a sample in zero field, one finds that the x-ray scattering is predicted to follow

$$S(0,0,q_{\parallel}) \sim (q_{\parallel}-q_0)^{-2+\eta} \tag{3a}$$

and

$$S(q_{\perp},0,q_0) \sim q_{\perp}^{-4+2\eta} \tag{3b}$$

with

$$\eta = kT(q_0^2/8\pi)(BK)^{-1/2} \ . \tag{3c}$$

We should emphasize that in x-ray scattering we measure the instantaneous correlations so that Eq. (3) represents quasielastic rather than true elastic scattering. Analogous results have been obtained for the two-dimensional harmonic crystal.[12-14] Near the nematic-smectic transition temperature T_c, B becomes very small[8] and the harmonic approximation

may be expected to fail; however there is as yet no theory for this region. As can be seen from Eq. (2), finite sample dimensions or a magnetic field remove the logarithmic singularity. In principle these will cause true Bragg scattering as well as the quasielastic scattering of Eq. (3); however this finite size scattering will be indistinguishable from Landau-Peierls scattering if the wave vector of the longest wave fluctuations is much less than the instrumental resolution.

II. EXPERIMENTAL

Experimentally it is not trivial to distinguish between the "Landau-Peierls" scattering of Eq. (3) and a combination of Bragg scattering plus thermal diffuse scattering from the acoustic modes. The instrumental resolution must be very good to distinguish between a δ function (Bragg) and a cusp (Landau-Peierls). High angular resolution can be obtained by using perfect crystals together with an essentially monochromatic x-ray line source; only x rays collimated within a few milliradians will fulfill the Bragg condition. In an earlier study we used the (111) reflection from perfect Ge crystals to define the direction of the incident and scattered x rays. In this study we did not fully appreciate the extent to which tails of the resolution function, arising primarily from extinction,[15] made it impossible to conclude unambiguously that the tails we observed were indeed caused by the Landau-Peierls scattering.

The solution to this problem is to use multiple Bragg reflections in the collimating crystals, since the resolution after *m* Bragg reflections is that for a single reflection raised to the *m*th power.[16] In the present paper we have investigated the resolution functions in great detail and are able to prove that the SmA line profile is consistent with Landau-Peierls scattering as described by Eq. (3) with essentially no adjustable parameters.

The high-resolution experimental setup at Risø is shown schematically in Fig. 3. The scattering plane is

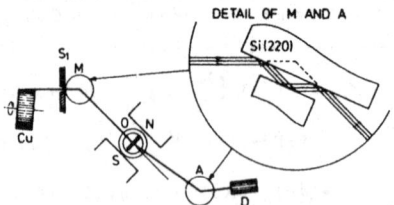

FIG. 3. Experimental setup using triple-reflection channel-cut crystals as monochromator (*M*) and analyzer (*A*) crystals. A single-domain smectic-*A* sample is obtained by aligning the director by a magnetic field. S_1 is the slit to separate $K\alpha_1$ from $K\alpha_2$ lines; *O* is a two-stage oven holding the sample; *D* is a scintillation detector.

horizontal and the x-ray source has dimensions 0.9 mm wide by 0.6 mm high seen from the monochromator M. The x-ray generator is a rotating-anode machine operating at 55 kV and 180 mA with a copper target. A 0.5-mm wide 3-mm high slit S_1, 65 cm from the source enables the monochromator to separate the $K\alpha_1$ and $K\alpha_2$ lines. Because the incident beam is depleted by scattering as it penetrates the monochromator crystal, only a finite number of crystal planes contribute to the Bragg scattering. Thus even a perfect crystal with a monochromatic x-ray beam will have a finite angular acceptance known as the Darwin width w_D; and the beam incident on the sample has an angular divergence equal to the Darwin width. The theoretical value of w_D for the Si (220) reflections we used is 0.0012°. The direction of x rays scattered by the sample is similarly determined by the analyzer A. A two-stage oven located at 0, 44 cm from the monochromator, contains the 1.5-mm-thick flat 8OCB sample between Be windows; to obtain a single domain sample a field of 4.8 kG was supplied by an electromagnet. The analyzer was 36 cm from the sample, and a slit (4 mm wide, 13 mm high) determined the illuminated area of the scintillation detector D. The channel-cut crystals used for M and A are discussed in more detail in an Appendix.

III. LINE PROFILE OF BRAGG SCATTERING

In this section we shall consider in some detail the Bragg-line shape one would have observed if the SmA phase had true long-range order. As we shall see one can, by simple scaling of a measured Bragg profile which in our case originated from a Si(111) reflection, simulate the Bragg profile of a crystal with the same planar spacing as 8OCB. Furthermore, we account in detail for the Si(111) profile from measured properties of the monochromator and analyzer crystals. In the end of the section we compare the simulated Bragg profile with the actual line profile from the SmA, and this comparison shows that the SmA phase exhibits anomalously intense diffuse scattering which is consistent with our expectations based on the harmonic theory but which appears to be much too strong to be accounted for on the basis of normal thermal diffuse scattering accompanying a true Bragg peak.

There are two contributions to the width, or in general the shape, of a Bragg profile. One is the Darwin width of the perfect monochromator and analyzer crystals as already discussed in the previous section; the other is the spectral width of the Cu $K\alpha_1$ line. In order to discuss the latter effect let us, as illustrated in Fig. 4, consider two wavelengths, λ and $\lambda + \Delta\lambda$, as they are Bragg scattered by the monochromator M, the sample S, and the analyzer A. The difference in Bragg angles between the two wave-

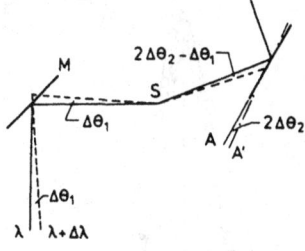

NONDISPERSIVE

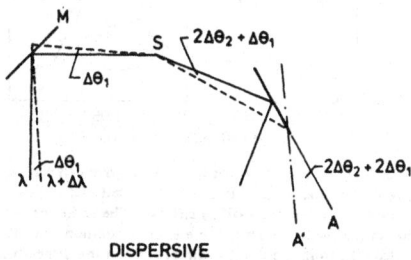

DISPERSIVE

FIG. 4. The geometry of the dispersive (bottom) and nondispersive (top) orientation of the monochromator (M) and analyzer (A) crystals and sample (S). Note that the dashed lines representing rays with wavelength $\lambda + \Delta\lambda$ will only be reflected by the anlayzer crystal if either $2\Delta\theta_2$ or $2\Delta\theta_2 + 2\Delta\theta_1$ is less than the Darwin width for the nondispersive and dispersive configurations, respectively. Therefore as the sample scattering angle $2\theta_2$ becomes small relative to the analyzer Bragg angle $2\theta_1$, the resolution function in the nondispersive configuration approaches the direct beam profile.

lengths is $\Delta\theta = (\Delta\lambda/\lambda)\tan\theta$. We use subscript 1 for Bragg angles in the identical monochromator and analyzer crystals, and subscript 2 for the Bragg angle in the sample. It is important to notice the difference between the two possible analyzer orientations as illustrated in the top part and in the bottom part of Fig. 4. In the first case the analyzer must be turned through the angle $2\Delta\theta_2$ counter clockwise in order to scatter the wavelength $\lambda + \Delta\lambda$, whereas it must be turned the angle $2\Delta\theta_2 + 2\Delta\theta_1$ clockwise in the latter case. Suppose there is no sample. In the top part of Fig. 4 the analyzer should then not be turned at all in order to scatter $\lambda + \Delta\lambda$, and scanning the analyzer would not yield any information about the wavelength distribution. But in the setting of the bottom part of Fig. 4 an analyzer scan determines directly the intrinsic linewidth of the Cu $K\alpha_1$ line, the angular width being $2(\Delta\lambda/\lambda)\tan\theta$, provided that the width is much larger than w_D. This orientation of monochromator and analyzer is called the dispersive orienta-

140

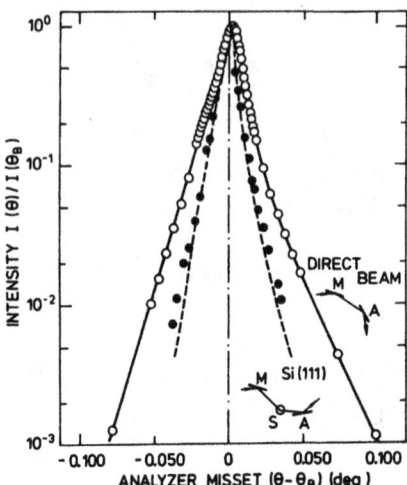

FIG. 5. Direct beam profile in the dispersive orientation (open circles) and the profile of a Si(111) reflection in the nondispersive orientation (filled circles). The solid line is a guide to the eye. The dashed line is the resolution function as calculated from the direct beam profiles in the dispersive and nondispersive orientation.

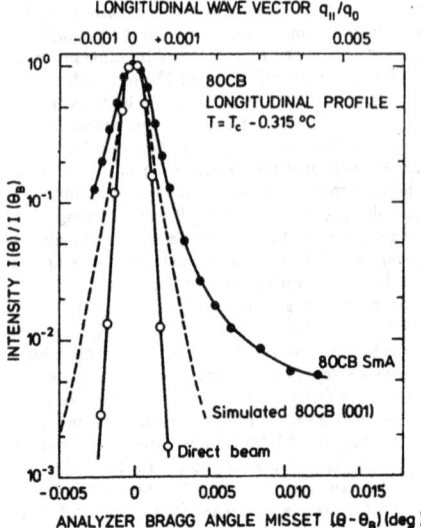

FIG. 6. The calculated and measured line profile of Si(111) gives by simple scaling a simulation of the 8OCB profile if the smectic-*A* phase had true long-range order (dashed line). This profile is somewhat broader than the direct beam profile (open circles) due to the linewidth of the Cu $K\alpha_1$ line. The measured profile from 8OCB (filled circles) exhibits pronounced wings, and the full line is a least-squares fit of the theoretical line shape discussed in the text folded with the experimental resolution.

tion. We have examined the resolution function of the apparatus by analyzer scans with no sample for both the dispersive and nondispersive orientations. With scattering through a finite angle by a sample, the x-ray linewidth broadens the resolution even for the nondispersive orientation. From the direct beam profile with the analyzer in the dispersive orientation (shown as open circles in Fig. 5) we find a full width at half maximum (FWHM) 0.015° or 0.262 mrad. With the Bragg angle $\theta_1 = 23.651°$ we find $\Delta\lambda/\lambda = 3.1 \times 10^{-4}$ for the Cu $K\alpha_1$ line. We can then use this result to calculate the line shape for Bragg scattering from a sample with nondispersive orientation. For Si(111), neglecting the finite Darwin width, we should expect to scale the abscissa by $\tan\theta_{111}/\tan\theta_{220} = 0.579$; we find $w = 0.00897°$ FWHM. Since $w^2 \gg w_D^2$, neglect of the Darwin width was a good approximation. The predicted Si(111) line (dashed curve) and the data points (solid circles) agree quite well. It should be noted that if the sample is a single crystal with a narrow mosaic spread, the sample must be rotated along with the analyzer to scan the Bragg peak. In the nondispersive orientation the ratio of sample to analyzer rotations is $-2\tan\theta_2/(\tan\theta_1 - \tan\theta_2) = -2.75$, the minus sign indicating rotations in opposite directions.

The direct beam profile for nondispersive orientations is shown (open circles) in Fig. 6. The FWHM is not far from the theoretical limit $w_D = 0.0012°$.

More important is the very rapid falloff in intensity as the angular misset exceeds w_D. This tailless resolution is the virtue of channel-cut crystals as discussed in the Appendix. We may calculate the expected line profile for a true Bragg peak corresponding to q_0 for 8OCB (which has $\theta_2 = 1.39°$). The result is obtained by scaling the abscissa for the Si(111) peak of Fig. 5 by $\tan(1.39°)/\tan(14.221°) = 0.096$, to obtain the width due to $\Delta\lambda$ and then convoluting this with the profile determined by scanning through the forward direction in the nondispersive orientation; the net result is shown as a dashed line in Fig. 6. The line profile observed at a reduced temperature $t = 1 - T/T_c = 9 \times 10^{-4}$ is shown as solid circles in Fig. 6. We see this differs significantly from what one would expect for a conventional Bragg peak, the difference being more than an order of magnitude at a misset of 0.005° from the peak. We emphasize that the expected profile for true Bragg scattering is obtained by direct scaling in angle of experimental data [for the Si(111) Bragg peak] that is, with this ultrahigh resolution, normal thermal diffuse scattering in the silicon is almost undetectable.

IV. QUANTITATIVE DATA ANALYSIS

In our experiment we measure the convolution of the scattering $S(\vec{q})$ with the spectrometer resolution function $R(\vec{q})$. Thus if the spectrometer is set for wave-vector transfer $\vec{q}$, a variety of scattering processes in the vicinity of $\vec{q}$ will be picked up as

$$S(\vec{q}) \sim \int S(\vec{q}\,')R(\vec{q}-\vec{q}\,')d\vec{q}\,' \ . \tag{4}$$

A conventional deconvolution by direct numerical integration is quite inefficient because $S(\vec{q})$ diverges at $(0,0,q_0)$. Fortunately we have an explicit theoretical expression for $S(\vec{r})$, the Fourier transform of $S(\vec{q})$, rather than for $S(\vec{q})$ itself. We may therefore use the folding theorem to calculate

$$S(\vec{q}) = (2\pi)^3 \int d\vec{r}\, S(\vec{r})F(\vec{r})e^{i\vec{q}\cdot\vec{r}} \ , \tag{5}$$

where $F(\vec{r})$ is the Fourier transform of $R(\vec{q})$ and the correlation function $S(\vec{r})$ was calculated by Caillé[11] (see Appendix) as

$$S(\vec{r}) \sim e^{-2\tau\eta}\left[\frac{1}{x^2+y^2}\right]^{\eta}\exp\left[-\left[\eta E_1\left[\frac{x^2+y^2}{4\lambda z}\right]\right]\right] \tag{6}$$

with $\lambda = (K/B)^{1/2}$ and η given by Eq. (3c). The resolution in the vertical (y) direction can be calculated from the geometry shown in Fig. 3. It is actually trapezoidal with base full width 20 mrad and peak full width 11 mrad; we approximate it for ease of calculation with a Gaussian of the same FWHM ($\sigma_y = 0.22q_0$, where $q_0 = 0.197$ Å^{-1}). The longitudinal resolution (along $q_z = q_{\parallel}$), is given by the dashed line in Fig. 6; it has tails which are very nearly exponential. We therefore represented it within the accuracy of our measurements by the best fit to the convolution of a Gaussian with an exponential; the functional form is given by Table I and the parameters are $\sigma_z = 2.2 \times 10^{-4}q_0$ with the momentum-space decay of the exponential tail being $\sigma\sigma_z = 1.54\sigma_z$. This resolution function has a FWHM of 0.0014°, only slightly larger than the theoretical Darwin width. The exponential wings of the longitudinal resolution

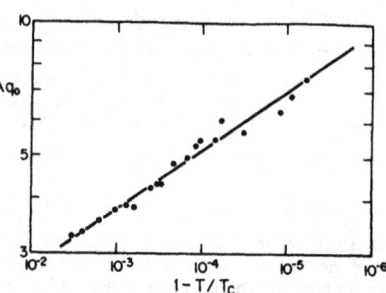

FIG. 7. Liquid-crystal penetration depth, $\lambda = (K/B)^{1/2}$, in the smectic-A phase of 8OCB. As T approaches T_c the layer stiffness constant B approaches zero. The solid line is the equation $\lambda q_0 = 1.55(1-T/T_c)^{-0.13}$ obtained from a least-squares fit to a power law. The data are from Ref. 17.

function are an experimental fact which we do not fully understand. Finally, the transverse in-plane resolution was determined by the mosaicity of the liquid-crystal sample, which showed some variation with temperature. We represented the mosaic distribution by a Gaussian whose width was determined by a fit to points above 20% peak intensity for sample rocking curves ($q_\perp = q_x$ scans) for each temperature. The results were: $\sigma_x/q_0 = 6.45 \times 10^{-3}$, 4.2×10^{-3}, and 4.9×10^{-3} for $t = 9.0 \times 10^{-4}$, 5.9×10^{-4}, and 4×10^{-6}, respectively. The reciprocal-space resolution function is given by $R(\vec{q}) = R_1(q_x)R_2(q_y)R_3(q_z)$, and its real-space transform is $F(\vec{r}) = F_1(x)F_2(y)F_3(z)$; all of these functions are given in Table I. The data analysis consisted of a nonlinear least-squares fit of the Fourier transform of $F(\vec{r})$ to the longitudinal scan data normalized to the peak height. Values of $q_{\parallel} < 0.999q_0$ were not used because of contaminant scattering from the $K\alpha_2$ line. $S(\vec{r})$ is determined by two parameters: λ and η. Since the penetration depth has been accurately measured by light scattering,[8,17] we used these results in our data analysis. The measured value of λq_0 for 8OCB is shown in Fig. 7. The fits were therefore carried out with only

TABLE I. The experimental resolution function (values for the parameters are given in the text).

Momentum space	Real space
$R(\vec{q}) = R_1(q_x)R_2(q_y)R_3(q_z)$	$F(\vec{r}) = F_1(x)F_2(y)F_3(z)$
$R_1(q_x) = \exp(-q_x^2/2\sigma_x^2)$	$F_1(x) = \exp(-x^2\sigma_x^2/2)$
$R_2(q_y) = \exp(-q_y^2/2\sigma_y^2)$	$F_2(y) = \exp(-y^2\sigma_y^2/2)$
$R_3(q_z) = \int \exp(-q_1^2/2\sigma_z^2)$	$F_3(z) = \dfrac{\exp(-z^2\sigma_z^2/2)}{(1+\sigma^2\sigma_z^2z^2/2)}$
$\quad \times \exp(\sqrt{2}|q_z-q_1|/\sigma\sigma_z)dq_1$	

TABLE II. Exponent η and elastic constant K.

t	9.0×10^{-4}	5.9×10^{-4}	4×10^{-6}
η	0.17 ± 0.02	0.23 ± 0.02	0.38 ± 0.06
$10^7 K$ (dyne)	8.4 ± 1	7.1 ± 0.6	7.7 ± 2

one adjustable parameter, the exponent η. The results of the fits are shown in Table II. The uncertainties are those which result in a doubling of χ^2 and include the effects of experimental errors in λ and the resolution function. The resulting fits for two longitudinal scans are shown along with the longitudinal resolution function in Fig. 8.

As a final check in our analysis, we give the elastic constant K deduced from each value of η in Table II. It has been established experimentally that K shows no anomalous behavior near the A-N transition; thus

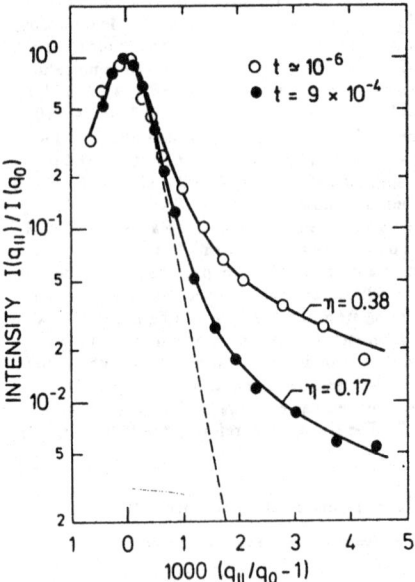

FIG. 8. Longitudinal line profiles at two reduced temperatures. The dashed line is the experimental resolution as shown and discussed in Fig. 6, the full line the best fit of the theoretical line shape $(q_{\parallel} - q_0)^{-2+\eta}$ folded with the three-dimensional resolution function using the exponent η and an overall scale factor as adjustable parameters. As discussed in the text the values of η so-obtained agree with those calculated from the empirical values of q_0, B, and K.

K should be constant over the narrow temperature range of our experiments. This is true within the experimental uncertainties. There are also independent measurements of K reported in the literature.[18] In these measurements a magnetic field induces a distortion in a liquid-crystal film (a Freedericks transition) of thickness l at a threshold field when $\xi_M = l/\pi$; thus a determination of K requires a knowledge of χ_a. Karat and Madhusudana[18] report $K \sim 1.8 \times 10^{-6}$ dyne, however they used an anisotropy in the 8OCB molar susceptibility $\Delta\chi_M \sim 118.6 \times 10^{-6}$ cm^3/mole. This figure was obtained by an incorrect average of the principal values for the biphenyl molecule. Using known anisotropies[19,20] for various elements of the 8OCB molecule we calculate $\Delta\chi_M \sim 44 \times 10^{-6}$ cm^3/mole. Including errors in $\Delta\chi_M$ and the nematic order parameter at the N-A transition, we find the correct value for K is $(6.8 \pm 0.7) \times 10^{-7}$ dyne. This is in excellent argeement with the values we obtained in Table II. Substituting our measured values into Eq. (2a) we obtain $\langle u^2 \rangle^{1/2} \sim 17$ Å at $t \sim 4 \times 10^{-6}$ for $L \sim 1$ cm. In a field of 4.8 kG at the same temperature we calculate $\xi_M \sim 6 \times 10^{-4}$ cm and $\langle u^2 \rangle^{1/2} \sim 14$ Å; this illustrates rather dramatically how slow is the logarithmic divergence of Eq. (2).

It should then be possible to use the parameters obtained from analysis of the longitudinal scans to calculate the transverse (q_x) scans. We find excellent agreement with no adjustable parameters down to 10% of peak height, but at 1% of peak height the experimental data are $\sim 60\%$ higher than we calculate. This is probably because the wings of the mosaic distribution fall off less rapidly than a Gaussian due to defects near the edges of the smectic sample. These defects (e.g., edge dislocations) are undoubtedly a small fraction of the total volume of the sample but can contribute to a broader mosaicity at the 1% level. A Gaussian mosaic spread is an excellent approximation to the transverse resolution used in deconvoluting longitudinal scans, but mosaicity must be very accurately known well below 1% of peak in order to analyze the transverse scans reliably.

In summary, then, the complete longitudinal profile can be accurately described using Caillé's harmonic theory with essentially no adjustable parameters. Further, the theory gives a good account of the transverse scans, although there is a slight discrepancy at the 1% level, presumably due to our incomplete knowledge of the sample mosaic distribution. As expected, the exponent η which characterizes the algebraic decay of the positional correlations increases markedly as one approaches T_c. Although we have not proven that the Caillé theory is unique, we have certainly proven that it is completely consistent with all of the data and we believe that this constitutes convincing evidence that the SmA phase does indeed exhibit the expected Landau-Peierls lack of true positional LRO.

V. CONCLUSIONS

The data of Fig. 6 show that the smectic-A state is not an infinite stacking of planes with a well-defined lattice spacing. This is consistent with the theoretical model of the SmA state as described by the free-energy density of Eq. (1). Furthermore, the x-ray line profiles can be evaluated in this model in the harmonic approximation in terms of independently determined parameters such as the penetration depth and the splay elastic constant. With no adjustable parameters, apart from an overall scale factor, we find excellent agreement between the theoretical line shape folded with the instrumental resolution and the experimental data within a temperature range where the exponent η describing the algebraic decay of correlations varies by a factor of two. This evidence for having observed the quasi-long-range order associated with divergent long-wavelength acoustic fluctuations could be further substantiated by examining the profile of higher-order reflections as well as transverse profiles and further work along these lines is in progress.

ACKNOWLEDGMENTS

We would like to acknowledge helpful discussions with Y. Imry, D. E. Moncton, and M. J. Stephen. The x-ray source and the spectrometer were granted by the Danish National Science Foundation. Work at M.I.T. was supported by the U. S. National Science Foundation Grant No. DMR-76-18035 and the Joint Services Electronics Program, Contract No. DAAG-29-78-C-0020.

APPENDIX A: THE CORRELALTION FUNCTION AND SCATTERING CROSS SECTION

The x-ray-scattering cross section is proportional to the Fourier transform of the correlation function

$$S(\vec{r}) = \langle \exp\{iq_0[u(r)-u(0)]\}\rangle \ .$$

In the harmonic approximation $u(r)$ is a Gaussian random variable so $S(\vec{r})$ is given by $\exp[-\frac{1}{2}q_0^2 \times \langle|u(r)-u(0)|^2\rangle]$. Thus we need to evaluate

$$\langle\tfrac{1}{2}q_0^2|u(\vec{r})-u(0)|^2\rangle$$

$$-\frac{kTq_0^2}{(2\pi)^3 B}\int\frac{[1-\cos(\vec{q}\cdot\vec{r})]d\vec{q}}{(q_{\shortparallel}^2+\lambda^2 q_{\perp}^4)}$$

with $\vec{r} = (\vec{\rho},z)$.

The integral

$$I(\vec{\rho},z) = \int\frac{1-\cos(\vec{q}\cdot\vec{r})}{q_{\shortparallel}^2+\lambda^2 q_{\perp}^4}d\vec{q}$$

is evaluated by first integrating over $q_{\shortparallel}$ using

$$\int_{-\infty}^{\infty}\frac{1-\cos[a(b-x)]}{x^2+c^2}dx = \frac{\pi}{c}[1-e^{-ac}\cos(ab)]$$

to obtain

$$\int_0^{\infty}\frac{1-\cos(q_{\shortparallel}z+\vec{q}_{\perp}\cdot\vec{\rho})}{q_{\shortparallel}^2+\lambda^2 q_{\perp}^4}dq_{\shortparallel}$$

$$-\frac{1}{2}\frac{\pi}{\lambda q_{\perp}^2}[1-\exp(-\lambda q_{\perp}^2 z)\cos(\vec{q}_{\perp}\cdot\vec{\rho})] \ .$$

Then with θ being the angle between $\vec{q}_{\perp}$ and $\vec{\rho}$

$$\int_0^{2\pi}[1-\exp(-\lambda q_{\perp}^2 z)\cos(\vec{q}_{\perp}\cdot\vec{\rho})]d\theta$$

$$-2\pi-\exp(-\lambda q_{\perp}^2 z)\int_0^{2\pi}\cos(q_{\perp}\rho\cos\theta)d\theta \ .$$

Using the integral representation of the Bessel function

$$J_0(x) = \frac{1}{\pi}\int_0^{\pi}\cos(x\cos\theta)d\theta$$

we obtain

$$I(\vec{\rho},z)$$

$$-\frac{1}{2}\pi 2\pi\int_{q_{min}}^{q_{max}}\frac{1-\exp(-\lambda q_{\perp}^2 z)J_0(q_{\perp}\rho)}{\lambda q_{\perp}^2}q_{\perp}dq_{\perp} \ .$$

With $q_{max} - 2\pi/d = q_0$ and $q_{min} = 0$ the integral is further evaluated as follows.

One inserts the series expansion form of $J_0(x)$

$$J_0(x) = \sum_{n=0}\frac{x^{2n}(-1)^n}{2^{2n}n!n!}$$

and integrates term by term. By comparing the result to the series expansion of the exponential integral $E_1(-x) = \int_x^{\infty} e^{-t}/t\,dt$

$$E_1(x) = -\gamma-\ln x-\sum_{n=1}^{\infty}\frac{(-1)^n x^n}{n(n!)}$$

(γ = Euler's const.), one finds

$$G(\vec{\rho},z) \sim e^{-2\eta\gamma}\left(\frac{4d^2}{\rho^2}\right)e^{-\eta}E_1(\rho^2/4\lambda z)$$

with $\eta = q_0^2 kT/(8\pi B\lambda)$. This expression for $G(\vec{\rho},z)$ has the asymptotic forms given in Eqs. (3a) and (3b).

APPENDIX B: SINGLE FACE VERSUS CHANNEL-CUT CRYSTALS

In this Appendix we compare for the sake of completeness single face crystals with channel-cut crystals. The conclusions are by and large in agreement with those obtained by Bonse and Hart in their pioneering paper on channel-cut crystals in 1968. In the left part of Fig. 9 is shown the direct beam profile in the nondispersive orientation (cf. Fig. 3) using the backside of the two channel-cut crystals. We notice that with the open detector geometry used in the experiment the direct beam has long tails falling off as the square of the misset angle, or as $q_\parallel^{-2}$. It would therefore not be possible to conclude from an observed $q_\parallel^{-2}$ falloff from an SmA sample whether it was a resolution effect or a true Landau-Peierls behavior. The $q_\parallel^{-2}$ dependence of the direct beam is a dynamical-diffraction effect due primarily to the finite penetration of the x-ray beam into the perfect crystals. The more of reciprocal space around the Bragg point is picked up by the detector, the larger will be the relative weight of the $q_\parallel^{-2}$ tails be. Indeed, as also shown in the left part of Fig. 8, when the vertical resolution is improved by about an order of magnitude by horizontal Soller splits in front of the detector, the relative weight of the $q_\parallel^{-2}$ component drops by about an order of magnitude.

However, when the Bragg scattering *and* the diffuse scattering is tripled in the channel-cut crystals,

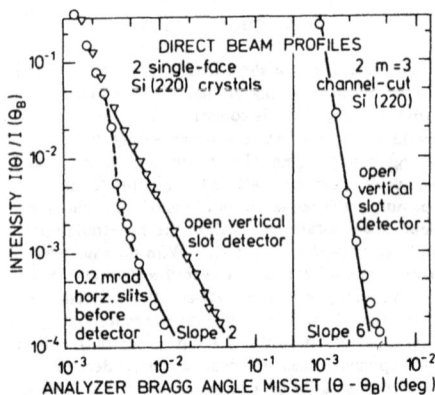

FIG. 9. Direct beam profiles of two Si(220) crystals in nondispersive orientation. With single reflections (left panel) tails fall off as $\Delta\theta^{-2}$, the tails being more pronounced with relaxed vertical resolution, whereas with triple reflections both in the monochromator and the analyzer channel-cut crystals the tails drop off very rapidly as $\Delta\theta^{-6}$.

the Bragg-peak intensity is only diminished to about 50%, but the diffuse $q_\parallel^{-2}$ wings become cubed to $q_\parallel^{-6}$ wings. Indeed, our experiment confirms this simple picture: The log-log plot gives approximately a slope of -6 for the direct beam profile of the channel-cut crystals.

*Permanent address: Dept. of Physics, Massachusetts Institute of Technology, Cambridge, Mass. 02139.

[1]R. E. Peierls, Helv. Phys. Acta Suppl. 7, 81 (1934).

[2]L. D. Landau, in *Collected Papers of L. D. Landau*, edited by D. ter Haar (Gordon and Breach, New York, 1965), p. 209.

[3]J. M. Kosterlitz and D. J. Thouless, J. Phys. C 6, 118 (1973).

[4]Very interesting related phenomena are, however, seen in helium films. See, for example, D.J. Bishop and J. D. Reppy, Phys. Rev. Lett. 40, 1727 (1978). Also recent experiments by D. E. Moncton and R. Pindak [Phys. Rev. Lett. 43, 701 (1979)] indicate that power-law decay of correlation functions may be observable in smectic-*B* films.

[5]L. D. Landau and E. M. Lifshitz, *Statistical Physics* (Addison-Wesley, Reading, Massachusetts, 1969), p. 403.

[6]P. G. de Gennes, *The Physics of Liquid Crystals* (Clarendon, Oxford, 1974).

[7]J. Als-Nielsen, R. J. Birgeneau, M. Kaplan, J. D. Litster, and C. R. Safinya, Phys. Rev. Lett. 39, 1668 (1977).

[8]H. Birecki, R. Schaetzing, F. Rondelez, and J. D. Litster, Phys. Rev. Lett. 36, 1376 (1976).

[9]C. A. Schantz and D. L. Johnson, Phys. Rev. A 17, 1504 (1978).

[10]B. I. Halperin and T. C. Lubensky, Solid State Commun. 14, 997 (1974).

[11]A. Caillé, C. R. Acad. Sci. Ser. B 274, 891 (1972).

[12]H. -J. Mikeska and H. Schmidt, J. Low Temp. Phys. 2, 371 (1970).

[13]Y. Imry and L. Gunther, Phys. Rev. B 3, 3939 (1971).

[14]Y. Imry, CRC Crit. Rev. Solid State and Mater. Sci. (USA) 8, 157 (1978).

[15]For a highly readable discussion of dynamical diffraction theory see B. E. Warren, *X-ray Diffraction* (Addison-Wesley, Reading, Massachusetts, 1969), Chap. 13.

[16]U. Bonse and M. Hart, Appl. Phys. Lett. 7, 238 (1965).

[17]J. D. Litster, J. Als-Nielsen, R. J. Birgeneau, S. S. Dana, D. Davidov, F. Garcia-Golding, M. Kaplan, C. R. Safinya, and R. Schaetzing, J. Phys. (Paris) 40, C3-339 (1979).

[18]P. P. Karat and N. V. Madhusudana, Mol. Cryst. Liq. Cryst. 47, 21 (1978).

[19]M. A. Laskeke, Philos. Trans. R. Soc. London, Ser. A 256, 357 (1963).

[20]I. H. Ibrahim and W. Haase, J. Phys. (Paris) 40, C6-167 (1979).

VOLUME 57, NUMBER 21   PHYSICAL REVIEW LETTERS   24 NOVEMBER 1986

Steric Interactions in a Model Multimembrane System: A Synchrotron X-Ray Study

C. R. Safinya,[1] D. Roux,[1],[3] G. S. Smith,[1],[2] S. K. Sinha,[1] P. Dimon,[1] N. A. Clark,[2] and A. M. Bellocq[3]

[1]Exxon Research and Engineering Company, Annandale, New Jersey 08801
[2]Department of Physics, University of Colorado, Boulder, Colorado 80309
[3]Centre de Recherche Paul Pascal and Groupement de Recherches Coordinées de Microemulsion,
Centre National de la Recherche Scientifique, 33405 Talence, Bordeaux, France

(Received 12 August 1986)

We report a high-resolution x-ray study of a model multilayer fluid membrane system in the lyotropic L_a phase of a quaternary microemulsion system. The structure factor exhibits power-law behavior characteristic of a Landau-Peierls system. As a function of intermembrane distance 3.8 nm $\leq d \leq$ 16.3 nm, the exponent $\eta(d)$ which describes the algebraic decay of layer correlations is predicted by the model of Helfrich where entropic steric repulsions dominate intermembrane interactions.

PACS numbers: 61.30.Eb, 82.70.Kj

The physical properties of multilayer membrane systems, which consist of equally spaced fluid bilayer lipid sheets embedded in water, have long been subjects of intense interest in biophysical research. In particular, much attention has been focused on the understanding of intermembrane interactions. The basic forces[1] between two parallel membrane sheets involve long-range van der Waals interactions, and short-range repulsive hydration and screened electrostatic forces. Recently, Helfrich[2] has proposed that large, thermally induced, out-of-plane layer fluctuations give rise to a repulsive interaction between membranes because of steric hindrance in multilayer systems. More generally, steric repulsions are known to be the dominant interactions associated with wandering walls of incommensurate phases.[3] This interaction has been calculated[2] for a multilayer system by use of the Landau–de Gennes[4,5] elastic theory of smectic-A liquid crystals with energy density

$$F/V = \{B(\partial u/\partial z)^2 + K[(\partial^2 u/\partial x^2) + (\partial^2 u/\partial y^2)]^2\}/2 , \quad (1)$$

where $u(\mathbf{r})$ is the layer displacement in the z direction normal to the layers, and B and K are the bulk moduli for layer compression (erg/cm^3) and layer curvature (erg/cm). The free energy of steric interaction per unit area for two membranes separated by a distance d is[2] $U_s/A = ak_BT/d^2$, where $a = 0.23k_BT/k_c$ and k_c is the elastic modulus for curvature for a single membrane ($K = k_cN$, where N is the number of layers per unit height). The van der Waals interaction is given by[1]

$$\frac{U_{vdW}}{A} \sim -\frac{H}{12\pi}\left\{\frac{1}{d^2} - \frac{2}{(d+\delta)^2} + \frac{1}{(d+2\delta)^2}\right\} ,$$

which for $d \gg \delta$ gives $\sim -H\delta^2/d^4$, where δ is the membrane thickness and $H \sim 0.75k_BT$ is the Hamaker constant. Thus, the Helfrich interaction competes in range with the van der Waals attraction and should dominate at large intermembrane separations.

The experimental relevance of this interaction remains unclear. To identify unambiguously the origin of intermembrane repulsion, it is crucial to carry out studies in uncharged systems. To date, studies[1] of neutral lipids have been limited to systems which exhibit a narrow range of intermembrane separations between ~ 1.0 and 2.0 nm, where the long-range van der Waals and steric interactions are both expected to fall off as $1/d^2$. Moreover, these studies have been associated with lipid systems in which the membrane curvature elasticity[6] $k_c > 20k_BT$, in which case van der Waals interactions are significantly larger. Thus, although the concept of this novel fluctuation-induced steric interaction has stimulated a large body of theoretical[7] and experimental[1,5] work, its significance in biological membranes remains largely controversial.

To elucidate the relevance of this interaction in multilayer systems, we report in this paper on a synchrotron diffuse x-ray scattering study in the lamellar L_a phase of the quaternary mixture of sodium dodecyl sulfate (SDS, the surfactant), pentanol (cosurfactant), water, and dodecane as a function of dodecane dilution. Figure 1 shows a cut of the phase diagram mapped out by Roux and Bellocq[8] represented on a standard triangular phase diagram in the plane with a constant water/SDS weight ratio equal to 1.55. This cut contains five one-phase regions: the microemulsion phases $\mu\varepsilon_1$ and $\mu\varepsilon_2$ and the liquid crystalline hexagonal (E), rectangular (R), and lamellar (L_a) phases.

The L_a phase consists of water layers (inverted membrane) embedded in oil, shown schematically in Fig. 1 with the fluid surfactant at the interface. This affords distinct advantages for studies of intermembrane interactions. First, the water layers are charge neutral so that there are no intermembrane electrostatic interactions. Second, and very significantly for this plane of the phase diagram, we are able to dilute over an unusually large oil range between 0 and 80 wt.%, which corresponds to intermembrane separations between ~ 2.0 and larger than

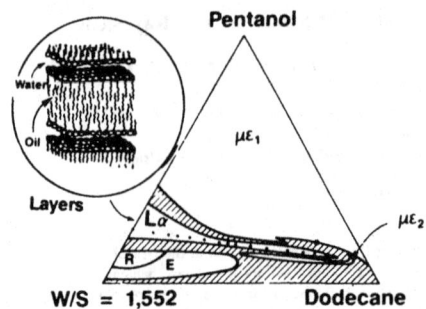

FIG. 1. A cut of the phase diagram of the quaternary mixture of SDS, pentanol, water, and dodecane shown in the plane with a constant water/SDS weight ratio $=1.55$ (Ref. 8). The dots in the lamellar phase labeled L_a correspond to the mixtures studied.

20.0 nm. Third, previous work[9] on the curvature elasticity in similar microemulsion L_a phases had indicated unusually small values of $k_c \sim k_B T$. These unique properties of the microemulsion L_a phases[10] thus allow for a comprehensive study of the long-range van der Waals and steric interactions.

Landau and Peierls[5] first demonstrated that the mean square layer displacements of a system of stacked fluid layers diverge logarithmically, $\langle u^2 \rangle \sim \ln(L/a)$, with sample size L, destroying conventional long-range order (a is of order of the intermolecular distance). For the x-ray structure factor, the consequences are dramatic. Caillé[11] has shown that for a smectic-A liquid-crystal phase, which has the identical elastic free energy [Eq. (1)] as that for the L_a phase, conventional δ-function Bragg peaks at $(0,0,q_m = mq_0 = m2\pi/d)$ (harmonic order $m = 1, 2, \ldots$) are replaced by singularities with asymptotic power-law behavior $S(0,0,q_z) \sim |q_z - q_m|^{-2+\eta_m}$ and $S(q_\perp, 0, q_m) \sim q_\perp^{-4+2\eta_m}$. Here, $\eta_m = m^2 q_0^2 k_B T/8\pi \times (BK)^{1/2}$ and q_z and $q_\perp$ are components of the wave vector normal and parallel to the layers. This power-law

behavior has been confirmed[12] for the first harmonic $q_z = q_0$ in the smectic-A phase of liquid crystals.

Here we find that the L_a phase of this SDS system corresponds to a Landau-Peierls system with characteristic power-law behavior for the first and second harmonics of the structure factor with η_m scaling as m^2. Most significantly we find that for intermembrane distance 3.8 nm $\leq d \leq 16.3$ nm, η_1 is accurately described by $\eta_1(d) = 1.33(1 - \delta/d)^2$ predicted by the Helfrich theory[2] (δ is the membrane thickness). This then is a direct experimental confirmation of the Helfrich steric interaction mechanism and its dominant behavior in appropriate model multimembrane systems. Additionally, our measurements confirm that the membrane curvature elasticity $k_c \sim k_B T$.

We carried out detailed studies for ten distinct mixtures: $x = 0, 0.07, 0.13, 0.18, 0.23, 0.29, 0.35, 0.47, 0.54$, and 0.62 (labeled by closed circles in Fig. 1). Here, x is the percentage of dodecane by weight of the mixtures. For $x = 0, 0.07$, and 0.13, the first and second harmonics of the structure factor were studied with the second harmonic not visible for dilutions x larger than 0.13. The experiments were carried out at the Stanford Synchrotron Radiation Laboratory on beam line VI-2 and at the National Synchrotron Light Source on the Exxon beam line X-10A. The monochromator and analyzer consisted of a double-bounce Si(111) and a triple-bounce Si(111) channel-cut crystal set at 8 keV in the nondispersive configuration, which yields a very sharp in-plane Gaussian resolution function with very weak tail scattering with half width at half maximum (HWHM) 8×10^{-5} Å^{-1}. A sharp Gaussian out-of-plane resolution function (HWHM $= 10^{-3}$ Å^{-1}) was achieved by use of extremely narrow slits. The mixtures were contained in sealed quartz capillaries with diameters of 1 and 2 mm, which yielded randomly oriented lamellar domains.

We show in Fig. 2 typical scattering profiles for longitudinal scans through the first harmonic for x between 0 and 0.54 in the L_a phase, where the total layer spacing $d = 2\pi/q_0$ increases from 3.82 to 11.5 nm. In the mixtures studied, the dilution corresponds to a path where water layers with approximately constant thickness d_w

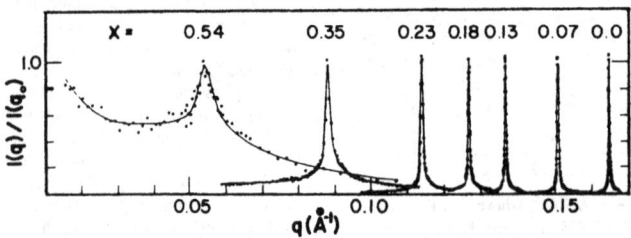

FIG. 2. Longitudinal profiles of the first harmonic of seven different mixtures along the dodecane dilution path. The percentage dodecane by weight of the mixtures (x) is indicated above each profile. All peak intensities are normalized to unity. The solid lines are fits by the Caillé power-law line shape [Eq. (2)].

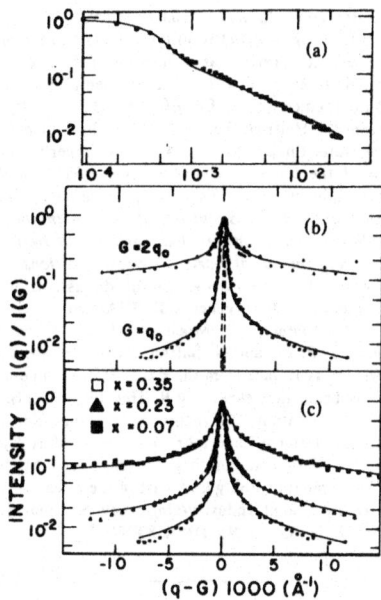

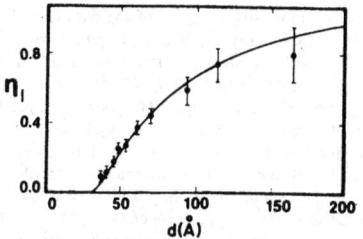

FIG. 4. Power-law exponent η_1 as a function of the inter-membrane distance for mixtures along the dilution path. The solid line is the prediction of the model of Helfrich of entropically driven steric interactions.

FIG. 3. (a) Profile of the first harmonic $(G = q_0)$ for the mixture $x = 0.23$ on a log-log scale which shows finite size rounding at small $q - G$ followed by power-law behavior at larger $q - G$. (b) Profile of the first and second harmonics for the mixture $x = 0.07$ on a logarithmic intensity scale. (c) Profile of the first harmonic $(G = q_0)$ for three mixtures on a logarithmic intensity scale. All peak intensities are normalized to unity. The solid lines are fits by the Caillé power-law line shape [Eq. (2)].

~ 18 Å are pushed apart with d_0/d_w varying from ~ 1 to ~ 7. ($d_0 = d - d_w$ is the oil thickness between water layers.) A striking feature of the profiles in Fig. 2 is the tail scattering which becomes dramatically more pronounced as d increases. This effect is further elucidated in Fig. 3(c), where we plot on a logarithmic intensity scale versus $q - q_0$ the scattering for three mixtures, $x = 0.07$, 0.23, and 0.35. The significant difference in the profiles over the entire dilution range is now immediately clear. It is qualitatively clear that $\eta_1(d)$, which characterizes the asymptotic scattering profile and which

is a measure of the ratio of the tail to peak intensity scattering, is increasing as d increases.

To analyze the profiles quantitatively, we begin with the Caillé[11] calculation. For a single crystal of infinite domain size in the L_a phase, the width of the measured profile at $q = q_0$ is primarily determined by the longitudinal resolution. Our very high-resolution setup can resolve length scales of 1.5 μm. The longitudinal resolution is shown dashed in Fig. 3(b). The widths of the profiles are normally 2 to 3 times larger than resolution. We find that for different preparations of the same mixture, the widths of the profiles may differ by as much as 50%. Thus, we attribute the broadening primarily to an extrinsic lamellar effect of finite lamellar domain sizes (associated with defects). The typical domain sizes are unusually large, between ~ 2500 and ~ 12500 Å, which allowed for meaningful analysis of the profiles.

We incorporate the finite-size effect in our analysis in a way analogous to that of Dutta and Sinha.[13] This modifies the structure factor $S(\mathbf{q})$ for an infinite single-crystal domain in a straightforward manner

$$S_{F.S.}(\mathbf{q}) \sim \int d^3R\, S(\mathbf{R})e^{-R^2\pi/L^2}e^{i(\mathbf{q}-\mathbf{G})\cdot\mathbf{R}},$$

where L^3 is the domain volume and $\mathbf{G} = mq_0\hat{z}$. The correlation function $S(\mathbf{R}) = S(z,\rho) \sim (1/\rho)^{2\eta}\exp\{-\eta \times [2\gamma + E_1(\rho^2/4\lambda z)]\}$ was calculated by Caillé.[11] Here $R^2 = z^2 + \rho^2$, γ is Euler's constant, $E_1(x)$ is the exponential integral function, and $\lambda = (K/B)^{1/2}$. Finally, because our samples consist of randomly oriented domains, we perform an exact powder average over all solid angles in reciprocal space:

$$\langle S(\mathbf{q})_{F.S.}\rangle \equiv S(q) \sim \int_{-\infty}^{\infty} dz \int_0^{\infty} d\rho\, S(z,\rho)e^{-R^2\pi/L^2}[(\sin qR)/qR]e^{-iGz}. \tag{2}$$

The analysis consists of least-squares fits of Eq. (2) convoluted with the resolution function to the observed profiles. We plot in Fig. 3(a) on a log-log scale the intensity versus $q - q_0$ for $x = 0.23$, where the solid line is a result of the fit yielding $\eta_1 = 0.25$ and $L = 8640$ Å. Two features in the scattering profile and the theoretical cross section are immediately apparent. While at large $q - q_0 \gtrsim 2\pi/L$ the scattering exhibits power-law behavior $S(q) \sim |q - q_0|^{-P}$ with $P \sim 1 - \eta$ at $q \simeq q_0$, the finite-size effects round off the observed profile with characteristic width $\sim 1/L$. (This effective power-law exponent P is due to the powder averaging.)

VOLUME 57, NUMBER 21   PHYSICAL REVIEW LETTERS   24 NOVEMBER 1986

A further, more subtle aspect of the data is evident in Fig. 3(c), where the profiles around $q = q_0$ are slightly asymmetric with the high-q scattering more intense than the low q. The parameter $\lambda \equiv (K/B)^{1/2}$, which enters the real-space correlation function $S(\mathbf{R})$, is a measure of the degree of anisotropy present in $S_{F.S.}(q_z, q_\perp)$.[11] The powder average of this anisotropic scattering cross section results in the observed asymmetric profile. For $\lambda > 2\pi/q_0$ the asymmetry is negligible and $S(q)$ is not very sensitive to the precise value of λ.

To summarize the essential points in the fits, the four parameters q_0, η, L, and λ are given respectively by the peak position, the power-law behavior away from the Bragg point, the central-peak width, and the asymmetry in the profile around the Bragg point. As is evident from the typical results of fits shown in Figs. 2 and 3, where η_1 varies by more than a factor of 7 (Fig. 4), the appropriately modified finite-size and powder-averaged Caillé cross section gives a quantitatively satisfactory description of the scattering. In contrast to the smectic-A system[12] where only the first harmonic was evident, we are able to confirm the scaling of η_m with m^2. We show in Fig. 3(b) the profiles and fits for the first and second harmonics for $x = 0.07$ on a normalized logarithmic intensity scale. We find $\eta_1(q_0) = 0.14 \pm 0.02$, $\lambda = 8.59 \pm 1$, $L = (10027 \pm 1000)$ Å for the first harmonic and $\eta_2(2q_0) = 0.575 \pm 0.02$, $\lambda = 8.13 \pm 2$, $L = (9115 \pm 1500)$ Å for the second harmonic. We also confirmed the scaling of η with m^2 for the mixtures $x = 0$ and $x = 0.13$.

For the last three dilutions ($x = 0.47$, 0.54, and 0.62) as seen in Fig. 2 for $x = 0.54$, in addition to the first harmonic, the data exhibited small-angle x-ray scattering (SAXS). For $0.8 > x > 0.62$ the SAXS dominates, although the existence of focal conics observed in optical microscopy indicates a lamellar structure. Although we do not understand the origin of the SAXS (possibly due to defects), we are able to fit the SAXS for samples in the range $0.8 > x > 0.62$ with a simple model[14] of polydisperse spheres with the same radius (~ 100 Å) but varying intensities. Accordingly, the fit to the profile for $x = 0.47$, 0.54, and 0.62 involves one more adjustable parameter which is the SAXS intensity.

We find that λ increases smoothly between 5 and 25 Å for $0 < x < 0.62$. The elastic constants B and k_c ($= Kd$) are derived from η and λ. The magnitude of B ranges from typical smectic-A values[12] (6×10^7 erg/cm^3) to more than 2 orders of magnitude less for large dilutions. In contrast k_c, which is the membrane-bending modulus, varies slowly. Very significantly, we find that $k_c \sim (0.5$ to $2.0) k_B T$ is about an order of magnitude smaller than that measured in other model membrane systems. Therefore, since the Helfrich mechanism scales as k_c^{-1}, we expect this interaction U_s/A, which ranges between $0.12 k_B T/d^2$ and $0.46 k_B T/d^2$, to dominate the van der

Waals interaction $U_{vdW}/A \sim -0.03 k_B T/d^2$.

We plot in Fig. 4 $\eta_1(d)$ resulting from fits to the profile at the first harmonic as a function of d. The solid line, which agrees well with the experimental data, is a plot of the predicted value for $\eta_1(d) = 1.33(1 - \delta/d)^2$ derived from the Helfrich theory. Here, we have taken the effective water thickness $\delta = 29$ Å slightly larger than d_w because of the known excluded volume effects[8] of the surfactant tails in the oil. *This then provides compelling evidence that in this SDS multimembrane system the intermembrane interactions are dominated by the Helfrich mechanism of entropically driven steric interactions.*

We gratefully acknowledge useful discussions with S. Alexander, R. J. Birgeneau, K. D'Amico, P. G. de Gennes, S. Leibler, S. Mochrie, D. E. Moncton, P. Pincus, J. Prost, and S. Safran. It is a pleasure to acknowledge assistance from the Exxon staff members at the Exxon beam line, especially R. Hewitt and M. Sansone. The National Synchrotron Light Source, Brookhaven National Laboratory, and the Stanford Synchrotron Radiation Laboratory are supported by the U. S. Department of Energy. A part of the research was supported by a joint Industry/University National Science Foundation Grant No. DMR-8307157.

[1]J. N. Israelachvili, *Intermolecular and Surface Forces* (Academic, Orlando, 1985); V. A. Parsegian, N. Fuller, R. P. Rand, Proc. Natl. Acad. Sci. 76, 2750 (1979).

[2]W. Helfrich, Z. Naturforsch. 33a, 305 (1978).

[3]S. G. J. Mochrie, A. R. Kortan, R. J. Birgeneau, and P. M. Horn, Z. Phys. B 62, 79 (1985).

[4]P. G. de Gennes, *The Physics of Liquid Crystals* (Clarendon, Oxford, 1974).

[5]L. D. Landau, in *Collected Papers of L. S. Landau*, edited by D. Ter Haar (Gordon and Breach, New York, 1965), p. 209; R. E. Peierls, Helv. Phys. Acta. 7, Suppl., 81 (1934).

[6]M. B. Schneider, J. T. Jenkins, and W. W. Webb, J. Phys. (Paris) 45, 1457 (1984).

[7]R. Lipowsky and S. Leibler, Phys. Rev. Lett. 56, 2561 (1986), and references therein.

[8]D. Roux and A. M. Bellocq, *Physics of Amphiphiles*, edited by V. DeGiorgio and M. Corti (North-Holland, Amsterdam, 1985).

[9]J. M. diMeglio, M. Dvolaitsky, and C. Taupin, J. Phys. Chem. 89, 871 (1985).

[10]F. C. Larché, J. Appell, G. Porte, P. Bassereau, and J. Marignan, Phys. Rev. Lett. 56, 1700 (1986).

[11]A. Caillé, C. R. Acad. Sci. Ser. B 274, 891 (1972).

[12]J. Als-Nielsen, J. D. Litster, R. J. Birgeneau, M. Kaplan, C. R. Safinya, A. Lindegaard-Andersen, and S. Mathiesen, Phys. Rev. B 22, 312 (1980).

[13]P. Dutta and S. K. Sinha, Phys. Rev. Lett. 47, 50 (1981).

[14]A. Guinier, *X-Ray Diffraction* (W. H. Freeman, New York, 1963).

3. NEMATIC, SMECTIC-A, SMECTIC C MULTI-CRITICAL POINT

3.1. Chen, J. H. and Lubensky, T. C. (1976). Landau-Ginzberg mean-field theory for the nematic to smectic-C and nematic to smectic-A phase transitions. *Phys. Rev.* **A14**, 1202–1207   151

3.2. Martinez-Miranda, L. J., Kortan, A. R. and Birgeneau, R. J. (1986). X-ray study of fluctuations near the nematic-smectic-A-smectic-C multicritical point. *Phys. Rev. Lett.* **56**, 2264–2267   157

PHYSICAL REVIEW A VOLUME 14, NUMBER 3 SEPTEMBER 1976

Landau-Ginzburg mean-field theory for the nematic to smectic-C and nematic to smectic-A phase transitions*

Jing-huei Chen and T. C. Lubensky[†]

Department of Physics and Laboratory for Research in the Structure of Matter, University of Pennsylvania, Philadelphia, Pennsylvania 19174
(Received 28 May 1976)

A Landau-Ginzburg free energy is presented that yields a nematic to smectic-A (NA) transition, a nematic to smectic-C (NC) transition, and a Lifshitz point (at the boundary between the NA and NC transitions). Fluctuation enhancements of the Frank elastic constants are calculated. For the NC transition, all three elastic constants K_1, K_2, and K_3 diverge as ξ^2, where ξ is the correlation length for fluctuations of the smectic order parameter. At the Lifshitz point, K_1 and K_2 diverge as $\ln\xi$ whereas K_3 diverges as ξ.

I. INTRODUCTION

In the nematic liquid crystalline state, long bar molecules are oriented on the average with their long axes along a preferred direction specified by a unit vector $\vec{n}$, called the director.[1,2] The molecular centers of mass are, however, free to diffuse throughout the system so that translational invariance is not destroyed. The nematic state is shown schematically in Fig. 1(a). In the smectic-A phase [shown in Fig. 1(b)] the molecules segregate into two-dimensional planes while maintaining translational invariance along the planes. In the smectic-C phase [shown in Fig. 1(c)] the molecules are tilted at an angle θ to the normal to the smectic planes. Phase transitions between all of these phases are possible. The nematic to smectic-A (NA) transition has been the most studied.[3-10] If director fluctuations are ignored, it can be a second-order transition with helium exponents.[3-6] Theoretical considerations indicate that director fluctuations cause the transition to be first order.[11,12] Though specific-heat and volumetric measurements[10] indicate a first-order transition, careful light scattering experiments[9] show no indication of a first-order transition. The smectic-A to smectic-C (AC) transition can be second order[6,13-15] and is believed to have helium exponents.[6] There can also be a direct nematic to smectic-C (NC) transition.[6,16] This has been the least studied of the transitions. In this paper, we present and study a model which has the potential to include all of these transitions. For compactness, we will refer to this as the nematic–smectic-A–smectic-C (NAC) model.

Our starting point will be the observation that the x-ray scattering in the nematic phase (single crystal) in the vicinity of the NA transition shows strong peaks at wave number $\vec{q}_A = \pm q_0\vec{n}$,[6,17] shown in Fig. 2(a). Near an NC transition, these two

peaks spread out into two rings at $\vec{q}_c = (\pm q_\parallel, q_\perp\cos\varphi, q_\perp\sin\varphi)$,[6,18] shown in Fig. 2(b). The latter observation prompted de Gennes[6] to introduce an infinite-dimensional order parameter $\Psi_\varphi = \rho(q_\parallel, q_\perp\cos\varphi, q_\perp\sin\varphi)$ for the smectic-C state where ρ is the center-of-mass density. In our model the NA and NC transitions can be described by the same free energy with different parameters. For the NA transition, the free energy is minimized if the center-of-mass density is periodic with wave number $\pm q_0\vec{n}$; for the NC transition, it is minimized if the center-of-mass density is periodic with wave number $\vec{q}_c$. The de Gennes and NAC models, though motivated by the same observation, predict different behavior. In particular, near the NC transition the de Gennes model predicts that the Frank elastic constants[6] K_1, K_2, and K_3 diverge as $\xi^{2/3}$, where ξ is the correlation length, while the NAC theory predicts that all three elastic constants diverge as ξ^2.

Models similar to the NAC model have appeared in other contexts. In particular, the NAC model is very similar to those used to describe transitions from the paramagnetic state to helical spin states[19,20] and to that used to describe the Benard instability in a cylindrical cavity.[21] All of these models have the common feature that fluctuations are a maximum at wave numbers $|\vec{q}_\perp| = q_c$, where $\vec{q}_\perp$ is an m-dimensional vector in a d-dimensional space. Mean-field theory predicts a second-order transition for these models. Fluctuations, however, are believed to lead to a first-order transition when $m = d$ or $m = d - 1$.[20,21] The NC transition has $d = 3$ and $m = 2$ so the transition is expected to be first order.[21] Nevertheless, pretransitional effects can be important, and the mean-field calculations presented in this paper are expected to be valid in some temperature range above the first-order transition. It is unclear whether a crossover from mean-field be-

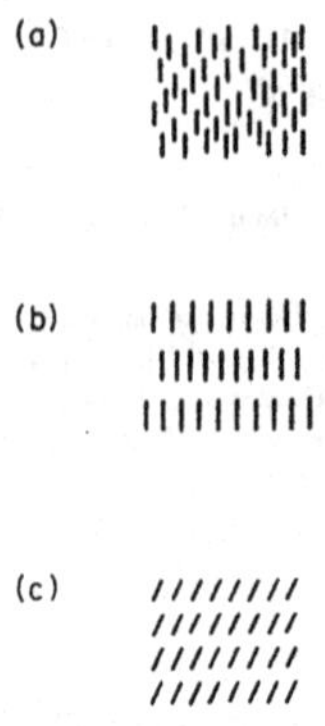

152

(a)

(b)

(c)

FIG. 1. Schematic representation of the position and orientation of molecules in (a) the nematic liquid phase, (b) the smectic-A phase, and (c) the smectic-C phase.

havior to some quasicritical behavior will occur before the first-order transition occurs. To date there is no satisfactory renormalization treatment of this model showing fixed points with first-order runaways. The boundary between the NA and NC transitions is a special point representative of a class of transitions recently considered by Hornreich, Luban, and Shtrikman.[22] This transition is second order with anisotropic scaling and can be treated using the renormalization group.

Section II introduces the model and Sec. III presents perturbation calculations of the elastic constants for the NA and NC transitions and at the Lifshitz point.

II. THE MODEL

In the smectic phases, the center-of-mass density $\rho(\vec{r})$ becomes periodic with fundamental wave number $\vec{q}_0 = (\pm q_{\parallel}, \vec{q}_{\perp})$ ($q_{\perp} = 0$ in the smectic-A phase). We will, therefore, take the part of ρ with wave numbers in the vicinity of $\vec{q}_0$ to be the order parameter $m(\vec{r})$ of our theory:

$$m(\vec{r}) = \int_D \frac{d^3k}{(2\pi)^3} e^{i\vec{k}\cdot\vec{r}} \rho(\vec{k}), \qquad (2.1)$$

where $\rho(\vec{k})$ is the Fourier transform of $\rho(\vec{r})$ and D is a two-part domain (excluding $\vec{k}=0$) centered around $(\pm q_{\parallel}, 0, 0)$ and large enough to contain the circles $k_{\parallel} = \pm q_{\parallel}$, $|k_{\perp}| = |\vec{q}_{\perp}|$. The model Landau-Ginzburg Hamiltonian can be written as a sum of three parts

$$\beta H = \beta H_m + \beta H_{e1} + \beta H_4, \qquad (2.2)$$

where $\beta = 1/kT$. H_m contains terms up to second order in the order parameter:

$$\beta H_m = \frac{1}{2} \int d^3r \left(am^2 + D_{\parallel}[(\vec{n}\cdot\nabla)^2 m]^2 \right.$$
$$- C_{\parallel}(\vec{n}\cdot\nabla m)^2 + \frac{C_{\parallel}^2}{4D_{\parallel}} m^2$$
$$\left. + C_{\perp}\delta_{ij}^T \nabla_i m \nabla_j m + D_{\perp}(\nabla_{\perp}^2 m)^2 \right), \qquad (2.3)$$

where $\vec{n}$ is the nematic director which may vary in space. The summation convention on repeated Cartesian indices is understood, $\delta_{ij}^T = \delta_{ij} - n_i n_j$ is the projection operator onto directions perpendicular to $\vec{n}$, and $\nabla_{\perp}^2 = \delta_{ij}^T \nabla_i \nabla_j$. $a = a'[(T - T_{NA})/T_{NA}]$ changes sign at the NA transition temperature T_{NA}. H_{e1} is the Frank free energy[1,2] for distortions in the nematic director:

$$\beta H_{e1} = \frac{\beta}{2} \int d^3r \left\{ K_1^0 (\nabla\cdot\vec{n})^2 + K_2^0 [\vec{n}\cdot(\nabla\times\vec{n})]^2 \right.$$
$$\left. + K_3^0 [\vec{n}\times(\nabla\times\vec{n})]^2 \right\}, \qquad (2.4)$$

where K_1^0, K_2^0, and K_3^0 are the unrenormalized Frank elastic constants. H_4 is the fourth-order term needed to stabilize the ordered phase:

$$\beta H_4 = u \int d^3r\, m^4(r). \qquad (2.5)$$

The partition function for this model is calculated in the usual way by taking the functional integral of $e^{-\beta H}$ over all configurations of $n_i(r)$ and $m(r)$:

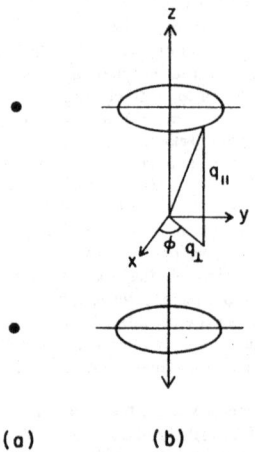

(a)   (b)

FIG. 2. X-ray intensity. Region of maximum x-ray scattering intensity in the vicinity of (a) the nematic to smectic-A transition and (b) the nematic to smectic-C transition.

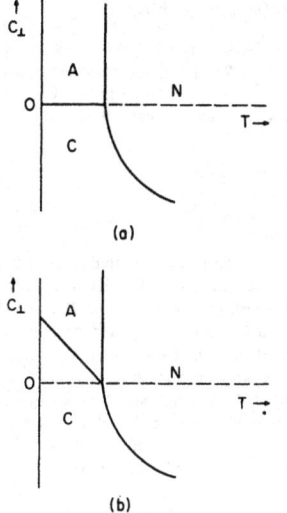

(a)

(b)

FIG. 3. (a) Phase diagram for the NAC model showing nematic (N), smectic-A (A) and smectic-C (C) phases. (b) Phase diagram for the modified NAC model presented in Appendix A.

$$Z = \int \mathfrak{D}n_t(r)\mathfrak{D}m(r)e^{-\beta H}. \tag{2.6}$$

In the nematic phase, H_4 can be neglected in the mean-field theory, this leads to an x-ray intensity in the vicinity of $\vec{q}_0$ of

$$I(k) \sim \langle \rho(k)\rho(-k)\rangle \equiv \langle m(k)m(-k)\rangle$$
$$= \frac{1}{a + D_{\parallel}(k_{\parallel}^2 - q_{\parallel}^2)^2 + C_{\perp}k_{\perp}^2 + D_{\perp}k_{\perp}^4}, \tag{2.7}$$

where $q_{\parallel}^2 = C_{\parallel}/2D_{\parallel}$. When $C_{\perp} > 0$, $I(k)$ has peaks at $k_{\parallel} = \pm q_{\parallel}$ corresponding to fluctuations into the smectic-A phase. When $C_{\perp}$ is negative Eq. (2.7) can be rewritten

$$I(k) \sim \frac{1}{\bar{a} + D_{\parallel}(k_{\parallel}^2 - q_{\parallel}^2)^2 + D_{\perp}(k_{\perp}^2 - q_{\perp}^2)^2}, \tag{2.8}$$

where $q_{\perp}^2 = |C_{\perp}|/2D_{\perp}$ and

$$\bar{a} = a'(T - T_{NC})/T_{NA},$$
$$T_{NC} = T_{NA} + C_{\perp}^2 T_{NA}/4D_{\perp}a'. \tag{2.9}$$

Thus for $C_{\perp} < 0$, $I(k)$ is a maximum on the two rings $(\pm q_{\parallel}, q_{\perp}\cos\varphi, q_{\perp}\sin\varphi)$ as required in the vicinity of the NC transition [cf., Fig. 2(b)]. Correlations in the directions parallel to $\vec{n}$ die off

with a correlation length

$$\xi_{\parallel}^2 = \begin{cases} 2C_{\parallel}/a \sim (T - T_{NA})^{-1} & \text{if } C_{\perp} > 0, \\ 2C_{\parallel}/a' \sim (T - T_{NC})^{-1} & \text{if } C_{\perp} < 0. \end{cases} \tag{2.10}$$

and correlations in directions perpendicular to $\vec{n}$ die off with length

$$\xi_{\perp}^2 = \begin{cases} C_{\perp}/a \sim (T - T_{NA})^{-1} & \text{if } C_{\perp} > 0, \\ 2|C_{\perp}|/a' \sim (T - T_{NC})^{-1} & \text{if } C_{\perp} < 0. \end{cases} \tag{2.11}$$

Equations (2.8), (2.9), and (2.11) yield the usual mean-field critical exponents $\gamma = 1$ and $\nu = \frac{1}{2}$. Equation (2.2) can also be minimized in the ordered phase in the usual way. The resulting phase diagram is shown in Fig. 3(a).

The model presented here does not show a smectic-A to smectic-C transition as a function of temperature. Such a transition is easily produced as shown in Appendix A by adding a term proportional to

$$\int m^2(r)\,\vec{\nabla}_{\perp}m(r)\cdot\vec{\nabla}_{\perp}m(r)\,d^3r.$$

Addition of this term, however, does not affect the elastic constant calculations presented in Sec. III.

Equation (2.2) reduces when $C_{\perp} > 0$ to the model introduced by de Gennes[5,6] to describe the nematic to smectic-A transition

$$\beta H_0 = \int d^3r \left(A|\psi|^2 + \frac{1}{2M_v}|\nabla_{\parallel}\psi|^2 \right.$$
$$\left. + \frac{1}{2M_T}|(\vec{\nabla}_{\perp} - q_{\parallel}\delta\vec{n})\psi|^2 \right), \tag{2.12}$$
$$\beta H_4 = \frac{3}{2}u\int|\psi|^4,$$

where $\delta\vec{n}$ is the deviation of $\vec{n}$ from its uniform equilibrium direction,

$$m(\vec{r}) = (2)^{-1/2}[e^{iq_{\parallel}z}\psi(r) + e^{-iq_{\parallel}z}\psi^*(r)], \tag{2.13}$$

where $z = \vec{n}\cdot\vec{r}$ and

$$A = \frac{1}{2}a, \quad C_{\perp} = 1/M_T, \quad C_{\parallel} = 1/2M_v. \tag{2.14}$$

III. ELASTIC CONSTANTS

In the smectic-A phase, bend and twist distortions of the director, even of small wave number, create large separations of the smectic planes. Thus these distortions have a finite rather than a vanishing energy at zero wave number in the A phase. Above T_{NA}, fluctuations into the A phase will, therefore, cause K_2 and K_3 to grow and finally diverge at T_{NA}. Splay deformations on the other hand have energy going to zero with wave number, even in the A phase. Therefore, no divergent anomalies are expected at T_{NA}. Calculations by de Gennes[6] and by Jähnig and Brochard,[23] in fact, predict that K_2 and K_3 diverge as the correlation length ξ and that K_1 undergoes no violent change at T_{NA}. A number of experiments[8,9] have verified that K_2 and K_3 diverge and that K_1 is relatively well behaved at T_{NA}, though agreement on the critical exponents for ξ has not been reached. In the smectic-C phase, all three director distortions lead to large separations of the smectic planes. One would, therefore, expect K_1, K_2, and K_3 to diverge near T_{NC}. There is some experimental evidence that this is the case.[16] The de Gennes theory[6] of the NC transition predicts that K_1, K_2, and K_3 diverge as $\xi^{3/2}$. The theory we present here predicts that all three will diverge as ξ^2 in three dimensions.

Our calculations are a straightforward perturbation expansion in the director–order-parameter couplings. In equilibrium, the director is uniform in space: $\vec{n}(r) = \vec{n}^0$. For convenience, we will take $\vec{n}^0$ (a unit vector) to point along the third axis. Deviations from equilibrium are expressed in terms of

$$\delta\vec{n} = (\delta n_1, \delta n_2, \{1 - [(\delta n_1)^2 + (\delta n_2)^2]\}^{1/2} - 1)$$
$$\cong (\delta n_1, \delta n_2, -\tfrac{1}{2}[(\delta n_1)^2 + (\delta n_2)^2]). \quad (3.1)$$

We can now expand βH_m in powers of $\delta\vec{n}$:

$$\beta H_m = \beta H_0 + \beta H_1 + \beta H_2 + O((\delta n)^3), \quad (3.2)$$

where βH_0 is given by Eq. (2.3) with $\vec{n}$ replaced by $\vec{n}^0$ and

$$\beta H_1 = \int \frac{d^3k}{(2\pi)^3} \int \frac{d^3q}{(2\pi)^3} \Gamma_i(\vec{k}, \vec{q}) n_i(-\vec{k}) m(-\vec{q}) m(\vec{q}+\vec{k}), \quad (3.3)$$

$$\beta H_2 = \int \frac{d^3k_1}{(2\pi)^3} \int \frac{d^3k_2}{(2\pi)^3} \int \frac{d^3q}{(2\pi)^3} \Gamma_{ij}^{(2)}(\vec{k}_1, \vec{k}_2, \vec{q})$$
$$\times n_i(-\vec{k}_1) n_j(-\vec{k}_2)$$
$$\times m(\vec{q}) m(-\vec{q}+\vec{k}_1+\vec{k}_2), \quad (3.4)$$

where

$$\Gamma_i(\vec{k}, \vec{q}) = D_{\shortparallel}[q_i q_{\shortparallel}(\vec{q}+\vec{k})_{\shortparallel}^2 + (\vec{q}+\vec{k})_i(\vec{q}+\vec{k})_{\shortparallel}q_{\shortparallel}^2]$$
$$- \tfrac{1}{2}(C_{\shortparallel}+C_{\perp})(2q_i q_{\shortparallel} + q_i k_{\shortparallel} + q_{\shortparallel}k_i)$$
$$- D_{\perp}[q_i q_{\shortparallel}(\vec{q}+\vec{k})_{\perp}^2 + (\vec{q}+\vec{k})_i(\vec{q}+\vec{k})_{\shortparallel}q_{\perp}^2]. \quad (3.5)$$

The expression for $\Gamma_{ij}^{(2)}$ is rather complicated. Since it will not be used directly in any calculations, we will not reproduce it here.

We now introduce propagators for the order parameter and the director:

$$G(\vec{k}) = \langle m(\vec{k}) m(-\vec{k}) \rangle,$$

$$D_{ij}(\vec{k}) = \langle \delta n_i(\vec{k}) \delta n_j(-\vec{k}) \rangle.$$

When H_m is replaced by H_0, $G(k)$ reduces to $G^0(k)$ given by Eq. (2.7). $D_{ij}(\vec{k})$ has two independent components. If $\vec{k}$ is chosen to be in the 1-3 plane and the coupling to m is ignored, they are

$$D_{11}^0(\vec{k}) = \frac{1}{\beta(K_1^0 k_1^2 + K_3^0 k_3^2)}, \quad (3.5a)$$

$$D_{22}^0(\vec{k}) = \frac{1}{\beta(K_2^0 k_1^2 + K_3^0 k_3^2)}. \quad (3.5b)$$

When the coupling to m is included, D_{ij} satisfies Dyson's equation $D_{ij}^{-1}(\vec{k}) = (D_{ij}^0)^{-1}(\vec{k}) - \pi_{ij}(\vec{k})$. The long-wavelength forms of D_{11} and D_{22} are the same as in Eqs. (3.5) with K_1^0, K_2^0, and K_3^0 replaced by the renormalized elastic constants K_1, K_2, and K_3. The lowest-order diagrams for $\pi_{ij}(\vec{k})$ are shown in Fig. 4. In Appendix B, we derive a Ward identity which shows that Fig. 4(b) exactly cancels the zero-momentum part of Fig. 4(a). The diagrams in Fig. 4, therefore, yield

$$\pi_{ij}(k) = -2 \int \frac{d^3q}{(2\pi)^3} \{\Gamma_i(\vec{k}, \vec{q}) \Gamma_j(\vec{k}, \vec{q}) G^0(\vec{q}) G^0(\vec{q}+\vec{k})$$
$$- \Gamma_i(\vec{0}, \vec{q}) \Gamma_j(\vec{0}, \vec{q}) [G^0(\vec{q})]^2\}. \quad (3.6)$$

This is in fact the most divergent contribution to π_{ij} when a (or $\tilde{a}$) goes to zero.

Evaluating Eq. (3.6) we obtain the elastic constant enhancement near the NA and NC transitions and near the Lifshitz point.

A. NA transition

$$\delta K_2 = \frac{kT}{24\pi} q_{\shortparallel}^2 \frac{C_{\perp}}{(2C_{\shortparallel}a)^{1/2}} = \frac{kT}{24\pi} q_{\shortparallel}^2 \frac{\xi_{\perp}^2}{\xi_{\shortparallel}},$$
$$\delta K_3 = \frac{kT}{24\pi} q_{\shortparallel}^2 \left(\frac{2C_{\shortparallel}}{a}\right)^{1/2} = \frac{kT}{24\pi} q_{\shortparallel}^2 \xi_{\shortparallel}. \quad (3.7)$$

There are no divergent contributions to K_1 as ex-

(a)

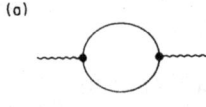

(b)

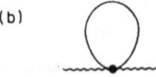

FIG. 4. Diagrams contributing to the fluctuation enhancement of the Frank elastic constants.

pected. Equations (3.7) are in agreement with the calculation of de Gennes[5,6] as corrected by Jähnig and Brochard.[23]

B. NC transition

$$\delta K_1 = \frac{9kT}{64\pi} \frac{q_{\shortparallel}}{(D_{\shortparallel}D_{\perp})^{1/2}} \frac{|C_{\perp}|^2}{\bar{a}}\left(1 + \frac{D_{\shortparallel}}{D_{\perp}}\frac{2}{9}\right)$$

$$= \frac{9kT}{64\pi}q_{\perp}\left(q_{\shortparallel}^2\frac{\xi_{\perp}^3}{\xi_{\shortparallel}} + \frac{2}{9}q_{\perp}^2\xi_{\shortparallel}\xi_{\perp}\right),$$

$$\delta K_2 = \frac{3kT}{64\pi} \frac{q_{\shortparallel}}{(D_{\shortparallel}D_{\perp})^{1/2}} \frac{|C_{\perp}|^2}{\bar{a}}\left(1 + \frac{2}{9}\frac{D_{\shortparallel}}{D_{\perp}}\right)\qquad(3.8)$$

$$= \frac{3kT}{64\pi}q_{\perp}\left(q_{\shortparallel}^2\frac{\xi_{\perp}^3}{\xi_{\shortparallel}} + \frac{2}{9}q_{\perp}^2\xi_{\shortparallel}\xi_{\perp}\right),$$

$$\delta K_3 = \frac{kT}{12\pi}q_{\shortparallel}^2\left(\frac{D_{\shortparallel}}{D_{\perp}}\right)^{1/2}\frac{|C_{\perp}|}{\bar{a}} = \frac{kT}{24\pi}q_{\shortparallel}^2q_{\perp}\xi_{\shortparallel}\xi_{\perp}.$$

All three elastic constants diverge as ξ^2 in three dimensions. $\delta K_1/\delta K_2 = 3$ is not special to the third dimension. It holds in any dimension near the NC transition.

C. Lifshitz point

$$\delta K_1 = \frac{kT}{4\pi}q_{\shortparallel}^2\left(\frac{D_{\perp}}{2C_{\shortparallel}}\right)^{1/2}\ln\frac{\Lambda^2}{a} + \frac{kT}{32\pi}\left(\frac{2C_{\shortparallel}}{D_{\perp}}\right)^{1/2}\ln\frac{\Lambda^2}{a},$$

$$\delta K_2 = \frac{kT}{2\pi}q_{\shortparallel}^2\left(\frac{D_{\perp}}{2C_{\shortparallel}}\right)^{1/2}\ln\frac{\Lambda^2}{a} + \frac{kT}{32\pi}\left(\frac{2C_{\shortparallel}}{D_{\perp}}\right)^{1/2}\ln\frac{\Lambda^2}{a},\qquad(3.9)$$

$$\delta K_3 = \frac{kT}{12\pi}q_{\shortparallel}^2\left(\frac{2C_{\shortparallel}}{a}\right)^{1/2} = \frac{kT}{12\pi}q_{\shortparallel}^2\xi_{\shortparallel}.$$

K_3 diverges as the correction length in the third direction, ξ_n, while K_1 and K_2 are only slightly divergent.

ACKNOWLEDGMENTS

We are grateful to P. C. Hohenberg and Jack Swift for bringing Ref. 20 to our attention, for communicating to us results of their work prior to publication, and for helpful discussions.

APPENDIX A: IMPROVEMENTS ON THE MODEL

In order to modify Eq. (2.3) to allow an AC transition we include a nonlocal term

$$\frac{1}{2}\int d^3r\, b\, m^2(r)\,\vec{\nabla}_{\perp}m(r)\cdot\vec{\nabla}_{\perp}m(r)$$

in βH_4 with $b < 0$. The effect of this term is to drive the coefficient of $k_{\perp}^2$ negative as the smectic order parameter m grows in the smectic-A phase as the temperature decreases. It thus has the potential to induce an AC transition. In terms of the de Gennes order parameter $\psi = |\psi|e^{i\varphi}$ the resulting free energy is

$$F = \frac{1}{2}\int d^3r\{a|\psi|^2 + C_{\shortparallel}|\nabla_{\shortparallel}\psi|^2 + C_{\perp}|\nabla_{\perp}\psi|^2 + D_{\perp}|\nabla_{\perp}^2\psi|^2$$

$$+ 3u|\psi|^4 + b[|\psi|^2(\nabla_{\perp}|\psi|)^2 + \tfrac{1}{2}|\psi|^2|\nabla_{\perp}\psi|^2]\},$$

$$\qquad(A1)$$

with $|\psi|$ constant and $\varphi = \vec{k}_{\perp}\cdot\vec{r}_{\perp}$. Equation (A1) is minimized when

$$|\psi|^2(C_{\perp} + b|\psi|^2 + D_{\perp}k_{\perp}^2)k_{\perp}^2 = 0.\qquad(A2)$$

Hence an A to C transition occurs when

$$C_{\perp} + b|\psi|^2 = 0,\qquad(A3)$$

where

$$|\psi|^2 = (a'/6u)(T_{NA} - T)/T_{NA}.\qquad(A4)$$

Equation (A3) is solved to give the phase boundary between the smectic-A phase and the smectic-C phase. For $b < 0$, we have

$$T_{AC} = T_{NA}[1 - (6u/a'|b|)C_{\perp}].\qquad(A5)$$

The resulting phase diagram is shown in Fig. 3(b).

APPENDIX B: DERIVATION OF WARD IDENTITIES

In this appendix we will derive Ward identities for the vertices coupling n_l to m which can be used to show that perturbation theory maintains rotational invariance. There are two vertices coupling n_l to m that are of interest:

$$\Gamma_l(\vec{x}_1,\vec{x}_2,\vec{x}_3) = \frac{1}{2}\frac{\delta G^{-1}(\vec{x}_1,\vec{x}_2)}{\delta n_l(\vec{x}_3)}\qquad(B1)$$

$$\Gamma_{lj}(\vec{x}_1,\vec{x}_2,\vec{x}_3,\vec{x}_4) = \frac{1}{4}\frac{\delta G^{-1}(\vec{x}_1,\vec{x}_2)}{\delta n_l(\vec{x}_3)\,\delta n_j(\vec{x}_4)}.\qquad(B2)$$

The vertex functions appearing in the text are related to the above via

$$\Gamma_l(\vec{k},\vec{q}) = \int d^3(x_1 - x_2)d^3x_3\, e^{-i\vec{q}\cdot(\vec{x}_1-\vec{x}_2)}$$

$$\times e^{-i\vec{k}\cdot\vec{x}_3}\Gamma_l(\vec{x}_1,\vec{x}_2,\vec{x}_3),\qquad(B3)$$

$$\Gamma_{lj}(\vec{k}_1,\vec{k}_2,\vec{q}) = \int d^3(x_1 - x_2)d^3x_3\,d^3x_4$$

$$\times [e^{-i\vec{q}\cdot(\vec{x}_1-\vec{x}_2)}e^{-i\vec{k}_1\cdot\vec{x}_3}$$

$$\times e^{-i\vec{k}_2\cdot\vec{x}_4}\Gamma_{lj}(\vec{x}_1,\vec{x}_2,\vec{x}_3,\vec{x}_4)].$$

$$\qquad(B4)$$

Combining Eqs. (B1)–(B4), we obtain

$$\lim_{k\to 0}\Gamma_l(\vec{k},\vec{q}) = \frac{1}{2}\frac{\partial G^{-1}(\vec{q})}{\partial n_l},\qquad(B5)$$

$$\lim_{k_1,k_2\to 0}\Gamma_{lj}(\vec{k}_1,\vec{k}_2,\vec{q}) = \frac{1}{4}\frac{\partial^2 G^{-1}(\vec{q})}{\partial n_l\,\partial n_j}.\qquad(B6)$$

But $G^{-1}(\vec{q})$ depends only on $q_{\shortparallel}^2 = (\hat{n}\cdot\vec{q})^2$ and $q_{\perp}^2 = q^2 - q_{\shortparallel}^2$. Therefore we have

$$\Gamma_i(0,\vec{q}) = \frac{1}{2}\left(q_i\frac{\partial G^{-1}}{\partial q_3} - q_3\frac{\partial G^{-1}}{\partial q_i}\right) = q_i q_3\left(\frac{\partial G^{-1}}{\partial q_3^2} - \frac{\partial G^{-1}}{\partial q_i^2}\right),$$

$$(B7)$$

$$\Gamma_{ij}(0,0,\vec{q}) = \frac{1}{2}\left(q_i\frac{\partial\Gamma_j}{\partial q_3} - q_3\frac{\partial\Gamma_j}{\partial q_i}\right).$$

$$(B8)$$

Equations (B7) and (B8) can be used to show that the contribution to π_{ij} from diagram 4(b) $(\pi_{ij}^{(2)})$ cancels the zero momentum contribution from diagram 4(a) $(\pi_{ij}^{(1)})$:

$$\pi_{ij}^{(2)} = \int \Gamma_{ij}(0,0,\vec{q})G(\vec{q}), \quad i,j = 1,2;$$

$$\pi_{ij}^{(2)} = \frac{1}{2}\int\left(q_i\frac{\partial\Gamma_j}{\partial q_3} - q_3\frac{\partial\Gamma_j}{\partial q_i}\right)G$$

$$= -\frac{1}{2}\int\left(q_i\frac{\partial G}{\partial q_3} - q_3\frac{\partial G}{\partial q_i}\right)\Gamma_j$$

$$= \frac{1}{2}\int\left(q_i\frac{\partial G^{-1}}{\partial q_3} - q_3\frac{\partial G^{-1}}{\partial q_i}\right)\Gamma_j G^2$$

$$= \int \Gamma_i(0,\vec{q})\Gamma_j(0,\vec{q})G^2(\vec{q}) = -\pi_{ij}^{(1)} \quad (\vec{q} = 0).$$

$$(B9)$$

This result is used in Eq. (3.6).

*Research was supported in part by the NSF and the Office of Naval Research.

†Alfred P. Sloan Research Fellow.

[1]C. W. Oseen, Trans. Faraday Soc. 29, 883 (1933); F. C. Frank, Disc. Faraday Soc. 25, 1 (1958).

[2]P. G. de Gennes, *The Physics of Liquid Crystals* (Oxford U. P., New York, 1974).

[3]W. L. McMillan, Phys. Rev. A 4, 1238 (1971); 6, 936 (1972).

[4]K. K. Kobayashi, Mol. Cryst. Liq. Cryst. 13, 137 (1971).

[5]P. G. de Gennes, Solid State Commun. 10, 753 (1972).

[6]P. G. de Gennes, Mol. Cryst. Liq. Cryst. 21, 49 (1973).

[7]J. W. Doane, R. S. Parker, B. Cuikl, D. L. Johnson, and D. L. Fishel, Phys. Rev. Lett. 28, 1694 (1972).

[8]L. Cheung, R. B. Meyer, and H. Gruler, Phys. Rev. Lett. 31, 349 (1973); M. Delaye, R. Ribotta, and G. Durand, *ibid.* 31, 443 (1973); H. Gruler, Z. Naturforsch. A 28, 996 (1973); L. Cheung and R. B. Meyer, Phys. Lett. A 43, 261 (1973); D. Salin, I. W. Smith, and G. Durand, J. Phys. (Paris) 35, L165 (1974); C. C. Huang, R. S. Pindak, P. J. Flanders, and J. T. Ho, Phys. Rev. Lett. 33, 400 (1974).

[9]K. C. Chu and W. L. McMillan, Phys. Rev. A 13, 1059 (1975); H. Birecki, R. Schaetzing, F. Rondelez, and J. D. Lifster, MIT report (unpublished).

[10]P. E. Cladis, Phys. Rev. Lett. 31, 1200 (1973); S. Torza and P. E. Cladis, *ibid.* 32, 1406 (1974); D. Djurek,

J. Bataric-Rubicic, and K. Franulovic, reported at the 1974 Stockholm Conference on Liquid Crystals.

[11]B. I. Halperin and T. C. Lubensky, Solid State Commun. 14, 997 (1974).

[12]B. I. Halperin, T. C. Lubensky, and Shang-Keng Ma, Phys. Rev. Lett. 32, 292 (1974).

[13]W. L. McMillan, Phys. Rev. A 8, 1921 (1973); R. J. Meyer and W. L. McMillan, *ibid.* 9, 899 (1974).

[14]A. Wulf, Phys. Rev. A 11, 365 (1975).

[15]R. G. Priest, J. Phys. (Paris) 36, 437 (1975).

[16]H. Gruler, Z. Naturforsch. A 28, 474 (1972).

[17]M. Alain Caille, C. R. Acad. Sci. (Paris) B 273, 891 (1972); W. L. McMillan, Phys. Rev. A 7, 1419 (1973).

[18]W. L. McMillan, Phys. Rev. A 8, 228 (1973).

[19]See for example, I. E. Dzyaloshinskii, Zh. Eksp. Teor. Fiz. 46, 1420 (1964); 47, 336 (1964); 47, 992 (1964) [Sov. Phys.-JETP 19, 960 (1964); 20, 223 (1965); 20, 665 (1965)].

[20]S. A. Brazovskii, Zh. Eksp. Teor. Fiz. 68, 175 (1975) [Sov. Phys.-JETP 41, 85 (1975)].

[21]P. C. Hohenberg and J. B. Swift, report of work prior to publication and private communication.

[22]R. M. Hornreich, M. Luban, and S. Shtrickman, Phys. Rev. Lett. 35, 1678 (1975).

[23]F. Jähnig and F. Brochard, J. Phys. (Paris) 35, 301 (1974).

X-Ray Study of Fluctuations near the Nematic–Smectic-A –Smectic-C Multicritical Point

L. J. Martínez-Miranda,[a] A. R. Kortan,[b] and R. J. Birgeneau

Department of Physics, Massachusetts Institute of Technology, Cambridge, Massachusetts 02139
(Received 19 February 1986)

We present the results of a high-resolution x-ray-scattering study of the pretransitional fluctuations above the nematic–to–smectic-C phase boundary in the $\overline{7}$S5/8OCB system. The fluctuations are well described by the Lifshitz-point model of Chen and Lubensky with minimal adjustable parameters. We have located the line along which the x-ray cross section $\sigma(\mathbf{q})$ exhibits pure $q_\perp^{-4}$ fluctuations. Close to the nematic–smectic-A –smectic-C point this line is simply related to the smectic-A –smectic-C transition boundary.

PACS numbers: 64.70.Md, 61.30.Eb

The phase-transition behavior of liquid-crystal systems exhibiting a confluence of nematic (N), smectic-A (S_A), and smectic-C (S_C) phases has been the subject of extensive investigation for the past decade.[1-5] In spite of this effort, a satisfactory experimental and theoretical description has not yet emerged. Experiments suggest that the N, S_A, and S_C phases come together at a single triple point, labeled the NAC multicritical point, with phase boundaries exhibiting universal geometrical behavior.[5] Much less is known about the nature of the mass-density fluctuations in the vicinity of the NAC point. This is an important shortcoming since the different theoretical models[1,2] make rather different predictions for the q dependence of the smectic susceptibility $\sigma(\mathbf{q})$.

Currently, the most extensive experimental information is available for the system $\overline{7}$S5$_x$-$\overline{8}$S5$_{1-x}$ (heptyloxy- and octyloxy-p'-pentylphenylthiol benzoate) which initially was viewed as prototypical.[3,4] However, recent work by Johnson and co-workers[4,5] indicates that the NAC multicritical region is unusually narrow in $\overline{7}$S5-$\overline{8}$S5. It is essential that detailed measurements be made on systems within the universal *NAC multicritical region* discovered by Brisbin et al.[5] and confirmed by Shashidhar, Ratna, and Krishna Prasad.[5] In this paper we report detailed microscope and high-resolution x-ray studies of mixtures of octyloxycyanobiphenyl (8OCB) and heptyloxypentylthiol benzoate ($\overline{7}$S5). We show from microscope studies that this system exhibits the "universal NAC diagram" of Brisbin et al. Further, from our x-ray studies in the nematic phase, we find that the smectic mass-density fluctuations are always well described by the simple Lifshitz form[6] proposed by Chen and Lubensky.[1] This is in partial contrast to the results of Ref. 3 where in the immediate vicinity of the NAC point the fluctuations are Lorentzian. We have located the Lifshitz line along which the coefficient of the transverse gradient squared term vanishes and we find that its locus mirrors that of the S_A-S_C boundary. Thus the NAC system is now experimentally the most completely characterized Lifshitz system; however, as

we shall discuss, important theoretical issues remain unresolved.

We discuss first the $\overline{7}$S5-8OCB phase diagram. Using a polarizing microscope we studied 32 different mixtures; the phase boundaries could be identified by an abrupt change in the texture. The nature of the phases and the locations of the phase boundaries were confirmed for eleven of the mixtures by use of x rays. The phase diagram so obtained is shown in Fig. 1. Comparison with the data in Ref. 5 shows that the $\overline{7}$S5-8OCB system exhibits the generic NAC phase diagram with an extended multicritical region. The NAC point occurs for a molar fraction of 8OCB of $X_{8OCB} = 0.0217$ at the temperature $T_{NAC} = 41.8\,°C$. Explicit fits by the crossover form suggested by Brisbin et

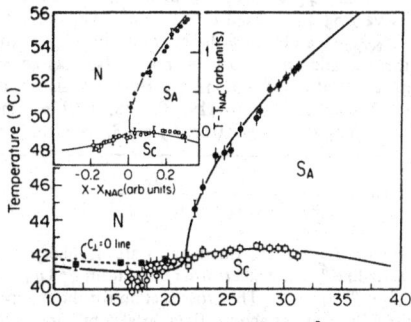

FIG. 1. NAC phase diagram obtained from the polarizing microscope studies. Solid circles, T_{NA}; open squares, T_{AC}; open circles, T_{NC}; and solid squares, $T(C_\perp = 0)$. The solid lines are the results of fits by Eq. (1) as discussed in the text. For the N-S_C transition which is hysteretic we show transition temperatures on both heating (upper) and cooling (lower). Inset: The data superimposed on the universal NAC curve of Ref. 5 with temperature and concentration rescaled by 0.124 and 0.033, respectively, and $B = 0$. Only the N-S_C transitions on heating are displayed in the inset.

158

$al.$[5]

$$T_\alpha - T_{NAC}$$

$$= A_\alpha |X - X_{NAC}|^{1/\phi_\alpha} + B(X - X_{NAC}), \qquad (1)$$

where α = NA, NC, and AC with $\phi_{NA} = \phi_{NC}$, give $1/\phi_{NA} = 0.5 \pm 0.05$, $1/\phi_{AC} = 1.6 \pm 0.1$, $A_{NA} = 80.5 \pm 10$, $A_{NA}/A_{NC} = -4 \pm 2$, $A_{AC} = 2240$, and $B = 174 \pm 130$. These agree well with the values obtained previously[5] in five different systems. The ratio A_{NA}/A_{NC} is poorly determined because of uncertainties connected with hysteresis at the N-C boundary. In the inset to Fig. 1 we show the "universal form" of Brisbin et $al.$[5] for the NAC boundaries with $B = 0$ superimposed on our data; clearly the agreement is good and could be improved by explicit fitting of B, etc. The exponents and amplitude ratio themselves, however, remain unexplained.

We carried out x-ray studies on eleven separate samples with molar concentrations $X_{8OCB} = 0$, 0.0069, 0.0097, 0.0119, 0.0158, 0.0176, 0.0197, 0.0217, 0.0219, 0.0261, and 0.0346. The first seven samples exhibit single N-S_C transitions. Experiments on each sample took about three weeks. The experimental configuration was essentially identical to that discussed by Safinya and co-workers.[3] For S_A pretransitional scattering one observes a single Lorentzian peak centered about $(0, 0, q_\parallel^0)$ while for S_C fluctuations one observes a diffuse ring centered about $(q_\perp^0 \cos\phi, q_\perp^0 \sin\phi, q_\parallel^0)$; the scattering is independent of the azimuthal angle ϕ. In the S_C fluctuation region we took four x-ray scans per temperature point, a transverse scan varying $q_\perp$ at fixed $q_\parallel^0$ and three longitudinal scans varying $q_\parallel$ at fixed $q_\perp = 0$, $\pm q_\perp^0$. For the samples $X_{8OCB} = 0$ and 0.0069 the pretransitional scattering in the nematic phase up to 8° above T_{NC} has an S_C character, that is, $q_\perp^0$ is nonzero. However, five samples, $X_{8OCB} = 0.0097$, 0.0119, 0.0158, 0.0176, and 0.0197, show a crossover from S_A- to S_C-type fluctua-

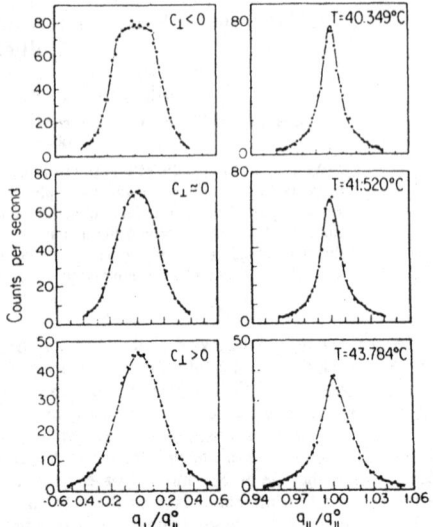

FIG. 2. Longitudinal and transverse x-ray scans in the crossover region for the $X_{8OCB} = 0.0158$ sample. The longitudinal scans are all through the position $q_\perp = 0.1$. The solid lines are the results of fits to the Chen-Lubensky form Eq. (2).

tions as the temperature is decreased. Representative scans for $X_{8OCB} = 0.0158$ spanning the crossover region in the nematic phase are shown in Fig. 2.

In order to discuss quantitatively the x-ray results we must first specify the appropriate form for the x-ray cross section $\sigma(\mathbf{q})$. In the Chen-Lubensky infinite-dimensional order-parameter model[1] the tilt $(q_\perp^0 \neq 0)$ enters through the gradient of the mass-density wave. The Ornstein-Zernike expression then is

$$\sigma(\mathbf{q}) = \frac{\sigma_0}{1 + \xi_\parallel^2 (1 + Kq_\perp^2)(q_\parallel - q_\parallel^0)^2 + C_\perp q_\perp^2 + D_\perp q_\perp^4}, \qquad (2)$$

where for $C_\perp > 0$ the fluctuations are S_A-like, for $C_\perp < 0$ the fluctuations are S_C-like, and $C_\perp = 0$ defines the Lifshitz line.[6] The higher-order cross term $q_\perp^2 (q_\parallel - q_\parallel^0)^2$ is necessitated by the data. The Lifshitz point is defined by $C_\perp = 0$, $\sigma_0 \to \infty$. The cross section for the independent tilt models is more complicated. For Lorentzian mass-density fluctuations about a fixed local configuration one obtains,[2] after averaging over the tilt azimuthal angle ϕ,

$$\sigma(\mathbf{q}) = \frac{\sigma_0}{[1 + \xi_\parallel^2 (q_\parallel - q_\parallel^0)^2 + \xi_\perp^2 (q_\perp - q_\perp^0)^2]^{1/2} [1 + \xi_\parallel^2 (q_\parallel - q_\parallel^0)^2 + \xi_\perp^2 (q_\perp + q_\perp^0)^2]^{1/2}}. \qquad (3)$$

If one includes a $D_\perp q_\perp^4$ term in the mass-density fluctuations one obtains an expression like (3) but algebraically more complicated. This form, which we do not quote explicitly here, was derived on a phenomenological basis by Safinya[3] and from a formal two–order-parameter theory by Andereck and Patton.[2] Here the adjustable parameters are σ_0, $\xi_\parallel$, $\xi_\perp$, $q_\perp^0$, and $D_\perp$.

The solid lines shown in Fig. 2 are the results of fits by Eq. (2). It is evident that this model, which has as adjustable parameters σ_0, $\xi_\parallel$, K, $C_\perp$, and $D_\perp$, works very well in the S_A, Lifshitz, and S_C regions. The goodness-of-fit parameter χ^2 is typically 1 for all temperatures and concentrations. Fits by Eq. (3) are completely unsuccessful in the S_C fluctuation region as found previously by Safinya and co-workers[3] in $\overline{7}S5$-$\overline{8}S5$. This is because the transverse scans fall off much faster than $q_\perp^2$. Fits by the Safinya-Andereck-Patton form which includes the $D_\perp q_\perp^4$ are more successful. Nevertheless, the χ^2 obtained from fits to this model are about three times larger than those found for the Chen-Lubensky model,[1] Eq. (2), for scans near the N-S_C boundary. Furthermore, near the Lifshitz point Eq. (3) with the $q_\perp^4$ correction fails completely while Eq. (2) describes the data quite well. This should be regarded as strong but not absolute evidence for the Chen-Lubensky approach since Eq. (3) and its isomorphs only represent one limit for the two-parameter models—that in which the tilt amplitude is fixed.

We now discuss the results obtained from the fits by the Chen-Lubensky form. For each N-S_C sample we obtained σ_0, K, $\xi_\parallel$, $C_\perp$, and $D_\perp$ over a range of temperatures down to the N-S_C boundary. For the samples $X_{8OCB}=0$ and 0.0069 the fits give $C_\perp < 0$ over the range studied ($\sim 8°$ above T_{NC}). For the remaining N-S_C samples we explicitly observe a continuous evolution of $C_\perp$ from positive to negative values with decreasing temperature. Results for the $X_{8OCB}=0.0158$ and 0.0197 samples are shown in Fig. 3. The data are most dramatic for the 0.0197 sample which is quite close to the NAC concentration. As is evident in Fig. 3, $C_\perp$ increases continuously as if one were headed towards a transition into an S_A phase ($C_\perp \sim \xi_\perp^2$).

However, approximately $1°$ above T_{NC}, $C_\perp$ reaches a maximum and then decreases rapidly through zero. There is then a weakly first-order transition into the S_C phase at $T_{NC}=41.53\,°C$.

The fitted transverse and longitudinal correlation lengths for the $X_{8OCB}=0.0197$ sample are shown in Fig. 4. Here $\xi_\perp$ is defined as the inverse half width at half maximum of the transverse scan while $\xi_\parallel$ is the inverse of the HWHM at $q_\perp = 0$ for $C_\perp \geq 0$ and $q_\perp = q_\perp^0$ for $C_\perp < 0$. It is notable that in the crossover region $\xi_\parallel / \xi_\perp \cong 50$ and the correlated regions are approximately $3000 \times 60 \times 60$ Å^3 in size. We do not know if this extreme anisotropy is generic. The temperatures at which $C_\perp = 0$ are shown in Fig. 1. The dashed line is just the inverse of the AC boundary, that is, Eq. (1) with $A_{C_\perp = 0} = -A_{AC}$; this is an interesting phenomenological observation which, hopefully, will stimulate theory for the Lifshitz line. We note that the $C_\perp = 0$ line rises dramatically for $X_{8OCB} < 0.01$ (not shown in Fig. 1) implying that one is departing from the N_{AC} multicritical region.

Before discussing the overall implications of these results we mention briefly the behavior on the N-S_A side of the NAC point. We find conventional N-S_A transitions with exponents $\gamma = 1.55 \pm 0.1$, $\nu_\parallel = 0.90 \pm 0.05$, and $\nu_\perp = 0.76 \pm 0.05$. Closely similar results for $\nu_\parallel$ have been independently obtained by Solomon and Litster[7] via light-scattering studies of the elastic response as discussed in the accompanying Letter. These exponents are close to but somewhat larger than those found in other N-S_A transitions.[8] The S_A-S_C transitions for the 0.0219, 0.0261, and 0.0346 samples are all consistent with mean-field behavior with a small

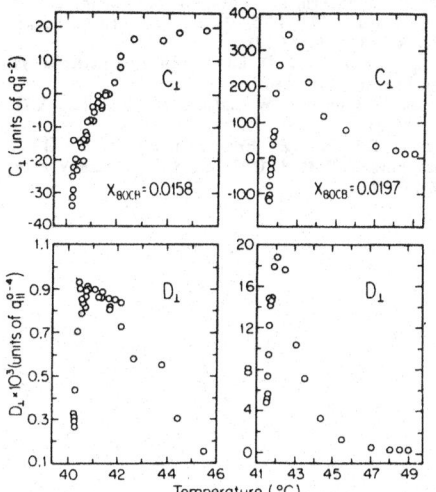

FIG. 3. Best-fit coefficients, $C_\perp$ and $D_\perp$ of Eq. (2), for the 0.0158 and 0.0197 samples.

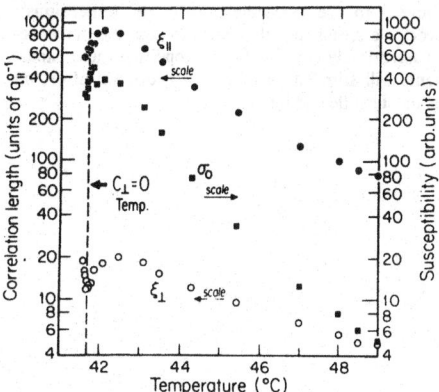

FIG. 4. Susceptibility and longitudinal and transverse correlation lengths as defined in the text for the $X_{8OCB}=0.0197$ sample; $q_\parallel^0 = 0.23195$ Å^{-1}

VOLUME 56, NUMBER 21   PHYSICAL REVIEW LETTERS   26 MAY 1986

sixth-order term in the tilt free energy; the effective tilt exponent for all three samples is 0.44 ± 0.04.

These experiments, therefore, have shown that in a prototypical system $\overline{7}$S5-8OCB which exhibits the universal NAC phase diagram[5] the smectic susceptibility in the nematic phase is always well described by the simple Chen-Lubensky Lifshitz form. Further, the Lifshitz line appears to be a simple extension of the AC boundary. Conventional two–order-parameter theories[2] are shown to be invalid both by the geometry of the phase boundaries[5] and by the nature of $\sigma(\mathbf{q})$. As noted by Brisbin *et al.*,[5] the fact that the nematic-to-smectic transition is suppressed as the fluctuations become C-like is suggestive of the Brazovskii model applied to the N-S_C transition by Swift.[9] This is further substantiated by our results near the NAC point where, as shown in Fig. 4, the longitudinal correlation length actually *decreases* as the fluctuations change from S_A-like to S_C-like. This decrease in $\xi_{\parallel}$ also manifests itself in the Solomon and Litster[7] light-scattering experiments.

An alternative approach to the NAC problem has been given by Grinstein and Toner,[2] who predict that in $d = 3$ the NAC point is a tetracritical point with decoupled mass-density and tilt orderings. Thus, close enough to the NAC point one should have the sequence of phases N $\rightarrow$ biaxial nematic $\rightarrow$ S_C. The x-ray cross section presumably should evolve from Eq. (2) to the Safinya-Andereck-Patton form. However, neither we nor other groups[5,7] find any evidence for the biaxial nematic phase nor for the anticipated evolution in the x-ray cross section. To make this more definite one requires explicit predictions for the magnitude of effects caused by the biaxial nematic ordering. The Chen-Lubensky model itself, however, presents the dilemma that an $m = 2$ Lifshitz point is believed to be unstable in three dimensions[6]; nevertheless, this model appears to describe the smectic susceptibility quite well. We hope that these experiments will stimulate a renewed theoretical effort on this most subtle and interesting problem.

We should like to thank A. Aharony, B. Andereck, P. Bak, C. W. Garland, G. Grinstein, D. L. Johnson, J. D. Litster, T. C. Lubensky, B. R. Patton, C. R. Safinya, and L. Solomon for stimulating discussions. This research was supported by the National Science Foundation Materials Research Program under Contract No. DMR84-18718.

[a]Current address: Department of Physics, University of California, Berkeley, Cal. 94720.

[b]Current address: AT&T Bell Laboratories, Murray Hill, N.J. 07974.

[1]J.-C. Chen and T. C. Lubensky, Phys. Rev. A 14, 1202 (1976).

[2]K. C. Chu and W. L. McMillan, Phys. Rev. A 15, 1181 (1977); L. Benguigui, J. Phys. (Paris), Colloq. 40, C3-222 (1979); C. C. Huang and S. C. Lien, Phys. Rev. Lett. 47, 1917 (1981); G. Grinstein and J. Toner, Phys. Rev. Lett. 51, 2386 (1983); B. S. Andereck and B. R. Patton, to be published.

[3]C. R. Safinya, R. J. Birgeneau, J. D. Litster, and M. E. Neubert, Phys. Rev. Lett. 47, 668 (1981); C. R. Safinya, L. J. Martinez-Miranda, M. Kaplan, J. D. Litster, and R. J. Birgeneau, Phys. Rev. Lett. 50, 56 (1983); C. R. Safinya, private communication.

[4]D. Johnson, D. Allender, R. DeHoff, C. Maxe, E. Oppenheim, and R. Reynolds, Phys. Rev. B 16, 470 (1977); R. DeHoff, R. Biggers, D. Brisbin, and D. L. Johnson, Phys. Rev. A 25, 472 (1982).

[5]D. Brisbin, D. L. Johnson, H. Fellner, and M. E. Neubert, Phys. Rev. Lett. 50, 178 (1983); R. Shashidar, B. K. Ratna, and S. Krishna Prasad, Phys. Rev. Lett. 53, 2141 (1984).

[6]R. M. Hornreich, M. Luban, and S. Shtrikman, Phys. Rev. Lett. 35, 1678 (1975); D. Mukamel and M. Luban, Phys. Rev. B 18, 3631 (1978).

[7]L. Solomon and J. D. Litster, following Letter [Phys. Rev. Lett. 56, 2268 (1986)].

[8]See, for example, C. W. Garland, et al., Phys. Rev. A 27. 3234 (1983).

[9]S. A. Brazovskii, Zh. Eksp. Teor. Fiz. 68, 175 (1975) [Sov. Phys. JETP 41, 85 (1975)]; J. Swift, Phys. Rev. A 14, 2274 (1976).

4. MULTIPLE SMECTIC-A PHASES

4.1. Levelut, A. M., Tarento, R. J., Hardouin, F., Achard, M. F. and Sigaud, G. (1981). Number of SA phases. *Phys. Rev.* **A24**, 2180–2186   163

4.2. Prost, J. (1984). The Smectic State. *Adv. Phys.* **33**, 1–46   170

4.3. Wang, J. and Lubensky, T. C. (1984). Theory of the $S_{A1}-S_{A2}$ phase transition in liquid crystal. *Phys. Rev,* **A29**, 2210–2217   216

4.4. Ratna, B. R., Shashidhar, R. and Raja, V. N. (1985). Smectic-A phase with two collinear incommensurate density modulations. *Phys. Rev. Lett.* **55**, 1476–1478   224

4.5. Chan, K. K., Pershan, P. S., Sorensen, L. B. and Hardouin, F. (1985). X-ray scattering study of the smectic-A_1 to smectic-A_2 transition. *Phys. Rev. Lett.* **54**, 1694–1697   227

4.6. Chan, K. K., Pershan, P. S., Sorensen, L. B. and Hardouin, F., (1986). X-ray studies of transitions between nematic, smectic-A_1, -A_2, and -A_d phases. *Phys. Rev.* **A34**, 1420–1433   231

PHYSICAL REVIEW A   VOLUME 24, NUMBER 4   OCTOBER 1981

Number of S_A phases

A. M. Levelut and R. J. Tarento

Laboratoire de Physique des Solides associé au Centre National de la Recherche Scientifique, Université Paris-Sud, Bâtiment 510, 91405 Orsay, Cédex, France

F. Hardouin, M. F. Achard, and G. Sigaud

Centre de Recherche Paul Pascal, Université de Bordeaux I, 33405 Talence, Cédex, France

(Received 17 February 1981)

We present here the phase diagram of a binary system of dipolar molecules in which five distinct smectic-A regions are put in evidence both by calorimetric and x-ray investigations, four of these smectic-A areas are contiguous and thus several S_A-S_A transition lines and two triple S_A points are present, and the fifth phase is only connected to the nematic phase. From x-ray measurements we are able to describe the molecular and the dipolar array of each of these five phases.

I. INTRODUCTION

For the first time a transition between two smectic-A phases has been discovered by Sigaud *et al.* in 1978 (Ref. 1) in a mixture of TBBA (terephtal-bis-p-butylaniline) with a nitrile derivative (DB$_5$). Further calorimetric investigations enable us to discover S_A-S_A transitions in other mixtures and even in pure compounds. All these mixtures, and *a fortiori* the pure compounds, are constituted by nitrile or nitro derivatives. Furthermore, another kind of nitrile molecule exhibits the reentrant polymorphism N-S_A-N-S_A (Ref. 2) and thus the uniqueness of the smectic-A phase is also questionable in this case. From x-ray investigations we can assert that the different S_A phases are layered phases in which the molecules are, on average, perpendicular to the layer plane and form a liquidlike array inside each layer. The smectic-A phases differ by their layer thickness compared to a mean molecular length: When the layer thickness is equal to the molecular length, one has an S_{A1} phase. A doubling of the layer thickness is induced by an antiferroelectric array of the dipolar molecules S_{A2} (Ref. 3) and the transition S_{A1}-S_{A2} occurs either directly or indirectly through an antiphase of the antiferroelectric order labeled $S_{\tilde{A}}$.[4] Another phase of intermediate layer thickness can exist: Schematically we can assume that the layer thickness comprised between one and two molecular lengths is due to a dipolar head to tail association of molecules with a more or less important overlapping of the molecules (Fig. 1). Let us call such a phase S_{Ad}, d for "dimers", keeping in mind that such dimers give a simplified description of the structures.[5]

An S_{Ad}-S_{A2} transition was observed in[6] DB$_7$ while in the double reentrant phenomena for which the two S_A phases correspond to an S_{Ad} and an S_{A1} phases, the transition from S_{Ad} to S_{A1} occurs through a reentrant nematic step. The thickness

is equal to $\simeq 1.2l$ (l is the molecular length) in the reentrant system and to $\simeq 1.7l$ in the case of the DB$_7$. In fact, S_{Ad} phases occur in many nitrile derivatives[5,7] such as CBOOA, 8OCB (p-cyanobenzylidene-p-octyloxyaniline and p-octyloxycyanobiphenyl), etc. The questions of the identification of the S_{Ad} phase with an S_{A1} phase or another S_{Ad}, and correlatively of the existence of transition lines between these phases, are important and have to be cleared up. Let us remark that investigations under a polarizing microscope are unable to distinguish the various S_A phases, except the two dimensional $S_{\tilde{A}}$ phase but only on free droplets.[4] Therefore careful calorimetric measurements together with x-ray diffraction studies are essential to clarify the matter. We report here our investigations upon binary mixtures of the "magic" DB$_5$ molecule ($N \leftrightarrow S_{A2}$) with the T_8 reentrant compound ($N \leftrightarrow S_{Ad} \leftrightarrow N \leftrightarrow S_{A1}$) (Ref. 8) (Fig. 2). The phase diagram of this binary system gives an idea of the great variety of the S_A-S_A transitions which could occur. We will describe successively the phase diagram and the x-ray pattern of each phase; after these descriptions we will give a discussion about the transition mechanism.

II. PHASE DIAGRAM

By contact method under the microscope we are able to distinguish three mesomorphic areas: the nematic one and two smectic-A areas, one S_A seems to exist for all concentrations below 130°–100 °C and above the crystallization (Fig. 3). Another smectic-A domain exists for 135 °C < T

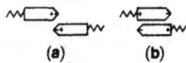

(a)   (b)

FIG. 1. Pair association with a more or less large overlap of the molecules.

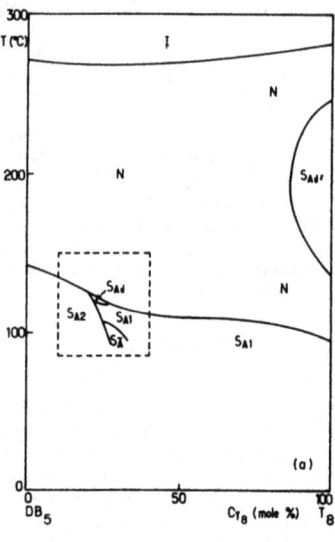

$$T_8 \begin{cases} C_8H_{17}O\text{-}\bigcirc\text{-}C\text{-}O\text{-}\bigcirc\text{-}CH=CH\text{-}\bigcirc\text{-}CN \\ \quad\quad\; \overset{\|}{O} \\ \underset{247\,°C}{\;}\;\;\underset{139\,°C}{\;}\;\;\underset{96\,°C}{\;} \\ N \longleftrightarrow S_{A4} \longleftrightarrow N \longleftrightarrow S_{A1} \end{cases}$$

$$DB_5 \begin{cases} C_9H_{11}\text{-}\bigcirc\text{-}O\text{-}C\text{-}\bigcirc\text{-}O\text{-}C\text{-}\bigcirc\text{-}CN \\ \quad\quad\;\; \overset{\|}{O}\quad\quad \overset{\|}{O} \\ \underset{139\,°C}{\;} \\ N \longleftrightarrow S_{A2} \end{cases}$$

FIG. 2. Longitudinal dipoles and transition temperatures for DB_5 and T_8 compounds.

$< 250\,°C$ and a concentration of T_8 more than 86 mole%. This domain is surrounded by a nematic zone and 86 mole% is the lower concentration limit of T_8 for the reentrant phenomenon. In the low-temperature smectic-A range, one expects to observe at least one transition line between the antiferroelectric S_{A2} phase of DB_5 and the "normal" S_{A1} phase of T_8, as in the binary diagram DB_5-TBBA.[3] Such a line can only be detected by calorimetric and x-ray investigations. We have performed enthalpic differential thermograms on several mixtures, and with x-ray measurements we have followed the evolution of the layer thickness. Only a comparison between the two methods allows us to establish the complete phase diagram (Fig. 3). Therefore S_A-S_A transitions take place for a concentration of T_8 greater than 20 mole% and smaller than 35 mole%. As a matter of fact, in this region three S_A-S_A transition lines are observed below the nematic-smectic line and two smectic-A triple points occur revealing four distinct smectic-A domains. The nature of each phase becomes clear if we refer to the x-ray patterns.

III. X-RAY PATTERNS

Samples are put in a Lindemann glass capillary and aligned in a 0.3-T magnetic field in the nematic range. The monochromatic $CuK\alpha$-x-ray beam reflected by a double bent pyrolitic graphite monochromator is perpendicular to the magnetic field and to the capillary axis. The diffracted light is collected on a photographic plate. Samples of various concentrations in T_8 have been studied (from 21 to 87 mole%). The large-angle region of the pattern consists of two broad crescents lying in a direction perpendicular to the director. The position and the broadness of these crescents remain rather invariant under temperature and concentration variations until the crystallization temperature is reached: The lateral liquidlike order remains apparently unchanged in the whole mesomorphic region. Therefore, our

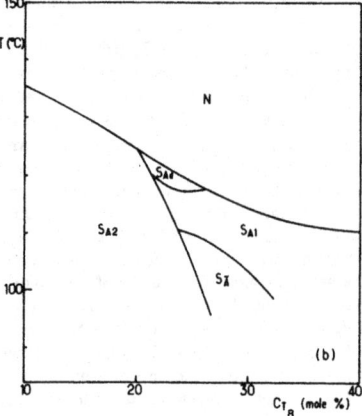

FIG. 3. (a) Binary isobaric diagram ($P = 1$ atm) between DB_5 (on left) and T_8 (on right). (b) Detail of the diagram around the $S_A - S_A$ phase boundaries.

interest will be focused on the small-angle part of the pattern; different examples are given on Fig. 4. Before a description of each phase, we will give the characteristics of the nematic patterns and define the mean molecular length for each mixture. For the small-angle region, the pattern is nearly symmetric with respect to the direct beam impact and our further description will concern only one side of the pattern.

A. Nematic phase and definition of the molecular length

Except for the domain of high concentration in T_8, the x-ray pattern of the nematic phase is similar at different concentrations. At low temperatures ($T < 150\,°C$), two diffuse maxima are seen along a direction parallel to the director [Fig. 4 (a)]. Far from the smectic-A phase, the two diffuse spots are similar and correspond to smectic-A fluctuations of very small size and of two different wave vectors. If we follow the position of the two diffuse maxima in the nematic phase from the pure DB_5 compound to the pure T_8, the ratio of the two wave vectors q_1/q_2 goes continuously

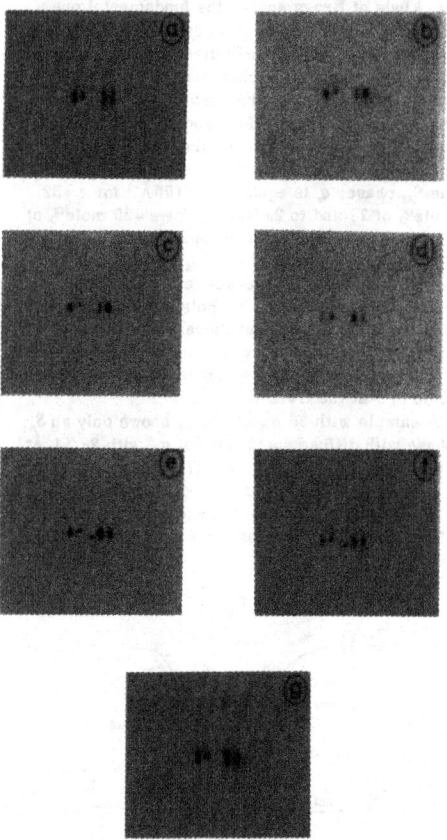

FIG. 4. X-ray diffraction photographs of some DB_5-T_8 mixtures: (a) $C_{T_8} = 31$ mole %, nematic phase, 116 °C; (b) $C_{T_8} = 31$ mole %, S_{A1}, 105 °C; (c) $C_{T_8} = 31$ mole %, $S_{\tilde{A}}$, 97 °C; (d) $C_{T_8} = 21$ mole %, S_{Ad}, 122 °C; (e) $C_{T_8} = 21$ mole %, S_{A2}, 116 °C; (f) $C_{T_8} = 23$ mole %, S_{Ad}, 120 °C; (g) $C_{T_8} = 23$ mole %, S_{A1}, 116 °C.

from 2 in the nematic phase in the pure DB_5 compound to a value of about 1.25 in the reentrant T_8 nematic phase. This evolution excludes the idea of an inhomogeneous mixture in which the local order of the pure components are seen simultaneously. Moreover, the existence of two wave-vector fluctuations have also been evidenced in pure compounds[6] such as DB_7. As we approach the S_A phase, one of these two spots becomes sharper and sharper: the outer one for a concentration of T_8 greater than 26 mole%, the inner one between 20 and 26 mole% T_8. The two spots condense together and simultaneously the two wave vectors become commensurable ($q_2 = 2q_1$) for a solution rich in DB_5 ($C_{T_8} < 20$ mole %). Whatever the position of the Bragg spots is in the smectic phase, the position of the outer spot (Bragg or diffuse) is independent of the temperature, while the position of the inner one, can vary in a large domain. The outer spot corresponds to a wave vector $q_2 = 2\pi/d_2$ with $26\,Å < d_2 < 31\,Å$ which is comparable to the molecular length of the two molecules (26 Å for DB_5, 31 Å for T_8). The concentration dependence of d_2 is not linear, nevertheless it is likely that d_2 is the wavelength of a density modulation corresponding to disordered single molecules. Therefore, we assume that d_2 corresponds to the apparent molecular length of the mixture. The other characteristic longitudinal wavelength will be compared to this d_2 value.

B. X-ray study of the T_8 rich samples

For a concentration of T_8 higher than 40 mole%, we observe only one smectic-A phase with a layer thickness equal to the apparent molecular length; two concentrations have been studied: 51 and 86 mole%. The inner diffuse line splits into two diffuse spots lying perpendicularly to the Z axis (parallel to the director). The corresponding wave vector in the Z direction is $2\pi/d_1$ with $40\,Å < d_1 < 44$–45 Å while the in-layer component $q_x = 2\pi/a$, $a \simeq 150$ Å. There is no temperature dependence on q_x while d_1 increases to 45 Å at low temperature for both samples. In fact, in patterns corresponding to high exposure times, we see a second-order Bragg spot lying at $2q_2 = 4\pi/d_2$ and below two diffuse spots with the same q_x and $2q_2 - q_x$ values as around the first-order Bragg spot. Therefore, in the smectic phases, diffuse spots are characteristic of a modulation of the density wave both in the z and x directions (x is perpendicular to z) and the period of the longitudinal modulation is equal to

$$\frac{1}{d_m} = \frac{1}{d_2} - \frac{1}{d_1} = \frac{q_x}{2\pi} .$$

At low temperature d_m varies between 70 and 80Å from 51 to 85 mole% of T_8 (2.5 to 2.8 layers). Similar modulations have been observed in T_8 (Ref. 8) and in other similar compounds in the low-temperature range of the smectic-A_1 phase.[6] For pure T_8, the period of the longitudinal modulation is 120Å ($\simeq 4$ layers). A model has already been proposed.[8]

C. X-ray diffraction of the DB₅ rich samples

Five different concentrations have been investigated between 20 and 36 mole% of T_8, in order to observe the different S_A phases. In this range of concentration the apparent molecular length (l) is equal to 26 ± 0.5Å and does not vary significantly. For concentrations of T_8 between 20 and 26 mole% the nematic phase transforms to a smectic-A phase by condensation of the fluctuations of the shorter wave vector q_1 and the layer thickness is equal to 47–48Å, i.e., 1.8 times the molecular length $(1.8l)$; the diffuse line at $q_2 = 2\pi/26$ Å^{-1} is still visible and a single order of reflexion is visible for the Bragg spot and for the diffuse line. Therefore, a fluctuation of period 26 Å seems to coexist in the sample besides the underlying layer modulation of 48 Å period. This smectic-A phase, which is an S_{Ad} phase [Fig. 4(d)] since the layer thickness is not commensurate with the apparent molecular length, is only stable over a few degrees. In a sample with 21 mole% of T_8, the S_{Ad} becomes smectic $A2$ [Fig. 4(e)] with a layer thickness of 52Å such that $q_2 = 2q_1$ and this kind of transition has been already observed in the pure compound DB₇.[6] For a sample with 23 mole% of T_8, the S_{Ad} phase [Fig 4(f)] transforms first to a smectic-A_1 phase [Fig. 4(g)]: The Bragg spot at $2\pi/48$ Å^{-1} becomes a diffuse line while the diffuse line lying at $2\pi/26$ Å^{-1} becomes a Bragg spot! During further cooling, the scattering vector for the diffuse line decreases and this line becomes a second Bragg spot leading to an S_{A2} transition. The transition S_{A1}-S_{A2} was observed in a DB₅-TBBA mixture, but we note that an S_{Ad}-S_{A1} transition is observed here for the first time. This transition exists until the concentration of T_8 reaches 26 mole%.

For a sample with 25 mole% T_8, the Bragg spot at $q_1 = 2\pi/48$ Å^{-1} splits into two diffuse spots out the z axis at the S_{Ad}-S_{A1} transition. These diffuse spots are precursors of a S_{A1}-$S_{\tilde{A}}$ transition, but in our x-ray experiments the crystallization of the sample occurs before the $S_{\tilde{A}}$ phase (i.e., the liquid antiphase) is reached. Differential scanning calorimetry (DSC) measurements confirm all these transitions (Fig. 5). In addition, the N-S_{Ad}, S_{A1}-S_{A2}, S_{A1}-$S_{\tilde{A}}$ transition enthalpies are weak

while the S_{Ad}-S_{A1} transition is clearly visible on DSC scans. Thus, this unexpected S_{Ad}-S_{A1} transition is a first order one in good agreement with the fact that it takes place without apparent symmetry change.

For concentration higher than 26 mole% the nematic phase [Fig 4(a)] transforms into an S_{A1} phase [Fig. 4(b)] and this S_{A1} phase transforms at a lower temperature into an $S_{\tilde{A}}$ phase [Fig. 4(c)] in which the dipoles undergo an antiferroelectric order along the director with π periodic changes of the phase of the antiferroelectric modulation.[4] We note that this N-S_{A1}-$S_{\tilde{A}}$ sequence was recently observed in a pure nitro compound.[8]

The diffraction pattern of the antiphase shows two kinds of Bragg spots: the fundamental one lies at $q_2 = 2\pi/l$ and the other lies at $q_{1s} = \pi/l$ along z and q_x in a perpendicular direction. As mentioned above the Bragg spots characteristic of the antiphase structure have precursor diffuse spots in the S_{A1} phase; these diffuse spots lie at $q_{1s} > \pi/l$ and q_x; q_x does not vary with the temperature through the S_{A1} and $S_{\tilde{A}}$ phase and we cannot reach the S_{A2} phase; q_x is equal to $2\pi/150$ Å^{-1} for $c = 32$ mole% of T_8 and to $2\pi/180$ Å^{-1} for $c = 25$ mole% of T_8. In Table I we give a schematic representation of the x-ray patterns of each S_A phase.

The variation of q_{1s} versus temperature is plotted in Fig. 6 for 23 and 32 mole% T_8 samples. The various S_A-S_A transitions appear to be of first order but the accuracy of our measurements of q is nearly of the same order of magnitude as the jump of q at the transition.

A sample with 36 mole% of T_8 shows only an S_{A1} phase with diffuse spots lying at q_{1s} with $2\pi/51$ Å^{-1} $< q_{1s} < 2\pi/45$ Å^{-1} while $q_2 = 2\pi/27$ Å^{-1}.

In fact, the double diffuse spots appear in the whole range of the S_{A1} phase except in the vicinity of the S_{A2} phase. With patterns of higher exposure

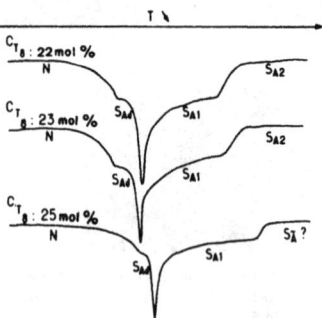

FIG. 5. Profile of DSC thermograms (Dupont 990) of some DB₅-T_8 mixtures.

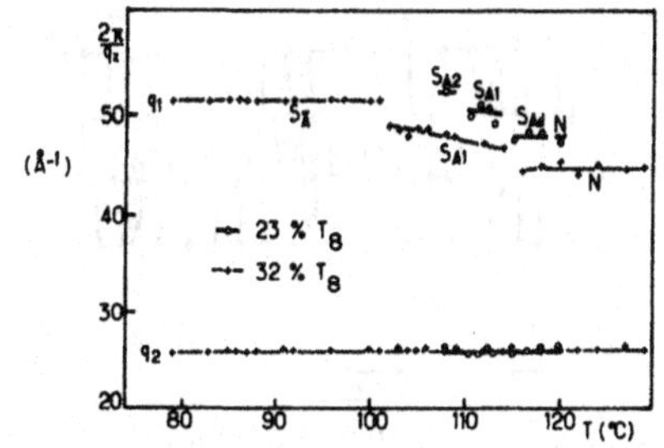

FIG. 6. Temperature dependence of the longitudinal modulations in two DB$_5$-T$_8$ mixtures, C_{T_8} = 23 mole % and C_{T_8} = 32 mole %.

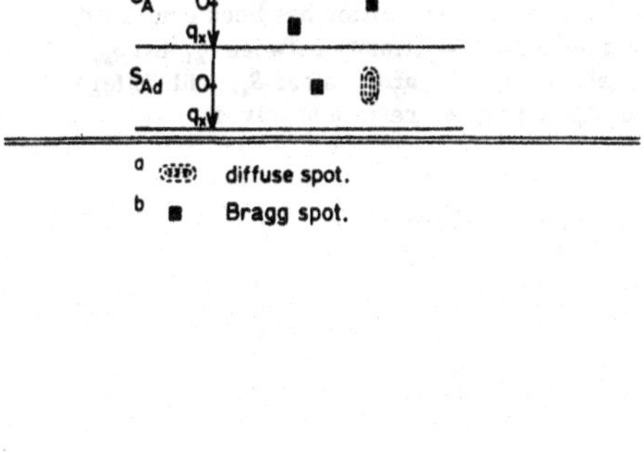

(Fig. 7). One expects a concentration between 36 and 50 mole% of T_8 to exist for which the period of the damped modulation does not vary with temperature. This concentration will be a kind of limit concentration for which the influence of DB and T_8 counterbalance each other.' Thus, as depicted on Fig. 7, the phase diagram is divided into two regions: the DB$_5$ rich region in which the head to head dipole array of DB$_5$ molecules governs the phase transitions and the T_8 rich region in which the molecular array is connected to the pure T_8 behavior. The "neutral" space between these two regions is less than a few mole%. A special limit role is devoted to the value $q = 2\pi$ 45 Å^{-1} and this wavelength seems also to act as a frontier.

From our x-ray and calorimetric studies, we are able to identify all the smectic-A areas of the phase diagram (Fig. 3). The reasons for having such a great variety of phases are not completely clear but we can try to give an idea from our established x-ray structures.

time we are able to point out at least two Bragg spots at q_2 and $2q_2$ and four diffuse spots at q_1, q_x (with $q_2 - q_x = q_1$) and $2q_2 - q_x$, q_x. The intensity of these second sets of diffuse spots is high compared to the intensity of the $2q_2$ Bragg spot and we exclude a mere multiple-scattering effect. We can assert that the diffuse spots do not come from independent fluctuations but originate in a modulated structure. We have two regimes for the thermal evolution of this modulation: For samples rich in T_8, the longitudinal period of the modulation d_m decreases from an upper limit of 200 Å towards 2.5 to 3 times the layer thickness. Indeed q_{1s} decreases with the temperature from $2\pi/35$ or $2\pi/40$ Å^{-1} towards a finite value of $2\pi/45$ Å^{-1}. On the contrary, for samples rich in DB$_5$, the modulation is equal to 2.5 times the molecular length at high temperatures and goes towards a limit value of two times the molecular length at low temperature

IV. DISCUSSION

In fact, our results show the complexity of the mixture DB$_5$-T$_8$ due to the opposite polar character of the two pure components. Let us recall that the terminal $C \equiv N$ group induces a strong longitudinal dipole on the molecule. The dipoles easily form pairs of antiparallel molecules and the properties of the mesophases are related to the length of such pairs. π bonds and other dipoles aligned in the same direction favor a delocalization of the dipole along the molecular core. In this case, the pair is not much greater than the length of a single molecule (e.g., T_8). On the contrary, the introduction of antiparallel dipoles in the core localizes the charges on the nitrile terminal group and favors a head to head antiparallel pair or at least a pair with a small overlap. Up to now, ex-

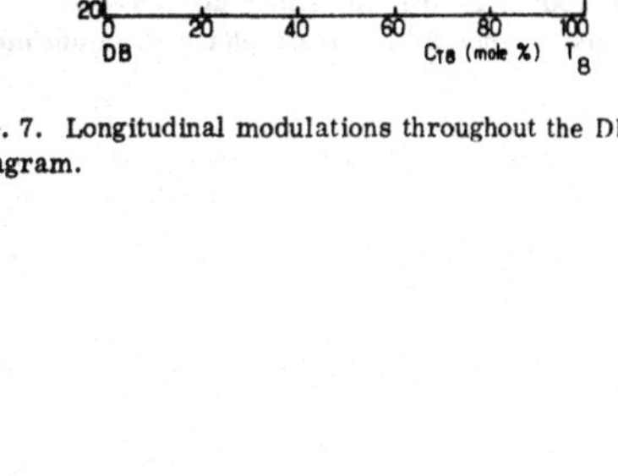

TABLE I. X-ray signatures of various S_A phases (note that the diffuse spots could be absent).

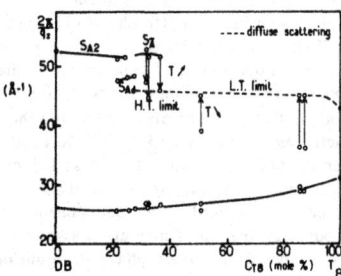

FIG. 7. Longitudinal modulations throughout the DB$_5$-T$_8$ diagram.

^{a} ⬚ diffuse spot.

^{b} ■ Bragg spot.

amples are only given by DB_5 and other nitrile
or nitrodibenzoate derivatives.[4,9]

We will exclude from our discussion the case of
molecules with cyclohexane rings for which the
area of the core is different from that of the chain.
In other rodlike compounds the two areas are
similar and a smectic-A phase corresponds to the
formation of a more or less defined planar surface of CH_3 terminal groups. The density through
the layer is nearly constant since the packing of
chains and of phenyl cores are equally dense (the
matching between the rather isotropic section of
the chains and the anisotropic phenyl one is en-
sured by the rotation of the phenyl group with a
local herring-bone array[10]). If we want to construct a smectic-A phase with antiparallel pairs
of polar molecules and if we want to keep the CH_3
planar surface, we obtain an S_{Ad} phase in which
the layer thickness is not a multiple of the molecular length. This kind of S_A phase is characterized by a rather dense molecular packing of the
molecular cores compared to the density of the chain packing. Moreover longitudinal fluctuations of the
center of mass are important and the second order
of Bragg reflection is extremely weak.[11] We can
assume that decreasing temperature or increasing
pressure favor a more uniform repartition of the
density leading to a decrease of the fluctuations.

When the overlap of two molecules of a dimer
is weak, the smectic-A fluctuations are damped
and low temperatures favor an antiferroelectric
S_{A2} phase. One can also have blocks of antiferroelectric layers with a periodic change of phase in
the layer. In this latter case, we suggest that the
periodicity in the layer plane is due to a compen-
sation between narrower aliphatic interfaces and
larger aromatic ones [Fig. 8(a)] and we obtain an
$S_{\tilde{A}}$ phase. A similar explanation was put forward
by Skoulios and Luzzati[12] in their description of
the ribbon structure of some soap mesophases.

When the layer thickness is near the molecular
length, a more uniform density is obtained at low
temperatures by the destruction of the CH_3 interfaces and a reentrant nematic phase can first
occur. Then, in the low temperature S_{A1} phase,
the competition between the "density wave" and
the "dipolar wave" can induce a modulation of the
density along the z axis in order to match the two
wavelengths as predicted by Prost.[13] Nevertheless such longitudinal modulations also induce
transverse modulations analogous to those seen
in the $S_{\tilde{A}}$ phase [Fig. 8(b)]. The correlation lengths
of the modulation remain finite and we have not
yet found an incommensurate phase in liquid meso-

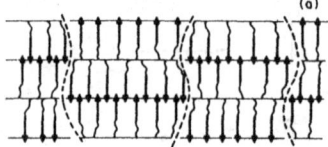

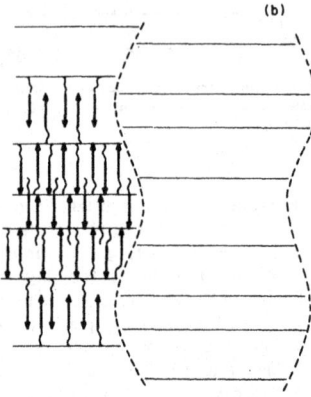

FIG. 8. Schematic representations of (a) $S_{\tilde{A}}$ antiphase, (b) low-temperature S_{A1} phase.

phases (although theoretically predicted[14]). In
mixtures $DB_5 - T_8$ (rich in T_8), we have no connection between the S_{Ad} and the S_{A1} domains, but such
a connection exists in another binary system including T_8 and it seems to be possible to go continuously from the S_{Ad} to S_{A1}.[15] Moreover, we have
not found a continuous variation of the layer thickness between once and twice the molecular length.
It is likely that in the case of cyano derivatives
with only two phenyl rings such as 8OCB or CBOOA
the dimer length ($1.4l$) is too far from a multiple
of the molecular length and a dense smectic-A
phase cannot be obtained, the increase of compactness favors only the nematic phase.

To conclude, we have to keep in mind that the
dimer is an image to describe an array in which
aromatic cores tend to overlap more or less.
Therefore the temperature dependence of the layer
thickness is complex and not easy to interpret.
The first S_{Ad}-S_{A1} transition has been evidenced,
but questions of continuity between S_{A1} and S_{Ad}
or between one S_{Ad} and another S_{Ad} with different
overlapping values remain unsolved.

[1]G. Sigaud, F. Hardouin, M. F. Achard, and H. Gasparoux, Proceedings of the Seventh International Liquid Crystal Conference, Bordeaux, 1978 [J. Phys. (Paris) 40, C3-356 (1979)].

[2]F. Hardouin, G. Sigaud, M. F. Achard, and H. Gasparoux, Phys. Lett. 71A, 347 (1979); Solid State Commun. 30, 265 (1979).

[3]F. Hardouin, A. M. Levelut, J. J. Benattar, and G. Sigaud, Solid State Commun. 33, 337 (1979).

[4]G. Sigaud, F. Hardouin, M. F. Achard, and A. M. Levelut, J. Phys. (Paris) 42, 107 (1981).

[5]P. E. Cladis, R. K. Bogardus, and D. Aadsen, Phys. Rev. A 18, 2292 (1978).

[6]F. Hardouin, A. M. Levelut, and G. Sigaud, J. Phys. (Paris) 42, 71 (1981).

[7]A. J. Leadbetter, J. C. Frost, J. P. Gaughan, G. W. Gray, and A. Mosley, J. Phys. 40, 375 (1979).

[8]F. Hardouin and A. M. Levelut, J. Phys. (Paris) 41, 41 (1980).

[9]F. Hardouin, G. Sigaud, Nguyen Huu Tinh, and M. F. Achard, J. Phys. Lett. (Paris) 42, L63 (1981).

[10]A. M. Levelut, J. Phys. (Paris) 37, C3-51 (1976).

[11]J. Als-Nielsen, R. J. Birgeneau, M. C. Kaplan, J. D. Lister, and C. Safinya, Phys. Rev. Lett. 39, 352 (1977).

[12]A. Skoulios and V. Luzzati, Acta Crystallogr. 14, 278 (1961).

[13]J. Prost, Liquid Crystal of One and Two Dimensional Order, edited by H. W. Helfrich and G. Hepke (Springer, Berlin, 1980), p. 125.

[14]P. Barois, C. Coulon, and J. Prost, J. Phys. Lett. 42, L107 (1981).

[15]Nguyen Huu Tinh, G. Sigaud, H. Gasparoux, and F. Hardouin, Proceedings of the Third Liquid Crystal Conference of the Socialist Countries, Budapest, 1979 (in press).

ADVANCES IN PHYSICS. 1984. VOL. 33, No. 1, 1 46

The smectic state

By J. PROST

Centre Paul Pascal, C.N.R.S., Domaine Universitaire,
33405 Talence, France

[Received 30 May 1984]

Abstract

The smectic state has been known for over a century but is currently being very actively investigated. We review the polymorphism associated with one-dimensional ordering, critical phenomena characteristic of the several phase transitions encountered in these systems, together with their very special dynamical properties.

Contents

	PAGE
1. Introduction	1
2. Smectic polymorphism and structures	7
2.1 Uniaxial phases	7
2.2. Biaxial phases	11
3. Phase transitions	14
3.1. Nematic–smectic A transition	14
3.2. Other phase transitions involving smectic phases	20
4. Frustrated smectics	26
5. Hydrodynamic properties	35
5.1. Conventional approach	35
5.2. Breakdown of conventional hydrodynamics	37
6. Conclusion	42
Acknowledgments	42
References	42

1. Introduction

Insert a few milligrammes of TBPA (i.e. terephthal-*bis*-pentylaniline $C_5H_{11}-\phi-N=CH-\phi-CH=N-\phi-C_4H_9$) between glass slides, and observe the preparation under a polarizing microscope as the temperature is changed. Above $232°C$ the sample will be totally isotropic, perfectly black between crossed polarizers, as for an ordinary liquid such as water or oil. But at temperatures just below $232°C$, and in the absence of any particular treatment of the glass slides, a bright threaded texture appears (figure 1). This is characteristic of the currently best-known liquid crystal state: the nematic. Such a phase is clearly liquid since it flows almost as well as water, but the strong birefringence revealed by the microscope observation thows the existence of orientational order. What is happening is that the elongated molecules are gliding freely over each other but stay parallel on the average (figure 2). Rotational symmetry has been broken whereas

2   *J. Prost*

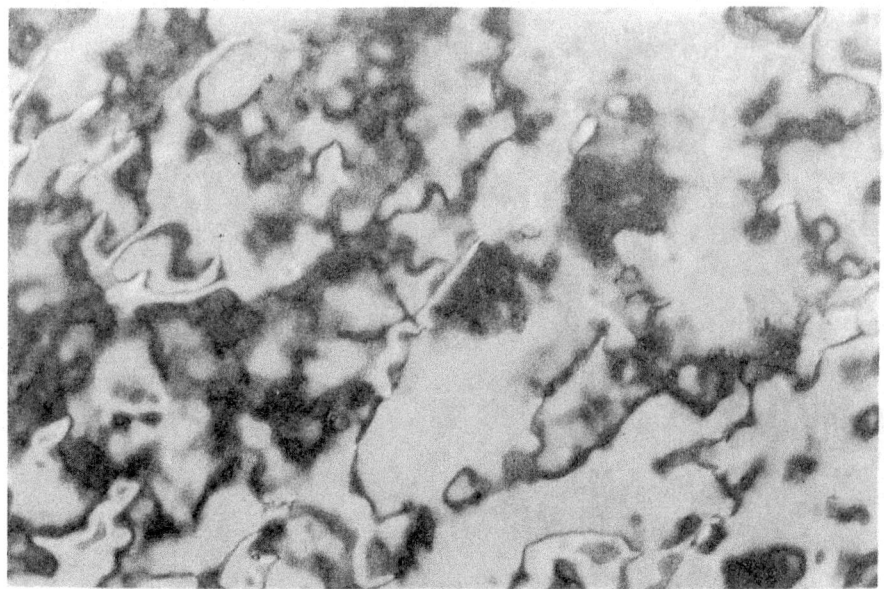

Figure 1.   Typical nematic texture as seen with a polarizing microscope.

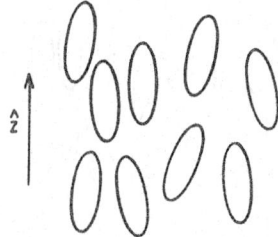

Figure 2.   Schematic representation of the nematic order: the long axes of the molecules are
on the average parallel to a macroscopic D_∞ symmetry axis, $\hat{z}$.

translational symmetry has been kept. A closer look at the sample shows that the
preferred direction varies from place to place; domains due to different boundary
conditions are separatec by thread-like lines, whence the name nematic. Some areas
remain black under the crossed polarizers, for any rotation angle of the stage: this
reveals a D_∞ symmetry axis perpendicular to the glass slide.

   If you cool the sample further its texture will again abruptly change, at 213·5°C.
The *threads* disappear and give way to the so-called focal conic texture (figure 3).
One sees arrays of ellipse/hyperbola couples, arranged in such a way that each curve
goes through one of the focal points of the other. These features were early
recognized as due to the existence of layers of constant thickness which could freely
glide over each other (for a review of the early work see Friedel 1922, Bragg 1934).
Indeed such ellipse/hyperbola couples correspond to the focal lines of Dupin cyclide
families, which are the mathematical objects that one gets by piling up surfaces at a
constant distance from each other. The regions which were previously black stay

Figure 3.   Focal conic smectic A texture as seen with a polarizing microscope.

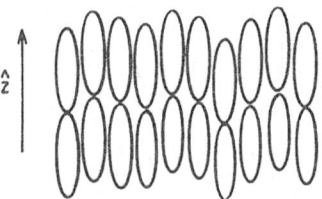

Figure 4.   Schematic representation of the smectic A order.

black between the crossed polarizers for any rotation of the stage which tells one that the molecules are perpendicular to the layers on the average (figure 4). The fact that the layers glide freely over each other indicates the absence of any genuine long-range translational order within the layers. Such a phase, called smectic A, is thus characterized by the breaking of symmetry in one direction of space.

This will be our general definition of the smectic state, i.e. a one-dimensionally ordered state. In the S_A case, the translational symmetry is broken along the optical axis. This need not always be the case. As a matter of fact if you keep on cooling the TBPA sample, below 182·5°C the texture changes again; the focal conics still exist, but the observation of biaxiality reveals that the optical axis is now tilted with respect to the layers (figure 6). Areas which exhibited a constant colour in the S_A phase show two types of domains in very thin samples. The regions which were dark in the S_A phase now exhibit a birefringent *schlieren texture* which in many ways resembles that obtained with nematics (figure 5). There is however a fundamental difference: half-integer disclinations (Kleman 1977) (i.e. in which the mean molecular direction, or its projection on the plane perpendicular to the disclination line,

4   *J. Prost*

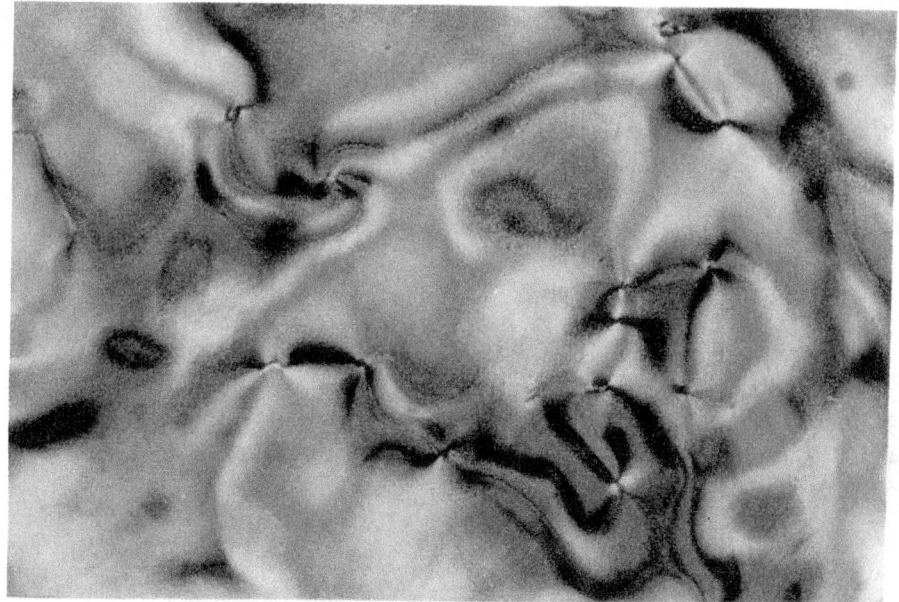

(a)

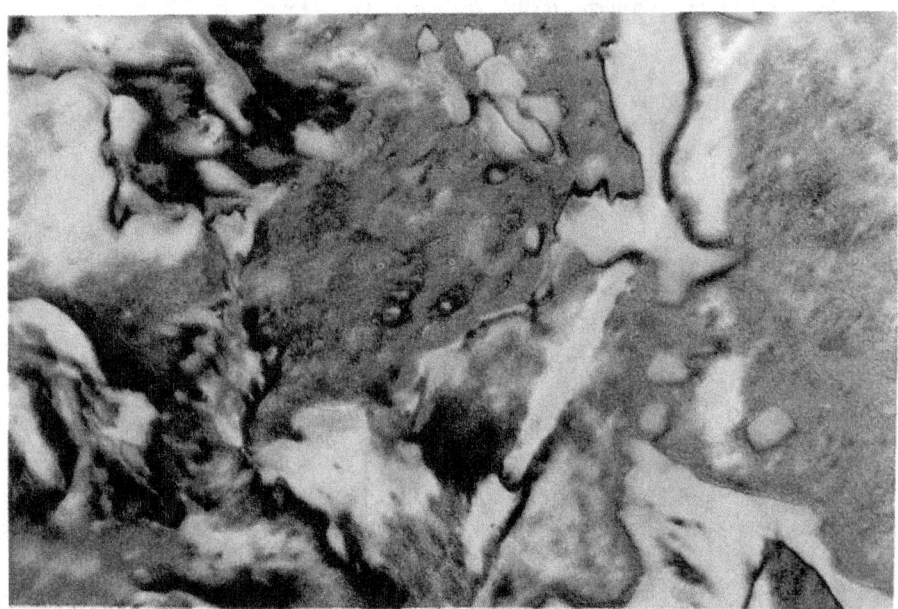

(b)

(*c*)

Figure 5. (*a*) Typical smectic C *schlieren* texture (smectic planes parallel to the microscope slides). (*b*) Smectic F *schlieren* texture. (*c*) Smectic G texture.

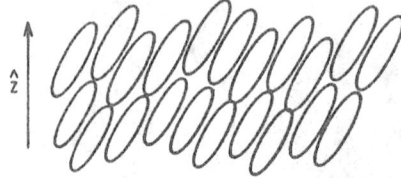

Figure 6. Schematic representation of the smectic C order.

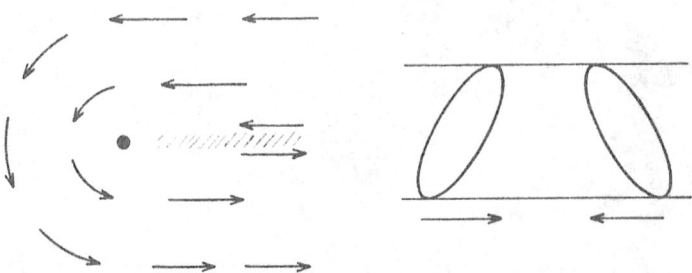

Figure 7. Topology of a *half-integer* disclination line. In nematics the sign of the arrows (which represent the local averaged molecular direction) has no physical meaning: such defects are topologically stable. In smectic C, as shown on the right side of the figure, the sign of the arrow indicates the tilt direction, and such defects are not topologically stable (i.e., the shaded area which is undistorted in a nematic, represents a splay-bent wall in a smectic C).

undergoes half a turn in any path encircling this line, see figure 7) are never observed in tilted smectic textures, since a π turn of the in-plane molecular direction does not leave the system invariant (figure 7). This phase, which is characterized by the C_{2v} point group symmetry plus one-dimensional translational order, is called smectic C (S_C).

The story does not stop here. Decrease the temperature below 153·5°C and a faint front sweeps through the sample suggesting another phase transition. In many cases the texture (figure 5(*b*)) resembles pretty much that of S_C. However the thermal fluctuations of the optical axis direction which were observable with the naked eye in both the nematic and smectic C phases, progressively vanish. Microscopic observations do not allow one to establish a clear-cut difference between this phase, called smectic F, and smectic C. The macroscopic symmetry has to be the same. Below 144°C the texture changes yet again, this time to a crystalline phase (three-dimensional translational order) which is called smectic G, a misnomer (see figure 5(*c*)).

Although a great deal of information can be obtained from simple observations under a polarizing microscope, we were unable to point to a clear difference between the S_C and S_F phases on this basis. One needs more experimental and conceptual tools to pin down the characteristics of these phases.

In §2 we give a concise description of the currently known smectic phases. The concepts which will allow a natural understanding of this polymorphism will be that of long-range, quasi-long-range and short-range order. As a matter of fact we will see that genuine long-range order never exists (Landau 1937, Peierls 1935) but that the new concept of bond orientational order introduced in two-dimensional melting (Halperin and Nelson 1978) is very useful. We will consider in another section the polymorphism arising from the *frustration* of smectics and the ferroelectric (chiral) smectic C (S_{C^*}). Our description of the smectic state excludes S_B, S_D, S_E, S_G and S_H phases, which all exhibit three-dimensional translational long-range order, and as such should be considered as crystals (Levelut *et al.* 1974, De Vries 1974, Doucet *et al.* 1978, Leadbetter *et al.* 1979 a, Pershan *et al.* 1981, Moncton and Pindak 1979).

The diversity of the smectic polymorphism allows the study of many different *phase transitions*. It is a very good field for testing universality concepts, but the absence of genuine long-range order may play a subtle role. We shall not describe all these transitions in detail but will concentrate on the NS_A transition in view of its historical and conceptual importance. An important parameter in modern theory is the dimension of the space of interaction: we will also briefly discuss theories and experiments concerned with two-dimensional systems. Indeed, the layered structure of smectics allows the stabilization of free-standing films from two to several hundred layers thick (Friedel 1922, Young *et al.* 1978). This provides the experimentalist with a powerful tool for varying the dimension from 3 to 2.

In our observation of the TBPA sample under the microscope, each phase transition obtained by lowering the temperature reduced the symmetry. This is often the case but does not need to be always true. If we had been looking at a sample of T8, i.e.

$$C_8H_{17}\text{-}O \ \phi \ OOC \ \phi \ CH=CH \cdot \phi \ CN$$

we would have observed the following sequence upon decreasing the temperature (Hardouin *et al.* 1979):

$$I \rightarrow N \rightarrow S_A \rightarrow N \rightarrow S_A \rightarrow \text{crystal.}$$

The even more surprising sequence would have been observed (Nguyen Huu Tinh *et al.* 1982)

$$I \rightarrow N \rightarrow S_A \rightarrow N \rightarrow S_A \rightarrow N \rightarrow S_A \rightarrow S_C \rightarrow S_C \rightarrow \text{crystal}$$

if we had been looking at a sample of DB90 NO_2, i.e.

$$C_9H_{19}O-\phi-OOC-\phi-OOC-\phi-NO_2.$$

Thus high-symmetry phases may *re-enter* at lower temperatures than low-symmetry phases in rather unexpected ways (Cladis 1975, Cladis *et al.* 1977). We will show in §4 that taking into account the existence of two fundamental lengths in the smectic problem allows one to understand most of the observed new polymorphism arising with highly polar molecules (Prost 1979, Prost and Barois 1983).

In the final Section we review the hydrodynamic properties of smectic systems. They have been extensively studied over the past 15 years both theoretically and experimentally (for a review see, de Gennes 1974a, Gasparoux and Prost 1976, Chandrasekhar 1977). However all these studies relied on the assumption that the absence of genuine long-range order would not invalidate the conventional *hydrodynamic* approach. More recently it has been shown (Grinstein and Pelcovits 1981, Mazenko *et al.* 1982) that the conventional concepts of elastic constants and viscosities breakdown in all one-dimensional ordered systems. Long wavelength fluctuations lead to elastic constants depending logarithmically on wavevector q (at small q) and viscosities diverging like the inverse frequency ω^{-1} (at small ω)!

2. Smectic polymorphisms and structures
2.1. *Uniaxial phases*

2.1.1. S_A

The simplest example is indeed the smectic A phase. One may think of it as a stack of two-dimensional liquid layers: the X-ray pattern reveals 00*l* sharp peaks as expected (quite often only 001 is observable), and a diffuse outer ring (isotropic when the incoming X-rays are perpendicular to the layers) showing that the in-plane correlations are indeed liquid-like, that is

$$\langle \rho(\mathbf{r})\rho(\mathbf{r}')\rangle \sim \exp\left(-|\mathbf{r}-\mathbf{r}'|/\xi\right) + \text{constant},$$

provided $\mathbf{r}$ and $\mathbf{r}'$ are taken in the same layer. ξ is found to be of the order of a few molecular diameters. For quite a long time the 00*l* spots have been considered as Bragg peaks: this is in fact not possible as had been recognized long ago by Landau and Peierls in the following way (Peierls 1935, Landau 1937).

Consider the layer displacement u, taken along the optic axis z. The excess free energy of a slightly deformed state can be written

$$F = \int_v \frac{1}{2}\left(B\left(\frac{\partial u}{\partial z}\right)^2 + K(\Delta_\perp u)^2\right) dv \tag{1}$$

$\Delta_\perp$ is the Laplacian in the xy plane; v is the sample volume; B is the compressional elastic constant, and has the dimensions of an energy per unit volume: it is the only first-order (or solid-like) term. K is a second-order (nematic-like) elastic constant which describes how much energy is required to bend the layers. It has the dimensions of an energy per unit length so that $\lambda = \sqrt{(K/B)}$ is a length (typically of the order of a molecular length). Note that there is no term involving the gradient of u in the xy plane corresponding to mere rotation of the layers.

In Fourier space

$$F = \sum_q \tfrac{1}{2}(Bq_z^2 + Kq_\perp^4)|\hat{u}^2(\mathbf{q})| \tag{2}$$

thus from the equipartition theorem

$$\langle |u^2(q)| \rangle = \frac{k_B T}{v(Bq_z^2 + Kq_\perp^4)} \tag{3}$$

and

$$\langle u^2(r) \rangle = \frac{k_B T}{(2\pi)^2 B} \int_{q_\perp} \int_{q_z} \frac{q_\perp \, dq_\perp \, dq_z}{q_z^2 + \lambda q_\perp^4} \tag{4}$$

$$\simeq \frac{k_B T}{2\pi (BK)^{1/2}} \ln (L/d) \tag{5}$$

in which L is the sample size ($r \sim L^3$) and d a molecular dimension. Thus the mean square displacement diverges with the size of the sample.

Within this level of approximation, the influence on the X-ray structure factor is fairly easily calculated (Caille 1972). It is given by the Fourier transform of the correlation function

$$\langle \exp[iq_0(u(\mathbf{r}) - u(0))] \rangle = \exp[-\tfrac{1}{2}q_0^2 \langle (u(\mathbf{r}) - u(0))^2 \rangle] \tag{6}$$

($q_0 = 2\pi/l$, the wavevector corresponding to the layer thickness l). Unlike $\langle u^2(\mathbf{r}) \rangle$ this quantity converges

$$\langle \exp[iq_0(u(\mathbf{r}) - u(0))] \rangle \propto \begin{cases} \left(\dfrac{d}{z}\right)^\eta, & \mathbf{r} = (0, 0, z), \\[2ex] \left(\dfrac{\lambda d}{r_\perp^2}\right)^\eta, & \mathbf{r} = \mathbf{r}_\perp = (x, y, 0), \end{cases} \tag{7}$$

with $\eta = q_0^2 kT/8\pi\sqrt{(KB)}$ and $d \ll |r| \ll L$. Thus, the X-ray peaks will not be delta functions

$$I(q) \begin{cases} \dfrac{1}{(q_z - q_0)^{2-\eta}}, & q_\perp = 0, \\[2ex] \dfrac{1}{q_\perp^{4-2\eta}}, & q_z = 0. \end{cases} \tag{8}$$

Experiments do confirm these expectations and even allow accurate measurements of η (Als Nielsen *et al.* 1977; for a review see Als Nielsen 1981). This absence of true long-range order has far-reaching consequences which are only just beginning to be understood. It results from the simultaneous existence in the free energy of a solid-like elastic term for wavevectors perpendicular to the layers and a nematic one for wavevectors parallel to them. Thus all smectic phases defined as one-dimensionally ordered will exhibit this feature. They are said to be at their lower marginal dimensionality in three dimensions (i.e. precisely that dimensionality below which long-range order no longer exists).

2.1.2. *Uniaxial hexatic phase S_{BH}*

Optical uniaxiality does not necessarily prove the smectic A character of the phase. Any n-fold symmetry axis with n larger than two would lead to the same result. One could thus be inclined to look for one-dimensional ordered phases having point group symmetries such as D_{4h}, D_{6h}, etc. To understand that it is indeed possible, it is useful to recall briefly some of the main results of melting theories in two dimensions (Kosterlitz and Thouless 1973, Halperin and Nelson 1978, Nelson and Halperin 1979). Let us for example consider the case of a triangular lattice. The energy excess associated with the introduction of one dislocation, of Burgers vector $b = a$ (the lattice parameter) can be written (Friedel 1964, Nabarro 1967)

$$W = a^2 (K/8\pi) \ln (L/a),\qquad(9)$$

with

$$K = \frac{4\mu\lambda}{2\mu + \lambda},\qquad(10)$$

μ and λ being Lamé coefficients. It is calculated from the elastic free energy

$$\delta F = \tfrac{1}{2} \int d^2 r [2\mu u_{ij}^2 + \lambda u_{ii}^2].\qquad(11)$$

(The summation convention is assumed and $u_{ij} = \tfrac{1}{2}(\nabla_i u_j + \nabla_j u_i)$ is the strain tensor.)

The logarithmic divergence of this excess energy with the size of the sample is a direct consequence of the bidimensionality of the system.

The free energy associated with the introduction of this extra dislocation is then

$$F = W - TS = [(a^2 K/8\pi) - 2kT] \ln (L/a).\qquad(12)$$

Thus below a critical temperature

$$T_{c1} = a^2 K/16\pi k\qquad(13)$$

the introduction of dislocations costs an infinite amount of energy, and the two-dimensional solid is stable with respect to that kind of excitation; above T_c the entropy term dominates, the system emits free dislocations spontaneously. The system can no longer sustain a shear, since the stress will be automatically relaxed by dislocation motion: it is now in a liquid state. If no strongly first-order transition occurs before, this process will provide a mechanism for melting. Renormalization-group analysis gives the same critical temperature, but provides a qualitatively new result, i.e. if the dislocation unbinding destroys the positional order, it does not destroy the bond order

but
$$\left.\begin{array}{l} \langle \rho(r)\rho(0)\rangle \sim \exp(-r/\xi), \quad (T > T_c), \\ \langle \psi_6^*(r)\psi_6(0)\rangle \sim (1/r)^{\eta_\psi}, \quad \text{with} \quad \psi_6(r) = \exp(6i\phi_6) \end{array}\right\}\qquad(14)$$

in which ϕ_6 is the angle of a bond with a reference axis (ϕ_6 is related to the displacement field by $\phi_6 = \tfrac{1}{2}\nabla \times \mathbf{u}$).

Equations (14) characterize a new phase, called the two-dimensional hexatic phase, which has been observed in free-standing films (Dawey *et al.* 1984).

The algebraic power-law decay of the bond order parameter ψ has the same origin as the one discussed for the displacement of layers in smectics

$$\delta F = \tfrac{1}{2} \int K_6 |\nabla \phi_6|^2 \, d^2 r\qquad(15)$$

10   *J. Prost*

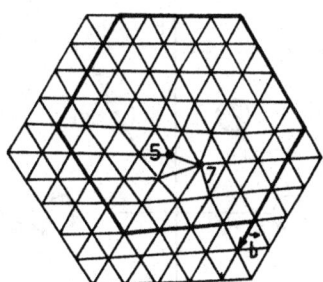

Figure 8.   Dislocation in a triangular lattice: note the very small perturbation of the bond
orientations.

so that in two dimensions

$$\langle \phi_6^2 \rangle \sim \ln (L/a) \tag{16}$$

and $\eta_\psi = 18kT/\pi K_A$. The angular elastic constant K_A has the same nature as the
second-order elastic constant K introduced in (1). The reason for destroying
positional correlations before the bond order is hinted at in figure 8; although a
dislocation dephases positional order very efficiently, it perturbs very little the bond
directions.

On the other hand, the elastic energies (11) and (15) have basically the same
structure, so that a defect-mediated orientational melting is expected at T_{c2}
$= (\pi K_A/72k)$. (Topological defects in that case are disclinations which are lines in
three dimensions and points in two dimensions.) Above T_{c2} a conventional liquid is
recovered; all correlation functions decay exponentially.

How can one adapt these results for the three-dimensional smectic case? One-
dimensional order implies that the layers are uncorrelated and one may hope to
obtain a stack of hexatic layers. Even if positional correlations from layer to layer
are absent, angular ones will always be present (because of anisotropic van der Waals
forces, for instance). As a consequence one can define a three-dimensional bond-
order parameter, and the corresponding free energy will be

$$\delta F = \tfrac{1}{2} \int K_A (\nabla \phi_6)^2 \, d^3r, \tag{17}$$

$$\langle \phi_6^2 \rangle = \frac{kT}{2\pi^2 K_A} \int_0^{q_c} \frac{q^2 \, dq}{q^2} = \frac{kT q_c}{2\pi^2 K_A}. \tag{18}$$

The mean square of the bond angle stays finite (no long wavelength divergence) and
thus an hexatic smectic exhibits conventional long-range bond order (Birgeneau and
Litster 1978). It is characterized by the D_{6h} point group symmetry.

Such a phase has been observed first in a thick film of n-hexyl-4-n
pentyloxybiphenyl-4-carboxylate (650 BC) (Pindack *et al.* 1981) (complete decoupling of the layers had been noticed by Leadbetter *et al.* 1979 a). The isotropic diffuse
outer ring of the smectic A (for incoming X-rays parallel to the optical axis) is
replaced by a six-fold modulated pattern (figure 9). The existence of this six-fold
symmetry confirms the existence of long-range bond order. The width of the diffuse
spots reveals an in-plane positional correlation length of about 100 Å. Locally, the
molecules are distributed on a triangular network but the number of defects is such
that the order does not propagate over distances larger than 100 Å.

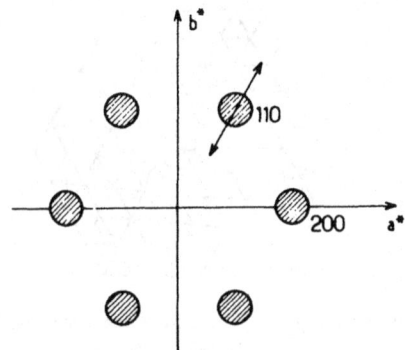

Figure 9. Typical X-ray pattern (incident beam perpendicular to the planes) revealing short-range positional and long-range bond orders, in a hexatic smectic B.

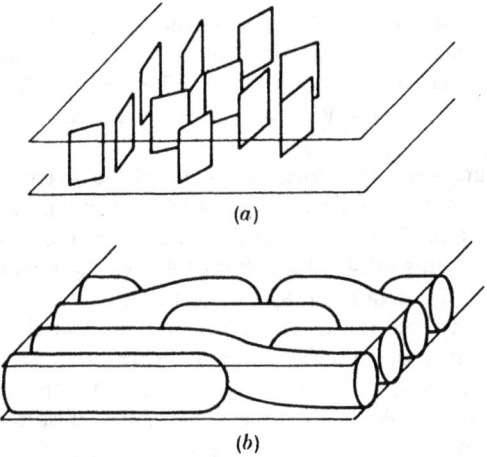

(a)

(b)

Figure 10. Schematic representation of a smectic phase characterized by D_{2h} symmetry. (a) Made of rectangular platelets. (b) Made of a partially melted columnar phase.

On decreasing temperature one obtains the sequence

$$I \xrightarrow{\;85°C\;} S_A \xrightarrow{\;68°C\;} \text{smectic hexatic} \xrightarrow{\;60°C\;} \text{crystal.}$$

Corresponding tetratic phases with D_{4h} symmetry have not yet been observed.

2.2. *Biaxial phases*

The simplest biaxial smectic phase would be characterized by a D_{2h} point-group symmetry, and the X-ray diffuse outer ring (again with incoming beam perpendicular to the layers) two-fold modulated. This would be the case if the molecules were more like rectangular platelets than cylinders (figure 10). For instance if the long sides of the rectangles contribute to the smectic layering the short sides could exhibit nematic like ordering. Another way of getting a smectic with D_{2h} symmetry could be by partial melting of a rectangular columnar phase: columnar phases are two-dimensionally ordered systems, liquid-like in a third direction (Chandrashekar 1977,

Destrade *et al.* 1983, Levelut 1983). Partial melting would indeed give one-dimensional order, and just as with hexatic phases, the column direction could well be kept, and give the D_{2h} symmetry. This phase has not been observed yet but recent observations (Davidson *et al.* 1983) on the charge transfer salt below correspond fairly closely to this picture (biaxiality has not yet been demonstrated).

$$A = \phi - OC_{12}H_{25}$$

Next to D_{2h} is C_{2h}. Phases corresponding to this symmetry do exist. The simplest is the smectic C already described in the Introduction (figure 6). Molecules are tilted within layers in which they assume a perfect liquid-like order. The projection of the molecular direction on the smectic planes constitutes a nematic-like variable which exhibits true long-range order, for the same reason as explained for the existence of three-dimensional bond ordering. An interesting complication is obtained when one considers optically active molecules (Meyer *et al.* 1975); the C_{2h} drops to a C_2 symmetry group and the only element left is a C_2 axis perpendicular to the molecular-tilt direction, and to the normal to the layers (figure 11). From the old Curie principle (Curie 1908) one can infer that there should be a macroscopic polarization along this C_2 axis- indeed no symmetry operation can suppress it. Chiral smectic C (S_{C*}) phases are thus ferroelectric. In fact, for the same reason, the elastic free energy for the molecular direction involves a Lifshitz invariant: in S_{C*} there is a helical precession of the tilt angle along the direction perpendicular to the layers. S_{C*} are thus helielectrics and on a macroscopic scale the D_∞ symmetry is recovered! Experiments do confirm these expectations; the average polarization corresponds to a few per cent of what one would expect if all transverse dipoles were aligned (Pieransky *et al.* 1975, Young *et al.* 1978, Martinot Lagarde *et al.* 1981, Beresnev and Blinov 1981). An interesting geometry is obtained in samples about $1\,\mu m$ thick, between glass slides treated for parallel alignment (figure 12) (Clark and Lagerwall 1980). Under these conditions the layers are perpendicular to the slides, and the helical precession of the tilt cannot develop. The polarization is uniform but can take either of two equivalent oppositely directed orientations, and one can switch from one to the other with an external electric field.

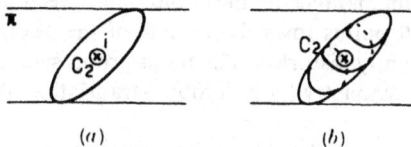

(*a*)   (*b*)

Figure 11.   (*a*) Symmetry elements of the C_{2h} point group characteristic of smectic C. π = symmetry plane perpendicular to the layers and parallel to the tilt; i = inversion point; C_2 two-fold axis perpendicular to π. (*b*) Introducing chirality suppresses the inversion point and the plane of symmetry. Only the C_2 is left.

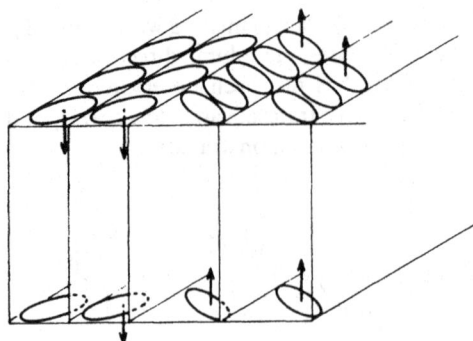

Figure 12.   Typical structure obtained with a thin smectic C* slab the boundaries of which impose parallel alignment: note that for a given orientation of the layers, two tilts are possible: since they correspond to opposite polarizations, they can be switched at will, with an external electric field.

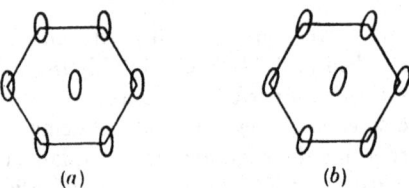

(a)   (b)

Figure 13.   (a) Short-range in-plane structure of a smectic F. (b) Short-range in-plane structure of a smectic I. (The ellipsoids represent the tilt direction.)

The tilt angle changes from $+\theta$ to $-\theta$ from one polarization to the other, and thus the coupling to light is excellent. This process could lead to switch times of about a microsecond which, though slow compared to solid-state devices, would be fast compared to most other liquid-crystal displays.

It seems natural that, if one can identify phases with in-plane short-range positional order and long-range bond order when molecules are perpendicular to the layers, the same should be true when the molecules are tilted. This is indeed the case. The S_F phase of TBPA which we could hardly distinguish from the S_C is such an example (Leadbetter *et al.* 1979 b, Benattar *et al.* 1979). As a matter of fact, S_F phases were synthesized before hexatic smectics (Demus *et al.* 1971). In a given layer molecules are distributed on a nearly triangular lattice and this order extends over a few hundred Ångströms as in the uniaxial hexatic case, but the bond ordering exhibits long-range order. X-ray patterns of single domain samples obtained by melting single crystals show six-fold symmetry (or *nearly* six-fold) in the diffuse outer ring (incoming X-ray beam parallel to the molecular direction). A more refined analysis shows that the tilt points towards the side of the hexagons (figure 13 (a)), which are consequently slightly distorted. The point-group symmetry is C_{2h}, as with S_C. The total macroscopic symmetry (point group + translational) is thus identical to that of smectic C.

If, instead of pointing towards the side of the hexagons, the tilt points toward the vertices (figure 13 (b)) one obtains a new phase, the S_I (Richter 1980, Benattar *et al.* 1981). The *nearly* six-fold outer ring appears much sharper than that of the hexatic phase which implies a much larger range of correlation for the in-plane translational

14   *J. Prost*

order. The line shape of the diffuse spot fits better the power law decay of a two-dimensional solid than the Lorentzian of a tilted hexatic phase (Benattar) *et al.* 1981, 1983). This might be an indication that there occurs in these systems melting mediated by grain boundaries, as suggested by computer simulations for two-dimensional solids.

3. Phase transitions

As stressed in the Introduction, one of the interesting features of liquid crystals is that they provide good model systems for testing universality concepts for phase transitions (Wilson and Kogut 1974, Ma 1976). Of course the first transition which comes in mind is that from nematic to smectic A. How can one describe the onset of one-dimensional order? Much of the following development is inspired from a recent paper by Lubensky (1983).

3.1. *Nematic–smectic A transition*

Over the past 10 years this phase transition has been very actively investigated but it is not yet possible to consider that we have a fully satisfactory understanding of it. The usual approach to describing a phase transformation is to define a suitable order parameter, to write down the Landau–Ginzburg free energy, and to try to solve the corresponding statistical problem. If it falls into an already known universality class, one is lucky and can draw on analogies with totally different systems. This procedure is valid provided the transition is continuous or nearly so. Microscopic models (Kobayashi 1970, MacMillan 1971) suggested that the nematic–smectic A transition could be second order. Symmetry leads to the same conclusion. How can one describe the appearance of layers? The existence of these layers implies that in S_A the mass density ρ is a periodic function in the direction z of the optical axis. If one expands ρ in Fourier series

$$\rho = \rho_0 + \frac{1}{\sqrt{2}} (\psi \exp(i\mathbf{q}_0 \cdot \mathbf{r}) + \mathrm{cc}) + \dots, \tag{19}$$

it will have a non-zero component at the wavevector $\mathbf{q}_0 = (2\pi/l)\mathbf{n}_0$ corresponding to the layer thickness l, $\mathbf{n}_0$ being the unit vector along the optic axis. In the nematic phase, on the other hand, translational invariance rules out any modulation of ρ; ψ is the appropriate order parameter, the amplitude and phase of which describe

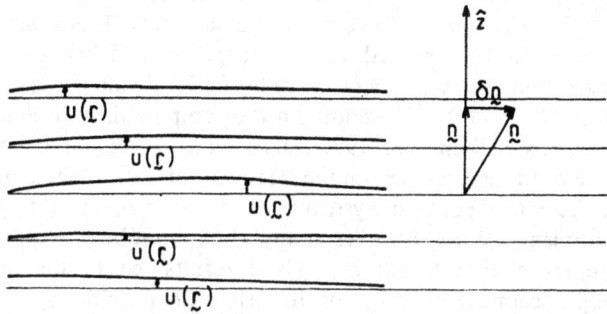

Figure 14. Displacing the layers over a distance u, changes the phase of the density modulation at q_0, of an amount $\phi = q_0 u(\mathbf{r})$. This can be included in (19) by choosing a complex order parameter $\psi = |\psi| \exp(iq_0 u)$.

respectively the degree of layering and the displacement of the layers—see figure 14 (de Gennes 1972). Requirement of rotational and translational invariance of the free energy leads to

$$F_S = \tfrac{1}{2} \int dr^d [A|\psi|^2 + C_\parallel |\nabla_\parallel \psi|^2 + C_\perp |(\nabla_\perp - iq_0\delta n)\psi|^2 + \tfrac{1}{2}B|\psi|^4]. \tag{20}$$

As usual $A = a(T - T_0)$, T_0 being the mean-field transition temperature. The dependence only on the modulus ensures translational invariance. Rotational invariance is taken care of by the $(\nabla_\perp - iq\delta n)$ term, to order δn^2, where $\delta n = n - n_0$.

Expression (20) reveals a remarkable analogy with the Landau–Ginzburg energy of the normal-superconductor transition (de Gennes 1972, 1973). ψ is analogous to the superconducting gap order parameter, and δn to the vector potential. Expression (20) is not sufficient, since it depends on the fluctuations of the nematic director n, and one has to add the nematic elastic energy (Oseen 1933, Frank 1958)

$$F_N = \tfrac{1}{2} \int K_1(\text{div } n)^2 + K_2(n \cdot \text{curl } n)^2 + K_3(n \times \text{curl } n^2)\, dr^d \tag{21}$$

and the total free energy is $F = F_S + F_N$. Although the analogy with superconductors is striking there are three potentially important differences (the anisotropy $C_\parallel \neq C_\perp$ can be easily removed by a volume-preserving scale transformation, Lubensky 1983)

whereas one has a complete gauge choice for the vector potential in electrodynamics, only the gauge $\delta n \cdot n_0 = 0$ has a physical meaning in the smectic case;

because of the existence of the splay term K_1, $F_S + F_N$ is not fully gauge invariant (as will be argued later, thermodynamical quantities are however);

as stressed in § 2, smectics do not possess long-range order.

The consequence is now to determine to what extent these differences are of consequence, and of course (if they are not) what the universality class of the superconductor problem is.

It was at first believed that these differences were irrelevant in the renormalization group sense and that the universality class should be that of an XY model (de Gennes 1973, Brochard 1973, Jähnig and Brochard 1974) the order parameter ψ having two components. Many experiments did support this point of view (Cheung *et al.* 1973, Leger 1973, Delaye *et al.* 1973 a, Pindak *et al.* 1974). There was however one major discrepancy: X-rays revealed an anisotropic critical behaviour of the correlation length exponents ($v_\parallel$ not very far from the $0.67 XY$ value, $v_\perp$ significantly smaller, being between 0.5 and 0.6 depending on the compound, MacMillan 1973, Birgeneau *et al.* 1981, Als Nielsen *et al.* 1977, Litster *et al.* 1974, 1980). This was a major point since one of the central ideas in the treatment of critical phenomena was that at or very near the critical point the system behaviour was governed by a unique length, the correlation length ξ (anisotropies, if any, should not be on exponents but rather on the bare correlation length ξ_0). These results led to the notion of generalized anisotropic scaling (Lubensky and Jing-Huei Chen 1978):

The correlation function

$$\langle \psi(r)\psi^*(0) \rangle = G(r); \qquad \langle \delta n_i(r)\,\delta n_j(0) \rangle = D_{ij}(r) \tag{22}$$

and the free energy f are supposed to obey the homogeneity relations:

$$
\begin{rcases}
G(q_\parallel, q_\perp, t, K_1) = \exp\left[(2-\eta_\perp)l\right]G[\exp(v_\parallel l/v_\perp)q_\parallel, \exp(l)q_\perp, \\
\qquad \exp(l\,v_\_)l, \exp(-\tau l)K_1], \\
D_{ij}(q_\parallel, q_\perp, t, K_1) = \exp\left[(2-\eta_n)l\right]D_{ij}[\exp(v_\parallel l/v_\perp)q, \exp(l)q_\perp, \\
\qquad \exp(l/v_\perp)t, \exp(-\tau l)K_1], \\
f(t, K_1) = \exp[-(d-1+v_\parallel/v_\perp)l]f[\exp(l/v_\perp)t, \exp(-\tau l)K_1],
\end{rcases} \tag{23}
$$

where $q_\parallel$ and $q_\perp$ are respectively the components of the wavevector parallel to, and perpendicular to, n_0. These relations imply that $\xi_\parallel \sim |t|^{-v_\parallel}$ and $\xi_\perp \sim |t|^{-v_\perp}$. According to equation (23), K_1 is a parameter which re-scales as

$$
K_1(l) = \exp(-\tau l)K_1, \tag{24}
$$

where K_1 is the elastic constant on the physical line and $K_1(l)$ is the iterated one. This procedure allows one to express the regularity of K_1 at the transition in the following way. If one considers fluctuations $\delta n_\perp$ of the director in the (n_0, q) plane and perpendicular to n_0, one expects from simple elastic theory

$$
\langle \delta n_\perp(q) \cdot \delta n_\perp(-q) \rangle = D_\perp(q_\parallel, q_\perp, t, K_1) = \frac{kT}{K_1 q_\perp^2 + K_3 q_\parallel^2}. \tag{25}
$$

From equation (23) this should also read

$$
D_\perp(q_\parallel, q_\perp, t, K_1) = \frac{kT \exp(2-\eta_n)l]}{\exp(2l)K_1(l)q_\perp^2 + \exp[2(1+\mu)l]K_3(l)q_\parallel^2}, \tag{26}
$$

where $\mu = (v_\parallel/v_\perp) - 1$. $K_1(l)$ is given by equation (24) and $K_3(l) = K_3^0$ is the bare bent elastic constant when $\exp(l) = \xi_\perp/\xi_{\perp 0}$, i.e. in the renormalization group transformation the system goes away from the critical point and when the re-scaled correlation length equals the bare one mean-field theory can apply.

The requirement that K_1 should be constant at the transition imposes, noting equations (25) and (26), the condition

$$
\eta_n = \tau
$$

(this is clear from the physics and can be shown to be valid to all orders of perturbation theory, Dunn and Lubensky 1981). The same identification leads to

$$
K_{33} = \exp\left[(2\mu + \eta_n)l\right]K_{33}^0 = \left(\frac{\xi_\perp}{\xi_{\perp 0}}\right)^{(2\mu + 3\eta_n)l}K_{33}^0.
$$

Adding the necessary independence of q_0 in the renormalization process one further obtains

and

$$
\begin{rcases}
\tau = \eta_n = 5 - d - \dfrac{v_\parallel}{v_\perp} \\[2mm]
2 - d = v_\parallel + (d-1)v_\perp.
\end{rcases} \tag{27}
$$

Similar procedures eventually give

$$\begin{aligned}
\delta K_3 &\sim t^{-(v_\parallel + (3-d)v_\perp)} = t^{-v_\parallel}, & (d=3), \\
\delta K_2 &\sim t^{v_\parallel - (5-d)v_\perp} = t^{v_\parallel - 2v_\perp}, & (d=3), \\
B &\sim t^{-(v_\parallel + (1-d)v_\perp)} = t^{2v_\perp - v_\parallel}, & (d=3), \\
D &\sim t^{v_\parallel - (3-d)v_\perp} = t^{v_\parallel}, & (d=3).
\end{aligned} \tag{28}$$

(D is the elastic constant describing how much energy is required to tilt the director away from the layers.)

Equation (24) can now be written in differential form

$$\frac{dK_1}{dl} = -\tau K_1 = (5 - d - v_\parallel/v_\perp)K_1. \tag{29}$$

Fixed points in the renormalization group flow must thus·satisfy one of the three conditions (if K_1^* is the fixed point value)

$$K_1^* = 0; \qquad K_1^* = \infty; \qquad v_\parallel = (5-d)v_\perp.$$

This considerably restricts the possibilities. All finite values of K_1^* necessarily imply anisotropic scaling with $v_\parallel = 2v_\perp$; the $K_1^* = 0$ value corresponds to the strict analogy with superconductors ($v_\parallel = v_\perp$). The $K_1^* = \infty$ value would also lead to anisotropic scaling, and be stable only if $v_\parallel > (5-d)v_\perp$. Current experimental data suggest that it *is* unstable. Possible renormalization group flows (Lubensky 1983) are displayed figure 15.

All this assumes that the phase change is second-order, but surprisingly, if fluctuations in $\delta \mathbf{n}$ are integrated out (in a mean-field approximation on ψ), one finds that the transition should be first-order. This would be valid if smectic A were equivalent to type 1 superconductors (Halperin *et al.* 1974). ε-expansion also suggests a discontinuous transition to first-order in $\varepsilon = 4 - d$, for n smaller than some

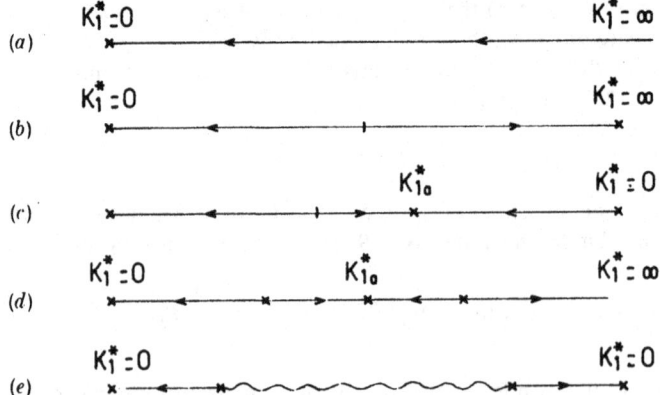

Figure 15. Possible renormalization flows according to Lubensky (1983). (*a*) One stable fixed point $K_1^* = 0$. (*b*) Two stable fixed points $K_1^* = 0$ and $K_1^* = \infty$; with such renormalization group flows, depending on the bare splay elastic constant, the NS_A transition can belong to two different universality classes. (*c*) Two stable fixed points: the $K_{1_0}^*$ fixed point implies necessarily $v_\parallel = 2v_\perp$. Again two possible universality classes. (*d*) Three stable fixed points: three possible universality classes. (*e*) Flow diagram involving a continuous line of fixed points (wavy-line) which cannot be *a priori* ruled out.

very large value (>200) depending on K_1 ($n=2$ in the liquid crystal case, Halperin *et al.* 1974). The $K_1^* = 0$ fixed point is reached for n values larger than this critical one. On the other hand, the $1/n$ expansion for $2 < d < 4$, predicts a second-order transition with only $K_1 = 0$, and $K_1^* = \infty$ fixed points (Dunn and Lubensky 1981). Dislocation theories have provided heuristic arguments in favour of anisotropic scaling (Helfrich 1978, Toner and Nelson 1981). In particular, the Toner–Nelson model suggests $v_\parallel = 2v_\perp$ in three dimensions which would correspond to the anisotropic fixed point $v_\parallel = (5-d)v_\perp$ already discussed. However, recent investigations of dislocation-mediated theories, in the ε-expansion (Toner 1982, Day *et al.* 1983) do not reveal the existence of such a fixed point and show that the relevant universality class should be that of an inverted XY model (like the normal conductor–superconductor transition; Dasgupta and Halperin 1981). Inverted means that high temperatures and low temperatures are exchanged as compared to a conventional XY model. This does not influence the exponent values, but exchanges the amplitudes c^+, c^- of the specific heat for instance. This result, based on the study of a generalized Villain model, strongly suggests that the transition is continuous. Thus, both dislocation-mediated theories and the original de Gennes formulation (the superconductor analogy) point towards the inverted XY universality class.

We have not yet specified the gauge in which calculations are made. Transforming on the isotropic form of the total free energy (after appropriate scale change) the transformation

$$\left.\begin{array}{l} \psi = \psi' \exp(-iq_0 L), \\[2mm] \delta n = \mathbf{A} - \nabla L, \end{array}\right\} \tag{30}$$

one finds (Jähnig and Brochard 1974, Halperin and Lubensky 1974, Halperin *et al.* 1974) that the expression is unchanged except for the splay term $K_1(\operatorname{div} \mathbf{n})^2$. Thus the elastic constants K_2, K_3, B, D should be gauge-invariant. Another interesting point is that the order parameter defined in the (non-physical) Coulomb gauge exhibits long-range order. The easiest way to understand this result is to consider the modulus of the order parameter as constant (as we have seen in §1, phase fluctuations alone are responsible for the divergence of $\langle u^2 \rangle$) with $\psi' = |\psi| \exp(iq_0 u')$ in which $q_0 u'$ is the phase of the transformed order parameter. The coupling term between $\mathbf{A}$ and ψ' then reads

$$\frac{C}{2} \int dr^d |(\nabla - iq_0 \mathbf{A})\psi')|^2 = \frac{C}{2} \int dr^d \{ |\nabla\psi'|^2 - 2q_0^2 |\psi|^2 \mathbf{A} \cdot \nabla u' + q_0^2 A^2 |\psi|^2 \}. \tag{31}$$

Integrating the second term on the right by parts, and making use of the Coulomb gauge $\operatorname{div} \mathbf{A} = 0$, completely decouples the fluctuations of $\mathbf{A}$ and of the phase of ψ'. The *elastic* or phase-dependent part of the free energy is simply

$$\tfrac{1}{2} C q_0^2 |\psi|^2 \int dr^d \nabla u'^2. \tag{32}$$

It has basically the same structure as that for a crystal, and indeed exhibits long-range order in three dimensions (i.e. a finite $\langle u^2 \rangle$; the lower marginal dimensionality is 2 for this transformed order parameter). In this gauge there is no thermodynamical singularity due to the lack of long-range order in the liquid-crystal phase. So the considerations developed in the preceding pages were implicitly carried out in the Coulomb gauge.

As argued previously, thermodynamical quantities are not expected to be gauge-dependent. On the other hand, X-ray scattering will be. Indeed, it depends on the physical correlation function

$$G(r) = \langle \psi(r)\psi^*(0) \rangle$$
$$= \langle \psi'(r)\psi'^*(0) \exp[-iq_0(L(r)-L(0))] \rangle \tag{33}$$

(Dunn and Lubensky 1981, Lubensky *et al.* 1981). And

$$G(r) \simeq \langle \psi'(r)\psi'^*(0) \rangle \langle \exp[-iq_0(L(r)-L(0))] \rangle. \tag{34}$$

This approximation is valid for $2 < d < 4$ when isotropic scaling holds but breaks down for the nematic phase and at the critical point for $d \leqslant 3$ (Lubensky and McKane 1984). In fact $\langle \psi'(r)\psi'^*(0) \rangle$ is the correlation function we have been speaking of previously. In particular equation (23) applies to it. The second term may be evaluated by much the same procedure as was used to derive equation (7). One finds

For $t < 0$:
$$\langle \exp[-iq_0(L(r)-L(0))] \rangle \sim \begin{cases} z^{-\eta}, & z \gg r_\perp, \\ |r_\perp|^{-2\eta}, & z \ll r_\perp, \end{cases} \tag{35}$$

which is another way of writing equation (7).

For $t > 0$:

$$\langle \exp[-iq_0(L(r)-L(0))] \rangle \sim \begin{cases} \exp(-z/l_\parallel), & z \gg |r_\perp| \\ \exp(-|r_\perp|/l_\perp), & z \ll |r_\perp| \end{cases} \tag{36}$$

$$\begin{cases} l_\parallel^{-1} \simeq \xi_\parallel^{-1} \ln \xi_\parallel, & t \to 0, \\ l_\perp^{-1} \simeq \xi_\perp^{-1/2}, & t \to 0. \end{cases} \tag{37}$$

Thus the correlation length measured by X-rays, ξ^{XR} will be

$$(\xi_\parallel^{XR})^{-1} = \xi_\parallel^{-1} + l_\parallel^{-1}; \qquad (\xi_\perp^{XR})^{-1} = \xi_\perp^{-1} + l_\perp^{-1}. \tag{38}$$

Close to $t = 0$, $l_\parallel^{-1}$ and $l_\perp^{-1}$ should win over $\xi_\parallel$ and $\xi_\perp$ (the *thermodynamical* correlation length defined by (23)), and one should always finish up with $v_\parallel^{XR} \simeq 2v_\perp^{XR}$. In practice, one can estimate that both terms contribute (Lubensky 1983), but the important result is that even with the isotropic $K_1^* = 0$ fixed points, X-rays will measure anisotropic apparent exponents.

As we have already noticed X-ray experiments favour anisotropic exponents but the ratio $v_\parallel^{XR} = 2v_\perp^{XR}$ is never reached. $v_\parallel^{XR}$ is typically of the order of what is expected from the XY model but slightly higher, typically 0·7 against 0·67, whereas $v_\perp^{XR}$ is significantly smaller, between 0·5 and 0·6 (Birgeneau *et al.* 1981, Litster *et al.* 1974). Thus they do not fit the defect-mediated theory, or any transition governed by a $K_1 \neq 0$ fixed point, but could fit relation (38) with the XY thermodynamical correlation length.

Calorimetric measurements show in general large values of the specific heat exponent (α ranging from 0 to 0·3 depending on the compound) which in some cases agree with the anisotropic hyperscaling relation (27) (Schantz and Johnson 1978, Kasting *et al.* 1980, Johnson *et al.* 1978, Hatta and Nokayama 1981, Lushington *et al.* 1980, Le Grange and Mochel 1980, Birgeneau *et al.* 1981). These results do not agree with the inverted XY model ($\alpha \simeq -0.02$). The proximity of a tricritical point

(MacMillan 1971, de Gennes 1972, Alben 1972) can explain the high values of α (Brisbin *et al.* 1979, Johnson 1983). Indeed, at a tricritical point, the exponents assume mean-field values ($\alpha = 0.5$; $\nu = 0.5$); note that the considerations concerning gauge transformations should still apply and apparent anisotropy could still be expected in X-ray experiments. The most recent experiments (Garland *et al.* 1983, Ocko *et al.* 1984 b, Thoen *et al.* 1984) show accurately the existence of a tricritical point and confirm the shift of the exponents from XY to harmonic values (for large nematic range, i.e. far from the tricritical point, $\alpha \simeq -0.03$, $\nu_\perp \simeq 0.68$, $\nu_\parallel \simeq 0.8$; for short nematic range, i.e. in the vicinity of the tricritical point, $\alpha \simeq 0.5$, $\nu_\perp \simeq 0.4$, $\nu_\parallel \simeq 0.6$). However, anisotropic scaling as given by (38) seems to hold everywhere (which is not expected with the isotropic XY or harmonic fixed points) and single power laws fit the data better than crossover functions in the intermediate regime.

Measured elastic constants are in some regards difficult to interpret. K_3 diverges with exponents in fairly good agreement with XY values (Cheung *et al.* 1973, Leger and Martinet 1976, Birecki *et al.* 1976, Birecki and Litster 1977, Von Känel and Litster 1981). The K_2 exponent is less well established, some experiments agreeing very well with an XY value (Pindak and Cheng-Cher Huang 1974), others giving smaller values which do not fit the $\nu_\perp = \nu_\parallel/2$ relation (Delaye *et al.* 1973 a). Measurements of the compressional elastic constant do not shed more light: low frequency and free surface measurements yield an exponent 0.3 (Birecki *et al.* 1976, Von Känel and Litster 1981, Clark 1976, Ribotta 1974, Fisch *et al.* 1981) which is significantly smaller than any theoretical prediction; second sound measurements agree better with tricritical behaviour (8CB : 0.49) or XY exponents (8CB : 0.62) (Ricard and Prost 1979, 1981). Low frequency measurements are probably perturbed by the presence of defects, the motion of which renormalizes the compressional elastic constant (Pershan and Prost 1975).

As a conclusion, one may wonder whether one is currently able to assign a universality class to the N–S_A transition. As we have seen the most recent theoretical and experimental data point towards the inverted XY model, with crossover towards a tricritical point when the nematic range is short. This picture describes rather well the current situation although some disagreements between experiment and theory persist. Whether the former or the latter should be improved is an open question.

3.2. *Other phase transitions involving smectic phases*

In this section we review the transitions smectic A to C, and smectic A to hexatic and the N–S_A–S_C multicritical point but not in as much detail.

The S_A–S_C transition is characterized by the onset of a tilt in a one-dimensional layered matrix. The angle of the director with the normal to the layers seems to be the relevant order parameter (De Gennes 1973). Since the tilt direction has azimuthal degeneracy (figure 16) one can again use a complex order parameter ψ_1 to describe the transition. If one neglects the coupling to the layers' undulation mode, obviously the transition is helium-like and should reveal XY exponents. Some light scattering, magnetic birefringence and tilt-angle measurements (Delaye 1979, Galerne 1981) are in fair agreement with these expectations ($\gamma \simeq 1.3$, $\beta \simeq 0.36$). Only the susceptibility exponent defined at the transition is at odds with this: $\delta = 3$, the mean-field value). Measurements on other compounds (high resolution calorimetry: Huang and Lien 1981, Schantz and Johnson 1978, Lim *et al.* 1980, Dehoff *et al.* 1981, 1982; X-rays: Safinya *et al.* 1980. Ocko *et al.* 1984 a, b, magnetic birefringence: Rosenblatt and Litster 1982) favour a mean-field behaviour in which the sixth-order term is unusually large.

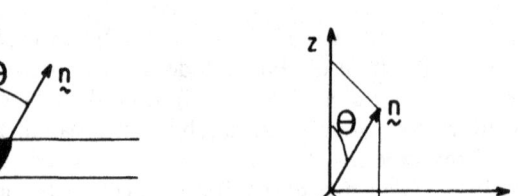

Figure 16. The tilt angle θ of the molecules can point in any azimuthal direction: there are thus two independent components: $\theta_x = \theta \cos \phi_1$ and $\theta_y = \theta \sin \phi_1$, which can be equivalently described by a complex order parameter $\psi_1 = \theta \exp (i\phi_1)$.

Recent mechanical measurements reveal an asymmetry between high and low temperature sides which would also agree with a strong ϕ^6 term (Bartolino and Durand 1984). The reason for mean-field behaviour in some systems is perhaps due to a large bare correlation length due to the proximity of a N-S_A—S_C point. One may also question the importance of the coupling to the layers' fluctuations since the absence of long-range order is a fundamental property of smectics. A model in which the director is fixed by a magnetic field and the layers allowed to undulate (Hornreich and Shtrikman 1977) shows that under those conditions the upper critical dimension is reduced to 3, justifying mean-field exponents (with a small logarithmic correction). The cross-over with the general case has been investigated (Grinstein and Pelcovits 1982).

On the other hand, if one is able to reduce the system to two dimensions, the problem of the layers' fluctuations no longer exists. This turns out to be possible with free-standing smectic films (Friedel 1922, Young *et al.* 1978, Pindak *et al.* 1980a, Rosenblatt *et al.* 1980). It is of some interest to understand the reasons for the films' stability. Breaking a thin film requires the creation of a hole, the energy of which will come essentially from two parts:

a positive line tension giving a contribution $2\pi N l \sigma R$ (N is the number of layers, l the layer thickness, σ the air/liquid crystal interfacial tension, R the radius of the hole);

a negative interfacial term $-2(\pi\sigma R^2)$ (the factor 2 comes from the two sides of the film. The minus sign comes from the fact that in the process of creating the hole, molecules are removed from the interface to the bulk where they cost no interfacial energy).

Thus the energy to create a hole of radius R is

$$W_R \simeq 2\pi\sigma(NlR - R^2) \qquad (39)$$

and there is an energy barrier

$$W_c \simeq \frac{\pi\sigma}{2}(Nl)^2 \qquad (40)$$

for nucleating such holes (it corresponds to a critical radius $R_c = Nl/2$ beyond which the hole grows spontaneously and the film breaks). Thus the nucleation frequency per unit area, will be

$$f \simeq f_0 \exp (- W_c/kT). \qquad (41)$$

Taking the value of f_0 to be of the order of the speed of sound divided by a molecular diameter, divided by a molecular area, and taking a value of σ typical of interfacial tensions ($\sigma \simeq 3 \times 10^{-2}$ N/m), we estimate

$$f \simeq 10^{28} \exp(-50N^2) \text{ per second.} \tag{42}$$

For $N = 1$, $f \simeq 2 \times 10^6 \text{ s}^{-1}$ while for $N = 2$, $f \simeq 10^{-59} \text{ s}^{-1}$. The N^2-dependence of the exponent has a tremendous consequence. Whereas one-layer films will almost never be stable, two-layer or thicker films should last for months when kept in suitable conditions. This stability provides very beautiful systems with which to probe two-dimensional behaviour.

In a smectic C film, and if the Kosterlitz–Thouless picture of disclination unbinding is correct, the projection $\mathbf{n}'$ of the molecular axis $\mathbf{n}$ on to the layer plane is the relevant order parameter. Again it is equivalent to a two-dimensional XY model (at least in respect of the number of components of the order parameter). Experiments proved to be more feasible with ferroelectric S_{C^*} smectics. The free energy then reads (Young *et al.* 1978)

$$F = \tfrac{1}{2} \int d^2r [K_1 \operatorname{div}^2 \mathbf{n}' + K_3(\mathbf{n}' \times \operatorname{curl} \mathbf{n}')^2 - 2\mathbf{P}_0 \cdot \mathbf{E}]$$

$$+ \tfrac{1}{2} \iint d^2r \, d^2r' \frac{\nabla \cdot \mathbf{P}_0 \nabla' \cdot \mathbf{P}_0}{|r - r'|}, \tag{43}$$

where $K_1 = NlK_{11} \sin^2 \theta$; $K_3 = Nl \sin^2 \theta \, (K_{22} \cos^2 \theta + K_{33} \sin^2 \theta)$; $\mathbf{P}_0$ is the permanent polarization and $\mathbf{P}_0 = P_0 \mathbf{n} \times \hat{z}$ where $\hat{z}$ is the unit vector normal to the film. There are two differences with the classical XY model. First the anisotropy of the elastic constant and second the existence of long-range interactions due to the polarization charges. Both have been proved to be unimportant near the transition (Pelcovits and Halperin 1979). The experiments agree fairly well with these expectations; in particular the existence of a disclination-unbinding temperature seems to be confirmed and agrees with the relation (Young *et al.* 1978, Pindak *et al.* 1980 a)

$$kT_c = \tfrac{1}{2}\pi K_c^*, \qquad K_c = \tfrac{1}{2}(K_1^* + K_3^*). \tag{44}$$

The multicritical point N–S_A–S_C on the other hand is not completely understood. Although several theoretical descriptions exist (Chen and Lubensky 1976, Chu and MacMillan 1977, Benguigui 1979, Huang and Viner 1982), we will describe only the first since it seems to correspond more closely to experiment. It is in some way a generalization of the de Gennes model of the NS_A and NS_C transitions (de Gennes 1973).

The free energy can be written directly in k-space

$$\int F d^3r = \tfrac{1}{2} \sum_{kk'k} (A + D_\parallel (k_\parallel^2 - q_0^2)^2 + C_\perp k_\perp^2 + D_\perp k_\perp^4) \rho(\mathbf{k}) \rho(-\mathbf{k})$$

$$+ \frac{B}{2} \rho(\mathbf{k}) \rho(\mathbf{k}') \rho(\mathbf{k}'') \rho(-\mathbf{k} - \mathbf{k}' - \mathbf{k}'') \tag{45}$$

the summation on $\mathbf{k}$, $\mathbf{k}'$, $\mathbf{k}''$ being performed on all wavevectors corresponding to smectic fluctuations, i.e. within a volume in k-space including the torus of diffuse scattering, defined in figure 17. $k_\parallel (k_\perp)$ refers to components of the vector taken parallel (perpendicular) to the director. A, $D_\parallel$, $D_\perp$, B are the usual Landau

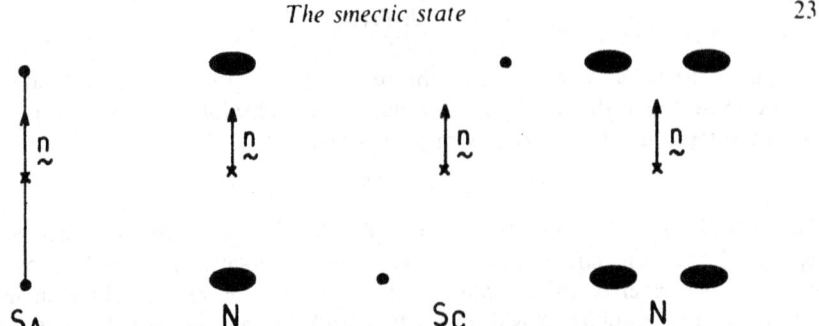

S_A N S_C N

Figure 17. Typical X-ray patterns for S_A, S_C and nematic phases (the cross shows the direct X-ray beam perpendicular to the director **n**). The integration domain of equation (45), corresponds to the interior of the surface obtained by rotating the rectangle of the far right figure around the director **n**.

coefficients. If $C_\perp > 0$, the minimum energy state is at $k_\| = q_0$, $k_\perp = 0$ and it corresponds to a smectic A phase (or nematic if $A > 0$). If $C_\perp < 0$, the minimum energy state is at $k_\|^2 = q_0^2$, $k_\perp^2 = (-C/2D_\perp)$ and this describes a smectic C phase (or nematic if $A - C^2/4D_\perp > 0$).

The fundamental ingredient of this theory is thus the sign change of C (and of course, as usual, of A). The N–S_A–S_C point is defined as a Lifshitz point (for renormalization group studies: see Mukamel and Hornreich 1980, Hornreich et al. 1975). NS_A, S_A, S_C transitions should be continuous, but the NS_C should be first-order. Perturbation theories also predict the first-order character of the NS_C transition (Brazovskii 1975, Swift 1976). (If d' is the dimension of the subspace in which the order parameter fluctuates, and d is the dimension of space, the transition is expected to be of first-order if $d' = d$ or $d' = d - 1$. In the NS_C case fluctuations occur in the nematic phase essentially on two rings defined by $k_\| = \pm q_0$ and $q_\perp^2 = -C_\perp/2D_\perp$).

The first-order character of this transition, and the vanishing of the enthalpy discontinuity at the Lifshitz point, have been beautifully confirmed experimentally (Johnson et al. 1977, Sigaud et al. 1977). The X-ray diffuse scattering pattern may be straightforwardly calculated from (45). Experiments appear to be in reasonably good agreement with the corresponding expression, but the vanishing of $C_\perp$ has not really been seen (Safinya et al. 1981). The same model predicts a divergence of the Frank elastic constants K_{11}, K_{22} and K_{33} as ζ^2 at the NS_C transition. Light scattering experiments far from the multicritical point agree with this ζ^2 divergence, but show the importance of both tilt and layer fluctuations close to the Lifshitz point (Witanachi et al. 1983). More recent theoretical treatments predict in fact that the N–S_A–S_C point should be a tetracritical one, and thus involve a biaxial nematic phase N_b as well (i.e. an N–N_b–S_A–S_C point, all phase boundaries being second-order, Grinstein and Toner 1984, Lubensky and Jiang Wang 1984). Biaxial nematics have not been reported yet in thermotropic systems, but their stability range might be small.

With the discoveries of the straight and tilted hexatic phases a whole class of new phase transitions became available to excite the curiosity of both experimentalists and theoreticians

$$(\text{Crystal } (S_B), S_{BHex}); \quad (S_{BHex}, S_A); \quad (\text{Crystal } (S_G), S_F);$$

$$(\text{Crystal}, S_I); \quad (S_F, S_I); \quad (S_F, S_C); \quad (S_I, S_C).$$

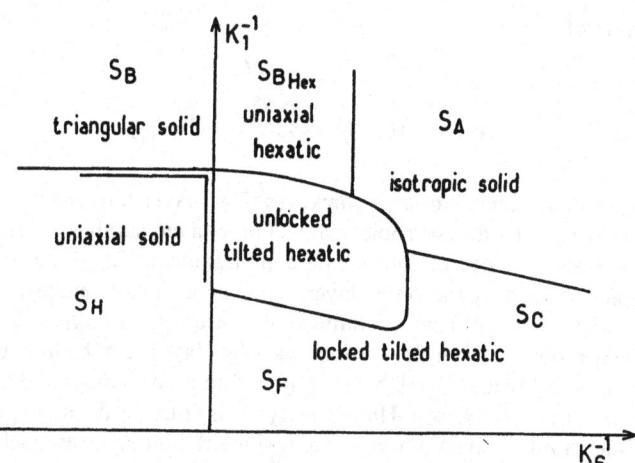

Figure 18. Phase diagram of two-dimensional tilted systems in the K_1^{-1}, K_6^{-1} plane according to Nelson and Halperin (1980); the symbols are given in accordance with the usual three-dimensional nomenclature.

One should note right away that, as already pointed out in §2 there is no symmetry difference between S_C, S_F and S_I phases (in fact the situation is qualitatively different for the S_I phase if the in plane positional correlations are quasi-long-range (Benattar *et al.* 1981)). Without any further consideration one can say that there can only be first-order transitions or no transition at all between these phases (Bruinsma and Nelson 1981, Prost 1981). It might be interesting to look for critical points in these systems.

The theoretical phase diagrams of two-dimensional tilted systems have been extensively studied (Nelson and Halperin 1980 and references therein). Not less than six phases and six Kosterlitz–Thouless types of phase boundary have been predicted (figure 18)! The starting free energy is rather simple

$$F = \tfrac{1}{2} \int \{ K_6 |\nabla \phi_6|^2 + K_1 |\nabla \phi_1|^2 + 2g \nabla \psi_6 \cdot \nabla \psi_1 - 2h \cos 6(\psi_6 - \psi_1) \} \, d^2 r. \qquad (46)$$

The first term corresponds exactly to expression (15) for hexatic order. The second is the splay term usual in smectic C liquid crystals and already written in equation (43). The third, allowed by symmetry, is in any case generated by renormalization procedures. The last term is the coupling that one expects from the six-fold symmetry of the bond orientational order. One can see the richness of dislocation-mediated theories by comparing (46) and figure 18. Most of the phases have three-dimensional counterparts displayed on the figure but an interesting unlocked tilted hexatic phase is expected to exist (i.e. the renormalized free energy has vanishingly small h) which has no three-dimensional counterpart for a reason which will soon be clear. The experimental situation is currently developing fast. As already mentioned, light scattering and tilt-angle measurements (Young *et al.* 1978, Pindak *et al.* 1980 b) seem consistent with the defect-mediated picture. On the other hand the melting of two-dimensional crystals is more complicated (Moncton *et al.* 1982). Films with from two to twelve layers of

$$\overline{14}S5(C_{14}H_{29}-O-\phi-\overset{\overset{\textstyle O}{\|}}{C}-S-\phi-C_5H_{11})$$

show a striking dependence on layer thickness. Two-layer films melt abruptly via a first-order transition to an isotropic state. Three-layer samples reveal a melting process in two steps, with an intermediate phase identified as hexatic. Data on thicker samples show that the outer layers melt at a higher temperature than the bulk, a result also obtained from mechanical measurements of amazing sensitivity in which the shear moduli of samples comprising a few layers can be measured (Pindak 1980 b, Tarczon and Miyano 1981). No evidence of an elastic constant discontinuity of the Kosterlitz–Thouless–Nelson–Halperin type is observed at the solid–liquid transition. One should remark however that, since the outer layers melt at a higher temperature through a direct first-order transition, the melting of the inner one is a new problem not included in the theoretical description. More recent experiments (Pindak 1983) show the existence of a two-layer hexatic phase with continuous transitions to the solid and isotropic phases, as expected from dislocation-mediated theories. Hexatic phases are found in films of heptyloxybenzylidene-heptylaniline having fewer than sixteen layers which are not found in bulk material (Collett *et al.* 1984).

Three-dimensional systems are also being actively investigated. A mean-field theory corrected for fluctuations in an $\varepsilon = (4-d)$ expansion gives a phase diagram reminiscent of that for two dimensions except for the disappearance of the *unlocked* phase. The order parameters of the hexatic and smectic C are again ψ_6 and ψ_1 respectively (as given by (14) and (39)) so that the $h\cos(6(\psi_6-\psi_1))$ in (46) is replaced by $(h/2(\psi_6\psi_1^{*6}+\psi_6^*\psi_1^6))$. In other words ψ_1^6 acts as an external field on ψ_6, and since it exhibits conventional long-range order in three dimensions, $\langle\psi_1\rangle \neq 0$ will always imply $\langle\psi_6\rangle \neq 0$. In three dimensions, tilted phases must exhibit long-range bond ordering (or there is no room for S_C–S_F transitions other than first-order, as already noticed). The S_A–S_C and S_A–S_{BHex} are expected to be XY like, and at the multicritical point the couplings between ψ_6 and ψ_1 are renormalized to zero: approached from the S_A side, the point corresponds to two decoupled XY models. As already mentioned, experimental results on the S_A–S_C transition conform either with the mean-field or with the XY expectations. Calorimetric measurements on the S_A–S_{BHex} transition show a critical exponent $\alpha = \alpha' = 0.64$ far from the expected value -0.02 (Huang *et al.* 1981, Viner *et al.* 1983, Rosenblatt and Ho 1982). In fact, X-ray measurements show the existence of an important *chevron* short-range order (Levelut 1976, Pindak *et al.* 1981). Taking account of this suggests that the high value of α might be linked to the proximity of a tricritical point (Bruinsma and Aeppli 1982). As far as the S_C–S_F case is concerned, both standard and high resolution calorimetry show first-order transitions (Kumar 1981, Huang *et al.* 1981). And X-ray measurements on TBDA (a compound similar to that quoted in the Introduction but with the pentyl group replaced by a decyl group) shows a sharp jump of the in-plane correlation length, showing the discontinuous character of the phase change (Benattar *et al.* 1983).

This brief review may seem somewhat confusing. One seldom deals with a simple *n* vector model in studies of smectic phase transitions. One is almost always near a

26 *J. Prost*

cross-over of some sort. This can be considered either as a drawback of the subject or (as we believe) an opportunity to deepen our understanding of phase transitions in general.

4. Frustrated smectics

Up to now we have considered that the phase characterized by the $D_{\infty h}$ point group and translational broken symmetry along the optic axis was a smectic A phase. We did not feel a need to specify anything more. However, inspection of the binary phase diagram of DB5 and TBBA (i.e. of 4'-pentylphenyl 4-cyanobenzoyloxy-benzoate, $C_5H_{11}-\phi-OOC-\phi-OOC-\phi-CN$, and terephthal-*bis*- 4 n-butylaniline $C_4H_9-\phi-N=CH-\phi-CH=N-\phi-C_4H_9$) reveals a first-order phase boundary separating two regions which under the microscope are identified as S_A (Sigaud *et al.* 1979) (figure 19). How can two S_A phases be different? Replace the DB5 by the DB7 (Hardouin *et al.* 1983) and a third S_A phase appears (figure 20)!

Replace TBBA by the more anisotropic three-benzene-ring molecule C_5 stilbene $(C_5H_{11}-\phi-CH=CH-\phi-OOC-\phi-CN)$ an entirely new phase called $S_{\tilde{A}}$ appears in a diagram having two S_A phases (Sigaud *et al.* 1981). Take instead T8 $(C_8H_{17}O-\phi-OOC-\phi-CH=N-\phi-CN)$ and three S_A-S_A boundaries, plus the $S_{\tilde{A}}$ phase, can be spotted (figure 21) (Levelut *et al.* 1981). The region of the phase diagram rich in T8 is also very interesting since it reveals a *re-entrant* nematic domain below a *high temperature* smectic pocket (Hardouin *et al.* 1979). Reverse one benzoate linkage in the DB_n series and place side by side the DB 7 8 and 8-9 diagrams and one obtains a very revealing picture (figure 22) (Hardouin *et al.* 1983). Again a re-entrant nematic domain appears, which has the typical topology of a bicritical point (Kosterlitz *et al.* 1976). Now change the cyano end-group of DB7 and

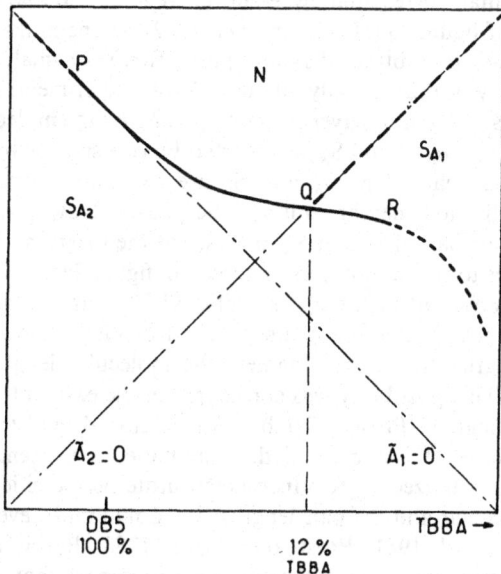

Figure 19. Binary phase diagram obtained by mixing DB5 and TBBA, or from the theory developed in Prost (1979). Note the existence of a monolayer-bilayer $S_{A_1}-S_{A_2}$ transition line, of the Ising type. The points P and R should be tricritical points (respectively helium-like and Ising-like).

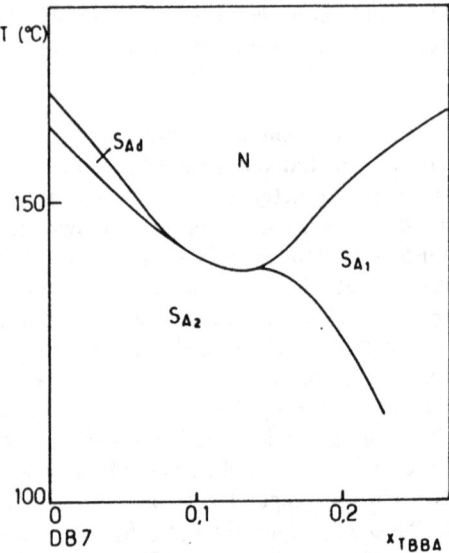

Figure 20. DB 7 TBBA phase diagram according to Hardouin *et al.* (1983). Note the existence of a new smectic A phase, separated from the bilayer S_{A_2}, by a first-order transition line. There are thus three different S_A phases and three NS_A lines in this diagram.

DB8 and change it to a nitro group NO_2: both are strongly polar, but NO_2 is somewhat more bulky. Again the binary diagram involves three smectic A phases, a *bicritical* point (of small corresponding re-entrant region) and one more new phase, $S_{\tilde{C}}$ on the diagram of figure 23 (Hardouin *et al.* 1982). If one looks at the molecular changes from one case to another, they are quite often very small even though the resulting diagram may be dramatically different. X-ray experiments (see Hardouin *et al.* 1983) show that S_{A_1} is a monolayer smectic ($q_0 \simeq 2\pi/l$; l the molecular length), S_{A_2} is a bilayer smectic ($q_1 = q_0/2$) and S_{A_d} is a partial bilayer smectic ($q_0 \simeq 2\pi/l'$; l' is the length of a molecular pair). The $S_{\tilde{A}}$ and $S_{\tilde{C}}$ are two-dimensionally ordered and according to our definition they are not smectic phases. However, locally they are very close to bilayer or partial bilayer S_A phases, but the location of the polar heads undulate with respect to the monolayers as shown in figure 24 (a).

Can one make sense out of these observations? It turns out that the answer is rather simple: one just needs to assume that, for some smectic systems, two incommensurate lengths are relevant namely the molecular length l and the pair length l' (figure 25). This possibility was considered in the early interpretation of the DB5 TBBA phase diagram (Prost 1979), but not developed until the X-ray analysis of the re-entrant phase of T8 revealed the simultaneous existence of fluctuating smectic domains characterized by two incommensurate periodicities, corresponding precisely to the molecular and the pair lengths (Hardouin and Levelut 1980 and see Prost 1980, Barois *et al.* 1981, Prost and Barois 1983, Barois *et al.* 1984). To understand the occurrence of two lengths one needs to remark that when asymmetric molecules are present one needs not only to specify a density modulation to define the existence of layers, but also where the heads of the molecules are with respect to the layers. In other words the de Gennes order parameter ρ is no longer sufficient. A

28   *J. Prost*

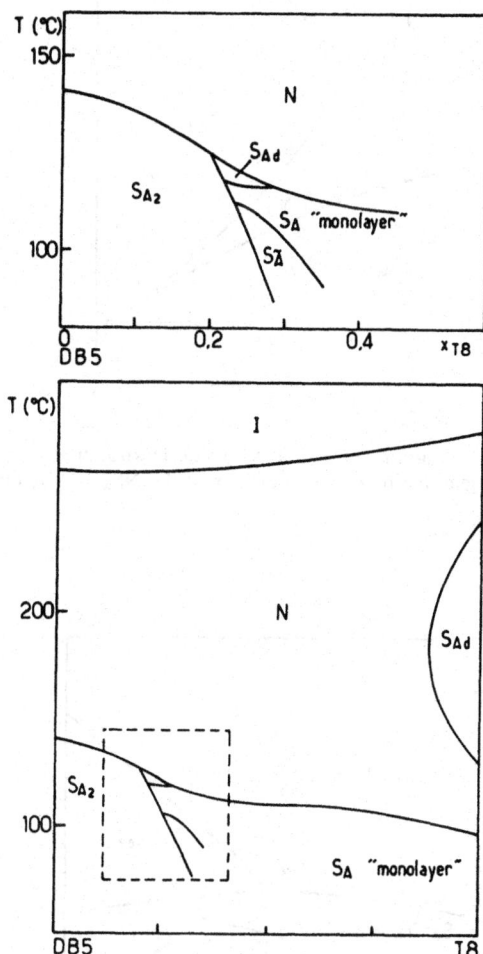

Figure 21. DB5 T8 phase diagram. The expanded (upper) part looks similar to figure 20. except for the appearance of a new phase: the $S_{\tilde{A}}$ antiphase (Levelut *et al.* 1981). In the region rich in TBBA, remark the wide re-entrant nematic domain.

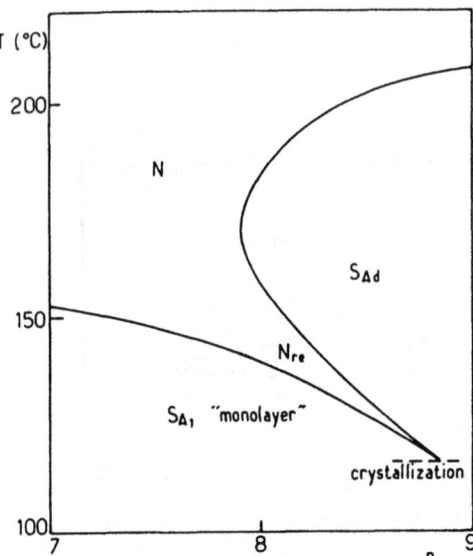

Figure 22. DB 7–8 and 8–9 diagrams placed side by side Hardouin *et al.* (1983): note the wide re-entrant nematic domain and the topology of the NS_{A_d}, NS_{A_1} lines, typical of a bicritical point.

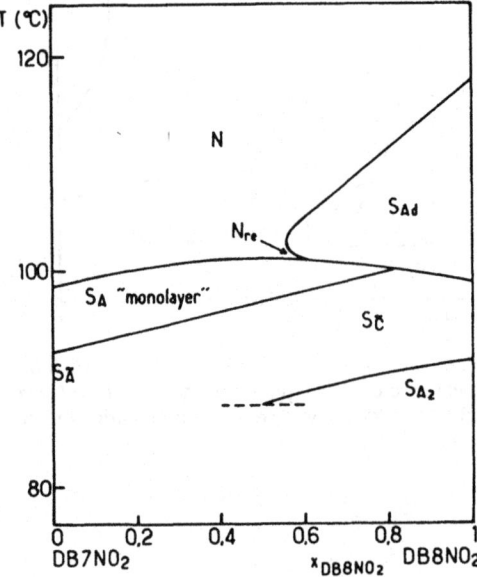

Figure 23. $DB7NO_2$–$DB8NO_2$ phase diagram (Hardouin *et al.* 1982). Note two S_A phases (S_{A_1} and S_{A_d}) separated by a first-order phase boundary. The NS_{A_1}, S_{A_d} point exhibits again the typical topology of a bicritical point, but with a much smaller range. For large $DB8NO_2$ concentrations, note the appearance of another new phase: the so-called ribbon $S_{\tilde{C}}$ phase.

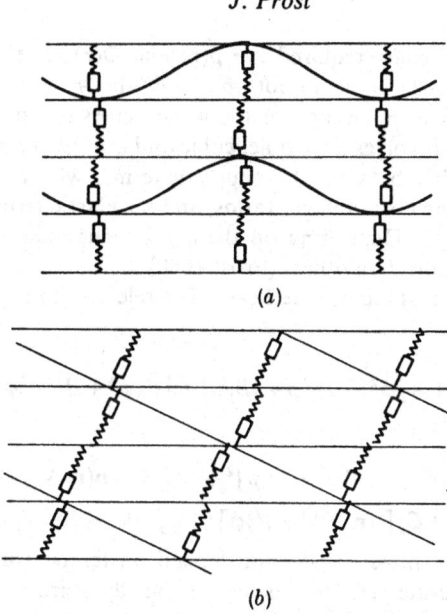

Figure 24.   (a) Structure of the $S_{\tilde{A}}$ antiphase: note a local bilayer smectic arrangement with the locations of the dipolar heads undulating periodically from one interface to the other. (b) Structure of the S_C phase: again there is a local bilayer arrangement but the planes of equal dipolar density make an angle with those of equal mass density. Furthermore, molecules are tilted with respect to both patterns.

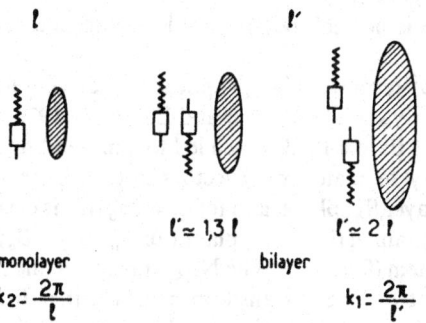

Figure 25.   The two important lengths in the smectic problem. Note that the pair length l' may range practically from l to $2l$.

parameter which unambiguously defines the relevant average vectorial symmetry is

$$P(r) = \frac{1}{V} \sum_i p_i \delta(r - r_i) \tag{47}$$

in which p_i is the molecular dipole moment. This order parameter, which will describe long-range antiferroelectric order, tends to condense at the pair-length, i.e. at wavevector $q'_0 = 2\pi/l'$. It is easier to use as order parameter the related potential Φ given by

$$P(r) = \frac{1}{4\pi} \nabla \Phi \tag{48}$$

Thus the two order parameters required are $\rho(\mathbf{r})$ and $\Phi(\mathbf{r})$. The free energy now involves harmonic and quartic terms in both ρ and Φ: the A_ρ and A_ϕ coefficients of ρ^2 and ϕ^2 are respectively at minimum for the wavevectors q_0 and q_0'. A priori the coupling between ρ and ϕ involves harmonic, cubic and quartic terms. Some original physics arise in the conflict between the coupling terms, which tend to lock the periods of the order parameters at integer ratios, and the elastic terms, which tend to enforce their natural length. Depending on the q_0/q_0' value, some of the coupling terms will be more important than others (Prost 1980).

For instance, let us discuss the $q_0/q_0' \simeq 2$ case. The relevant couplings are $\Phi^2\rho$ and $\Phi^2\rho^2$.

$$F = \tfrac{1}{2} \int \{A_\rho\rho^2 + A_\phi\phi^2 - D\phi^2\rho + \tfrac{1}{2}B_\rho\rho^4 + \tfrac{1}{2}B_\phi\phi^4 + B_{\rho\phi}\rho^2\phi^2\} \, dv, \tag{49}$$

with

$$\left. \begin{aligned} A_\rho\rho^2 &= a_\rho(T-T_\rho)\rho^2 + C_\rho^\|[((\mathbf{n}\cdot\mathbf{\nabla})^2 + q_0^2)\rho]^2 + C_\rho^\perp[(\mathbf{\nabla}-\mathbf{n}(\mathbf{n}\cdot\mathbf{\nabla}))\rho]^2, \\ A_\phi\Phi^2 &= a_\phi(T-T_\phi)\Phi^2 + C_\phi^\|[((\mathbf{n}\cdot\mathbf{\nabla})^2 + q_0^2)\phi]^2 + C_\phi^\perp[(\mathbf{\nabla}-\mathbf{n}(\mathbf{n}\cdot\mathbf{\nabla}))\rho]^2. \end{aligned} \right\} \tag{50}$$

The coefficients in these expressions do not depend on temperature. The gradient terms favour incommensurate periodicities l and l', but the third-order term appears as a lock-in contribution which will gain energy whenever wavevectors are matched in $\mathbf{k}$ space. The Fourier transform of this term reads

$$-D \sum_{k_1 k_2 k_3} \Phi(\mathbf{k}_1)\Phi(\mathbf{k}_2)\rho(\mathbf{k}_3)\delta_{\mathbf{k}_1 + \mathbf{k}_2 + \mathbf{k}_3}.$$

The different ways of satisfying this wavevector conservation law, yield all the phases observed—see figure 26.

First, if $\phi = 0$, there is no real lock-in problem, and one gets the monolayer S_{A_1} phase.

But if $\Phi \neq 0$, the $D\phi^2$ term will always appear as a driving field and induce a non-zero ρ. (This point, clearly stated in Meyer and Lubensky 1976, Prost 1979, 1980, was overlooked in Prost and Barois 1983, which led to a misleading description of the S_{A_d} phase). The easiest way to match wavevectors is to choose $\rho = \rho_0 \cos 2k_1 z$ and $\phi = \phi_0 \cos k_1 z$, i.e. the bilayer S_{A_2} phase, and for $q_0 = 2q_0'$ this is clearly the least energetic solution. The phase diagram in the A_ρ, A_ϕ plane (for $B_\rho = B_\phi = B_{\rho\phi}$) is exactly that found for the DB5–TBBA system (figure 19). The NS_{A_1} transition line is continuous far from the isotropic phase, but the NS_{A_2} transition can be either first or second-order, a tricritical point separating the two regimes. It had been previously thought that the S_{A_1}–S_{A_2} line should belong to the Ising universality class (Prost 1979), but the coupling to the layers' displacement field (Jiang Wang and Lubensky 1984) should lead to gauge corrections similar to those described in the previous section for the NS_A case. Recent experimental studies (Chen et al. 1984) yield results close to the Ising universality class but not in complete agreement.

Let us now introduce incommensurability, i.e. $q_0 \neq 2q_0'$. For $(q_0 - 2q_0')/(q_0 + 2q_0')$ small, nothing dramatic happens: k_1 is a trade-off between q_0' and $q_0/2$, but the topology of the diagram is still the same. However the layer spacing, which in a conventional theory of the NS_A and S_{A_1}–S_{A_2} transitions is independent of temperature, should vary critically close to the phase changes. With greater incommensurability the tricritical point R disappears leading the way to a critical end point R'. Simultaneously a new phase boundary between two S_{A_2} phases appears (figure 20). The phases differ in that one displays very little harmonic, and

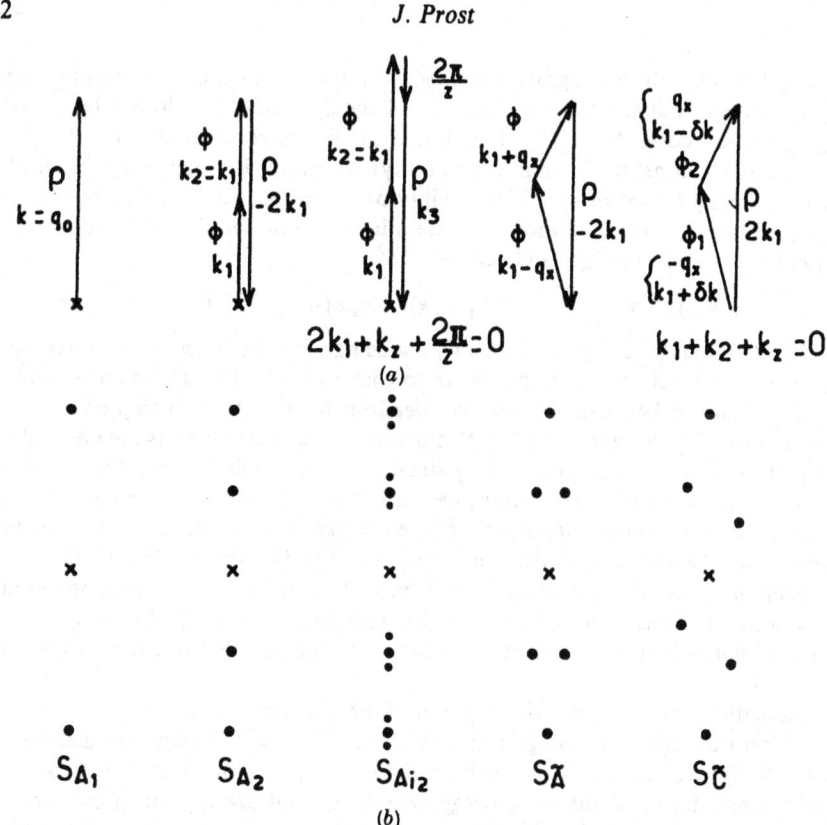

Figure 26.   (a) Different ways of matching wavevectors to satisfy the cubic term $\phi^2 l$. Each of
     them implies a well-defined symmetry, which corresponds to the phases described in the
     text and outlined on the figure. (b) Corresponding X-ray patterns.

has a layer spacing centred essentially on the natural pair length $k_1 \simeq q_0'$, while the
other reveals a sizeable amount of harmonic and its layer spacing compromises
between l and $l'(q_0' > k_1 > q_0/2)$. The topology agrees well with the DB7–TBBA
diagram if one identifies the S_{A_d} with the S_{A_2} having small harmonic modulation.
Experimentally the diffuse scattering corresponding to ϕ-fluctuations is often larger
than the driven harmonic (which led to the misconception of Prost and Barois 1983).
Furthermore, as there is no symmetry difference between S_{A_d} and S_{A_2}, one can go
continuously from one phase to the other and one can predict the existence of a
critical point ending the line of the first-order transition (Barois *et al.* 1984). Recent
experiments (Hardouin *et al.* 1984) seem to corroborate this expectation.

Increasing incommensurability still further results in R' moving towards Q, the
N–S_{A_1}–S_{A_2} triple point, until a S_{A_d}–S_{A_1} phase boundary opens up. This corresponds
to the blown-up part of the DB5 T8 diagram. On the other hand, the S_{A_2} type of
solution costs a lot of elastic energy. If it is easier to glide molecular pairs past each
other rather than change their length, the $S_{\tilde{A}}$ phase will be favoured (Barois *et al.*
1981), i.e. $\rho = \rho_0 \cos 2k_1 z$ and $\Phi = \Phi_0 \cos k_1 z \cos q_x x$. Mean-field theory predicts a
transition fairly similar to the S_{A_1}–S_{A_2} one, but since the order parameter fluctuations
occur on a cone in k space it should, like the NS_C transition, be first-order. Recent
experiments show indeed a weak first-order character, and confirm the theoretical
interpretation of the $S_{\tilde{A}}$ phase as a two-dimensional escape from incommensurability

(Safinya 1983). The phase diagram obtained is not far from the experimental one (figure 21), except that the existence domain of the S_{A_d} phase should not be closed according to the theory, the S_{A_2}–S_{A_d} line should end before reaching the NS_{A_2} line (at least in a plot at constant incommensurability). It would be interesting to check whether it is really closed or not. If the ability of molecular pairs to glide past each other is even larger, and if the molecules are allowed to tilt within the layers, an $S_{\tilde{C}}$ stability domain opens up. $S_{\tilde{C}}$ is defined by

$$\rho = \rho_0 \cos(2k_1 z); \quad \phi(r) = \phi_1 \cos[(k_1 + \delta k)z + q_x x] + \phi_2 \cos[(k_1 - \delta k)z - q_x x],$$

or $\Phi(r) = \Phi_0(x, z) \cos[k_1 z + \phi(x,z)]$ with Φ_0, ϕ being periodic functions of wavevector $(q_x, \delta k)$. A theoretical diagram similar to that of figure 23 can be conjectured since the unbalance between Φ_1 and Φ_2 (leading to $\delta k \neq 0$) is prompted by the proximity of the S_{A_d} phase (i.e. the S_{A_d}–$S_{\tilde{C}}$ transition is characterized as the onset of a (small) ρ, Φ_2 pair in a matrix of a strong already condensed Φ_1). Again fluctuations should render this transition first-order, because of the Brazovskii argument. On the other hand the $S_{\tilde{C}}$–S_{A_1} boundary should be clearly first-order even in the mean-field approximation. The incommensurate phase $S_{A_{i2}}$ predicted by theory (Prost 1980) has not yet been observed. An incommensurate phase of a similar type has been observed in the smectic E phase but this is, strictly speaking, a crystal (Brownsey and Leadbetter 1980) and it corresponds more closely to the $S_{A_{i1}}$ that one can predict in the case $l' \simeq l$.

The nematic re-entrant behaviour seen in figures 22 and 23 is clearly connected with the bicritical type of topology near the N–S_{A_d}–S_{A_1} point. Mean-field diagrams involving N–S_{A_d}–S_{A_1} points do not exhibit re-entrant behaviour but they do show that the relevant terms of the free energy near that point are typical of two-order parameters coupled only at fourth order

$$F' = \tfrac{1}{2} \int \{ A'_\rho \rho'^2 + A'_\Phi \Phi'^2 + \tfrac{1}{2} B_\rho \rho'^4 + \tfrac{1}{2} B_\phi \phi'^4 + B_{\rho\phi} \rho'^2 \phi'^2 \} \, dv \tag{51}$$

Such a free energy has been extensively studied in d dimensions, ρ' and ϕ' being considered as vectorial order parameters with n components. In the smectic case, ρ' and Φ' are two-component order parameters,

$$\rho' \begin{cases} \rho_0 \cos \psi_\rho \\ \rho_0 \sin \psi_\rho, \end{cases} \quad \Phi' \begin{cases} \phi_0 \cos \psi_\phi \\ \phi \sin \psi_\phi, \end{cases}$$

ψ_ρ and ψ_ϕ being the phase as described in §2, and A'_ρ and A'_ϕ are given by (50) but centred at q_0 and $q'_0 = 0$. Equation (51) neglects the coupling with the director fluctuations. Hand-waving arguments (Prost 1980) suggest that re-entrance may occur when $A'_\rho \simeq A'_\phi < 0$, but $A'_\rho + B_{\rho\phi} \langle \phi^2 \rangle$, and $A'_\phi + B_{\rho\phi} \langle \rho'^2 \rangle$, are greater than 0. In other words, re-entrance is induced by the fluctuations of the competing order parameters. Solutions for quasi-one-dimensional Ising systems involving the transfer integral technique (Coulon and Prost 1981), the harmonic approximation for two coupled XY parameters (Prost and Barois 1983), together with an exact solution of the $n = \infty$ case do reveal a re-entrant behaviour connected to the existence of a multicritical point (bicritical or tetracritical). The general argument can be cast in a simple form for any n and $2 < d < 4$. Let us express the temperature dependence of A'_ρ and A'_ϕ explicitly, i.e.

$$\left. \begin{array}{l} A'_\rho = a_\rho(T - T_0 + g) + \text{gradient terms,} \\ A'_\phi = a_\phi(T - T_0 - g) + \text{gradient terms.} \end{array} \right\} \tag{52}$$

J. Prost

The existence of a multicritical point governed by (51) implies (Kosterlitz *et al.* 1976) that the correlation functions $G_{\rho\rho}$ and $G_{\phi\phi}$ obey scaling relations (t=reduced temperature)

$$G(t,g) = \bar{t}^{-\gamma} f\left(\frac{\tilde{g}}{\bar{t}^{\psi}}\right) \tag{53}$$

in which the scaling axes $\tilde{g}$ and $\bar{t}$ are given by linear combinations of g and t

$$\left. \begin{array}{l} \bar{t} = t - pg, \\ \tilde{g} = g + qt. \end{array} \right\} \tag{54}$$

Equation (53) implies that the phase boundaries obey the relation (figure 26)

$$\tilde{g} = \pm W_{\pm} \bar{t}^{\psi}. \tag{55}$$

As figure 27 makes clear, it is then straightforward to show that there exists a range of g-values for which one observes a re-entrant behaviour. One should remark that such an interpretation of re-entrance is very general and does not depend on the particular system under study (magnetic systems should also exhibit re-entrance at the spin-flop bicritical point). Models based on in-plane frustration (Berker and Walker 1981) also lead to re-entrance because of the existence of a zero-temperature bicritical point. However, as is clear from a comparison between figures 22 and 23, the amplitude of the phenomenon is not universal. Note that diagrams such as the one displayed in figure 27 lead to a fundamentally doubly re-entrant phenomenon such as observed with T8 or in figures 22 and 23 (this may be experimentally hidden if crystallization occurs before the smectic re-entrance). The triple re-entrance referred to in the Introduction has not yet been interpreted.

If many experimental systems support this interpretation (Hardouin and Levelut 1980, Hardouin *et al.* 1982, 1983, Engelen *et al.* 1979) it is not clear that the re-entrance obtained with two benzene ring compounds (Cladis 1975, Cladis *et al.* 1977) is open to the same type of explanation. Many microscopic models, or *ad hoc* Landau theories, can give single re-entrance in agreement with these early experiments (Pershan and Prost 1979, Cladis 1981, Longa and De Jeu 1982, Hida 1981, Dowel 1983). The real problem is to find which one is the more physically relevant.

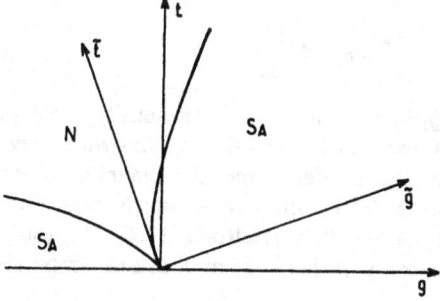

Figure 27. Topology near a multicritical point: the two boundaries separating the high symmetry phase from the low symmetry ones are tangent to the $\bar{t}$ scaling axis. However the physical t axis is rotated from it by a non-universal amount. Thus varying temperatures will always lead to re-entrant behaviour for well chosen parameters.

Finally we remark that if one approaches tangentially from the curved boundaries, the exponents γ and ν should be doubled in the high symmetry phase (Prost 1980). This point, known in other systems (Griffiths and Wheeler 1970, Tufeu *et al.* 1975) has been accurately checked in the N–S_A case (Kortan *et al.* 1981, 1984).

5. Hydrodynamic properties

5.1. *Conventional approach*

The understanding of the hydrodynamic properties of smectics has improved strikingly in recent years, but in order to assess correctly the importance of these developments it is necessary to recall the more conventional theories which were used earlier. By definition, hydrodynamic theories are preoccupied with the long wavelength, low frequency, behaviour of systems. Hydrodynamic variables are such that the lifetime of a given mode goes to infinity with its wavelength. They can be of two different origins: conserved quantities or continuous broken symmetry. In smectic A, we count, as in any system, five conserved quantities (mass, energy, momentum) plus one variable associated with the translational broken symmetry (de Gennes 1968, Martin *et al.* 1972), the layer displacement u. From the linearity of hydrodynamic equations, one can infer that the number of normal modes equals that of the variables. Thus in S_A liquid crystals we should have six normal modes: since, because of time reversal symmetry, propagating modes occur in pairs, we expect on general grounds either one propagating with four diffusive modes or two propagating with two diffusive modes (since one always expects ordinary or first sound and thermal diffusion to exist, we do not consider the extreme cases of six diffusive and of three propagative modes).

With S_{BHex} or S_C, we have one more variable and hence one more mode. In both cases this extra mode is expected to be nematic-like, that is diffusive. Hydrodynamic equations of both phases should be very similar, the lower symmetry of S_C allowing, however, for a larger number of cross-terms. On the other hand, the hydrodynamics of S_F should be identical to that of S_C. In the following we will concentrate on hydrodynamic properties of smectic A. The equations read (Martin *et al.* 1972)

$$\left.\begin{aligned}
&\frac{\partial \rho}{\partial t}+\nabla_i(\rho v_i)=0,\\[2mm]
&\frac{dg_i}{dt}=-\nabla_i p+\nabla_j\sigma_{ij}-\delta_{iz}\frac{\delta F_0}{\delta u},\\[2mm]
&\frac{\partial u}{\partial t}-v_z=\lambda_p\frac{\delta F}{\delta u}
\end{aligned}\right\} \tag{56}$$

(ρ os the mass density, $g_i=\rho v_i$ is the momentum density, p the pressure, and σ_{ij} the stress tensor). The first two equations are well known. They express mass and momentum conservation. The last determines the dynamics of the layer displacement variable. Note that $\partial u/\partial t$ is not identical to v_z which implies that matter can flow through the layers, i.e. permeation (Helfrich 1968). In (56), F_0 is the energy considered at constant entropy, rather than at constant temperature as was the case for equation (1).

Including the bulk compressibility we obtain

$$F_0=\frac{1}{2}\int\left[B_0\left(\frac{\partial u}{\partial z}\right)^2+2C_0\frac{\partial u}{\partial z}\left(\frac{\delta\rho}{\rho}\right)+A_0\left(\frac{\delta\rho}{\rho}\right)^2+K_1(\Delta_\perp u)^2\right]dv \tag{57}$$

(The zero subscript denotes that the moduli are to be taken at constant entropy.)
$(-\delta\rho/\rho)$ is the bulk dilation. B_0 is the layer-compressibility elastic constant, A_0 the
bulk modulus, and C_0 describes the layer-density coupling. K_1 is again the splay
Frank–Oseen elastic constant.

Solving these equations shows that, depending on the direction of the wave-
vector with respect to the optic axis, one can indeed have first and second sound, plus
two diffusive modes, or first sound only and four diffusive modes (de Gennes 1968,
Martin et al. 1972). They are identified as follows:

$(\mathbf{k}\cdot\mathbf{n}_0)$ and $(\mathbf{k}\times\mathbf{n}_0)\neq 0$:	first sound (essentially density modulation) second sound (layer shear wave) vorticity diffusion (polarized in a direction perpendicular to the $(\mathbf{k},\mathbf{n}_0)$ plane) thermal diffusion
$\mathbf{k}\cdot\mathbf{n}_0=0$:	first sound overdamped nematic-like, layer undulation mode two vorticity diffusion modes (two orthogonal polarizations) thermal diffusion
$\mathbf{k}\times\mathbf{n}_0=0$:	first sound permeation mode two vorticity diffusion modes (two orthogonal polarizations) thermal diffusion

Experiments have largely supported this approach to the dynamical behaviour of
smectics. The anisotropy of the first sound has been evidenced by Brillouin scattering
as well as by ultrasonic techniques (Liao et al. 1973, Miyano and Ketterson 1973,
Kiry and Martinoty 1976, 1978, Bradberry and Vaughan 1977). Second sound has
also been evidenced with these two techniques (Liao et al. 1973, Bacri 1973) but the
slow propagation velocity renders its observation very difficult. Furthermore, the
frequency range in which Brillouin scattering and ultrasound measurements were
made is clearly outside the hydrodynamic regime ($\sim$gigahertz). Hence specific
experiments had to be designed to study second sound in a forced Rayleigh
scattering technique using interdigitated electrodes to excite the layer shear mode
(Ricard and Prost 1979, 1981) and later a surface wave technique (Fisch et al. 1981).
The continuous shift from propagating to overdamped behaviour has been clearly
observed for $\mathbf{k}\cdot\mathbf{n}_0\simeq 0$. The purely overdamped $\mathbf{k}\cdot\mathbf{n}_0\simeq 0$, nematic-like mode had
been previously observed (Ribotta et al. 1974, Birecki et al. 1976). Permeation has
also been indirectly evidenced by the observation of the flow profile around obstacles
(Clark 1978) as suggested by de Gennes (1974 b). Thus, most of the experimental
observations agree with theory. However, one can observe a fairly large scatter in the
data obtained for B, depending on whether it is measured at low ($\lesssim 10^3$ Hz), or high
(10^9 Hz) frequencies. There is typically a factor of 10 to 20 between the results of
each type of experiment. This can be rationalized if one remarks that, at low
frequencies, dislocations may move and reduce the apparent elastic constant
(Pershan and Prost 1975, Ricard and Prost 1981) and that above 10^7 Hz one leaves
the hydrodynamic regime (i.e. the characteristic time of the molecular tilt with

respect to the layers is $\sim 10^{-7}$ s). There is however no formal proof that this way of thinking is correct. Another point concerns ultrasound attenuation. As previously stated, equations (56) predict two modes propagating according to a dispersion relation ($i = 1, 2$, i.e. first, and second sounds)

$$\left.\begin{array}{l} q_1 = \omega/C_1(\alpha) + i\Gamma_1(\alpha, \omega), \\ q_2 = \omega/C_2(\alpha) + i\Gamma_2(\alpha, \omega) \end{array}\right\} \tag{58}$$

($\mathbf{q}$ is the wavevector, $\omega/2\pi$ the frequency, $C(\alpha)$ the propagation velocity which depends on α the angle $(\mathbf{q}, \mathbf{n}_0)$). Within this type of treatment, Γ_i scales like ω^2. Some experiments in smectics do not follow this square dependence (Battacharya *et al.* 1979, 1981) but others are in good agreement with conventional hydrodynamic theories (Martinoty 1979), provided molecular relaxation processes are taken into account.

5.2. *Breakdown of conventional hydrodynamics*

In their general formulation of the hydrodynamics of crystals, liquid crystals and liquids, Martin *et al.* (1972) pointed out that, because of the lack of long-range order, it was not clear that the usual expansion in terms of gradients would be valid for smectics. Until recently, this remark has generally been considered as a purist word of caution and nothing more. Recent work concerning the statics (Grinstein and Pelcovits 1981, 1982) and dynamics (Mazenko *et al.* 1982, 1983) profoundly change the picture by showing that the conventional notions of elastic constants and viscosities breakdown at long wavelength. We will try to illustrate their results in a simple-minded fashion. The first point is that since there is no long-range order in the harmonic approximation, fluctuations are large and anharmonic corrections to the smectic free energy should be taken into account. As written in equation (1), the

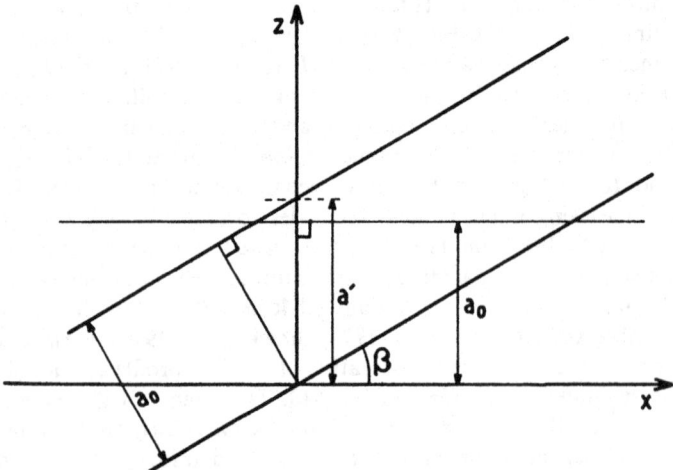

Figure 28. Nature of the corrective term $\frac{1}{2}(\nabla_\perp u)^2$: the layers rotated by an angle $\beta = |\nabla_\perp u|$, at constant layer spacing a_0, appear to have a dilated thickness $a' = (a_0/\cos\beta) \simeq a[1 + (\beta^2/2)] \simeq a_0\{1 + [(\nabla_\perp u)^2/2]\}$ in the z direction in which by convention it is measured. This additive term must be subtracted off since it does not correspond to a physical dilation.

J. Prost

expression is rotationally invariant only to lowest order; as shown in figure 28 rotating the layers from an angle $\nabla_\perp u$ leads to an apparent dilation $\frac{1}{2}(\nabla_\perp u)^2$ which must be subtracted from the genuine physical dilation. Hence the compressional term reads

$$\int \frac{B}{2}(\nabla_z u - \tfrac{1}{2}(\nabla_\perp u)^2)^2 \, dv. \tag{59}$$

This modification of the free energy had been previously introduced to successfully explain the spectacular undulation instability, obtained when one pulls the boundaries of a smectic confined between two parallel flat plates in the homeotropic geometry (Delaye *et al.* 1973 b, Clark and Meyer 1973, Ribotta 1974, Clark 1976). The renormalization of B, due to the layer fluctuations can be estimated in a way similar to that used to discuss the influence of impurities in smectics (Pershan and Prost 1975).

According to (59) the stress field ϕ_z reads

$$\Phi_z = B(\nabla_z u - \tfrac{1}{2}(\nabla_\perp u)^2). \tag{60}$$

We want now to find the relation between an external stress $\Phi_{z(q)}^{ext}$, and $|\nabla_z u|$. Because of the coupling between Φ_z^{ext} and $(\nabla_\perp u)^2$ this last quantity will be polarized

$$\tfrac{1}{2}\langle(\nabla_\perp u)^2(\mathbf{q})\rangle_{\Phi_z} - \tfrac{1}{2}\langle(\nabla_\perp u)^2\rangle_{\Phi_z = 0} = \chi_{(q)}\Phi_{z(q)}^{ext} \tag{61}$$

hence

$$\Phi_z^{ext}(1 + B\chi(\mathbf{q})) = B\langle\nabla_z u - \tfrac{1}{2}\langle(\nabla_\perp u)^2\rangle_{\Phi_z = 0}\rangle, \tag{62}$$

or with a suitable redefinition of the equilibrium layer spacing

$$\Phi_z^{ext} = \frac{B}{1 + B\chi(\mathbf{q})}\langle\nabla_z\tilde{u}\rangle \tag{63}$$

that is

$$B^R(q) = B/(1 + B\chi(\mathbf{q})). \tag{64}$$

$\chi(\mathbf{q})$ may be estimated via the fluctuation-dissipation theorem

$$\chi(\mathbf{q}) = \frac{\beta}{4}\{\langle(\nabla_\perp u)^2(q)(\nabla_\perp u)^2(-q)\rangle_{\Phi_z = 0} - \langle(\nabla_\perp u)^2\rangle_{\Phi_z = 0}^2 \delta_q\}. \tag{65}$$

The last term on the right-hand side of (65) does not bring anything remarkable but the first, when estimated in the harmonic approximation, gives a $\ln q$ dependence

$$\langle(\nabla u)^2(q)(\nabla u)^2(-q)\rangle_{\Phi_z = 0} = \frac{k^2 T^2}{8B^2}\int \frac{d^3k}{8\pi^3}\frac{[\mathbf{k}_\perp \cdot (\mathbf{q}_\perp - \mathbf{k}_\perp)]^2}{(k_z^2 + \lambda^2 k_\perp^4)(q_z - k_z)^2 + \lambda^2(\mathbf{q} - \mathbf{k}_\perp)^4)},$$

$$(\lambda = \sqrt{(K/B)}). \tag{66}$$

Taking $k_\perp$ as $O(\varepsilon)$ and k_z as $O(\varepsilon^2)$ shows that the integrand for $q = 0$ behaves as $(\varepsilon^8/\varepsilon^8) = \varepsilon^0$, that is it diverges logarithmically. One finds

$$\chi(\mathbf{q})\alpha\frac{kT}{B^2\lambda^3}\ln\left[\frac{k_c^4}{\max(\lambda^{-2}q_z^2, q_\perp^4)}\right] \tag{67}$$

and

$$B^{R}_{(q)} = B \left[1 + xW \ln \frac{k_c^2}{\max(\lambda^{-1}q_z, q_\perp^2)} \right]^{-1}, \tag{68}$$

x is a numerical factor. $W = kT/B\lambda^3 = kTB^{1/2}K_1^{-3/2}$ is a dimensionless number, crucial for estimation of the cross-over regime beyond which the logarithmic behaviour should show up. Because of the exponential dependence of this length on W, its value is poorly known. If instead of applying a stress at a wavevector $\mathbf{q}$, one applies an homogeneous stress Φ_0, the cut-off is no longer governed by a wavevector but by the quantity $\Phi_0/\lambda^2 B$; combining (63) and (68) one easily sees that Hooke's law is no longer satisfied at small stresses which is usually the range in which it should apply!

The quick derivation used here gives the essential characteristics of the conventional elastic-theory breakdown. However it is not rigorous: indeed the elastic constant B appears on both sides of equation (68), which shows clearly that the correct approach would be a renormalization procedure. Such a procedure has been developed (Grinstein and Pelcovitz 1981, 1982) in a formalism using the Callan Zymanzig equations (see, for example, Brezin *et al.* 1976) directly in three dimensions. One finds

$$\left. \begin{aligned} B^{R}(q) &\simeq B \left(1 + \frac{5W}{128\pi} \ln \frac{k_c^2}{\max(\lambda^{-1}q_z, q_\perp^2)} \right)^{-4/5}, \\[2ex] K_1(q) &= K_1 \left(1 + \frac{5w}{128\pi} \ln \frac{k_c^2}{\max(\lambda^{-1}q_z, q_\perp^2)} \right)^{2/5}. \end{aligned} \right\} \tag{69}$$

The rigorous result (69) is not strikingly different from the approximate one we derived. It is conceptually very important, and not only shows the limits of the *gradient expansion assumption* (i.e. dispersion of Bq_z^2) but also gives a rigorous measure of the importance of the Landau–Peierls instability (i.e. divergence of $\langle u^2 \rangle$). Note that B is renormalized to zero at long wavelength whereas K_1 goes to infinity. According to Grinstein and Pelcovitz this effect could be measurable although not spectacular, in the currently available experimental range. Their estimates are based on bare elastic constants extracted from second sound measurements. Considering bare elastic constants close to crystalline values would increase drastically the validity range of the logarithmic behaviour but by no means explain the large dispersion observed between low and high-frequency measurements. As proposed in the preceding subsection dislocation and impurity motion on the one hand, intrinsic non-hydrodynamic modes on the other certainly play an important role. Direct mechanical measurements (Bartolino and Durand 1977) could in principle reveal non-hookean behaviour, provided finite size effects could be avoided. An extension of this theory, including that of bulk compressibility, shows that A does not diverge at small q (although slightly renormalized) and that C behaves like B (Mazenko *et al.* 1983).

More spectacular results are obtained in the dynamical case. Three viscosities are found to diverge as $1/\omega$, which is a very strong divergence (Mazenko *et al.* 1982, 1983). Again we find it useful to give a simple derivation of results that have been obtained in a more rigorous way. The difference between static and dynamic results will appear as a simple consequence of the Kramers–Kronig relations. Consider the field h conjugated to the layer displacement u. The local expression of the equations of motion may be written

$$h(\mathbf{q}, \omega) = L(q, \omega)u(\mathbf{q}, \omega) + NL(u)(\mathbf{q}, \omega), \tag{70}$$

$L(q, \omega)$ is the linear operator that we get upon solving equations (56). $NL(u)$ is the non-linear part in which for simplicity we keep only the static non-linearity. The average relation between $\langle h \rangle$ and $\langle u \rangle$ should obey the linear response theory with

$$\left. \begin{aligned} \langle h(\mathbf{q}, \omega) \rangle &= \chi(q, \omega)\langle u(\mathbf{q}, \omega) \rangle, \\ \chi''(q, \omega) &= \frac{\omega}{2kT} \langle h(\mathbf{q}, \omega)h(-\mathbf{q}, -\omega) \rangle. \end{aligned} \right\} \tag{71}$$

Replacing h by (70) and remarking that $\langle LuNL(u) \rangle = 0$, one finds

$$\chi''(q, \omega) = L''(q, \omega) + \frac{\omega}{2kT} \langle NL(u)(q, \omega)NL(u)(-q, -\omega) \rangle, \tag{72}$$

with

$$NL(u)(q, \omega) = \frac{B}{2} \int \frac{d^3k}{8\pi^3} \frac{d\Omega}{2\pi} u(\mathbf{k}, \Omega)u(\mathbf{q} - \mathbf{k}, \omega - \Omega)$$

$$\times [jq_z\mathbf{k}_\perp \cdot (\mathbf{q} - \mathbf{k}_\perp) \mapsto 2jq_\perp k_z(\mathbf{q}_\perp - \mathbf{k}_\perp)] \tag{73}$$

(in which for simplicity we have assumed $C = 0$).

Thus, evaluating the average in the gaussian approximation, one finds

$$\chi''(q, \omega) = L''(q, \omega) + \frac{\omega B^2}{8kT} \int \frac{d^3k}{8\pi^3} \frac{d\Omega}{2\pi} \langle u(\mathbf{k}, \Omega)u(-\mathbf{k}, -\Omega) \rangle$$

$$\times \langle u(\mathbf{q} - \mathbf{k}, \omega - \Omega)u(\mathbf{k} - \mathbf{q}, \Omega - \omega) \rangle \Gamma(\mathbf{k}, \mathbf{q}), \tag{74}$$

with

$$\Gamma(\mathbf{k}, \mathbf{q}) = -(q_z\mathbf{k}_\perp \cdot (\mathbf{q}_\perp - \mathbf{k}_\perp) + 2q_\perp k_z(\mathbf{q}_\perp - \mathbf{k}_\perp))$$

$$\times [2q_z\mathbf{k}_\perp \cdot (\mathbf{k}_\perp(\mathbf{k}_\perp - \mathbf{q}_\perp)) + 2q_\perp \cdot (k_z(\mathbf{k}_\perp - \mathbf{q}_\perp) + (k_z - q_z)\mathbf{k}_\perp].$$

Keeping the leading terms only

$$\Gamma(\mathbf{k}, \mathbf{q}) \simeq -2[q_z k_\perp^2 + 2(\mathbf{q}_\perp \cdot \mathbf{k}_\perp)k_z]^2. \tag{75}$$

Equations (74) and (75) correspond exactly to the results found by Mazenko *et al.* (1982, 1983) (in the case $C = 0$). Performing the integrals, and identifying the expressions obtained with the structure expected for χ'', one finds that the viscosity η_1 diverges

$$\eta_1 \simeq \eta_1^0 + \frac{BW}{64|\omega|}, \qquad (\text{for } |\omega| \gg |\mathbf{q}|), \tag{76}$$

or

$$\Delta\eta_1 = \eta_1 - \eta_1^0 \alpha \frac{W\eta_3^0}{\lambda^2 \max(\lambda^{-1}q_z, q_\perp^2)} \qquad (\text{for } |\mathbf{q}| \gg |\omega|). \tag{77}$$

Removing the hypothesis $C = 0$ leads to a similar divergence of η_4 and η_5 with the relation $\Delta\eta_1\Delta\eta_4 = (\Delta\eta_5)^2$. The exact formulae are given in Mazenko *et al.* (1983)

(note that the relations here correspond to unit-dimensions $ML^{-1}T^{-2}$ compared to L^2T^{-1} in their paper). Higher-order corrections lead also to the divergence of η_2.

It is interesting to understand the origin of the difference between the dynamic and static effects. They are both governed by very similar integrals over correlation functions:

$$\left.\begin{array}{ll} \text{Static} & \int d^3k k_\perp^4 \langle u(k)u(-k)\rangle^2, \\[2em] \text{Dynamic} & \int d^3k k_\perp^4 d\Omega \langle u(k,\Omega)u(-k,-\Omega)\rangle^2. \end{array}\right\} \qquad (78)$$

Taking $k_\perp = O(\varepsilon)$, $k_z = O(\varepsilon^2)$, $\Omega = O(\varepsilon^2)$, one finds by simple power counting that the integrand behaves at small $\mathbf{k}$ and Ω like ε^0 in the static case (i.e. a logarithmic divergence), like ε^{-2} in the dynamical one (i.e. a q^{-2} or q_z^{-1} or ω^{-1} divergence). Indeed knowing from (1) that $\langle (u(k)u(-k)\rangle = [O(\varepsilon^4)]^{-1}$, one deduces right away, from the Kramers–Kronig relations, that $\langle u(k,\Omega)u(-k,-\Omega)\rangle = [O(\varepsilon^6)]^{-1}$ which is a much more *dangerous* divergence.

The behaviour of the η_3 viscosity coefficient, which governs the damping of the undulation mode, is different. Indeed, it is obtained from (75) in the case $q_z = 0$, $q_\perp \neq 0$. The integrand now does not read $\int d^3k k^4$ but $\int d^3k k_z^2 k^2$, and power counting gives ε^0 as in the static case. Thus η_3 exhibits only a weak logarithmic divergence (this is fortunate since equations such as (77) contain η_3 as an input, and if it were strongly modified, one would need a full renormalization procedure).

The experimental consequences of the divergences of the viscosities are probably not yet all known. The most direct is perhaps that the damping terms in ultrasonic attenuation are not expected to scale like ω^2 but like $(\omega + a\omega^2)$. Experimental data seem to agree with these theories (Battacharya and Ketterson 1982); indeed a linear contribution may provide a good fit to the experiment, and the bare viscosities extracted from the fitting procedures agree fairly well with data obtained in the nematic phase (another favourable point is that the linear term is not found in the nematic phase). On the other hand one may question the role of non-hydrodynamic modes in these experiments (Martinoty 1979) and it is valuable to confirm this theory with different techniques. Recent measurements with ultrasonic techniques (Gallani and Martinoty 1984) confirm the diverging character of viscosities for mono-layer smectics, but also show that finite-size effects may be significant. Forced Rayleigh scattering experiments on a Lecithin-water system do give results which suggest diverging viscosities (Marcerou *et al.* 1983). The most surprising consequence originates probably from the divergence of η_2. When sheared *inside* the layers, smectics A should reveal their liquid-like nature in directions perpendicular to the optic axis and flow like a liquid and this is indeed the case according to conventional hydrodynamic theories. However, the divergence of η_2 shows that one has to exceed a yield stress before flow can start! Indeed, in an experiment at constant shear rate (dimension as for ω) the cut-off in integrals such as (74) will be governed by the shear rate itself the resulting *viscous stress* $\sigma = s\eta_2 = s\eta_2^0 + s\delta\eta_2 = s\eta_2^0 + \text{constant}$ exhibits a value independent of s (Ramaswamy 1983). Experiments seem again to support these expectations (Kim *et al.* 1976), Battacharya and Letcher 1980). Last but not least, the existence of an imaginary part in the diverging term of η_2 leads to the prediction of a new propagating pair of modes, replacing the single vorticity diffusion mode.

J. Prost

Even the equality between the number of modes and the number of hydrodynamic variables breaks down!

6. Conclusion

After reading this review, one may have the impression that a great deal of effort is involved in understanding these systems which are rather exotic. Layered structures are in fact frequently encountered states of matter. The fastest liquid-crystal display devices are currently based on the smectic state (the smectic A–nematic–isotropic transitions: Hareng and Leberre 1978, Hareng *et al.* 1983; ferroelectric S_{C^*}: Clark and Lagerwall 1980, Clark *et al.* 1983). They have response times compatible with TV. Smectics have also been shown to be excellent lubricants (Oswald and Kleman 1982). The development of modern technology, in particular miniaturization, requires an extensive understanding of two-dimensional physics and smectics provide excellent model systems for testing our knowledge in that domain. Far-from-equilibrium spatial instabilities often start as roll patterns which mimic smectics: one can use the elastic and hydrodynamic theories of (V, A) to discuss the long wavelength behaviour, and the defect patterns in such systems (Guazzelli *et al.* 1983, Dubois-Violette *et al.* 1983). One can even construct the equivalent of frustrated smectics with *Williams domains* submitted to an externally periodic field (Lowe *et al.* 1983). A long-known field of application of smectics is lyotropic systems (Charvolin 1983, Pershan 1979). What will be the relevance to biological systems is not clear yet, but the myelin in nerves is made of cylindrically stacked layers and the rods and cones of the eyes and chloroplasts of plants have structures similar to that of smectics; membrane cells are in many ways similar to thin films, and so on. Smectics also play an important role in polymer physics (Samulski and Dupre 1983, Finkelmann *et al.* 1983), and appear in microemulsion-related phase diagrams (Di Meglio *et al.* 1983). Perhaps even in astrophysics smectics may play a role: it is currently believed that pion condensation in neutron stars assumes a layered structure (Baym 1978)!

Acknowledgments

It is a pleasure to thank P. Barois, F. Hardouin, G. Sigaud and N. A. Clark for their critical reading of the manuscript, C. Destrade and Nguyen Huu Tinh for providing us with the pictures of different smectic phases and T. C. Lubensky for the illuminating discussions we had during our stay at the University of Pennsylvania and his help in the scaling-law formulation of re-entrance.

References

ALBEN, R., 1972, *Solid St. Commun.*, **13**, 1783.
ALS NIELSEN, J., 1981, *Symmetries and Broken Symmetries*, edited by N. Bocarra (Paris: IDSET), p. 107.
ALS NIELSEN, J., BIRGENEAU, R. J., KAPLAN, M., LITSTER, J. D., and SAFINAYA, C. R., 1977, *Phys. Rev. Lett.*, **39**, 352.
BACRI, J. C., 1973, *J. Phys., Paris*, **37**, C3.
BAROIS, P., COULON, C., and PROST, J., 1981, *J. Phys. Lett., Paris*, **42**, L107.
BAROIS, P., PROST, J., and LUBENSKY, T. C., 1984 (to be published).
BARTOLINO, R., and DURAND, G., 1977, *Phys. Rev. Lett.*, **39**, 1346; 1984, *J. Phys., Paris*, **45**, 889.
BATTACHARYA, S., and KETTERSON, J. B., 1982, *Phys. Rev. Lett.*, **49**, 997.
BATTACHARYA, S., and LETCHER, S. V., 1980, *Phys. Rev. Lett.*, **44**, 414.
BATTACHARYA, S., SARMA, B. K., and KETTERSON, J. B., 1981, *Phys. Rev. B*, **23**, 2397.

BATTACHARYA, S., SHEN, S., and KETTERSON, J. B., 1979, *Phys. Rev.* A, **19**, 1219.

BAYM, 1978, *Proceedings of Les Houches Summer Institute 1977* (Amsterdam: North Holland).

BENATTAR, J. J., DOUCET, J., and LAMBERT, M., 1979, *Phys. Rev.* A, **20**, 2505.

BENATTAR, J. J., MOUSSA, F., and LAMBERT, M., 1981, *J. Phys. Lett., Paris*, **42**, L.67; 1983, *J. Chim. phys.*, **80**, 99.

BENGUIGUI, L., 1979, *J. Phys., Paris*, **40**, C3–419.

BERESNEV, L. A., and BLINOV, L. M., 1981, *Ferroelectrics*, **33**, 129.

BIRKER, P. N., and WALKER, J. S., 1981, *Phys. Rev. Lett.*, **47**, 1469.

BIRECKI, H., and LITSTER, J. D., 1977, *Molec. Crystals liq. Crystals*, **42**, 33.

BIRECKI, H., SCHAETZING, R., RONDELEZ, F., and LITSTER, J. D., 1976, *Phys. Rev. Lett.*, **36**, 1376.

BIRGENEAU, R. J., GARLAND, C. W., KASTING, G. V., and OCKO, B. M., 1981, *Phys. Rev. A*, **24**, 2624.

BIRGENEAU, R. J., and LITSTER, J. D., 1978, *J. Phys. Lett., Paris*, **39**, 399.

BRADBERRY, G. W., and VAUGHAN, J. M., 1977, *Physics Lett.*, **62**, 225.

BRAGG, W. H., 1934, *Nature, Lond.*, **133**, 445.

BRAZOVSKII, S. A., 1975, *Zh. ésp. teor. Fiz.*, **68**, 175 (*Soviet Phys. JETP*, **41**, 85).

BREZIN, E., LE GUILLOU, J. C., and ZINN-JUSTIN, J., 1976, *Phase Transitions and Critical Phenomena*, Vol. 6, edited by C. Domb and M. S. Green (London: Academic).

BRISBIN, D., DE HOFF, R., LOCKHART, T. E., and JOHNSON, D. L., 1979, *Phys. Rev. Lett.*, **43**, 1171.

BROCHARD, F., 1973, *J. Phys., Paris*, **34**, 411.

BRUINSMA, R., and AEPPLI, G., 1982, *Phys. Rev. Lett.*, **48**, 1039.

BRUINSMA, R., and NELSON, D. R., 1981, *Phys. Rev. B*, **23**, 335.

CAILLE, A., 1972, *C. r. hebd. Séanc. Acad. Sci., Paris* B, **274**, 891.

CHANDRASEKHAR, S., 1977, *Liquid Crystals* (Cambridge Monographs on Physics) (Cambridge, London, New York, Melbourne: Cambridge University Press).

CHARVOLIN, J., 1983, *J. Chim. phys.*, **80**, 15.

CHEN, J. H., and LUBENSKY, T. C., 1976, *Phys. Rev. A*, **14**, 1202.

CHEN, K., SORENSEN, L., HARDOUIN, F., and PERSHAN, P. S., 1984 (to be published).

CHEUNG, L., MEYER, R. B., and GRULER, H., 1973, *Phys. Rev. Lett.*, **31**, 349.

CHU, K. C., and MACMILLAN, W. L., 1977, *Phys. Rev. A*, **15**, 1181.

CLADIS, P. E., 1975, *Phys. Rev. Lett.*, **35**, 48; 1981, *Molec. Crystals liq. Crystals*, **67**, 833.

CLADIS, P. E., BOGARDUS, R. K., DANIELS, W. B., and TAYLOR, G. N., 1977, *Phys. Rev. Lett.*, **39**, 720.

CLARK, N. A., 1976, *Phys. Rev. A*, **14**, 1551; 1978, *Phys. Rev. Lett.*, **40**, 1663.

CLARK, N. A., and LAGERWALL, 1980, *Appl. Phys. Lett.*, **36**, 899.

CLARK, N. A., and MEYER, R. B., 1973, *Appl. Phys. Lett.*, **30**, 3.

CLARK, N. A., HANDSCHY, M. A., LAGERWALL, S. T., 1983, *Molec. Crystals liq. Crystals*, **94**, 213.

COLLETT, J., PERSHAN, P. S., SIROTA, E. B., and SORENSEN, L. B., 1984, *Phys. Rev. Lett.*, **52**, 356.

COULON, C., and PROST, J., 1981, *J. Phys. Lett., Paris*, **42**, L241.

CURIE, P., 1908, *Oeuvres* (Gauthier-Villars).

DAVIDSON, P., LEVELUT, A. M., STRZELECKA, H., and GIONIS, V., 1983, *J. Phys. Lett., Paris*, **44**, L823.

DASGUPTA, C., and HALPERIN, B. I., 1981, *Phys. Rev. Lett.*, **47**, 1556.

DAY, A. R., LUBENSKY, T. C., and MCKANE, A. J., 1983, *Phys. Rev. A*, **27**, 1461.

DAWEY, S. C., BUDAI, J., GOODBY, J. W., and PINDAK, R., 1984, *Phys. Rev. Lett.* (submitted).

DE GENNES, P. G., 1968, *J. Phys., Paris*, **30**, C4; 1972, *Solid St. Commun.*, **10**, 753; 1973, *Molec. Crystals liq. Crystals*, **21**, 49; 1974 a, *The Physics of Liquid Crystals* (Oxford: Clarendon Press); 1974 b, *Physics Fluids*, **17**, 1645.

DEHOFF, R., BIGGERS, R., BRISBIN, D., and JOHNSON, D. L., 1982, *Phys. Rev. A*, **25**, 472.

DEHOFF, R., BIGGERS, R., BRISBIN, D., MAHMOOD, R., GOODEN, C., and JOHNSON, D. L., 1981, *Phys. Rev. Lett.*, **47**, 664.

DELAYE, M., 1979, *J. Phys., Paris*, **40**, C3–350.

DELAYE, M., RIBOTTA, R., and DURAND, G., 1973 a, *Phys. Rev. Lett.*, **31**, 443; 1973 b, *Physics Lett. A*, **44**, 139.

DEMUS, D., DIELE, S., KLAPPERSTÜCK, M., LINK, V., and ZASCHKE, H., 1971, *Molec. Crystals liq. Crystals*, **15**, 161.

DESTRADE, C., GASPAROUX, H., FOUCHER, P., NGUYEN HUU TINH, MALTHETE, J., and JACQUES, J., 1983, *J. Chim. Phys.*, **80**, 138.

DE VRIES, A., 1974, *Chem. Phys. Lett.*, **28**, 252.

DI MEGLIO, J. M., DVOLAITZKY, M., OBER, R., and TAUPIN, C., 1983, *J. Phys. Lett., Paris*, **44**, 2229.

DOUCET, J., LEVELUT, A. M., and LAMBERT, M., 1978 *Annls Phys.*, **3**, 157.

DOWEL, F., 1983, *Phys. Rev. A*, **28**, 3526.

DUBOIS-VIOLETTE, E., GUAZZELLI, E., and PROST, J., 1983, *Phil. Mag. A*, **48**, 727.

DUNN, S. G., and LUBENSKY, T. C., 1981, *J. Phys., Paris*, **42**, 1201.

ENGELEN, B., HEPPKE, G., HOPF, R., and SCHNEIDER, F., 1979, *Molec. Crystals liq. Crystals*, **49**, 193.

FINKELMANN, H., BENTHACK, H., and REHAGE, G., 1983, *J. Chim. Phys.*, **80**, 163.

FISCH, M. R., SORENSEN, I. B., and PERSHAN, P. S., 1981, *Phys. Rev. Lett.*, **47**, 43.

FRANK, F. C., 1958, *Discuss. Faraday Soc.*, **25**, 19.

FRIEDEL, G., 1922, *Annls Phys.*, **18**, 273.

FRIEDEL, J., 1964, *Dislocations* (Oxford: Pergamon Press).

GALERNE, Y., 1981, *Phys. Rev. A*, **24**, 2284.

GALLANI, J. C., and MARTINOTY, P., 1984, *Fourth European Winter Conf. on Liquid Crystals*, Bovec, Yugoslavia.

GARLAND, C. W., MEICHLE, M., OCKO, B. M., KORTAN, A. R., SAFINYA, C. R., YU, L. J., LITSTER, J. D., and BIRGENEAU, R. J., 1983, *Phys. Rev. A*, **27**, 3234.

GASPAROUX, H., and PROST, J., 1976, *An. Rev. phys. Chem.*, p. 175.

GRIFFITHS, R. B., and WHEELER, J. C., 1970, *Phys. Rev. A*, **2**, 1047.

GRINSTEIN, G., and PELCOVITS, R., 1981, *Phys. Rev. Lett.*, **47**, 856; 1982, *Phys. Rev. A*, **26**, 915, 2196.

GRINSTEIN, G., and TONER, D., 1984, *Phys. Rev. Lett.*, **51**, 2386.

GUAZZELLI, E., GUYON, E., and WESFREID, E., 1983, *Phil. Mag. A*, **48**, 709.

HALPERIN, B. I., and LUBENSKY, T. C., 1974, *Solid St. Commun.*, **14**, 997.

HALPERIN, B. I., LUBENSKY, T. C., and SHANG-KENG MA, 1974, *Phys. Rev. Lett.*, **32**, 292.

HALPERIN, B., and NELSON, D. R., 1978, *Phys. Rev. Lett.*, **41**, 21.

HARDOUIN, F., ACHARD, M. F., DESTRADE, C., and NGUYEN HUU TINH, 1984, *J. Phys. Lett., Paris*, **45**, 765.

HARDOUIN, F., and LEVELUT, A. M., 1980, *J. Phys., Paris*, **41**, 41.

HARDOUIN, F., LEVELUT, A. M., ACHARD, M. F., and SIGAUD, G., 1983, *J. Chim. phys.*, **80**, 53.

HARDOUIN, F., NGUYEN HUU TINH, ACHARD, M. F., and LEVELUT, A. M., 1982, *J. Phys. Lett., Paris*, **43**, L327.

HARDOUIN, F., SIGAUD, G., ACHARD, M. F., and GASPAROUX, H., 1979, *Physics Lett.*, **71**, 347.

HARENG, M., and LEBERRE, S., 1978, *I.E.D.M. Tech. Dig.*, p. 258.

HARENG, M., LEBERRE, S., MOURET, B., MOUTOU, P. C., PERBET, J. N., and THIRANT, L., 1983, *I.E.E.E. Trans. Electron Devices*, **30**, 507.

HATTA, I., and NOKAYAMA, T., 1981, *Molec. Crystals liq. Crystals*, **66**, 417.

HELFRICH, W., 1978, *J. Phys., Paris*, **39**, 1199.

HIDA, K., 1981, *J. phys. Soc. Japan*, **50**, 3869.

HORNREICH, R. M., and SHTRICKMAN, S., 1977, *Physics Lett. A*, **63**, 39.

HORNREICH, R. M., LUBAN, M., and SHTRIKMAN, S., 1975, *Phys. Rev. Lett.*, **35**, 1678.

HUANG, C. C., and LIEN, S. C., 1981, *Phys. Rev. Lett.*, **47**, 1917.

HUANG, C. C., and VINER, J. M., 1982, *Phys. Rev. A*, **25**, 3385.

HUANG, C. C., VINER, J. M., PINDAK, R., and GOODBY, J. W., 1981, *Phys. Rev. Lett.*, **46**, 1289.

JÄHNIG, F., and BROCHARD, F., 1974, *J. Phys., Paris*, **35**, 301.

JIANG WANG, and LUBENSKY, T. C., 1984, *Phys. Rev. A*, **29**, 2210.

JOHNSON, D. L., 1983, *J. Chim. phys.*, **80**, 45.

JOHNSON, D. L., ALLENDER, D., DE HOFF, R., MAZE, C., OPPENHEIM, E., and REYNOLDS, R., 1977, *Phys. Rev. B*, **16**, 470.

JOHNSON, D. L., HAYES, C. F., DE HOFF, R. J., and SCHANTZ, C. A., 1978, *Phys. Rev. B*, **18**, 4902.

KASTING, G. B., LUSHINGTON, K. J., and GARLAND, C. W., 1980, *Phys. Rev. B*, **22**, 321.

KIM, M. G., PARK, S., COOPER, M., and LETCHER, S. V., 1976, *Molec. Crystals liq. Crystals*, **36**, 143.

KIRY, F., and MARTINOTY, P., 1976, *J. Phys., Paris*, C3–37, 113; 1978, *Ibid.*, **39**, 1019.

KLEMAN, M., 1977, *Points, Lignes, Parois* (Orsay: Editions de Physique).

KOBAYASHI, K. K., 1970, *Physics Lett.* A, **31**, 125.

KORTAN, A. R., KÄNEL, H. V., BIRGENEAU, R. J., and LITSTER, J. D., 1981, *Phys. Rev. Lett.*, **47**, 1206.

KORTAN, A. R., KÄNEL, H. V., BIRGENEAU, R. J., LITSTER, J. D., 1984, *J. Phys., Paris*, **45**, 529.

KOSTERLITZ, J. M., NELSON, D. R., and FISHER, M. E., 1976, *Phys. Rev.* B, **13**, 412.

KOSTERLITZ, J. M., and THOULESS, D. J., 1973, *J. Phys.* C, **6**, 1181.

KUMAR, S., 1981, *Phys. Rev.* A, **23**, 3207.

LANDAU, L. D., 1937, *Phys. Z. SowjUn.*, **2**, 26.

LEADBETTER, A. J., FROST, J. C., and MAZID, M. A., 1979 a, *J. Phys. Lett., Paris*, **40**, L325.

LEADBETTER, A. J., GAUGHAN, J. P., KELLY, B., GRAY, G. W., and GOODBY, J. W., 1979 b, *J. Phys., Paris*, **40**, C3–178.

LEADBETTER, A. M., MAZID, M. A., KELLEY, B. A., GOODBY, J., and GRAY, W., 1979 c, *Phys. Rev. Lett.*, **43**, 630.

LEGER, L., 1973, *Physics Lett.* A, **44**, 535.

LEGER, L., and MARTINET, A., 1976, *J. Phys., Paris*, **37**, C3–89.

LE GRANGE, J. D., and MOCHEL, J., 1980, *Phys. Rev. Lett.*, **45**, 35.

LEVELUT, A. M., 1976, *J. Phys., Paris*, **37**, C3–51; 1983, *J. Chim. phys.*, **80**, 149.

LEVELUT, A. M., DOUCET, J., and LAMBERT, M., 1974, *J. Phys., Paris*, **35**, 773.

LEVELUT, A. M., TARENTO, R. J., HARDOUIN, F., ACHARD, M. F., and SIGAUD, G., 1981, *Phys. Rev.* A, **24**, 2180.

LIAO, Y., CLARK, N. A., and PERSHAN, P. S., 1973, *Phys. Rev. Lett.*, **30**, 639.

LIM, K. C., HO, J. T., and NEUBERT, M. E., 1980, *Molec. Crystals liq. Crystals*, **58**, 245.

LITSTER, J. D., ALS-NIELSEN, J., BIRGENEAU, R. J., DANA, S. S., DAVIDOV, D., GARCIA-GOLDING, F., KAPLAN, M., SAFINYA, C. R., and SCHAETZING, R., 1974, *J. Phys., Paris*, **40**, C3–339.

LITSTER, J. D., BIRGENEAU, R. J., KAPLAN, M., SAFINYA, C. R., and ALS-NIELSEN, J., 1980, *Ordering in Strongly Fluctuating Condensed Matter Systems*, edited by T. Riste (New York: Plenum), p. 357.

LONGA, L., and DE JEU, W., 1982, *Phys. Rev.*, **26**, 1632.

LOWE, M., GOLLUB, J. P., and LUBENSKY, T. C., 1983, *Phys. Rev. Lett.*, **51**, 786.

LUBENSKY, T. C., 1983, *J. Chim. phys.*, **80**, 31.

LUBENSKY, T. C., DUNN, S. G., and ISAACSON, J., 1981, *Phys. Rev. Lett.*, **22**, 1609.

LUBENSKY, T. C., and JING-HUEI CHEN, 1978, *Phys. Rev.*, **17**, 366.

LUBENSKY, T. C., and JIANG WANG, 1984 (private communication).

LUBENSKY, T. C., and McKANE, A. J., 1984, *Phys. Rev.* A, **29**, 317.

LUSHINGTON, K., KASTING, G. B., and GARLAND, C. W., 1980, *Phys. Rev.* B, **22**, 2569.

MA, S. K., 1976, *Modern Theory of Critical Phenomena* (Frontiers in Physics) (Benjamin).

MACMILLAN, W. L., 1971, *Phys. Rev.* A, **4**, 1238; 1973, *Ibid.*, **7**, 1419.

MARCEROU, J. P., ROUILLON, J. C., and PROST, J., 1983, *Molec. Crystals liq. Crystals* (submitted).

MARTIN, P., PARODI, O., and PERSHAN, P. S., 1972, *Phys. Rev.* A, **6**, 2401.

MARTINOT-LAGARDE, P., DUKE, R., and DURAND, G., 1981, *Molec. Crystals liq. Crystals*, **75**, 249.

MARTINOTY, P., 1979, *J. Phys. Lett., Paris*, **40**, L291.

MAZENKO, G. F., RAMASWAMY, S., and TONER, J., 1982, *Phys. Rev. Lett.*, **49**, 51; 1983, *Ibid.*, **28**, 1618.

MEYER, R. B., LIEBERT, L., STRZELECKI, L., and KELLER, P., 1975, *J. Phys. Lett., Paris*, **36**, L.69.

MEYER, R. B., and LUBENSKY, T. C., 1976, *Phys. Rev.* A, **14**, 2307.

MIYANO, K., and KETTERSON, J. B., 1973, *Phys. Rev. Lett.*, **31**, 639.

MONCTON, D. E., and PINDAK, R., 1979, *Phys. Rev. Lett.*, **43**, 701.

MONCTON, D. E., PINDAK, R., DAVEY, S. C., and BROWN, G. S., 1982, *Phys. Rev. Lett.*, **49**, 1865.

MUKAMEL, D., and HORNREICH, R. M., 1980, *J. Phys.* C, **13**, 161.

NABARRO, F. R. N., 1967, *Theory of Dislocations* (New York: Clarendon).

NELSON, D. R., and HALPERIN, B. I., 1979, *Phys. Rev.* B, **19**, 2456; 1980, *Ibid.*, **21**, 5312.

NGUYEN HUU TINH, HARDOUIN, F., and DESTRADE, C., 1982, *J. Phys., Paris*, **43**, 1127.

OCKO, B. M., BIRGENEAU, R. J., and LITSTER, J. D., 1984 b, *Phys. Rev. Lett.*, **52**, 208.

Ocko, B. M., Kortan, A. R., Birgeneau, R. J., and Goodby, J. W., 1984 a, *J. Phys., Paris,* Oseen, C. W., 1933, *Trans. Faraday Soc.,* **29,** 883.
Oswald, P., and Kleman, M., 1982, *J. Phys. Lett., Paris,* **43,** L411.
Peierls, R. E., 1935, *Annls Inst. Henri Poincaré,* **5,** 177.
Pelcovits, R. I., and Halperin, B. I., 1979, *Phys. Rev. B,* **19,** 4614.
Pershan, P. S., 1979, *J. Phys., Paris,* **40,** C3–423.
Pershan, P. S., Aeppli, G., Litster, J. D., and Birgeneau, R. J., 1981, *Molec. Crystals liq. Crystals,* **70C,** 861.
Pershan, P. S., and Prost, J., 1975, *J. appl. Phys.,* **46,** 2343; 1979, *J. Phys. Lett., Paris,* **40,** L27.
Pieransky, P., Guyon, E., Keller, P., 1975, *J. Phys. Lett., Paris,* **36,** L69.
Pindak, R., 1983 (private communication).
Pindak, R., Bishop, D. J., and Sprenger, W. O., 1980 b, *Phys. Rev. Lett.,* **44,** 1461.
Pindak, R., Cheng-Cher Huang, and Jo, J. T., 1974, *Phys. Rev. Lett.,* **32,** 43.
Pindak, R., Moncton, D. E., Davey, S. C., and Goodby, J. W., 1981, *Phys. Rev. Lett.,* **46,** 1135.
Pindak, R., Young, C. Y., Meyer, R. B., and Clark, N. A., 1980 a, *Phys. Rev. Lett.,* **45,** 1193.
Prost, J., 1979, *J. Phys., Paris,* **40,** 581; 1980, *Proceedings of the Conference on Liquid Crystals of One and Two Dimensional Order,* Garmisch Partenkirchen (Berlin, Heidelberg, New York: Sprinter-Verlag), p. 125; 1981, *Symmetries and Broken Symmetries,* edited by N. Boccara (Paris: IDSET), p. 159.
Prost, J., and Barois, P., 1983, *J. Chim. phys.,* **80,** 65.
Ramaswamy, S., 1984, *Phys. Rev. A,* **29,** 1506.
Ramaswamy, S., and Toner, J., 1983, *Phys. Rev. A,* **28,** 3159.
Ribotta, R., 1974, *C. r. hebd. Séanc. Acad. Sci., Paris B,* **279,** 295.
Ribotta, R., Salin, R., and Durand, G., 1974, *Phys. Rev. Lett.,* **32,** 6.
Ricard, L., and Prost, J., 1979, *J. Phys., Paris,* **40,** C–3; 1981, *Ibid.,* **42,** 861.
Richter, L., 1980, Dissertation, Halle/S.
Rosenblatt, C., and Ho, J. T., 1982, *Phys. Rev. A,* **26,** 2293.
Rosenblatt, C., and Litster, J. D., 1982, *Phys. Rev. A,* **26,** 1809.
Rosenblatt, C., Meyer, R. B., Pindak, R., and Clark, N. A., 1980, *Phys. Rev. Lett.,* **21,** 140.
Safinya, S., 1983 (private communication).
Safinya, C., Birgeneau, R., Litster, J. D., and Neubert, M. E., 1981, *Phys. Rev. Lett.,* **47,** 668.
Safinya, C. R., Kaplan, M., Als-Nielsen, J., Birgeneau, R. J., Davidov, D., Litster, J. D., Johnson, D. L., and Neubert, M. E., 1980, *Phys. Rev. B,* **21,** 4149.
Samulsky, E., and Dupre, D. B., 1983, *J. Chim. phys.,* **80,** 25.
Sigaud, G., Hardouin, F., and Achard, M. F., 1977, *Solid St. Commun.,* **23,** 35.
Sigaud, G., Hardouin, F., Achard, M. F., and Gasparoux, H., 1979, *J. Phys., Paris,* **40,** C3–356.
Sigaud, G., Hardouin, F., Achard, M. F., and Levelut, A. M., 1981, *J. Phys., Paris,* **42,** 107.
Schantz, C. A., and Johnson, D. L., 1978, *Phys. Rev. A,* **17,** 1504.
Swift, J., 1976, *Phys. Rev. A,* **14,** 2274.
Tarczon, J. C., and Miyano, K., 1981, *Phys. Rev. Lett.,* **46,** 119.
Thoen, J., Marynissen, H., and Van Dael, W., 1984, *Phys. Rev. Lett.,* **52,** 204.
Toner, J., 1982, *Phys. Rev. B,* **26,** 462.
Toner, J., Nelson, D. R., 1981, *Phys. Rev. B,* **23,** 363.
Tufeu, R. J., Keyes, P. H., and Daniels, W. B., 1975, *Phys. Rev. Lett.,* **35,** 1004.
Viner, J. M., Lamey, D. Huang, C. C., Pindack, R., and Goodby, J. W., 1983, *Phys. Rev. A,* **28,** 2433.
Von Känel, H., and Litster, J. D., 1981, *Phys. Rev. B,* **23,** 3251.
Wilson, K., and Kogut, J., 1974, *Phys. Rep.,* **12,** 77.
Witanachi, S., Huang, J., and Ho, J. T., 1983, *Phys. Rev. Lett.,* **50,** 594.
Young, C., Pindak, R., Clark, N. A., and Meyer, R. B., 1978, *Phys. Rev. Lett.,* **40,** 773.

PHYSICAL REVIEW A   VOLUME 29, NUMBER 4   APRIL 1984

Theory of the S_{A_1}-S_{A_2} phase transition in liquid crystals

Jiang Wang and T. C. Lubensky

Department of Physics, University of Pennsylvania, Philadelphia, Pennsylvania 19104

(Received 3 August 1983)

We present a phenomenological model for the phase transition between the monolayer (S_{A_1}) and the bilayer (S_{A_2}) phases of smectic liquid crystals. This model contains all relevant symmetries and Goldstone modes. We study the transition using the ϵ expansion and find it to be in the same universality class as the Ising model to first order in ϵ. In three dimensions, the physical correlation function, however, exhibits nonuniversal power-law behavior leading to a nonuniversal susceptibility exponent γ.

I. INTRODUCTION

In the smectic-A (S_A) phase of liquid crystals,[1] oriented barlike molecules of length l segregate into stacks of structureless two-dimensional planes. It is now clear that there are several types of S_A phases [2,3] characterized by the ratio of the interplanar spacing d to l. Conceptually, the simplest phase is the monolayer phase with $d = l$ denoted by S_{A_1} and depicted schematically in Fig. 1(a). In systems composed of polar molecules, a bilayer phase with $d = 2l$ and an incommensurate phase with $d = sl$ with $1 < s < 2$ can occur as well. These phases are depicted schematically in Figs. 1(b) and 1(c). Phase transitions between all pairs of the above phases have been observed,[2] indicating that they are indeed distinct phases.

In this paper, we will be concerned with the S_{A_1}-S_{A_2} transition. The x-ray diffusion pattern for these two phases is shown in Fig. 2. In the S_{A_1} phase, there is a quasi-Bragg peak at wave number $q = 2q_0 = 2\pi/l$ and a diffuse spot at wave number $q_0 = 2\pi/2l$. In the S_{A_2} phase, there are quasi-Bragg peaks at both q_0 and $2q_0$. The order parameter for these two phases can be constructed by expanding the center-of-mass density ρ in a Fourier series

$$\rho = \rho_0 + \sum (\psi_n e^{inq_0 z} + \text{c.c.}) . \tag{1.1}$$

The order parameter of the S_{A_1} phase is ψ_2, the complex amplitude of the mass-density wave at $2q_0$. The S_{A_2} phase has an additional order parameter ψ_1, the complex amplitude of the mass-density wave at q_0. Thus to study the S_{A_1}-S_{A_2} transition, we need to study the fluctuations of ψ_1 in the presence of a nonvanishing $\langle \psi_2 \rangle$.

It is clear that there is no symmetry argument that would indicate that the S_{A_1}-S_{A_2} transition has to be first order, and our primary concern will be to identify the universality class of this transition when it is second order. Since ψ_2 is a complex order parameter, a naive argument would predict the universality class of the xy model. This, however, neglects the important coupling between ψ_1 and ψ_2. There is a preferred relative phase of the two order parameters so that only the amplitude of ψ_1 is critical once ψ_2 has ordered, indicating as pointed out by Prost[4]

that the universality class should be that of the Ising model. This argument is again incomplete because it neglects the coupling between ψ_1 and the hydrodynamic phase mode (Goldstone boson) associated with the nonzero ψ_2 of the S_{A_1} phase. In this paper, we will develop a model which takes into account all of these couplings.

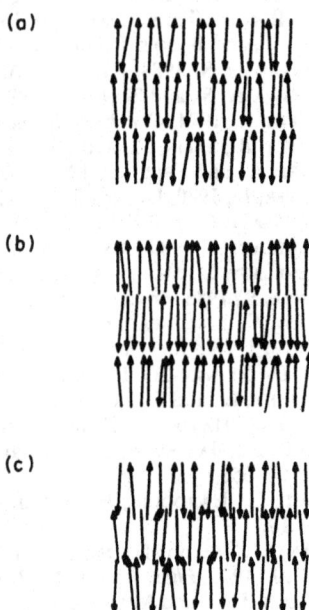

FIG. 1. Schematic representation of the three types of smectic-A phases: (a) the monolayer S_{A_1} phase, (b) the bilayer S_{A_2} phase, and (c) the S_{A_d} phase with layer spacing intermediate between one and two molecular lengths.

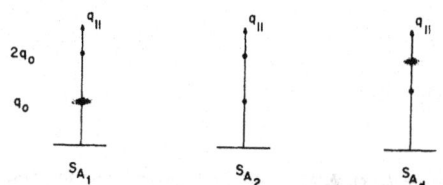

FIG. 2. X-ray diffraction patterns for the three smectic-A phases: (a) the S_{A_1} phase with a quasi-Bragg peak at $2q_0$ and a diffuse spot at q_0, (b) the S_{A_2} phase with quasi-Bragg peaks at both q_0 and $2q_0$, and (c) the S_{A_d} phase with a quasi-Bragg peak at a wave number between q_0 and $2q_0$ and a diffuse spot in the vicinity of $2q_0$.

Our conclusion is that the S_{A_1}-S_{A_2} transition is in the Ising universality class (at least near four dimensions) with isotropic correlation length critical exponents in agreement with the simple argument of Prost.[4] The physical correlation function $G(\vec{x},0)$ does not, however, behave simply as the spin correlation function in the Ising model. Interactions between the order parameter and the director lead to nonuniversal power-law behavior in $G(\vec{x},0)$ at the critical point and a resulting nonuniversal susceptibility exponent γ. This nonuniversality is simply a reflection of the well-known[5] lack of long-range order in ψ_2 in the S_{A_2} phase.

The outline of this paper is as follows. In Sec. II, we formulate an isotropic model for the S_{A_1}-S_{A_2} transition that neglects fluctuations of the Frank director, and we discuss its symmetries and associated Ward identities. In Sec. III, we develop momentum-shell recursion relations and study this model using the ϵ expansion. In Sec. IV, we formulate the full model for the S_{A_1}-S_{A_2} transition including anisotropy and couplings to the director and show that it is in the same universality class as the isotropic model of Sec. II. We then discuss the physical correlation function and derive its nonuniversal power-law behavior. Finally, there are two appendices presenting a derivation of the fundamental Ward identity and calculations of vertex functions to one-loop order in perturbation theory.

II. ISOTROPIC MODEL

The simplest model Hamiltonian $\bar{H}$ capable of describing the nematic (N), S_{A_1}, and S_{A_2} phases is a functional of the fields ψ_1 and ψ_2 only. In general, of course, all ψ_n are needed, but they can be expressed as functions of ψ_1 and ψ_2. The Hamiltonian must be invariant under uniform translations of the system. As can be seen from Eq. (1.1), a uniform translation by z along the z axis changes ψ_n to

$\psi_n e^{inq_0 z}$. Thus, $\bar{H}$ must be invariant under $\psi_1 \to \psi_1 e^{i\phi}$, $\psi_2 \to \psi_2 e^{2i\phi}$. The simplest Hamiltonian[6] incorporating this invariance is

$$\bar{H} = \bar{H}_0 + \bar{H}_{\text{int}} , \tag{2.1}$$

where

$$\bar{H}_0 = \int d^d x \,\tfrac{1}{2} [r_1 |\psi_1|^2 + r_2 |\psi_2|^2 + |\nabla\psi_1|^2 + |\nabla\psi_2|^2] \tag{2.2}$$

is the noninteracting part and

$$\bar{H}_{\text{int}} = \int d^d x \left[\frac{u_1}{4!} |\psi_1|^4 + \frac{u_2}{4!} |\psi_2|^4 + \frac{u_{12}}{12} |\psi_1|^2 |\psi_2|^2 - \tfrac{1}{2}\bar{w}\,\text{Re}(\psi_1^2 \psi_2^*) \right]. \tag{2.3}$$

In the above, $r_1 \sim (T-T_1)$ and $r_2 \sim (T-T_2)$ where T is the temperature and T_1 and T_2 are, respectively, the mean-field N-S_{A_1} and N-S_{A_2} transition temperatures. Couplings of ψ_1 and ψ_2 to the Frank director $\vec{n}$ and spatial anisotropies are not included in this equation. They will be treated in Sec. IV. The phases of ψ_1 and ψ_2 have been chosen so that the potential $\bar{w}$ is real and a relative phase of zero is favored.

The Hamiltonian Eq. (2.1) has been used to study the N-S_{A_1}, N-S_{A_2} and S_{A_1}-S_{A_2} transitions in mean-field theory.[3,4,6] A renormalized perturbation theory and renormalization-group analysis can be applied directly to this model to study the S_{A_1}-S_{A_2} transition. Such an analysis is complicated, however, by the requirement that the equation of state for $\langle\psi_2\rangle$ be calculated consistently at each order in perturbation theory. Near the S_{A_1}-S_{A_2} transition, fluctuations in the amplitude of ψ_2 are unimportant compared to those of the phase of ψ_2. We, therefore, replace ψ_2 by $|\psi_2| e^{i\phi}$ and treat $|\psi_2|$ as a constant to obtain the modified model

$$H = H_0 + H_{\text{int}} , \tag{2.4}$$

where

$$H_0 = \int d^d x \left[\tfrac{1}{2} r |\psi_1|^2 + \tfrac{1}{2} |\nabla\psi_1|^2 + \frac{K}{2} |\nabla\phi|^2 \right], \tag{2.5}$$

and

$$H_{\text{int}} = \int d^d x \left[\frac{u}{4!} |\psi_1|^4 - \frac{w}{2} \text{Re}(e^{-i\phi}\psi_1^2) \right], \tag{2.6}$$

where $r = r_1 + u_{12} |\psi_2|^2/6$, $K = |\psi_2|^2$, and $w = \bar{w} |\psi_2|$. In what follows, we will find it convenient to express H in terms of the variables $\psi_x = \text{Re}\psi_1$ and $\psi_y = \text{Im}\psi_1$:

$$H = \int d^d x \left[\tfrac{1}{2} r(\psi_x^2 + \psi_y^2) + \tfrac{1}{2} K (\nabla\phi)^2 + \tfrac{1}{2}(\nabla\psi_x)^2 + \tfrac{1}{2}(\nabla\psi_y)^2 - w\,[\sin\phi\,\psi_x\psi_y + \tfrac{1}{2}\cos\phi(\psi_x^2 - \psi_y^2)] + \frac{u}{4!}(\psi_x^2 + \psi_y^2)^2 \right]. \tag{2.7}$$

The phase translational invariance of Eq. (2.1) now takes the form of an invariance of Eq. (2.7) with respect to the transformations

$$\phi \rightarrow \phi + 2\theta, \quad \begin{bmatrix} \psi_x \\ \psi_y \end{bmatrix} \rightarrow \begin{bmatrix} \cos\theta & -\sin\theta \\ \sin\theta & \cos\theta \end{bmatrix} \begin{bmatrix} \psi_x \\ \psi_y \end{bmatrix}. \tag{2.8}$$

This invariance leads to a series of Ward identities for the vertex functions

$$\Gamma^{(n)}_{a_1 \cdots a_n}(x_1, \ldots, x_n) = \frac{\delta^n \Gamma}{\delta a_1(x_1) \cdots \delta a_n(x_n)},$$

where Γ is the Legendre-transformed free energy and $a_i(x)$ can be $\psi_x(x)$, $\psi_y(x)$, or $\phi(x)$. The general form for these Ward identities is derived in Appendix A. Of particular interest in what follows are the relations, valid in the S_{A_1} phase,

$$\lim_{q \to 0} \Gamma^{(2)}_{\phi\phi}(\vec{q}) = 0, \tag{2.9a}$$

$$\lim_{q \to 0} [\Gamma^{(2)}_{xx}(\vec{q}) - \Gamma^{(2)}_{yy}(\vec{q})] = \lim_{q \to 0} 2\Gamma^{(3)}_{\phi xy}(\vec{q}, \vec{q}, 0), \tag{2.9b}$$

$$\lim_{q \to 0} \Gamma^{(4)}_{\phi\phi xx}(\vec{q}, \vec{q}, 0, 0) = \lim_{q \to 0} -2\Gamma^{(3)}_{\phi xy}(\vec{q}, \vec{q}, 0), \tag{2.9c}$$

where it is understood that x and y stand for ψ_x and ψ_y. Equation (2.9a) states the obvious that ϕ is a hydrodynamical variable in the S_{A_1} phase.

In order to develop a systematic loop expansion and renormalization-group recursion relations for functions arising from Eq. (2.7), it is convenient to expand in a power series in ϕ and to introduce a more general space of potentials. We, therefore, write

$$H = \int d^d x \left[\frac{1}{2} r_x \psi_x^2 + \frac{1}{2} r_y \psi_y^2 + \frac{1}{2} c_x (\nabla \psi_x)^2 + \frac{1}{2} c_y (\nabla \psi_y)^2 + \frac{1}{2} K (\nabla \phi)^2 + \frac{1}{4!} u_x \psi_x^4 + \frac{1}{12} u_{xy} \psi_x^2 \psi_y^2 + \frac{1}{4!} u_y \psi_y^4 \right.$$

$$\left. - w_1 \phi \psi_x \psi_y + \frac{1}{3!} w_3 \phi^3 \psi_x \psi_y + \frac{1}{4} w_{2x} \phi^2 \psi_x^2 - w_{2y} \phi^2 \psi_y^2 - \frac{w_{4x}}{2 \cdot 4!} \phi^4 \psi_x^2 + \frac{w_{4y}}{2 \cdot 4!} \phi^4 \psi_y^2 \right]. \tag{2.10}$$

To make Eq. (2.10) equivalent to Eq. (2.7), we must have $r_x = r - w$, $r_y = r + w$, $w_1 = w_3 = w_{2x} = w_{2y} = w$, $c_x = c_y = 1$, and $u_x = u_y = u_{xy} = u$. In this case the Ward identities of Eq. (2.9) are easily seen to be valid to lowest order in a loop expansion where $\Gamma^{(2)}_{xx}(0) = r_x$, $\Gamma^{(2)}_{yy}(0) = r_y$, $\Gamma^{(3)}_{\phi xy}(0,0,0) = w_1$, $\Gamma^{(4)}_{\phi\phi xx} = -w_{2x}$. We will use the Hamiltonian of Eq. (2.10) in subsequent sections, always remembering the relations among the bare potentials imposed by Eq. (2.9).

III. ϵ EXPANSION

The momentum-shell recursion relations of the ϵ expansion map the original Hamiltonian H with a spherical Brillouin zone of unit radius onto a Hamiltonian $H' \equiv RH$ by first removing degrees of freedom with wave number $\vec{q}$ between b^{-1} and 1 and rescaling the fields: $\psi_x \rightarrow \zeta_x \psi_x$, $\psi_y \rightarrow \zeta_y \psi_y$, and $\phi \rightarrow \zeta_\phi \phi$. Vertex and correlation functions can be expressed in terms of H or H'. For example,

$$\Gamma^{(2)}_{\phi\phi}(q, H) = b^{-d} \zeta_\phi^2 \Gamma^{(2)}_{\phi\phi}(q/b, H'), \tag{3.1}$$

$$\Gamma^{(2)}_{xx}(q, H) = b^{-d} \zeta_x^2 \Gamma^{(2)}_{xx}(q/b, H').$$

The Ward identities discussed in the previous section are satisfied by vertex functions of the original Hamiltonian; they are not, however, satisfied by those of the rescaled Hamiltonian H'. This is because H' is not invariant with respect to the simple transformations of Eq. (2.8) but rather to more general transformations involving rescaling of the fields. The Ward identity Eq. (2.9a) is, however, valid

in all rescaled Hamiltonians $R^n H$ along the renormalization-group trajectory including any fixed-point Hamiltonian H^*. This provides a valuable check of the validity of our recursion relations.

The important vertex functions for this problem are evaluated in Appendix B to one-loop order for arbitrary potentials of Eq. (2.10) using the diagrams of Fig. 3. They satisfy the Ward identities as required. In this section, we will need to consider the vertex function $\Gamma^{(2)}_{\phi\phi}$ carefully. To one-loop order, it satisfies

$$\Gamma^{(2)}_{\phi\phi}(k) = Kk^2 - w_1^2 \int_q \frac{1}{r_x + c_x q^2} \frac{1}{r_y + c_y (\vec{k} + \vec{q})^2}$$

$$+ \frac{1}{2} w_{2x} \int_q \frac{1}{r_x + c_x q^2}$$

$$- \frac{1}{2} w_{2y} \int_q \frac{1}{r_y + c_y q^2}, \tag{3.2}$$

where

$$\int_q = \int \frac{d^d q}{(2\pi)^d}.$$

As just discussed, $\lim_{k \to 0} \Gamma^{(2)}_{\phi\phi}(k) = 0$ for all Hamiltonians along the RG trajectory.

To derive the momentum-shell recursion relations, we note that ψ_x is critical whereas ψ_y is non-critical. Usually in situations like this, the non-critical field is simply removed to produce a new effective Hamiltonian that is a function of the critical fields only. We will discuss this approach at the end of the section. Since ψ_y plays an

For $\Gamma_{\phi\phi}^{(2)}$

(a)

For $\Gamma_{xx}^{(2)}$

(b)

For $\Gamma_{yy}^{(2)}$

(c)

For $\Gamma_{\phi xy}^{(3)}$

(d)

For $\Gamma_{xx\phi\phi}^{(4)}$

(e)

FIG. 3. Diagrams for some vertex functions. The unbroken lines represent G_{xx}, the broken lines G_{yy}, and the wavy lines $G_{\phi\phi}$. (a), (b), (c), (d), and (e), respectively, show diagrams for $\Gamma_{\phi\phi}^{(2)}$, $\Gamma_{xx}^{(2)}$, $\Gamma_{yy}^{(2)}$, $\Gamma_{\phi xy}^{(3)}$, and $\Gamma_{xx\phi\phi}^{(4)}$.

essential role in maintaining Ward identities, we find it convenient, for the moment, not to remove it. We will, however, rescale ψ_x and ψ_y differently. We set

$$\zeta_x^2 = b^{d+2-\eta} \, ,$$
$$\zeta_y^2 = b^d \, , \tag{3.3}$$
$$\zeta_\phi^2 = b^{d+2-\eta_\phi} \, ,$$

where η and η_ϕ are chosen to keep the coefficients c_x and K constant and equal to unity. With this choice of rescaling, all potentials involving ψ_y except w_1 and r_y are irrelevant. Furthermore, the potentials r_y and w_1 always occur in the combination $g = w_1^2/r_y$ since c_y is irrelevant. The differential recursion relations for $b = e^{\delta l}$ to first order in ϵ for the potentials of interest are therefore

$$\frac{du_x}{dl} = \epsilon u_x - \tfrac{3}{2}K_d u_x^2 - \tfrac{3}{2}K_d w_{2x}^2$$
$$+ 6K_d g w_{2x} - 6K_d g^2 \, ,$$

$$\frac{dw_{2x}}{dl} = \epsilon w_{2x} - 2K_d g u_x + 4K_d g w_{2x}$$
$$- \tfrac{1}{2}K_d u_x w_{2x} - 2K_d w_{2x}^2 \, ,$$

$$\tag{3.4}$$

$$\frac{dg}{dl} = \epsilon g + 3K_d g^2 - 2K_d w_{2x} g \, ,$$

$$\frac{dr_x}{dl} = 2r_x + \tfrac{1}{2}K_d w_{2x} + \tfrac{1}{2}K_d \frac{u_x}{1+r_x} - K_d g \, ,$$

where $K_d = \Omega_d/(2\pi)^d$ where Ω_d is the solid angle subtended by the unit sphere in d-dimension. η_ϕ and η are zero to first order in ϵ. The general fixed point structure of these equations is quite complex. We know, however, from Eq. (3.2) that at the fixed point $g = g^* = w_1^{*2}/r_y^* = w_{2x}^*/2$ in order to ensure that $\Gamma_{\phi\phi}^{(2)}(q) \sim q^2$ at small q because the potentials w_y and c_y are zero. Equations (3.4) have a stable fixed point satisfying these conditions with

$$K_d u_x^* = \tfrac{3}{2}\epsilon \, , \quad K_d g^* = \tfrac{1}{2}\epsilon \, , \quad K_d w_{2x}^* = \epsilon \tag{3.5}$$

and

$$\nu = \tfrac{1}{2} + \tfrac{1}{12}\epsilon \, , \quad \eta = 0 \, . \tag{3.6}$$

It is clear that this is a fixed point with Ising symmetry since the equations for u_x and r_x decouple completely from those for g and w_{2x} if $g = w_{2x}/2$. In other words, the fixed-point Hamiltonian consists of two disjoint parts for the critical fields ψ_x and ϕ: the Ising Hamiltonian for the field ψ_x and a non-interacting Gaussian spin-wave Hamiltonian for the field ϕ.

Another way of seeing this result is to consider the effective Hamiltonian for ψ_x and ϕ with the non-critical field ψ_y removed. We find

$$H_{\text{eff}} = \int d^d x [\tfrac{1}{2} r_x \psi_x^2 + \tfrac{1}{2}(\nabla \psi_x)^2 + u_x \psi_x^4] + \int d^d x \left[\frac{r_\phi}{2}\phi^2 + \tfrac{1}{2}K(\nabla\phi)^2 \right]$$

$$- \int d^d x \int d^d x' u_{\phi x}(x,x')\phi(x)\psi_x(x)\phi(x')\psi_x(x') \, , \tag{3.7}$$

where

$$u_{\phi x}(x,x') = 1/4[w_{2x}\delta(\vec{x}-\vec{x}') - 2w_1^2 G_{yy}(\vec{x},\vec{x}')]$$

and $r_\phi = -\frac{1}{2} w_{2y} G_{yy}(\vec{x}, \vec{x})$ where $G_{yy}(\vec{x}, \vec{x}') = \langle \psi_y(\vec{x})\psi_y^*(\vec{x}')\rangle$. To one-loop order, the contribution from $u_{\phi x}$ cancels the r_ϕ term keeping $\Gamma_{\phi\phi}^{(2)}(q=0)=0$ as required. At the fixed point, $u_{\phi x}=0$ and $r_\phi^*=0$. It seems likely to us that the decoupling of ψ_x and ϕ remains true to higher order in ϵ. We have, however, not found a general symmetry argument leading to this result nor have we carried out the rather tedious calculations needed to verify this to second order in ϵ.

IV. GENERAL MODEL

The model presented in Sec. II is incomplete in two respects: it does not include anisotropy and coupling of the order parameters ψ_1 and ψ_2 to the Frank director $\vec{n}$. The invariance of the Hamiltonian with respect to simultaneous rotations of the director and the smectic planes leads to a gaugelike coupling between $\delta\vec{n} = \vec{n} - \vec{n}_0$ (where $\vec{n}_0$ is the uniform equilibrium director) and both ψ_1 and ψ_2 to quadratic order in $\delta\vec{n}$. The non-interacting Hamiltonian of Eq. (2.2) is thus replaced by

$$\bar{H}_0 = \int d^dx \tfrac{1}{2}[r_1|\psi_1|^2 + r_2|\psi_2|^2 + |\nabla_\parallel\psi_1|^2 + c_\parallel|\nabla_\parallel\psi_2|^2 + |(\nabla_\perp - iq_0\delta\vec{n})\psi_1|^2 + c_\perp|(\nabla_\perp - 2iq_0\delta\vec{n})\psi_2|^2]\,, \quad (4.1)$$

where we have chosen length scales so that the coefficients of $|\nabla_\parallel\psi_1|^2$ and $|\nabla_\perp\psi_1|^2$ are unity. The full Hamiltonian is

$$H = \bar{H}_0 + \bar{H}_{int} + H_F \,, \qquad (4.2)$$

where $\bar{H}_{int}$ is given in Eq. (2.3) and H_F is the usual Frank Hamiltonian

$$H_F = \int d^dx \tfrac{1}{2}\{K_1(\nabla\cdot\vec{n})^2 + K_2(\vec{n}\cdot\nabla\times\vec{n})^2 + K_3[\vec{n}\times(\nabla\times\vec{n})]^2\}\,. \qquad (4.3)$$

This Hamiltonian leaves out non-linear terms needed to ensure global rotational invariance.[7] Though these terms do affect the elastic properties of the smectic phases, we do not believe they have an important effect on the S_{A_1}-S_{A_2} transition, and we will ignore them.

Equation (4.2) is the analog of Eq. (2.1). To obtain the analog of Eq. (2.4), we set $\psi_2 = e^{i\phi}$ as before to obtain

$$H_0 = \int d^dx \tfrac{1}{2}[r|\psi_1|^2 + |\nabla_\parallel\psi_1|^2 + |(\nabla_\perp - iq_0\delta\vec{n})\psi_1|^2 + K_\parallel(\nabla_\parallel\phi')^2 + K_\perp(\nabla_\perp\phi' - 2q_0\delta\vec{n})^2] \qquad (4.4)$$

and

$$H = H_0 + H_{int} + H_F \,, \qquad (4.5)$$

where H_{int} is expressed in Eq. (2.6). The simplest way to study the effects of couplings to the director is to perform a change of variables to decouple ϕ' from $\delta\vec{n}$. This change of variables has the appearance of gauge transformation [8,9]

$$\delta\vec{n} = \vec{A} + \nabla L \,,$$
$$\phi' = \phi + 2q_0 L \,, \qquad (4.6)$$
$$\psi_1 = \psi e^{iq_0 L} \,.$$

After this transformation, H_{int} is unchanged,

$$H_0 = \int d^dx \tfrac{1}{2}[r|\psi|^2 + |(\nabla_\parallel - iq_0 A_\parallel)\psi|^2 + |(\nabla_\perp - iq_0\vec{A}_\perp)\psi|^2 + K_\parallel(\nabla_\parallel\phi - 2q_0 A_\parallel)^2 + K_\perp(\nabla_\perp\phi - 2q_0\vec{A}_\perp)^2]\,, \qquad (4.7)$$

and H_F becomes a function of both L and $\vec{A}$. To decouple $\vec{A}$ and ϕ, we choose

$$K_\parallel\nabla_\parallel A_\parallel + K_\perp\nabla_\perp\cdot\vec{A}_\perp = 0 \,. \qquad (4.8)$$

When $K_\parallel = K_\perp$, this is the usual transverse constraint of the "SC" gauge.[8,9] Equation (4.8) determines L in terms of $\delta\vec{n}$ via Eq. (4.6a) and allows $\delta\vec{n}$ and thus H_F to be expressed in terms of $\vec{A}$ only. Defining $\vec{A}_s$ to be the component of $\vec{A}$ in the $\vec{n}_0$-$\vec{q}$ plane and A_t the component of $\vec{A}$ in the space perpendicular to the $\vec{n}_0$-$\vec{q}$ plane, we obtain

$$H = H_0' + H_{int} + H_A \,, \qquad (4.9)$$

where

$$H_0' = \int d^dx \tfrac{1}{2}[r|\psi|^2 + |(\nabla - iq_0\vec{A})\psi|^2] + \int d^dx \tfrac{1}{2}[K_\parallel(\nabla_\parallel\phi)^2 + K_\perp(\nabla_\perp\phi)^2] \qquad (4.10)$$

and

$$H_A = \int_q \frac{1}{2}\left\{\left[4q_0^2 K_\parallel K_\perp + \left[K_3 + K_1\frac{q_\perp^2}{q_\parallel^2}\right]K(q)\right]\left[\frac{K(q)}{K_\parallel^2 q_\parallel^2 + K_\perp^2 q_\perp^2}\right]\vec{A}_s^2 + (4q_0^2 K_\perp + K_2 q_\perp^2 + K_3 q_\parallel^2)A_t^2\right\}\,, \qquad (4.11)$$

where

$$K(q) = K_{\parallel} q_{\parallel}^2 + K_{\perp} q_{\perp}^2 .$$

We are now in a position to analyze the effect of anisotropy and couplings to $\delta \vec{n}$ on the S_{A_1}-S_{A_2} transition. The first thing to note is that $\vec{A}$ is not a hydrodynamic variable: its independent components have "masses" determined by $K_{\parallel}$ and $K_{\perp}$ and the constraint, Eq. (4.8). Thus $\vec{A}$ can be removed to produce an effective Hamiltonian that is identical in form to that of Eq. (2.2) except for the anisotropy imposed by $K_{\parallel} \neq K_{\perp}$. At the Ising fixed point discussed in the previous sections, the critical fields ψ_x and ϕ decouple completely. Since anisotropy has no effect on the critical exponents of the Ising fixed point associated with an independent ψ_x or the Gaussian fixed point associated with an independent ϕ, it is clear that introduction of anisotropy in the original Hamiltonian will have no effect on critical exponents. This has been explicitly verified to first order in ϵ using recursion relations appropriate to the anisotropic system.

Though the S_{A_1}-S_{A_2} transition is in the same universality class as the Ising model, the coupling of ψ to $\delta \vec{n}$ leads to anomalous behavior for the physical correlation function $G(\vec{x}, \vec{x}') = \langle \psi_1(\vec{x}) \psi_1^*(\vec{x}') \rangle$. The correlation function in the "SC" gauge is, apart from dependence on irrelevant variables, identical to the spin-correlation functions of the Ising model. Correlations in the physical field ψ_1 can be related to correlations in ψ_{SC} via

$$G_{SC}(\vec{x}, 0) = \langle \psi_{SC}(\vec{x}) \psi_{SC}^*(0) e^{-iq_0[L(\vec{x}) - L(0)]} \rangle$$

$$\equiv \widetilde{G}_{SC}(\vec{x}, 0) C(\vec{x}) , \qquad (4.12)$$

where

$$\widetilde{G}_{SC}(\vec{x}, 0) = \frac{\langle \psi_{SC}(\vec{x}) \psi_{SC}^*(0) e^{-iq_0[L(\vec{x}) - L(0)]} \rangle}{\langle e^{-iq_0[L(\vec{x}) - L(0)]} \rangle} \qquad (4.13)$$

and

$$C(\vec{x}) = \langle e^{-iq_0[L(\vec{x}) - L(0)]} \rangle . \qquad (4.14)$$

The large-$|\vec{x}|$ behavior[8,10] of $C(\vec{x})$ is dominated by the cumulant $\langle [L(\vec{x}) - L(0)]^2 \rangle$ which can be calculated using

$$\langle |L(\vec{q})|^2 \rangle = \frac{K_{\perp} q_{\perp}^2}{K^2(q)} D_{\perp}(\vec{q}) , \qquad (4.15)$$

where

$$D_{\perp}(\vec{q}) = (K_{\perp} q_{\perp}^2 + B q_{\parallel}^2 / q_{\perp}^2)^{-1} , \qquad (4.16)$$

where B is the usual compression modulus for smectics. In the present model, $B = 4 q_0^2 K_{\perp}$ to lowest order in perturbation theory. Using Eqs. (4.14) and (4.15), it is easy to verify that at large $|\vec{x}|$ $C(\vec{x})$ is identical to the function calculated by Caillé:[5]

$$C(\vec{x}) = \begin{cases} x_{\parallel}^{-\eta_c} & x_{\perp} = 0 \\ x_{\perp}^{-2\eta_c} & x_{\parallel} = 0 \end{cases} \qquad (4.17)$$

where

$$\eta_c = \frac{q_0^2 k_B T}{8\pi \sqrt{K_{\perp} B}} . \qquad (4.18)$$

Note that $C(\vec{x})$ dies off algebraically rather than exponentially as it does in the nematic phase and at the nematic to smectic-A critical point.[8-10] To complete the calculation of $G(\vec{x}, 0)$, we need to consider $\widetilde{G}_{SC}(\vec{x}, 0)$ which can be expanded in a power series[10] in q_0. Low-order diagrams in this expansion are shown in Fig. 4. The first term is precisely $G_{SC}(\vec{x}, 0)$ which has Ising critical behavior. Higher-order diagrams [Figs. 4(b) and 4(c)] involve both $G_{SC}(\vec{x}, 0)$ and $D_{LA_i}(\vec{x}, \vec{x}') = \langle L(\vec{x}) A_i(\vec{x}') \rangle$ which die off, respectively, with characteristic lengths $\xi \sim t^{-\nu}$ and $\lambda = (K_{\perp}/B)^{1/2}$ where t is the reduced temperature. It is tedious but straightforward to verify that there are no infrared singularities in these diagrams[10] so that they die off with a length $l \sim \min(\lambda, \xi)$. λ is not a critical function of temperature in the S_{A_1} phase as the S_{A_2} line is approached so that $l \sim \lambda$ as $t \to 0$. Thus $\widetilde{G}_{SC}(\vec{x}, 0) \sim G_{SC}(\vec{x}, 0)$ as $t \to 0$ and

$$G(\vec{x}, 0) = C(\vec{x}) |\vec{x}|^{-(d-2+\eta)} f\left(\frac{|\vec{x}|}{\xi} \right) , \qquad (4.19)$$

where we have inserted the isotropic scaling form for $G_{SC}(\vec{x}, 0)$ near the critical point. This form implies that $G(\vec{x}, 0)$ has anisotropic power-law behavior with critical exponents varying continuously along the S_{A_1}-S_{A_2} critical line:

$$G(\vec{x}, 0) \sim \begin{cases} x_{\parallel}^{-(d-2+\eta+\eta_c)} & x_{\perp} = 0 \\ |x_{\perp}|^{-(d-2+\eta+2\eta_c)} & x_{\parallel} = 0 \end{cases} . \qquad (4.20)$$

Equations (4.18) and (4.19) imply that the susceptibility χ satisfies

(a)　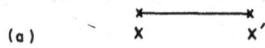

(b)　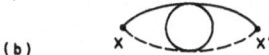

(c)　

FIG. 4. Diagrams contributing to $\widetilde{G}_{SC}(\vec{x}, \vec{x}')$. The unbroken line represents $G_{SC}(\vec{x}, \vec{x}')$ and the broken line D_{LA_i}. The circles in (b) and (c) represent vertex functions. (a) is simply G_{SC} and is the dominant contribution to $\widetilde{G}_{SC}$ near $t = 0$. The D_{LA_i} propagators in (b) and (c) cause these diagrams to decay to zero for $|\vec{x} - \vec{x}'| > \lambda$.

$$\chi = \int d^d x \, G(\vec{x},0) \sim \xi^{2-\eta-2\eta_c} \tag{4.21}$$

so that

$$\gamma = (2-\eta-2\eta_c)\nu \tag{4.22}$$

is not universal but depends on the values of B and K_1 at the critical point. η_c can be calculated using Eq. (4.18). Using $q_0 = 0.222 \times 10^8 \text{cm}^{-1}$, $K_1 = 10^{-6}$ dynes, $B = 10^8$ dynes/cm^2, and $T = 360$ K, we obtain $\eta_c \sim 0.1$. Thus we predict that γ is of order 0.1 to 0.2 less than the Ising value of 1.242 (Ref. 11) that is presumed to be appropriate for the SC gauge in three dimensions.

ACKNOWLEDGMENT

This work was supported in part by the National Science Foundation under Grant No. DMR 82-19216 and by the Office of Naval Research under Grant No. 0158.

APPENDIX A

To derive the Ward identities discussed in the text, we consider the partition function

$$Z(h_x,h_y,h_\phi) = \int D\psi_x D\psi_y D\phi \, e^{-H-H_{ext}} , \tag{A1}$$

where H is given by Eqs. (2.4) to (2.6) and

$$H_{ext} = -\int d^d x [h_x(x)\psi_x(x) + h_y(x)\psi_y(x) + h_\phi\phi(x)] . \tag{A2}$$

The invariance of H under the transformations of Eq. (2.8) implies that

$$Z(h_x,h_y,h_\phi) = \exp\left[\int d^d x \, 2h_\phi(x)\theta\right] Z(h_x\cos\theta + h_y\sin\theta, \, -h_x\sin\theta + h_y\cos\theta, \, h_\phi) \tag{A3}$$

or that

$$\frac{\partial \ln Z}{\partial \theta} = \int d^d x \, [2h_\phi(x) + \langle\psi_x(x)\rangle h_y(x) - \langle\psi_y(x)\rangle h_x(x)] , \tag{A4}$$

where $\langle\psi_x(x)\rangle$ and $\langle\psi_y(x)\rangle$ are the averages of $\psi_x(x)$ and $\psi_y(x)$ with respect at $H + H_{ext}$. Introducing the Legendre-transformed free energy

$$\Gamma(\langle\psi_x\rangle,\langle\psi_y\rangle) = -\ln Z + \int d^d x \, [h_x(x)\langle\psi_x(x)\rangle + h_y(x)\langle\psi_y(x)\rangle + h_\phi(x)\langle\phi(x)\rangle] , \tag{A5}$$

Eq. (A3) can be reexpressed as

$$\int d^d x \, [2\Gamma_\phi^{(1)}(x) + \langle\psi_x(x)\rangle\Gamma_y^{(1)}(x) - \langle\psi_y(x)\rangle\Gamma_x^{(1)}(x)] = 0 , \tag{A6}$$

where these vertex functions are defined as

$$\Gamma_{a_1\cdots a_n}^{(n)}(x_1,\ldots,x_n) = \frac{\delta^n \Gamma}{\delta a_1(x_1)\cdots\delta a_n(x_n)} , \tag{A7}$$

where a_i can be ψ_x, ψ_y, or ϕ with the convention that x and y rather than ψ_x and ψ_y be used as subscripts on the left-hand side of the equation. Successive differentiation of Eq. (A5) yields the Ward identities of Eqs. (2.9) where

$$\Gamma_{ab}^{(2)}(q) = \int d^d x \, e^{-i\vec{q}\cdot\vec{x}} \, \Gamma_{ab}^{(2)}(\vec{x},0) \tag{A8}$$

and

$$\Gamma_{a_1\cdots a_n}^{(n)}(k_1,\ldots,k_n) \, \delta^{(d)}(\vec{k}_1 + \cdots + \vec{k}_n) = \int d^d x_1 \cdots d^d x_n e^{-i\vec{k}_1\cdot\vec{x}_1} \cdots e^{-i\vec{k}_n\cdot\vec{x}_n} \Gamma_{a_1\cdots a_n}^{(n)}(\vec{x}_1,\ldots,\vec{x}_n) . \tag{A9}$$

APPENDIX B

Perturbation expressions for the vertex functions are easily derived from the general Hamiltonian of Eq. (2.10). One-loop diagrams for some vertex functions are shown in Fig. 3. From these, we obtain the following one-loop expressions for $\Gamma_{xx}^{(2)}$, $\Gamma_{yy}^{(2)}$, $\Gamma_{\phi xy}^{(3)}$, and $\Gamma_{\phi\phi xx}^{(4)}$:

$$\Gamma_{xx}^{(2)}(\vec{q}=0) = r_x + \tfrac{1}{4}w_{2x}\int_q G_{\phi\phi}(\vec{q}) + \tfrac{1}{2}u_x\int_q G_{xx}(\vec{q}) + \tfrac{1}{6}u_{xy}\int_q G_{yy}(\vec{q}) - w_1^2\int_q G_{\phi\phi}(\vec{q})G_{yy}(\vec{q}) , \tag{B1}$$

$$\Gamma_{yy}^{(2)} = r_y - \tfrac{1}{2}w_{2y}\int_q G_{\phi\phi}(\vec{q}) + \tfrac{1}{2}u_y\int_q G_{yy}(\vec{q}) + \tfrac{1}{6}u_{xy}\int_q G_{xx}(\vec{q}) - w_1^2\int_q G_{\phi\phi}(\vec{q})G_{xx}(\vec{q}) , \tag{B2}$$

$$\Gamma_{\phi xy}^{(3)} = w_1 - \tfrac{1}{2}w_3\int_q G_{\phi\phi}(\vec{q}) + w_1 w_{2y}\int_q G_{\phi\phi}(\vec{q})G_{yy}(\vec{q}) - w_1 w_{2x}\int_q G_{\phi\phi}(\vec{q})G_{xx}(\vec{q})$$
$$+ w_1^3\int_q G_{\phi\phi}(\vec{q})G_{xx}(\vec{q})G_{yy}(\vec{q}) - \tfrac{1}{3}u_{xy}w_1\int_q G_{xx}(\vec{q})G_{yy}(\vec{q}) , \tag{B3}$$

$$\Gamma^{(4)}_{\phi\phi xx} = w_{2x} - \tfrac{1}{2} w_{4x} \int_q G_{\phi\phi}(\vec{q}) - 2w_{2x}^2 \int_q G_{\phi\phi}(\vec{q}) G_{xx}(\vec{q}) + w_1 w_3 \int_q G_{\phi\phi}(\vec{q}) G_{yy}(\vec{q}) + \tfrac{1}{6} u_{xy} w_{2y} \int_q G_{yy}^2(\vec{q})$$

$$- \tfrac{1}{2} w_{2x} u_x \int_q G_{xx}^2(\vec{q}) - w_1 w_{2y} \int_q G_{\phi\phi}(\vec{q}) G_{yy}^2(\vec{q}) + \tfrac{1}{3} w_1^2 u_{xy} \int_q G_{xx}(\vec{q}) G_{yy}^2(\vec{q})$$

$$+ 4 w_{2x} w_1 \int_q G_{\phi\phi}(\vec{q}) G_{xx}(\vec{q}) G_{yy}(\vec{q}) + w_1^2 u_x \int_q G_{xx}^2(\vec{q}) G_{yy}(\vec{q}) - 2 w_1^2 \int_q G_{\phi\phi}(\vec{q}) G_{xx}(\vec{q}) G_{yy}^2(\vec{q}) \ . \tag{B4}$$

One can easily see that the Ward identities of Eqs (2.9) are satisfied when $w_1 = w_{2x} = w_{2y} = w_3 = w_{4x} = w_{4y}$ and $u_x = u_y = u_{xy}$.

---

[1] P.G. de Gennes, *The Physics of Liquid Crystals* (Clarendon, Oxford, 1975).

[2] F. Hardouin, A.M. Levelut, M.F. Achard, and G. Sigaud, J. Chim. Phys. **80**, 53 (1983).

[3] J. Prost and P. Barois, J. Chim. Phys. **80**, 65 (1983).

[4] J. Prost, J. Phys. (Paris) **40**, 581 (1979).

[5] A. Caillé, C.R. Acad. Sci. (Paris) **B274**, 891 (1972).

[6] R.B. Meyer and T.C. Lubensky, Phys. Rev. A **14**, 2307 (1976).

[7] G. Grinstein and R.A. Pelcovits, Phys. Rev. Lett. **47**, 856 (1981).

[8] T.C. Lubensky, S.G. Dunn, and Joel Issacson, Phys. Rev. Lett. **22**, 1609 (1981).

[9] T.C. Lubensky, J. Chim. Phys. **80**, 31 (1983).

[10] T.C. Lubensky and A.J. McKane, Phys. Rev. A **29**, 317 (1984).

[11] J. C. Le Guillou and J. Zinn Justin, Phys. Rev. B **21**, 3976 (1980).

224

VOLUME 55, NUMBER 14   PHYSICAL REVIEW LETTERS   30 SEPTEMBER 1985

Smectic-A Phase with Two Collinear Incommensurate Density Modulations

B. R. Ratna, R. Shashidhar, and V. N. Raja

Raman Research Institute, Bangalore 560080, India

(Received 8 January 1985)

The first observation of a smectic-A phase with two collinear incommensurate density modulations is reported. X-ray diffraction studies of this incommensurate phase reveal reflections corresponding to both partially bilayer (A_d) and bilayer (A_2) periodicities, the relative intensities of the two reflections being strongly temperature dependent.

PACS numbers: 61.30.Eb, 64.70.Ew

A smectic-A liquid crystal may be described as an orientationally ordered fluid with a one-dimensional mass-density wave along the optic axis.[1] When the constituent molecules possess a strongly polar end group several types of smectic-A phases exist, which can be characterized unambiguously by x-ray diffraction.[2] The monolayer A_1 phase exhibits a peak at a wave vector $2q_0 = 2\pi/l$, where l is approximately equal to the molecular length; in addition there may be diffuse scattering centered at a wave vector intermediate between q_0 and $2q_0$. The bilayer phase (A_2) is characterized by two reflections, the fundamental at q_0 and its second harmonic at $2q_0$. In the case of the partially bilayer A_d phase, there is a reflection at $q_0' = 2\pi/l'$, where $l < l' < 2l$, and, generally, a diffuse maximum centered at $2q_0$. There is also the fluid antiphase $\tilde{A}$ whose characteristic x-ray pattern consists of a spot at $2q_0$ and two spots displaced from the Z axis (optic axis) in a perpendicular direction situated symmetrically about the q_0 position.[3] Recent high-resolution x-ray studies[4] have shown the existence of two types of $\tilde{A}$ phases with rectangular lattices of different symmetries.

Although theoretically predicted,[5,6] a smectic-A phase with two incommensurate collinear periodicities has never been observed so far. The only case reported to date of a smectic with two coexistent incommensurate density modulations is the three dimensionally ordered smectic-E phase of 4-octyl-4′-cyanoterphenyl.[7] We present here the results of our x-ray studies on binary mixtures of 4-octyloxy-4′-cyanobiphenyl (8OCB) and 4-n-heptyloxyphenyl-4′-cyanobenzoyloxybenzoate (DB7OCN). The studies have revealed a new type of A phase (referred to here as A_{ic}) which intervenes between the A_d and A_2 phases and which has *two collinear incommensurate density modulations*, one of wavelength $2\pi/q_0'$ and the other of wavelength $2\pi/q_0$. The amplitude of the former modulation decreases with decreasing temperature while that of the latter increases, leading finally to a lockin transition to the A_2 phase.

The phase diagram of the binary system (obtained by a combination of optical and x-ray diffraction techniques) in the region of existence of the A_{ic} phase is shown in Fig. 1. For 8OCB molar concentrations

(X) < 24%, there is only the A_d-A_2 transition which is clearly seen as an abrupt change in the slope of the curve of layer spacing (d) versus temperature (Fig. 2). We have taken x-ray diffraction photographs[8] for several concentrations as a function of temperature in the A_d, A_{ic}, and A_2 phases. The sample had to be cooled extremely slowly (less than 1 °C per hour) in order to obtain a monodomain of the A_{ic} phase. Microdensitometer traces of a series of representative photographs scanned along the Z axis (parallel to the director) for the $X = 34.8\%$ mixture are given in Figs. 3(a)–3(f). Starting from the A_d phase at 119 °C [Fig. 3(a)], we see a sharp peak at q_0'. On cooling, a second sharp peak is seen at q_0, in addition to that q_0' [Fig. 3(b)]. This signifies the onset of the incommensurate phase. On further decrease of temperature the intensity of the reflection at q_0' decreases while that at q_0 increases with an accompanying increase in the intensity of the second harmonic at $2q_0$. The switchover of the relative strengths of the q_0' and q_0 reflections is clearly seen in Figs. 3(c) and 3(d). Finally at 108 °C the peak at q_0' has disappeared altogether [Fig. 3(f)] leaving a

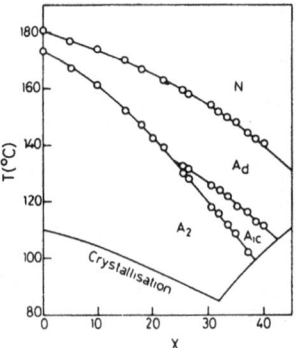

FIG. 1. Partial temperature-concentration (T-X) diagram for mixtures of 8OCB and DB7OCN. X is the mole percent of 8OCB in the mixture. The incommensurate A_{ic} phase intervenes between the partially bilayer (A_d) and bilayer (A_2) phases.

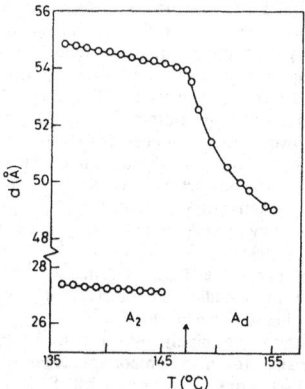

FIG. 2. Temperature variation of the layer spacing (d) in the A_d and A_2 phases of the $X = 18\%$ mixture. The arrow represents the A_d-A_2 transition temperature.

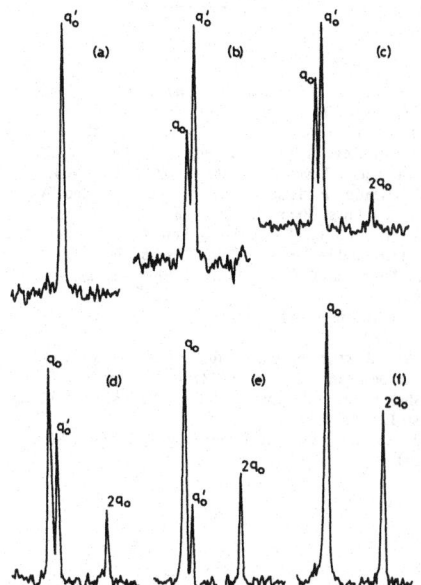

FIG. 3. Raw microdensitometer scans of the x-ray diffraction photographs taken for the $X = 34.8\%$ mixture at different temperatures: (a) A_d phase, 119 °C; (b)–(e) A_{ic} phase, 117, 116, 114.5, and 112 °C, respectively; (f) A_2 phase, 108 °C. The wave vector corresponding to each reflection is marked. The direction of the scan is along the Z axis (optic axis) for all the photographs.

clear signature of the A_2 phase—strong reflections at q_0 and $2q_0$. It must be emphasized that regardless of their amplitudes, the sharpness of these reflections remains the same throughout at all temperatures. No reflections corresponding to combinations of q_0 and q_0' were recorded even with long exposures. [The setup did not allow very-low-angle reflections ($\theta < 0.5°$) to be recorded.] We have also verified from the high-angle diffraction maximum that the in-plane order is liquidlike in all the three phases.

Figure 4 gives the intensity contour diagram (obtained with an X-Y microdensitometer—Joyce-Loebl Scandig 3—in conjunction with an on-line computer) of the photograph taken for the $X = 34.8\%$ mixture at 115.5 °C. Typical widths of the spots are 0.8×10^{-2} Å^{-1} in the Z direction and 1.7×10^{-2} Å^{-1} in the X direction. The larger width in the X direction arises from the geometry of the x-ray monochromator setup. However, it is evident that any displacement of the reflections along the X axis arising from a lateral periodicity of several hundred angstroms would at once be revealed in the contours. We therefore conclude that the three wave vectors are collinear along the Z axis.

The thermal evolution of the layer spacing corresponding to the different modulations in the A phases of the same mixture is shown in Fig. 5. The variations in the A_d and A_2 phases are similar to those seen for the 18% mixture (Fig. 2). In the A_{ic} phase, $2\pi/q_0'$ shows a marked decrease with decrease of temperature. Measurements of the layer spacing for a number of concentrations in the region of existence of the A_{ic}

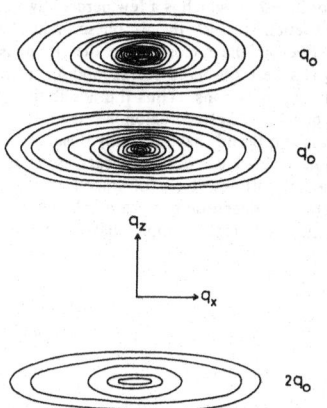

FIG. 4. Intensity contour diagram of a photograph taken for the $X = 34.8\%$ mixture at 115.5 °C. Typical widths of the spots are $q_z = 0.8 \times 10^{-2}$ Å^{-1} and $q_x = 1.7 \times 10^{-2}$ Å^{-1}. The spot at $2q_0$ has been displaced (along Z) closer to the other spots for convenience of representation.

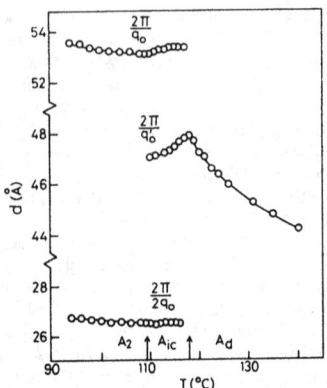

FIG. 5. Temperature variation of the layer spacing (d) in the A_d, A_{ic}, and A_2 phases of the $X = 34.8\%$ mixture. The wave vectors corresponding to the different periodicities are discussed in the text. The arrows indicate the temperatures of transition between the A phases.

phase have confirmed this behavior. This decreasing trend is exactly opposite to what one gets from a simple calculation of the layer spacing variation in the A_{ic} phase assuming it to be a two-phase region. Also, differential scanning calorimetry runs (taken with a Perkin-Elmer DSC-4 in conjunction with the Thermal Analysis Data Station) of the A_d-A_2 transition show a rapid decrease in the strength of the signal with increasing 8OCB concentration, and no signal is observable for $X \sim 20\%$ which is a few percent away from the concentration at which the A_{ic} phase makes its appearance. Even in the region where the A_{ic} phase exists, no signals were observed corresponding to the A_d-A_{ic} and A_{ic}-A_2 transitions. Thus the possibility of the A_{ic} phase being a two-phase region is ruled out.

Cladis and Brand[9] have recently observed an inverted cholesteric phase which appears as an island surrounded by different types of smectic-A phases in the temperature-concentration plane. These authors argued that their results imply that the coexistence of two percolating collinear density waves with different periodicities in a smectic-A phase is incompatible with the fluidity and order-parameter rigidity of the A phase. The present study shows that the A phase can support two incommensurate collinear periodicities over a range of temperature. Prost and Barois[5] have suggested two molecular models for the incommensurate A phase: Depending on the relative strengths of the elastic and lockin terms in the free-energy expansion, the incommensurate density modulations can coexist either by percolating through each other or as a multisoliton regime. Which of the two molecular models represents the A_{ic} phase remains to be settled. Clearly, further studies are needed for a complete understanding of this new phase.

The authors are highly indebted to Professor S. Chandrasekhar for many important suggestions and valuable discussions. The help of Mr. S. Somasekhar during the early part of this work and the cooperation of Dr. Kalyani Vijayan in obtaining the x-ray intensity contour diagrams are gratefully acknowledged.

[1]See, e.g., J. D. Litster, in *Liquid Crystals of One- and Two-Dimensional Order,* edited by W. Helfrich and G. Heppke (Springer-Verlag, Berlin, 1980), p. 65.

[2]A. M. Levelut, R. J. Tarento, F. Hardouin, M. F. Achard, and G. Sigaud, Phys. Rev. A **24**, 2180 (1981).

[3]G. Sigaud, F. Hardouin, M. F. Achard, and A. M. Levelut, J. Phys. (Paris) **42**, 107 (1981).

[4]A. M. Levelut, J. Phys. (Paris), Lett. **45**, L-603 (1984).

[5]J. Prost and P. Barois, J. Chim. Phys. **80**, 65 (1983).

[6]J. Wang and T. C. Lubensky, J. Phys. (Paris) **45**, 1653 (1984).

[7]G. J. Brownsey and A. J. Leadbetter, Phys. Rev. Lett. **44**, 1608 (1980).

[8]For a description of the setup used in the x-ray diffraction experiments see B. R. Ratna, S. Krishna Prasad, R. Shashidhar, G. Heppke, and S. Pfeiffer, Mol. Cryst. Liq. Cryst. **124**, 21 (1985).

[9]P. E. Cladis and H. R. Brand, Phys. Rev. Lett. **52**, 2261 (1984).

X-Ray Scattering Study of the Smectic-A_1 to Smectic-A_2 Transition

K. K. Chan,[a] P. S. Pershan, and L. B. Sorensen[b]

Division of Applied Sciences, Harvard University, Cambridge, Massachusetts 02138

and

F. Hardouin

Centre de Recherche Paul Pascal, Université de Bordeaux I, 33405 Talence Cédex, France

(Received 22 January 1985)

X-ray scattering measurements are reported for critical smectic-A_2 fluctuations along a line of second-order transitions between the smectic-A_1 and smectic-A_2 phases in mixtures of hexylphenyl cyanobenzoyloxy benzoate (DB$_6$) and terephthal-*bis*-butylaniline (TBBA). The measured exponents $\gamma = 1.46 \pm 0.05$ and $\nu_\| = \nu_\perp = 0.74 \pm 0.03$ are constant along the second-order line and agree with recent heat-capacity measurements and the scaling law, $3\nu + \alpha = 2$. They disagree with current theoretical expectations.

PACS numbers: 64.70.−p, 61.30.Eb

The smectic-A phase, which corresponds to the establishment of a one-dimensional density wave in a three-dimensional (3D) liquid, is a system precisely at its lower marginal dimension and consequently does not have true long-range order (LRO) but instead has only quasi long-range order (QLRO).[1] The concept of QLRO in a correlation function has broad importance in equilibrium statistical physics and interest is particularly acute for the smectic-A phase.[2] This phase continues to be one of the most important systems in which QLRO can be studied. The one-dimensional density wave can result in either a monolayer smectic-A_1 phase where the periodicity corresponds to the molecular length, l, or a bilayer smectic-A_2 phase where the periodicity corresponds to twice the molecular length, $2l$.[3] Very recently systems with a smectic-A_1 to smectic-A_2 (A_1A_2) transition were discovered by Hardouin and co-workers.[4,5] This transition is of great intrinsic interest since it corresponds to a transition between two systems with QLRO.[6-8]

The A_1A_2 transition is also relevant to the nematic to smectic-A (NA) transition which, despite considerable experimental and theoretical effort, remains one of the principal unsolved problems in equilibrium statistical physics.[9-11] Complications in the NA transition include the divergent phase fluctuations of the smectic-A order parameter which produce QLRO, strong coupling between the nematic director and the smectic layering, and the anisotropy ($\nu_\| \neq \nu_\perp$) of divergences in the correlation lengths parallel ($\xi_\| \sim t^{-\nu_\|}$) and perpendicular ($\xi_\perp \sim t^{-\nu_\perp}$) to the nematic director. All materials which have been studied appear to obey the anisotropic hyperscaling relation, $\nu_\| + 2\nu_\perp + \alpha = 2$ (where α is the heat-capacity exponent).[10] However, there is no viable explanation for the anisotropy ($\nu_\| - \nu_\perp \sim 0.1$–$0.2$ for all measured materials) or for the increase in the observed values of $\nu_\|$ ($\nu_\| \sim 0.7$–0.8 for the transitions which are currently thought to be representative of NA critical behavior)

relative to the theoretically predicted value ($\nu_\| \sim 0.66$). The origin of this puzzling anisotropy has been widely attributed to the divergent phase fluctuations[9] but it has not been previously possible to test this hypothesis directly. The A_1A_2 transition allows an indirect test since the smectic-A_2 phase fluctuations are quenched by the presence of the established smectic-A_1 density wave.

In this Letter we report the first high-resolution x-ray scattering studies of the A_1A_2 transition. We measured the susceptibility and correlation lengths for a series of mixtures of hexylphenyl cyanobenzoyloxy benzoate (DB$_6$) and terephtal-*bis*-butylaniline (TBBA) with A_1A_2 transitions and found that these transitions are isotropic with correlation-length exponents, $\nu_\| = \nu_\perp = 0.74 \pm 0.03$, elevated compared to the theoretically predicted value ($\nu = 0.63$).[8] This strongly supports the hypothesis that the phase fluctuations are responsible for the anisotropy in the NA transition.

Prost developed a mean field model for the A_1A_2 transition which introduces complex order parameters ψ_1 and ψ_2 corresponding to density fluctuations at $q_0 = (2\pi/2l)\hat{z}$ and $2q_0 = (2\pi/l)\hat{z}$, respectively.[6,7] In this model the NA$_1$ transition is a normal NA transition in which ψ_2 develops QLRO. At the A_1A_2 transition ψ_1 develops in the presence of a nonzero ψ_2. Because of the coupling $\psi_1^\dagger\psi_2^2$ between ψ_1 and ψ_2, the phase of ψ_1 is locked to that of ψ_2. Consequently only the amplitude $|\psi_1|$ has critical behavior at the A_1A_2 transition, and the mean-field model predicts that the transition will be in the 3D Ising class. A more accurate description notes that the "Ising" transition occurs on a smectic-A_1 "lattice" (with QLRO) rather than on a true 3D crystal lattice (with LRO). This presents all the complications inherent in the smectic-A phase: QLRO, anisotropy, and coupling between smectic layering and nematic director. Wang and Lubensky[8] have studied a model which includes all of these effects. To first order in the ϵ expansion they

228

find the A_1A_2 transition to be in the same universality class as the Ising model with isotropic correlation length exponents $\nu_{\parallel} = \nu_{\perp} = 0.63$. However, a nonuniversal susceptibility exponent (γ) of order 0.1 to 0.2 less than the Ising value[12] of 1.242 is predicted.

The x-ray source was a Rigaku RU-200 rotating-anode generator with a 0.3×0.3-mm^2 effective source size. Germanium (220) monochromator and analyzer crystals provided a longitudinal resolution of 1.6×10^{-4} Å^{-1} [half width at half maximum (HWHM)] in the scattering plane. The transverse in-plane resolution was effectively perfect while the transverse out-of-plane resolution was set by slits to be 0.06 Å^{-1} (HWHM). The dispersion of the monochromator induced a spatial separation of the Cu $K\alpha$ doublet, which allowed the $K\alpha_2$ ($\lambda = 1.55439$ Å) line to be eliminated by a slit. Only the $K\alpha_1$ ($\lambda = 1.54056$ Å) component was incident on the sample. The scattered intensity was normalized to the signal from a beam monitor between the monochromator and sample. The sample was contained in a beryllium cell inside a two-stage oven with a dry nitrogen atmosphere. The measured temperature uniformity over the illuminated sample volume (typically 1–2-mm linear dimensions) was better than 2 mK and the stability was about 0.5 mK over a period of several hours. The nematic director was aligned with a 4.3-kG field provided by samarium-cobalt permanent magnets. Typical sample mosaics in the smectic-A_1 phase were about 0.6° FWHM.

Pretransitional fluctuations above T_c (T_c is the A_1A_2 transition temperature) give rise to critical scattering which can be described by an x-ray structure factor

$$S(\mathbf{q}) = \sigma/[1 + \xi_{\parallel}^2 (q_{\parallel} - q_0)^2 + \xi_{\perp}^2 q_{\perp}^2], \quad (1)$$

where $q_0 = 2\pi/2l$ is the wave number corresponding to the bilayer spacing ($2l$) and $\parallel$ and $\perp$ indicate directions with respect to the nematic director. Longitudinal scans (vary $q_{\parallel}$ at $q_{\perp} = 0$) and transverse scans (vary $q_{\perp}$ at $q_{\parallel} = q_0$) were done at a series of temperatures close to the transition. Simultaneous fits of the longitudinal and transverse scans by Eq. (1) (appropriately convolved with the instrumental resolution and corrected for sample mosaic which was directly measured using the monolayer peak at $2q_0$) yielded σ, $\xi_{\parallel}$, and $\xi_{\perp}$. The results of the structure factor fits were then fitted by the power laws $\sigma(t) = \sigma^0 t^{-\gamma}$, $\xi_{\parallel}(t) = \xi_{\parallel}^0 t^{-\nu_{\parallel}}$, $\xi_{\perp}(t) = \xi_{\perp}^0 t^{-\nu_{\perp}}$ [where $t = (T - T_c)/T_c$] to obtain the critical exponents. The transition temperature, T_c, was determined to within ± 3 mK as the temperature below which the widths of longitudinal and transverse scans were resolution and mosaic limited.

The phase diagram for the DB$_6$-TBBA system is shown in Fig. 1. The A_1A_2 transitions, for five mixtures chosen to span the second-order to first-order crossover, were studied and are reported here. A complete study including the NA$_1$, NA$_2$, NA$_d$, and A$_d$A$_2$ transitions will be published separately. The 18.0-, 16.4-, and 13.2-mol% TBBA mixtures had second-order transitions while the 12.1- and 11.9-mol% mix-

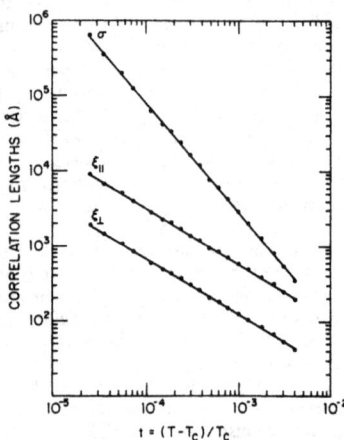

FIG. 2. Measured evolution of the A_2 fluctuations in the A_1 phase for the 16.4-mol% mixture. The divergence of the susceptibility ($\sigma = \sigma^0 t^{-\gamma}$), longitudinal correlation length ($\xi_{\parallel} = \xi_{\parallel}^0 t^{-\nu_{\parallel}}$), and transverse correlation length ($\xi_{\perp} = \xi_{\perp}^0 t^{-\nu_{\perp}}$) is shown. The units for the susceptibility are arbitrary and $q_0 = 0.1171$ Å^{-1}. The exponents determined by a simultaneous power-law fit for γ, $\nu_{\parallel}$, and $\nu_{\perp}$ are $\gamma = 1.46 \pm 0.05$, $\nu_{\parallel} = 0.75 \pm 0.03$, and $\nu_{\perp} = 0.74 \pm 0.03$. The transition temperature T_c (~ 127 °C) was allowed to vary in the fits.

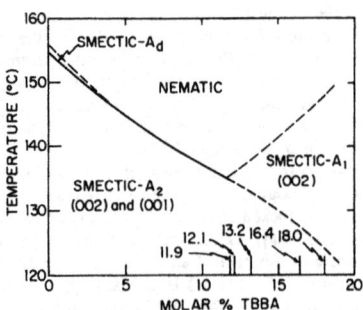

FIG. 1. Phase diagram for mixtures of DB$_6$ with TBBA. The quasi-Bragg reflection from the bilayer ordering is indexed (001) and the monolayer ordering is indexed (002). The dashed lines indicate second-order phase transitions and the solid lines indicate first-order transitions. The order of the A_dA_2 phase boundary has not been determined. The five mixtures reported in this study are indicated.

1695

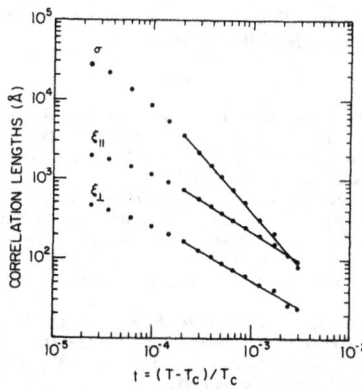

FIG. 3. Measured evolution of the A_2 fluctuations in the A_1 phase for the 12.1-mol% mixture. The susceptibility (σ), longitudinal correlation length ($\xi_{\parallel}$), and transverse correlation length ($\xi_{\perp}$) are shown. The units for the susceptibility are arbitrary and $q_0 = 0.1185$ Å^{-1}. Power-law fits over the range $t > 2 \times 10^{-4}$ yielded $\gamma = 1.41$, $\nu_{\parallel} = 0.76$, and $\nu_{\perp} = 0.72$.

tures had first-order transitions.

Figure 2 shows the measured susceptibility and correlation lengths for the 16.4-mol% sample, together with the power-law fits. In the fits the transition temperature was allowed to vary as a free parameter; it never deviated by more than 3 mK from the value determined from simple inspection of the linewidths of the raw data. For this mixture the resulting critical exponents are $\gamma = 1.46 \pm 0.05$, $\nu_{\parallel} = 0.75 \pm 0.03$, and $\nu_{\perp} = 0.74 \pm 0.03$. The quoted errors are almost entirely systematic, reflecting uncertainties in the corrections for mosaic, background, and instrumental resolution. Table I lists the results for all of the mixtures.

The results for the 12.1-mol% sample are shown in Fig. 3. The data could not be fitted with simple power laws using the transition temperature identified by linewidth inspection. The presence, near T_c, of a sharp resolution-limited peak superposed on diffuse fluctuation scattering indicated a first-order transition with a small coexistence region. An extrapolation of the data yielded an asymptotic transition temperature about 20 mK below the beginning of the coexistence. Power-law fits to the data over the reduced temperature range $t > 2 \times 10^{-4}$ yielded exponents comparable to those obtained for the other mixtures (see Table I). The 11.9-mol% mixture was unambiguously first order; there was a sudden transition (within 3 mK) and no observed coexistence.

We tested the effect of including a fourth-order term of the form $c \xi_{\perp}^4 q_{\perp}^4$ in the denominator of the structure factor in Eq. (1). Other workers have found this to be necessary to account for anomalous transverse tails observed in the NA transition.[13] We found no measurable effect on the quality of the line-shape fits (the χ^2 were typically 1.0 to 1.5) and in all cases the coefficient c was consistent with zero within a statistical uncertainty many times larger than the best-fit value.

Our data show critical exponents that are invariant along the second-order $A_1 A_2$ line. The correlation-length exponents are isotropic ($\nu_{\parallel} = \nu_{\perp}$) within the errors. The average values are $\gamma = 1.46 \pm 0.05$ and $\nu_{\parallel} = \nu_{\perp} = 0.74 \pm 0.03$. Although there is a change in the absolute lengths $\xi_{\parallel}^0$ and $\xi_{\perp}^0$ with concentration (with smaller values closer to the tricritical point), within experimental uncertainty $\xi_{\parallel}^0 / \xi_{\perp}^0 \sim 4$ is invariant. This appears explicitly if the data for different concentrations are fitted with $\nu_{\parallel} = \nu_{\perp}$ held fixed at 0.74. Over the range of length scales and reduced temperatures measured, we find no indication of critical to tricritical crossover effects, as reported for the NA transition.[10,11]

Our exponents do not agree with the 3D Ising values

TABLE I. Summary of the power-law fits for the susceptibility ($\sigma = \sigma^0 t^{-\gamma}$), longitudinal correlation length ($\xi_{\parallel} = \xi_{\parallel}^0 t^{-\nu_{\parallel}}$), and transverse correlation length ($\xi_{\perp} = \xi_{\perp}^0 t^{-\nu_{\perp}}$) for each mixture. The units for the susceptibility are arbitrary. Except for the 12.1-mol% sample the reduced temperature range used in the fits was $2 \times 10^{-5} \leq t \leq 8 \times 10^{-3}$; for the 12.1-mol% sample $t > 2 \times 10^{-4}$.

mol% TBBA	σ^0	γ	$\xi_{\parallel}^0$ (Å)	$\nu_{\parallel}$	$\xi_{\perp}^0$ (Å)	$\nu_{\perp}$
18.0	0.14	1.46[a]	3.5	0.75[b]	0.91	0.73[b]
16.4	0.12	1.46	3.3	0.75	0.77	0.74
13.2	0.062	1.45	2.5	0.74	0.58	0.73
12.1	0.024	1.41	1.2	0.76	0.39	0.72

[a]The error on the susceptibility exponent, γ, is ± 0.05.
[b]The error on the correlation length exponents, $\nu_{\parallel}$ and $\nu_{\perp}$, is ± 0.03.

($\gamma = 1.24$, $\nu = 0.63$) but are consistent with the recent heat-capacity measurements by Chiang and Garland[14] and the scaling law $3\nu + \alpha = 2$. Combining their value of $\alpha = -0.14 \pm 0.06$ with our values yields $3\nu + \alpha = 2.08 \pm 0.15$ which agrees with the scaling law. Since the experiments do not agree with theoretical expectations, it is important to determine if the observed exponents are universal. Experiments on other systems with an A_1A_2 transition are clearly needed to answer this question.

In conclusion, we note that the isotropic correlation-length exponents observed in these A_1A_2 transitions provide the first experimental support for the widespread hypothesis that the divergent phase fluctuations of the smectic order parameter are responsible for the anisotropy in the NA transition. In addition, our observation that the pretransitional A_2 fluctuations are well described by a pure Lorentzian form in both the longitudinal and transverse directions suggests that the inherent anisotropy in the NA transition is related to the anomalous q^4 transverse tails. There is no understanding at the present time of the increase in the correlation-length exponents compared to the theoretically expected value, but it is suggestive that this behavior is found in both the NA ($\nu_\parallel$) and A_1A_2 transitions. The theoretically predicted decrease in the susceptibility exponent relative to the Ising value is also in disagreement with the observed increase. Clearly, continued experimental and theoretical work on the A_1A_2 transition is needed.

The work reported here was supported by the National Science Foundation under Grants No. DMR-82-12189 and No. MRL-80-20247.

(a)Present address: National Synchrotron Light Source, Brookhaven National Laboratory, Upton, N.Y. 11973.

(b)Present address: Department of Physics, University of Washington, Seattle, Wash. 98195.

[1]A. Caille, C. R. Acad. Sci. Ser. B274, 891 (1972).

[2]J. Als-Nielsen, J. D. Litster, R. J. Birgeneau, M. Kaplan, C. R. Safinya, A. Lindegaard-Andersen, and S. Mathiesen, Phys. Rev. B 22, 312 (1980).

[3]R. B. Meyer and T. C. Lubensky, Phys. Rev. A 14, 2307 (1976).

[4]G. Sigaud, F. Hardouin, M. F. Archard, and H. Gasparoux, J. Phys. (Paris) Colloq. 40, C3-356 (1979).

[5]F. Hardouin, A. M. Levelut, M. F. Archard, and G. Sigaud, J. Chim. Phys. 80, 53 (1983).

[6]J. Prost, J. Phys. 40, 581 (1979).

[7]J. Prost and P. Barois, J. Chim. Phys. 80, 65 (1983).

[8]J. Wang and T. C. Lubensky, Phys. Rev. A 29, 2210 (1984).

[9]T. C. Lubensky, J. Chim. Phys. 80, 31 (1983).

[10]B. M. Ocko, R. J. Birgeneau, J. D. Litster, and M. E. Neubert, Phys. Rev. Lett. 52, 208 (1984).

[11]J. Thoen, H. Marynissen, and W. Van Dael, Phys. Rev. Lett. 52, 204 (1984).

[12]J. C. LeGuillou and J. Zinn-Justin, Phys. Rev. B 21, 3976 (1980).

[13]D. Davidov, C. R. Safinya, M. Kaplan, S. S. Dana, R. Schaetzing, R. J. Birgeneau, and J. D. Litster, Phys. Rev. B 19, 1657 (1979).

[14]C. Chiang and C. W. Garland, to be published.

PHYSICAL REVIEW A   VOLUME 34, NUMBER 2   AUGUST 1986

X-ray studies of transitions between nematic, smectic-A_1, -A_2, and -A_d phases

K. K. Chan,[*] P. S. Pershan, and L. B. Sorensen[†]

Division of Applied Sciences, Harvard University, Cambridge, Massachusetts 02138

F. Hardouin

Centre de Recherche Paul Pascal, Université de Bordeaux I, 33405 Talence Cédex, France

(Received 24 January 1986)

We report high-resolution x-ray scattering measurements of the critical fluctuations in mixtures of hexylphenyl cyanobenzoyloxy benzoate (DB_6) and terephtal-bis-butylaniline (TBBA). The phase sequence exhibited on cooling pure DB_6 (or mixtures with a low concentration of TBBA), is nematic (N) to smectic-A_d (A_d) to smectic-A_2 (A_2). Mixtures with ≥ 12 molar percent (mol%) of TBBA have a smectic-A_1 (A_1) phase between the nematic and smectic-A_2 phases. For each of the second-order transitions the critical-temperature dependence of the susceptibility and correlation lengths are fit to power laws of the form t^x where $t = (T - T_c)/T_c$. For the N-A_d transition in pure DB_6 the susceptibility exponent $\gamma = 1.29 \pm 0.05$ and the parallel and perpendicular correlation-length exponents are $\nu_{\parallel} = 0.67 \pm 0.03$ and $\nu_{\perp} = 0.52 \pm 0.03$, respectively. Close to the multicritical point (12 mol% TBBA) where the second-order N-A_1 line meets the first-order portion of the A_1-A_2 line, the N-A_1 exponents are $\gamma = 1.09 \pm 0.05$, $\nu_{\parallel} = 0.57 \pm 0.03$, and $\nu_{\perp} = 0.43 \pm 0.03$. The correlation length anisotropy ($\nu_{\parallel} \neq \nu_{\perp}$) persists along the entire N-A_1 line, with the observed exponents decreasing as the concentration approaches the multicritical point. The A_1-A_2 line has both first-order and second-order regions. All the measured exponents ($\gamma, \nu_{\parallel}, \nu_{\perp}$) were invariant along the second-order portion of the A_1-A_2 line and the correlation-length exponents were isotropic ($\nu_{\parallel} = \nu_{\perp}$). The measured exponents were $\gamma = 1.46 \pm 0.05$, and $\nu_{\parallel} = \nu_{\perp} = 0.74 \pm 0.03$. These numbers also held close to the tricritical point where the A_1-A_2 transition became first order.

I. INTRODUCTION

At the nematic–to–smectic-A phase transition the continuous translational symmetry of the nematic phase is spontaneously broken by the appearance of a one-dimensional density wave in the smectic-A phase. Although this transition would naively appear to be the simplest solidification and/or melting transition which can occur in nature, its physics has been found to be remarkably subtle. Consequently it has been a rich subject for numerous experimental and theoretical studies. Despite these considerable efforts, it remains an unsolved problem.

The conventional smectic-A phases have been classified according to the periodicity of their one-dimensional waves. When the periodicity corresponds to the molecular length, l, the phase is called a monolayer or smectic-A_1 phase and when the periodicity corresponds to twice the molecular length, $2l$, the phase is called a bilayer or smectic-A_2 phase. There are numerous "conventional" examples of both monolayer and bilayer smectic systems that undergo continuous, or second-order, transitions between the nematic and smectic-A phases for which the critical smectic density fluctuations exhibit only a single wave vector corresponding to either the monolayer or bilayer periodicity.

Recently, a new class of polar liquid-crystal molecules were discovered[1] which exhibit a variety of new smectic phases including polar monolayer and bilayer phases and a smectic-A_d phase with a periodicity d intermediate between the monolayer and bilayer periodicity; $l < d < 2l$.

These remarkable polar smectic-A phases melt into polar nematic phases which exhibit two simultaneous fluctuations with either commensurate $d_1 = 2d_2$ or incommensurate $d_1 \neq 2d_2$ periodicities.

In this paper we report the results of a comprehensive high-resolution x-ray scattering study of these remarkable polar nematic and smectic-A phases and the transitions between them for a series of mixtures of hexylphenyl cyanobenzoyloxy benzoate (DB_6) and terephtal-bis-butylaniline (TBBA). We were motivated to study these new polar phases both because of their intrinsic interest and because they have properties that provide insight into the unsolved nematic–to–smectic-A transition. We have found that the nematic–to–smectic-A transitions in this polar system (with two simultaneous smectic fluctuations) are quantitatively different from the conventional nematic–to–smectic-A transition (in systems with a single density wave). In addition, we have found that the correlation-length exponents at the smectic-A_1 to smectic-A_2 transitions in these mixtures are isotropic, providing the first experimental support for the hypothesis that the divergent phase fluctuations of the smectic order parameter are responsible for the observed anisotropy in the nematic–to–smectic-A transition.

The next section of this paper provides a review of the experimental and theoretical background to the present work. Section III describes the experimental apparatus and techniques. Section IV presents the details of the data analysis and the experimental results. Section V provides a discussion and conclusion.

II. BACKGROUND

A. Polar smectic-A phases

The loss of translational invariance in the conventional (nonpolar) smectic-A phases is fully characterized by a single complex order parameter, $\Psi = \psi(\mathbf{r})e^{i\phi(\mathbf{r})}$, which specifies the amplitude and the phase of the one-dimensional density wave. When the constituent molecules contain strong dipole moments, the dipolar ordering can induce a second spatial period. It is convenient to describe these two possible density waves by introducing two complex order parameters. Much of the interesting physics in these systems arises from the competition between these two order parameters which, in general, have incommensurate periodicities.

Figure 1 schematically illustrates the four liquid-crystalline phases described in this study. The molecules are represented as thin rods, with an arrow at one end to denote the lack of inversion symmetry introduced by an off-center dipole. In the nematic phase the centers of mass of the molecules have fluidlike order. The molecules tend to orient with their long axis parallel to a common director $\hat{\mathbf{n}}$, and there is a long-range molecular-orientational order. Since the nematic phase has inversion symmetry there are equal numbers of molecules pointing along $\hat{\mathbf{n}}$ and opposite to $\hat{\mathbf{n}}$. The smectic-A phases have a periodic density modulation parallel to the nematic director. Roughly, this can be thought of as the formation of "statistical layers" in the material. Three possible types of smectic-A phases are shown in Fig. 1. For each of

these, there is a quasi-long-range correlation between the positions of the layers. Within each layer, the molecules act like a two-dimensional fluid.

These smectic-A phases are distinguished by their layer spacings and by their dipolar order. The smectic-A_1 is a monolayer phase with the dipolar heads oriented randomly within each layer. This produces a periodicity d equal to l, the molecular length. The smectic-A_2 phase is a bilayer phase. This is attributed to a preferential up-down ordering of the dipolar heads within each layer, which alternates from layer to layer producing a periodicity d equal to twice the molecular length, $2l$. There are other systems, not studied here, that develop a further modulation parallel to the layers. This is attributed to further modulations of the head directions within the layers.[1] The smectic-A_d phase is an intermediate phase characterized by $l < d < 2l$. The dipolar heads are preferentially ordered with a slight overlap rather than end to end as in the smectic-A_2 phase. The macroscopic symmetries of the smectic-A_d and smectic-A_2 phases are indistinguishable.

The x-ray diffraction patterns of the polar nematic and polar smectic-A phases are shown in Fig. 2. The scattering for the nematic phase has two diffuse spots with wave vectors $\mathbf{q}_2 = (2\pi/l)\hat{\mathbf{n}}$ and $\mathbf{q}_1 = q_1\hat{\mathbf{n}}$ ($2\pi/2l \leq q_1 < 2\pi/l$) corresponding to the preferred layer spacings for monolayer and dipolar ordering, respectively. Since the wave numbers are in general incommensurate, the temperature dependence of q_1 and q_2 can be different. The presence of two strong diffuse spots in the nematic phase distinguishes these materials from the nonpolar nematic materials which have only a single diffuse spot corresponding either to the monolayer or bilayer periodicity. The smectic-A_1 phase has a Bragg-like peak at q_2 corresponding to monolayer ordering and diffuse scattering at wave

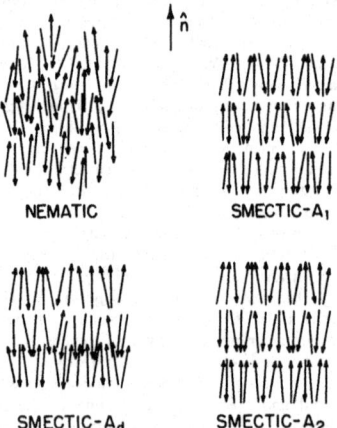

NEMATIC   **SMECTIC-A_1**

SMECTIC-A_d   **SMECTIC-A_2**

FIG. 1. Schematic representation of the polar nematic and smectic-A phases. The molecules are drawn as thin rods with an arrow on one end to denote the lack of inversion symmetry produced by an off-center dipole.

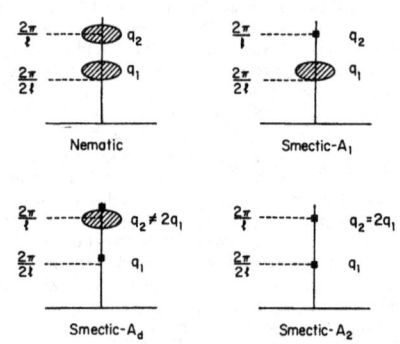

FIG. 2. Diffraction patterns of the polar nematic and smectic-A phases. The vertical axis indicates the q_z direction and the horizontal axis indicates the $q_x q_y$ plane. The small solid squares denote Bragg-like spots and the large ellipses denote diffuse spots. The monolayer ordering ($d = l$) produces the scattering at q_2 and the dipolar ordering ($l < d \leq 2l$) produces the scattering at q_1.

vector q_1. The smectic-A_2 and -A_d have Bragg-like peaks at both q_1 and $2q_1$. The smectic-A_d phase also has a diffuse peak at q_2 corresponding to smectic fluctuations at the monolayer periodicity.

The first observation of a smectic-A_1–smectic-A_2 transition was made by Sigaud et al.[2] and DB$_5$-TBBA mixtures. In this paper we report studies of a similar system consisting of mixtures of DB$_6$ and TBBA. The phase diagram of DB$_6$-TBBA is shown in Fig. 3. All four of the phases discussed above are present. DB$_6$ has a strong dipolar end group and forms a smectic-A_d phase over a narrow temperature range, but primarily forms a smectic-A_2 phase. TBBA is a nonpolar molecule that forms a smectic-A_1 phase. The addition of sufficient TBBA [> 12 mole percent (mol %)] to DB$_6$ leads to a smectic-A_1 phase intermediate between the nematic and smectic-A_2 phases.

Prost and co-workers[3-6] have constructed mean-field descriptions of these phases and the corresponding phase diagrams for numerous polar smectic systems. The Prost model introduces two complex fields, Ψ_1 and Ψ_2, that order preferentially with wave numbers q_1 and q_2, respectively. This model predicts that the N-A_1 transition is a conventional N-A transition in which Ψ_2 develops quasi-long-range order (QLRO) at the transition. The model also predicts that at the A_1-A_2 transition, Ψ_1 develops in the presence of a nonzero Ψ_2 and because of a $\Psi_1^2\Psi_2^*$ coupling, the phase of Ψ_1 is locked to that of Ψ_2. Consequently, only the amplitude $|\Psi_1|$ is predicted to have critical fluctuations at the A_1-A_2 transition. The N-A_d transition is also predicted to be a conventional N-A transition where Ψ_1 develops QLRO and $|\Psi_2| / |\Psi_1|$ is small. Finally, the A_d-A_2 transition is predicted to be first-order and is characterized by a sudden change in the ratio $|\Psi_2| / |\Psi_1|$.

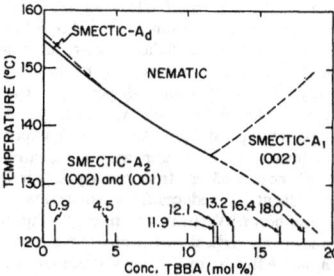

FIG. 3. Schematic phase diagram for mixtures of DB$_6$ and TBBA. The reflection at q_1 from the bilayer ordering is indexed (001) and the reflection at q_2 from the monolayer ordering is indexed (002). The dashed lines indicate second-order phase boundaries and the solid lines indicate first-order boundaries. The order of the $A_d A_2$ phase boundary has not been determined. The mixtures which were measured during this study are indicated at the bottom of the diagram.

B. Nematic-to-smectic-A transition

Since the earliest work by Kobayashi[7,8] and McMillan,[9] considerable effort has been devoted to the nematic-to-smectic-A transition. Present theories[10,11] based on the de Gennes free energy[12,13] propose two possibilities. The first is that the N-A transition is in the inverted XY class with critical exponents:[14] $\alpha = -0.02$ (specific heat); $\nu_{||} = \nu_\perp = 0.67$ (correlations parallel and perpendicular to the nematic director); and $\gamma = 1.32$ (susceptibility). The second possibility is an anisotropic class where $\nu_{||} = 2\nu_\perp$. Unfortunately, direct comparison of the predicted XY model correlation length exponents with experiment is complicated by the absence of long-range order in the smectic-A phase.[10] Lubensky et al.[15] predict that asymptotically close to the transition, $\nu_\perp = \nu_{||}/2$ regardless of the universality class. Toner[16] predicts modified XY values of the form $\nu_{||} = \frac{6}{5}\nu_{XY} = 0.80$ and $\nu_\perp = \frac{4}{5}\nu_{XY} = 0.53$.

The experimentally observed correlation-lengths exponents are anisotropic with $\nu_{||} - \nu_\perp \sim 0.1-0.2$ for all measured materials, and in addition are nonuniversal with $\nu_{||}$ ranging from $\sim 0.6-0.8$ for different materials. Brisbin et al.[17] proposed that the apparent nonuniversality of the critical exponents is due to the presence of a nearby tricritical point. This suggestion has been confirmed by a number of studies which show that the measured critical exponents cross over from a set of "saturated values," when the nematic range is large, to a set of "tricritical values" when the nematic range is small.[18-22] Such a crossover was originally predicted by McMillan[9] to occur as the result of coupling between the nematic director fluctuations and the smectic ordering when the ratio T_{NA}/T_{NI} (called the McMillan ratio) became larger than about 0.99.

The clearest evidence for this crossover was provided by combined specific-heat and x-ray scattering studies of the critical exponents for the two homologous series, 4-n-pentylbenzenethio-4-alkoxybenzoate ($\overline{n}$S5) and alkylcyanobiphenyl (nCB). The crossover occurred at $n = 10$ for $\overline{n}$S5 and $n = 9$ for nCB. For $\overline{10}$S5 ($T_{NA}/T_{NI} = 0.983$) $\alpha = 0.45 \pm 0.05$,[17] $\gamma = 1.10 \pm 0.05$, $\nu_{||} = 0.61 \pm 0.03$, and $\nu_\perp = 0.51 \pm 0.03$.[21] For 9CB ($T_{NA}/T_{NI} = 0.994$) $\alpha = 0.50 \pm 0.05$,[22] and $\gamma = 1.10 \pm 0.05$, $\nu_{||} = 0.57 \pm 0.03$, $\nu_\perp = 0.37$ (+ 0.07, −0.03).[21] For these materials the specific-heat exponents approach the tricritical value,[23] $\alpha = 0.5$, and the susceptibility exponent γ drops to 1.10 which is slightly above the tricritical value of 1.0. In addition, an average correlation length exponent $\overline{\nu} = \frac{1}{2}(\nu_{||} + \nu_\perp)$ approaches the corresponding isotropic tricritical value of 0.50. However, there is no existing theory for an anisotropic tricritical point.

The materials which are currently thought to be the most representative of true saturated N-A critical behavior have large nematic ranges (with correspondingly low McMillan ratios) and specific-heat exponents which are consistent with the predicted XY values. Two of these saturated materials are $\overline{8}$S5 and 4O.7. For $\overline{8}$S5 ($T_{NA}/T_{NI} = 0.936$) $\alpha = 0.0 \pm 0.02$, $\gamma = 1.53 \pm 0.02$, $\nu_{||} = 0.83 \pm 0.01$ and $\nu_\perp = 0.68 \pm 0.02$.[17,20] For 4O.7 ($T_{NA}/T_{NI} = 0.926$) $\alpha = -0.03 \pm 0.04$, $\gamma = 1.46 \pm 0.03$, $\nu_{||} = 0.78 \pm 0.02$ and $\nu_\perp = 0.65 \pm 0.02$. [$n$O.7 is N-(4-n-alkoxybenzylidene)-4-heptylaniline.] However, as originally pointed out by Garland et al.,[19] neither compound has correlation-length exponents that can be reconciled with theory.

Recently, measurements have been made of the mean-square amplitude of the order parameter near the N-A transition.[24] By fitting the experimental data to the scaling form $\langle |\Psi|^2 \rangle = L \mp M^{\pm} |t|^x$ (here $M^{\pm}$ are constants above and below the transition temperatures and t is the reduced temperature) it was demonstrated that x is not equal to $1-\alpha$ as is predicted for the Landau–de Gennes free energy.[25] In addition, the discrepancy is largest for materials with long nematic ranges, which are expected to be described best by the de Gennes model. These results suggest that the Landau–de Gennes free energy must be inadequate for quantitative predictions.

C. Smectic-A_1-to–smectic-A_2 transition

Considerably less work has been done on the A_1-A_2 transition. Cursory inspection based only on the dimensionality of the system and the number of components in the order parameter would place the A_1-A_2 transition in the three-dimensional (3D) Ising class.[3] The actual situation is however more subtle, since the A_1-A_2 transition presents the novel case of a transition between two phases with QLRO. The "Ising" transition occurs on a smectic-A_1 "lattice" (with QLRO) rather than a true 3D crystal lattice (with LRO). This presents all the complications inherent in the theoretical descriptions of the smectic-A phase: QLRO, anisotropy, and coupling between the smectic layering and the nematic director. Wang and Lubensky[26] have done a renormalization-group calculation which includes all of these effects. The first order in the ϵ expansion they found that the correlation-length exponents are equal to the three-dimensional Ising value, $\nu_{||} = \nu_{\perp} = 0.63$, and the susceptibility exponent γ is nonuniversal and is of order 0.1–0.2 less than the Ising value of 1.242.

Experimentally, x-ray scattering measurements on DB$_6$-TBBA mixtures[27] have found $\gamma = 1.46 \pm 0.05$ and $\nu_{||} = \nu_{\perp} = 0.74 \pm 0.03$. Heat-capacity measurements on the same system[28] found $\alpha = -0.14 \pm 0.06$. The two measurements are clearly consistent with the appropriate isotropic scaling law $3\nu + \alpha = 2$ (experimental value 2.08 ± 0.15), but disagree with the theoretically predicted critical exponents. Recently Huse[29] proposed that the discrepancy arises because the data were taken at fixed concentration, while the theoretical values correspond to fixed chemical potential. Using the technique of Fisher renormalization[30] the expected experimental values are given in terms of the theoretical Ising values by $\gamma = \gamma_I/(1-\alpha_I) = 1.24/(1-0.11) = 1.39$, and $\nu = \nu_I/(1-\alpha_I) = 0.63/(1-0.11) = 0.71$ where the subscript I denotes the Ising values and the theoretical value[31] $\alpha_I = 0.11$ is used.

III. EXPERIMENTAL

Figure 4 shows a top view of the scattering geometry and apparatus. The axes q_x and q_z define the horizontal

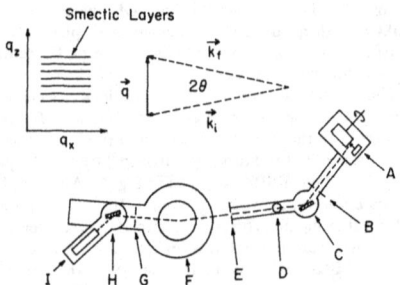

FIG. 4. Schematic diagram showing the x-ray scattering apparatus and the corresponding scattering geometry. A, rotating anode x-ray generator; B, monochromator input slits; C, germanium (220) monochromator crystal; D, incident beam intensity monitor; E, monochromator output slits; F, Huber 420 two-circle goniometer; G, analyzer input slits; H, germanium (220) analyzer crystal; I, NaI scintillation detector. The q_z direction is normal to the smectic layers and the momentum transfer is $\mathbf{q} = \mathbf{k}_f - \mathbf{k}_i$.

scattering plane. The third axis, q_y, out of this plane is referred to as the vertical direction.

The x-ray source (A) is a Rigaku RU-200 rotating anode generator operated at 4.5 kW. The monochromator input slit (B) is 2 mm wide and 25 cm from the source providing an angular collimation of about 0.5° at an angle of roughly 6° from the copper anode surface. With a 3.0 mm wide by 0.3 mm high electron focal spot, this provides an effective x-ray source size of 0.3×0.3 mm. A germanium monochromator crystal (C) cut for a (220) reflection is oriented to reflect the K_α lines of copper ($2\theta = 45.2°$). For each wavelength, the Darwin width of the reflection results in an angular divergence of 3.4 mdeg full width at half maximum (FWHM) in the horizontal plane. Following the monochromator is a sheet of kapton (D) which scatters a portion of the beam vertically into a scintillation detector. This serves as a monitor for the beam intensity. The 0.06° angular dispersion of the K_{α_1} and K_{α_2} components ($\lambda_1 = 1.5406$ Å, $\lambda_2 = 1.5444$ Å) results in a spatial separation of these two components. Horizontal adjustment of a 1 mm wide monochromator output slit (E) located 65 cm from the source eliminates the K_{α_2} component and reduces the bremsstrahlung contribution to the beam. Vertical adjustment of this slit (2 mm) determines the vertical extent of the beam incident on the sample. After exiting from the output slit, the beam passes over a Huber model 420, two-circle goniometer (F). The rotation axes of the two circles are coincident and are centered on the beam. The sample oven (90 cm from source) is mounted on one Huber circle. A rail assembly that is mounted on the second Huber circle holds the analyzer input slits (G) 20 cm from sample, a matching analyzer crystal (H) 10 cm from analyzer slit, and a scintillation detector (I) for counting the scattered x-ray photons.

The in-plane resolution of the spectrometer is determined primarily by the monochromator and analyzer crystals. The rocking curve of a pair of Ge(220) reflections has an angular width of 4.5 mdeg (FWHM). At small scattering angles, the rocking curve can be used as a first approximation for the longitudinal (along q_z) resolution. This approximation is valid only when looking at imperfectly aligned samples or diffusion scattering, as is always the case in this experiment. In addition, there is a correction for the dispersion of the natural linewidths of the copper lines. The corresponding reciprocal space resolution is 3.2×10^{-4} Å^{-1} (FWHM) in q_z, and essentially infinitely sharp in q_x. The transverse out-of-plane or vertical resolution is determined by the height of the beam at the sample (3 mm) and the analyzer input slit (8 mm vertical). These produced a 0.12-Å^{-1} (FWHM) resolution in q_y. The implications of the resolution are discussed in detail, in Sec. IV.

The liquid-crystal samples are held in a rectangular cell machined from a solid piece of beryllium (see Fig. 5). The sample thickness is 2 mm, and the illuminated scattering volume is approximately 0.5 mm wide by 3 mm high. Scattering is done in transmission. A Teflon cap is pressed against the cell to provide a seal. The cell is held in a vacuum tight, beryllium-windowed oven with two independent heating stages. The temperatures are sensed by thermistors and regulation is provided by a computer-interfaced temperature controller. The absolute temperature calibration was about 0.5 K. Typically the oven was operated with a 3–5 K temperature difference between inner and outer stages. The temperature gradient over the illuminated sample volume was measured to be less than 2 mK. Temperature stability over a period of hours was better than 1 mK. Two 2 in. diameter, 1 in. thick samarium-cobalt magnets spaced 1 in. apart in a magnetic yoke provide a 4.5-kG field for aligning the nematic

director. All measurements were done with a dry nitrogen atmosphere in the oven.

A Digital Equipment Corporation PDP 11/34 minicomputer was used to control the spectrometer angles, timing, counting, and setting and measurement of the sample temperature. The interfacing was done with a combination of CAMAC (IEEE-583) modules and GPIB (IEEE-488) instruments.

IV. EXPERIMENTAL RESULTS AND DATA ANALYSIS

Measurements were done on pure DB$_6$ and seven DB$_6$-TBBA mixtures. The concentrations studied are indicated on the phase diagram shown in Fig. 3. The five mixtures, 18.0, 16.4, 13.2, 12.1, and 11.9 mol% TBBA in DB$_6$ are to the right of the NA_1A_2 triple point in the phase diagram, and second-order N-A_1 transitions were observed. The 18.0, 16.4, and 13.2 m% mixtures are to the right of the A_1A_2 tricritical-point and second-order A_1-A_2 transitions were observed. The 12.1 and 11.0 mol% mixtures had first-order A_1-A_2 transitions. The 0.9 and 4.5 mol% mixtures were not studied in detail, but were examined to verify the topology of the phase diagram.

The organization of this section is as follows. We first present an overview of the principal structural features of the various phases of pure DB$_6$. This is followed by detailed discussions of the fitting procedures that were used to extract the correlation-length and susceptibility exponents at the N-A_d transition. After a brief discussion of the N-A_1 transition the section concludes with a longer discussion of the A_1-A_2 transition.

A. Results for pure DB$_6$

Pure DB$_6$ has a smectic-A_d phase over a narrow temperature range ($\sim$0.5 K) between the nematic and

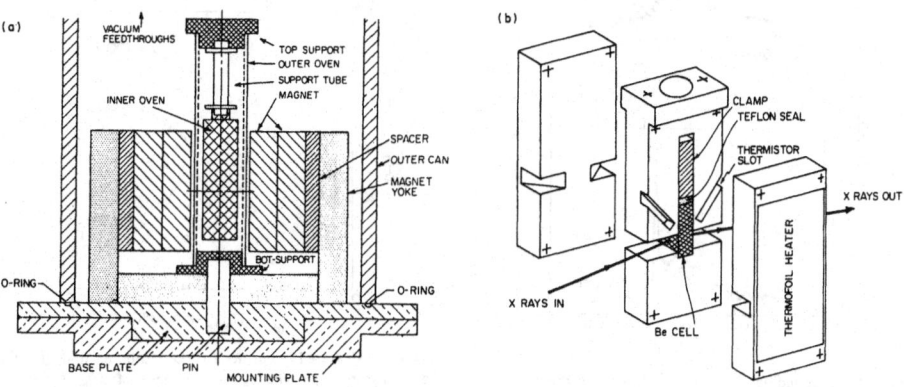

FIG. 5. Schematic diagram of the two-stage oven used in these studies. (a) Cross-sectional view of the complete oven. The magnetic field is provided by four samarium-cobalt magnets mounted in a return yoke. The outer oven is held 3–5 K below the inside oven. (b) Detail of the inner oven. The sample is sealed in a solid beryllium cell with a Teflon seal.

smectic-A_2 phases. Several degrees above the N-A_d transition, the scattering from DB$_6$ shows two diffuse incommensurate peaks. An (001) peak corresponding to dipolar ordering is centered at 0.1277 ± 0.0002 Å^{-1}, and an (002) peak corresponding to the monolayer ordering is centered at 0.241 ± 0.001 Å^{-1}, slightly less than twice the wave vector of the (001) peak. The width of the (002) scattering indicates longitudinal correlations of about 60 Å, or between two or three molecular lengths. The diffuse (002) scattering shows no change in shape or intensity when the temperature is lowered below the N-A_d transition temperature. The (001) peak is due to pretransitional smectic order-parameter fluctuations. These fluctuations show critical behavior and are discussed in detail below.

Below the N-A_d transition (T_{NA}), the (001) scattering is resolution limited in the longitudinal direction and mosaic limited in the transverse direction. The original diffuse (002) scattering shows no change, but a resolution-limited (002) peak appears off to one side of the diffuse scattering (Fig. 6). This resolution-limited (002) peak is at exactly twice the wave vector of the (001) smectic peak, and is first observable about 20 mK below T_{NA}. The data suggest that it is present immediately below T_{NA}, but is too weak to be seen above the diffuse scattering and background. Mosaic scans of the (001) and (002) resolution-limited scattering always showed identical structures, precluding any appreciable contribution from multiple scattering effects.[32]

In the smectic-A_d phase, the (001) wave vector q_1 changes dramatically with temperature as shown in Fig. 7. The wave vector of the resolution-limited (002) peak remains commensurate with the changing (001) wave vector at all temperatures below the N-A_d transition. The

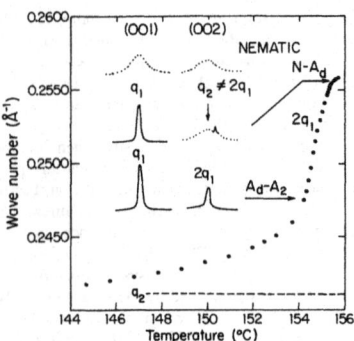

FIG. 7. Temperature dependence of the wave vectors in DB$_6$. The (001) peak condenses at the N-A_d transition and just below the N-A_d transition the sharp second harmonic of the (001) peak appears slightly above the center of the diffuse (002) peak. The value of $2q_1$ has been plotted in the figure to emphasize its lock in ($2q_1 \rightarrow q_2$) at low temperature.

wave vector of the (002) diffuse scattering q_2 is temperature independent. Initially, the intensity of the resolution limited (002) peak grows slowly, as the temperature is decreased below the N-A_d transition. Then about 0.5 K below the N-A_d transition it undergoes a dramatic increase. This rapid increase is the signature of the A_d-A_2 transition. Figure 8 shows the measured intensity as a function of temperature. The sample mosaics vary rapidly with temperature due to the large changes in the smectic lattice constants. A typical mosaic width 1 K below the N-A_d transition is 4° FWHM. This variation makes it difficult to directly measure the intensity. To provide re-

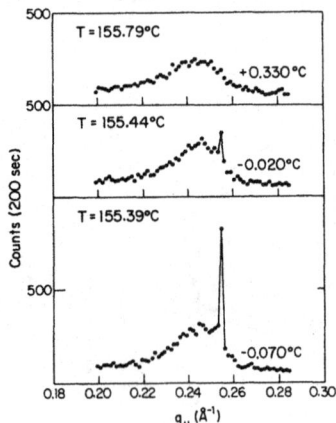

FIG. 6. Longitudinal (q_z) scans showing the diffuse (002) scattering and the sharp $2q_1$ scattering near the N-A_d transition in DB$_6$. Finer scans of the sharp peak show that it is resolution limited. The nematic-to-smectic-A_d transition occurs at ~ 155.46 °C.

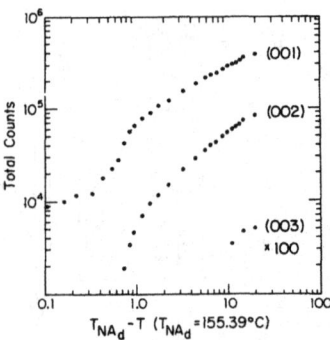

FIG. 8. Measured intensities for the resolution limited (001) and (002) scattering in the A_d and A_2 phases of DB$_6$. The rapid growth of the (002) peak indicates the transition to the A_2 phase. The temperatures are given relative to the N-A_d transition. The intensities were integrated over the mosaic. Deep in the A_2 phase a resolution-limited (003) peak is visible.

liable intensity data we integrated over the sample mosaic for each 2θ value. The data shown in Fig. 8 are the integrated results and were taken by heating up from the A_2 phase. Although the mosaics at low temperatures were poor, they remained relatively stable while the sample was heated, except very close (~ 50 mk) to the N-A_d transition. The data obtained from cool-down sequences showed qualitatively similar results. Deep in the A_2 phase, a weak resolution limited peak at the (003) position can be seen. This is shown in Fig. 8, magnified by a factor of 100.

In the 0.5-K region of the A_d phase, the system chooses a wave vector for the QLRO that is a compromise between the two competing wave vectors. The periodicity of the first wave vector q_1 is given by the center of the (001) fluctuations in the nematic phase which corresponds to the preferred spacing for dipolar order. The periodicity of the second wave vector q_2 is given by the center of the (003) diffuse peak, corresponding to the monolayer order. The A_d wave vector is intermediate between these two possible choices. As q_1 decreases with decreasing temperature it approaches $q_2/2$, and as the incommensuration in the system decreases, there is a transition to the A_2 phase.[6] In the A_2 phase, q_1 actually continues to change with temperature, but at a much slower rate as it approaches its low-temperature equilibrium value with $q_1 = q_2/2$ as shown in Fig. 7.

For contrast, Fig. 9 shows similar intensity measurements for DB$_5$, which has a N-A_2 transition. Calorimetry results indicate the transition is first order.[2] The x-ray data show no pretransitional fluctuation scattering on the nematic side of the transition. As the temperature is lowered two commensurate resolution-limited peaks suddenly appear and grow continuously

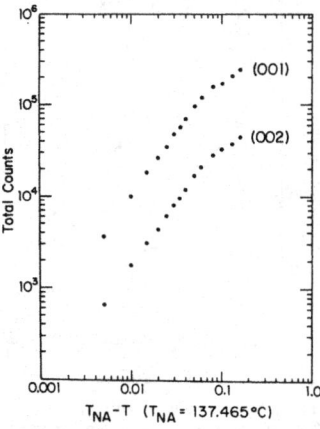

FIG. 9. Measured intensities for the resolution limited (001) and (002) scattering in the smectic-A_2 phase of DB$_5$. The intensities were integrated over the mosaic. The temperatures are given relative to the N-A_2 transition.

from zero intensity at the N-A_2 transition. This transition displayed a 200 mK hysteresis as expected for a first-order transition. The data shown in Fig. 9 were collected by cooling down from the nematic phase. Unlike DB$_6$, DB$_5$ does not show any dramatic changes in lattice constant below the N-A transition.

B. Data analysis and the nematic—to—smectic-A_d transition in DB$_6$

Close to the N-A_d transition, the nematic phase develops smectic order-parameter fluctuations. The scattering produced by these fluctuations are described by a modified Ornstein-Zernike expression

$$S(\mathbf{q}) = \frac{\sigma}{1 + \xi_\parallel^2 (q_\parallel - q_0)^2 + \xi_\perp^2 q_\perp^2 + c\xi_\perp^4 q_\perp^4} . \quad (1)$$

The addition of the $q_\perp^4$ term in the denominator is purely phenomenological and will be discussed further below. In the smectic-A phase, the liquid crystal has Bragg-like scattering. Our instrumental resolution is unable to distinguish the subtle differences[33] between true Bragg scattering and the algebraic line shapes predicted for smectic-A liquid crystals. Consequently, the longitudinal scans are resolution limited in the smectic-A phase. As the temperature is lowered into the smectic-A phase the elastic constants increase and produce long healing lengths. As a result, competition between the alignment effects of the magnetic field and boundary conditions imposed by the walls of the beryllium cell create a polycrystalline sample. The distribution in the orientation of these macroscopic crystallites over the illuminated sample volume is the observed mosaic. The transverse scans in the smectic-A phase are limited in their width by the sample mosaic, which is always much broader than the transverse resolution. As T_{NA} is approached from below, the effect of the cell walls decreases and the alignment improves. If the magnetic field is grossly misaligned with the rectangular walls of the cell, this can lead to sudden shifts in the mosaic. Typically the mosaic widths show a minimum close to T_{NA}. The mosaics are not always smooth. Some samples displayed sharp or erratic structures such as double peaks.

In the nematic phase, the rapidly decreasing correlation lengths lead to a broadening of the scattering profile as the temperature is increased. The longitudinal scans are no longer resolution limited and the transverse widths are now dominated by fluctuation scattering instead of the sample mosaic. The transition temperature can be identified from the behavior of these linewidths. Figure 10 shows the measured linewidths for a DB$_6$ sample close to the N-A_d transition. In this case, the widths for both the longitudinal and transverse scans show no variation in the small temperature range below T_{NA}. At T_{NA} both the widths exhibit a slope discontinuity. In some samples, the mosaic width broadens slightly below T_{NA}. This leads to a cusp in the temperature dependence of the transverse FWHM's at T_{NA}. Either temperature or concentration inhomogeneities can obscure the sharpness of the cusp by

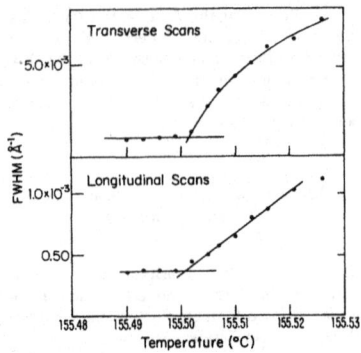

FIG. 10. The observed longitudinal (q_z) and transverse (q_x) linewidths (FWHM) near the N-A_d transition in DB$_6$. The transition temperature estimate from these linewidths measurements is $\sim 155.502\,°C$.

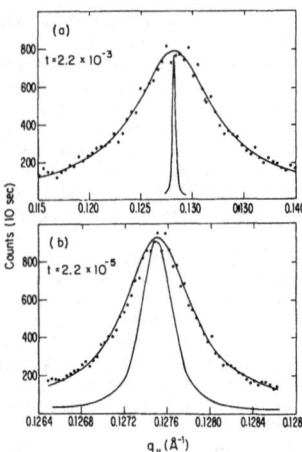

FIG. 11. Representative longitudinal (001) scans near and far from the N-A_d transition in DB$_6$. The line through the data point is the result of the least-square fits to Eq. (1) convolved with the instrumental resolution. The narrow solid line is a resolution-limited smectic-A scan.

increasing the width of the narrowest transverse scan observed. Nonetheless, as shown in Fig. 10, the transition temperature can be identified to within a few mK.

The data were fitted to Eq. (1) convolved with the instrumental resolution. Longitudinal and transverse scans at each temperature were fitted together as a pair since the coarse vertical resolution (necessary to obtain adequate signal) couples the longitudinal and transverse widths. A Marquardt[34] nonlinear least-squares-fitting routine adapted from Bevington[35] was used. The free parameters are the correlation lengths $\xi_{\|}$ and $\xi_{\perp}$, the susceptibility σ, the wave number q_0 and, in some cases, the quartic coefficient c.

A small absorption correction was applied to the data. In the wings of the widest transverse scans ($\pm 25°$), this amounted to an additional 10% attenuation. The thickness of the liquid-crystal sample is roughly one e length. The absorption has little effect on any of the results. A small background correction (0.15/sec) was also made.

The deviation of the structure factor from a pure Lorentzian form in the transverse direction has been reported by previous investigators. The earliest measurement was by McMillan[36] who proposed the modified form

$$S(\mathbf{q}) = 1 + \left[\beta_0 t^\gamma + \alpha_{\|} \left[\frac{q_{\|} - q_0}{q_0} \right]^2 + \alpha_{\perp} \left[\frac{q_{\perp}}{q_0} \right]^n \right]^{-1}. \quad (2)$$

Equation (1) follows later workers[37] in the addition of a $q_{\perp}^4$ term to the denominator. Its inclusion is necessary to obtain adequate fits to the data. For a transverse scan at a reduced temperature of 2×10^{-5}, it reduces χ^2 from 8 to 1.5. The best-fit value of c is generally in the range 0.05–0.20. Figure 11 shows least-squares fits using Eq. (1) at reduced temperatures close to and far away from the transition.

Figure 12 shows the results of the nematic correlation-length measurements for DB$_6$. The longitudinal and transverse scans at a series of temperatures were fitted in

pairs using Eq. (1). The typical values for χ^2 were 1.0–1.3. The resulting values for σ, $\xi_{\|}$, and $\xi_{\perp}$ at each temperature were fitted to the power laws

$$\sigma = \sigma^0 t^{-\gamma}, \quad \xi_{\|} = \xi_{\|}^0 t^{-\nu_{\|}}, \quad \xi_{\perp} = \xi_{\perp}^0 t^{-\nu_{\perp}} \quad (3)$$

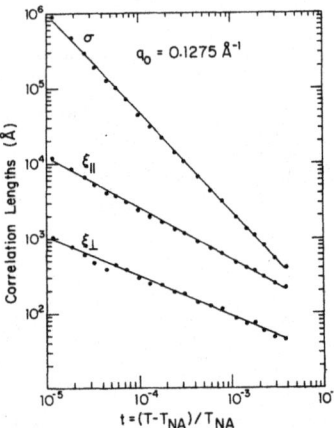

FIG. 12. Power-law fits $\sigma = \sigma^0 t^{-\gamma}$, $\xi_{\|} = \xi_{\|}^0 t^{-\nu_{\|}}$, and $\xi_{\perp} = \xi_{\perp}^0 t^{-\nu_{\perp}}$ for the N-A_d transition in DB$_6$. The units for the susceptibility are arbitrary. For these fits T_{NA_d} was fixed at the value determined from the linewidth inspection, $T_{NA_d} = 155.502\,°C$.

[t is the reduced temperature $(T - T_{NA})/T_{NA}$]. The error bars used in the power-law fits are the statistical errors produced by the line-shape fitting program. One-standard-deviation statistical errors on the exponents were about 0.005.

One defined of Eq. (1) is that the fourth-order coefficient is not well determined. This problem is compounded by the vertical resolution. Integration over the vertical direction raises the tails observed in the in-plane scans. The fourth-order correction compensates by reducing the tails. However, χ^2 has a broad and ill-defined minimum with respect to variation in c. Thus the best-fit value varies drastically, depending on the wave-vector range covered by the transverse scan and in turn affects the best-fit value of $\xi_\perp$. The wiggles in the plot of the transverse correlation lengths (Fig. 12) are the result of this coupling.

For comparison purposes, the raw data were also fit to a variation of McMillan's form appropriately convolved with the resolution

$$S(q) = \frac{\sigma}{1 + \xi_\parallel^2 (q_\parallel - q_0)^2 + (\xi_\perp q_\perp)^{2-\eta}} , \quad (4)$$

here both $\xi_\perp$ and $q_\perp$ are raised to $2 - \eta$. The χ^2 values for individual scan pairs remained essentially unchanged. The primary effect was on the transverse correlation lengths. At $t = 10^{-5}$, $\xi_\perp$ decreased by 20%. The effect grew progressively less at larger reduced temperatures, becoming totally negligible for $t > 10^{-3}$. The resulting effect on $\nu_\perp$ was a change from 0.51 to 0.48. The spurious oscillations in $\xi_\perp$ were reduced, but not eliminated.

Finally, the data were refit using a fixed value of c. This value was determined by averaging the values obtained for all reduced temperatures when c was left as a free parameter. This produced an increase in χ^2 for a pair of scans by at most 10% and produced no statistically significant increase in $\nu_\perp$. This effectively eliminated the troublesome oscillations. The effect on $\nu_\perp$ was about 0.01.

The data is most susceptible to artifacts at temperatures close to the transition ($t < 5 \times 10^{-5}$). Subtle errors in modeling the instrumental resolution or in locating the transition temperature can have drastic effects on these points. These systematic errors are not accounted for by the statistical weights used in the power-law fits. To estimate this systematic error, the transition temperature was varied as a free parameter. This moved the transition temperature up 1 mK from that estimated by the line-shape inspection (see Fig. 10), and caused the points closest to the transition to be consistent with the simple power-law form. This, of course, hides any crossover physics that may occur in the range $t < 5 \times 10^{-5}$. However, the readjustment of T_{NA} leaves it within the uncertainty range bracketed by the direct line-shape inspection. Consequently, the data is consistent with, and well described by, a simple power law over the entire measured temperature range.

Finally, the result of holding the value of c fixed at 0.05 and treating the transition temperature as a free parameter is shown in Fig. 13. Note that a single transition temperature results in straight lines for all three quantities

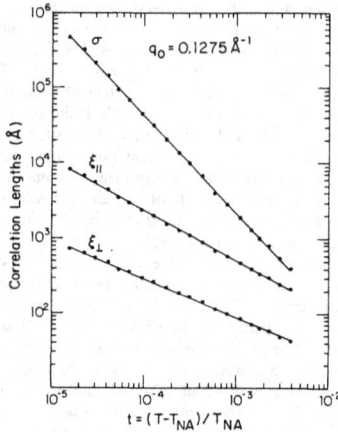

FIG. 13. Power-law fits $\sigma = \sigma^0 t^{-\gamma}$, $\xi_\parallel = \xi_\parallel^0 t^{-\nu_\parallel}$, and $\xi_\perp = \xi_\perp^0 t^{-\nu_\perp}$ for the N-A_d transition in DB$_6$ with $c = 0.05$ and T_{NA_d} as a free parameter. The units for the susceptibility are arbitrary.

plotted, σ, $\xi_\parallel$, and $\xi_\perp$

Pure DB$_6$ was found to be quite stable. The N-A_d transition temperature drifted downwards by 0.5 mK. A typical scattering measurement took about 30 h with most of the time spent far away from the transition (where the signal is weaker). The total time spent in the reduced temperature decade $t < 10^{-4}$ ($\Delta T < 40$ mK) was less than a few hours (about 20 min per temperature). The scan times for temperatures farthest away from T_{NA} were a few hours per temperature. No correction was applied for the drift in T_{NA}, but it was carefully measured at the beginning of each run. Small impurities in the sample are not believed to be important. If they are distributed homogeneously, they should simply vary the coefficients in the free energy, producing a different transition temperature. Since the basic symmetry properties of the free energy will remain unchanged, the critical exponents should remain the same. Although the N-A transition has been observed to be nonuniversal, small impurity concentrations do not produce a measurable effect on the critical exponents. The measured exponents for a given sample are always found to be independent of the small drifts in transition temperature due to sample decomposition.

Equation (1) assumes a perfectly aligned sample. If the sample has an orientational distribution in the nematic phase, the scattering from the different orientations must be included in the analysis. The mosaic distribution can be measured in the smectic-A phase by rocking the sample (mosaic scan). If the mosaic inferred from the narrowest observed mosaic scan in the smectic-A phase is assumed to hold in the nematic phase, it can be used to estimate the mosaic corrected scattering. Although the precise sample alignment at or above the transition temperature is unknown, it should depend only on the uniformity of the magnetic field. An attempt was made to include a Gaussian mosaic correction with FWHM equivalent to the narrowest smectic-A transverse scan ($\sim 0.4°$). This affected points where $1/(q_0\xi_\perp)$ is less than or comparable to the mosaic angular spread (in radians). It did not improve the power-law fits, and in fact made them slightly worse. In addition, when the N-A_d transition temperature was allowed to vary, the resulting exponents became identical to the values obtained without mosaic corrections. We conclude that mosaic corrections are smaller than the precision of the present experiments.

Reasonable variations in the background and vertical resolution were also made, to estimate the corresponding systematic errors. The critical exponents for the N-A_d transition are sensitive at the level of about 0.01 to these variations. The largest variations resulted from floating the transition temperature and changing the scaling form. A total uncertainty of about 0.03 in $\nu_{||}$ and $\nu_\perp$, and 0.05 in γ accounts for the observed range of values produced by all these different assumptions. We use these values as our systematic error estimate. The statistical errors are negligible.

To summarize, the observed critical exponents for the N-A_d transition in DB$_6$ are

$$\gamma = 1.29 \pm 0.05 \ ,$$

$$\nu_{||} = 0.67 \pm 0.03 \ ,$$

$$\nu_\perp = 0.52 \pm 0.03 \ .$$

Although the combined effect of all the contributing factors affects the exponents at the level of 0.03, they are remarkably immune to larger changes. In fact, a crude estimate was done using only a pencil and a ruler and the plots of the raw data. This analysis ignored the complications due to the vertical resolution, the choice of the scaling forms, etc., and simply measured the FWHM's. The result was exponents at the bounds of the error estimates quoted above.

C. Nematic—to—smectic-A_1 transition

Correlation length measurements were done on five DB$_6$-TBBA mixtures with an N-A_1 transition (see Fig. 3). Each mixture exhibited different stability problems. They all showed N-A_1 transition temperatures that drifted downwards at typically a few mK/h. A complete characterization of the N-A_1 transition typically took from 12 to 18 hours. Short scan times were always used at the temperatures closest to the transition, where the signal is strong; 10 min per temperature was typical. The scans were started slightly below the transition temperture, and the temperature was increased in small increments past the transition. The temperature steps were then increased according to distance from the transition. Most of the data-taking time was spent far away from the transition where the signal is weak and the transition-temperature drifts are relatively unimportant.

The raw data were analyzed as described for the N-A_d transition in DB$_6$. The data were first fit to Eq. (1) with c, the coefficient of the $q_\perp^4$ term free. The data were then refit with c fixed at the average value determined by the free fits. No mosaic correction was applied. All five mixtures showed second-order transitions. The correlation lengths obtained for each mixture are well described by power-law behavior over the entire measured temperature range. The transition temperature was treated as a free parameter in the power-law fits. It never varied by more than 3 mK from the value determined by direct line-shape inspection. No correction was made for the drifting transition temperature. Figure 14 shows the power-law fits for the 18.0 and 11.9 mol% samples. The resulting exponents for the N-A_1 transition are summarized in Table I.

D. Smectic-A_1—to—smectic-A_2 transition

Correlation-length measurements at the A_1-A_2 transition were made on the same five mixtures used for the N-A_1 transition and were reported briefly in an earlier publication.[27] The temperatures for the A_1-A_2 transition were much less susceptible to drifts than the N-A_1 transitions. The 18.0, 16.4, and 13.2 mol% mixtures exhibited second-order transitions with transition-temperature drifts of less than 1 mK/h. The 12.1 mol% mixture exhibited a weakly first-order transition and also had drifts less than 1 mK/h. The 11.0 mol% mixture had a first-order transition with about 60 mK hysteresis.

The analysis procedure for the A_1-A_2 transition differs from the N-A_1 analysis in two important ways. These are the mosaic correction and the scaling form. Unlike the N-A_1 analysis which required no mosaic correction, the A_1-A_2 definitely requires one. The A_1-A_2 transition

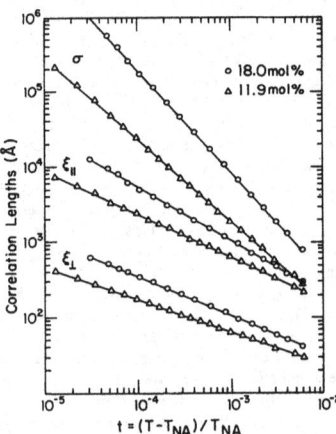

FIG. 14.   Power-law fits $\sigma = \sigma^0 t^{-\gamma}$, $\xi_{||} = \xi_{||}^0 t^{-\nu_{||}}$ and $\xi_\perp = \xi_\perp^0 t^{-\nu_\perp}$ for the N-A_1 transition in two mixtures. For the 18 mol% mixture $q_0 = 0.2332$ Å^{-1} and for the 11.9 mol% mixture $q_0 = 0.2371$ Å^{-1}. The units for the susceptibility are arbitrary.

CHAN, PERSHAN, SORENSEN, AND HARDOUIN

TABLE I. Summary of the N-A_1 power-law fits for the susceptibility $(\sigma=\sigma^0 t^{-\gamma})$, longitudinal correlation length $(\xi_\parallel=\xi_\parallel^0 t^{-\nu_\parallel})$, and transverse correlation length $(\xi_\perp=\xi_\perp^0 t^{-\nu_\perp})$ for each mixture. The units for the susceptibility are arbitrary. The reduced temperature range used in the fits was $2\times10^{-5} \leq t \leq 8\times10^{-3}$.

mol % TBBA	σ^0	γ	$\xi_\parallel^0$ (Å)	$\nu_\parallel$	$\xi_\perp^0$ (Å)	$\nu_\perp$	c
18.0	0.71	1.36[a]	7.4	0.72[b]	3.0	0.52[b]	0.05
14.4	0.84	1.30	8.6	0.69	3.1	0.49	0.08
13.2	0.69	1.27	8.5	0.67	3.0	0.48	0.09
12.1	0.93	1.16	10.4	0.62	3.5	0.44	0.09
11.9	1.0	1.09	11.8	0.57	3.3	0.43	0.09

[a]The error on the susceptibility exponent γ is ±0.05.
[b]The error on the correlation length exponents $\nu_\parallel$ and $\nu_\perp$ is ±0.03.

occurs in the presence of the ordering field of a well-developed smectic-A_1 lattice. Because of this, the sample alignment could be directly measured with a mosaic scan of the corresponding (002) smectic peak. For each sample, the smectic-A_1 crystal was "grown" by slow cooling from the nematic phase. The typical procedure was to lower the temperature several hundred mK across the N-A_1 transition over the period of an hour. The dominant features of the mosaic were found to be fully formed by about 100 mK below the N-A_1 transition. The temperature was then lowered the rest of the way to the A_1-A_2 transition (as much as 20 K lower) over a period of several hours. This could be accomplished with no serious degradation of the mosaic. Finally, cooling the temperature past the A_1-A_2 transition, into the A_2 phase, produced small changes in the mosaic distribution resulting in a growth in the tails relative to the center.

A mosaic scan for the 16.4 mol % sample is shown in Fig. 15. In the data analysis, these mosaics were modeled as Lorentzians (indicated by the solid line in the figure) out to ±8 HWHM (half width at half maximum). The

fits of the data to Eq. (1) convolved with the resolution and corrected for the mosaic gave typical χ^2 values from 1.0 to 1.5. In sharp contrast to the N-A_1 transition which always required a nonzero fourth-order coefficient, the coefficient c for the A_1-A_2 transition was always driven to zero within a statistical uncertainty many times larger than the best-fit value. Consequently, the A_1-A_2 data were analyzed without any fourth-order corrections $(c=0)$.

Figure 16 shows the results for the 18.0 and 13.2 mol % samples. Again, the transition temperature was left as a free parameter. It never varied by more than 3 mK. A 20% change in the FWHM of the mosaic distribution changed $\nu_\parallel$ and $\nu_\perp$ by about 0.01. Different assumptions about the background and vertical resolution also have ef-

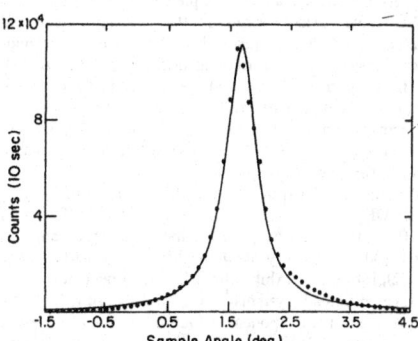

FIG. 15. Typical mosaic profile in the smectic-A_1 phase. The data points are for the 16.4 mol % mixture. The solid line indicates a Lorentzian model for the mosaic distribution.

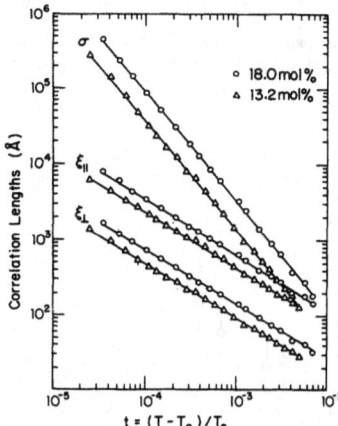

FIG. 16. Power-law fits $\sigma=\sigma^0 t^{-\gamma}$, $\xi_\parallel=\xi_\parallel^0 t^{-\nu_\parallel}$, and $\xi_\perp=\xi_\perp^0 t^{-\nu_\perp}$ for the A_1-A_2 transition in two mixtures. For the 18 mol % mixture $q_0=0.1168$ Å^{-1} and for the 13.2 mol % mixture for $q_0=0.1179$ Å^{-1}. The units for the susceptibility are arbitrary.

TABLE II. Summary of the A_1-A_2 power-law fits for the susceptibility ($\sigma = \sigma^0 t^{-\gamma}$), longitudinal correlation length ($\xi_{\parallel} = \xi_{\parallel}^0 t^{-\nu_{\parallel}}$), and transverse correlation length ($\xi_{\perp} = \xi_{\perp}^0 t^{-\nu_{\perp}}$) for each mixture. The units for the susceptibility are arbitrary. Except for the 12.1 mol% sample, the reduced temperature range used in the fits was $2 \times 10^{-5} \le t \le 8 \times 10^{-3}$. For the 12.1 mol% sample (which was weakly first order) $t > 2 \times 10^{-4}$.

mol% TBBA	σ^0	γ	$\xi_{\parallel}^0$ (Å)	$\nu_{\parallel}$	$\xi_{\perp}^0$ (Å)	$\nu_{\perp}$
18.0	0.14	1.46[a]	3.5	0.75[b]	0.91	0.73[b]
16.4	0.12	1.46	3.3	0.75	0.77	0.74
13.2	0.062	1.45	2.5	0.74	0.58	0.73
12.1	(0.024)	(1.41)	(1.2)	(0.76)	(0.39)	(0.72)

[a]The error on the susceptibility exponent γ is ± 0.05.
[b]The error on the correlation length exponents, $\nu_{\parallel}$ and $\nu_{\perp}$ is ± 0.03.

fects at this level. The results for the measured A_1-A_2 exponents are summarized in Table II. When the data for different concentrations were fit with the correlation-length exponents held fixed at the overall average value $\nu_{\parallel} = \nu_{\perp} = 0.74$, the ratio of the bare correlation lengths ($\xi_{\parallel}^0$ and $\xi_{\perp}^0$) were found to be independent of the TBBA concentration with $\xi_{\parallel}^0 / \xi_{\perp}^0 \sim 4$.

In the A_2 phase, the (001) and (002) Bragg-like spots were always commensurate within the instrumental resolution. This was not true in the A_1 phase where the (001) scattering was still diffuse and a slight temperature-dependent incommensuration was observed. This is shown for the 18.0 mol% sample in Fig. 17. The measured incommensuration was about 0.5%, 1 K above the A_1-A_2 transition temperature ($\xi_{\parallel} \sim 330$ Å, $\xi_{\perp} \sim 75$ Å), and gradually decreased to zero at the transition.

V. DISCUSSION AND CONCLUSIONS

In this paper we have presented the results of a comprehensive high-resolution x-ray scattering measure-

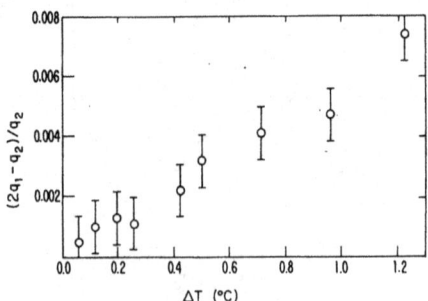

FIG. 17. Plot of the incommensurability of the (001) diffuse scattering with respect to the resolution limited (002) scattering just above the A_1-A_2 transition. The data shown are for the 18.0 mol% sample. At the transition $q_1 = 0.1168 \pm 0.0001$ Å^{-1}

ments of the phases and phase transitions in mixtures of DB_6 and TBBA. The physics of this system is unusually rich. It contains multicritical points, commensurate-incommensurate transitions, critical phenomena, and phases with QLRO.

Our measurements on pure DB_6 have revealed a previously unobserved feature of the A_d phase—the simultaneous presence of a Bragg-like reflection at $2q_1$ and diffuse scattering at q_2 as shown in Fig. 6. This observation confirms the mean-field prediction of Barois et al.[6] which places the A_d phase in the symmetry class of the A_2 phase. The dramatic wave-vector shifts in the A_d phase and the subsequent growth in the intensity of the second harmonic at the A_d-A_2 transition is also in qualitative agreement with the mean-field-theory predictions. We also found a slower temperature-dependent shift of the wave vector in the A_2 phase as the system continued to evolve towards the original monolayer fluctuation wave vector q_2. At low temperature $2q_1 \rightarrow q_2$. Taken all together, our observations lend strong support to the Prost model[5] which describes the A_d-A_2 transition in terms of the two competing order parameters. Heat-capacity measurements are now needed for this transition to confirm its first-order nature and to complete the description.

We have also measured the susceptibility and correlation-length exponents for the different second-order transitions which occur in this system. For the N-A_d transition in pure DB_6 and for the N-A_1 transition in mixtures, the exponents are for smectic-A fluctuations in the nematic phase. For the A_1-A_2 transitions, the exponents are for smectic-A_2 critical fluctuations which occur in the smectic-A_1 phase.

The measured exponents for the N-A_d transition in pure DB_6 are $\gamma = 1.29 \pm 0.05$, $\nu_{\parallel} = 0.67 \pm 0.03$, and $\nu_{\perp} = 0.52 \pm 0.03$. Although DB_6 has a nematic range of about 100 C, and very small McMillan ratio ($T_{NA}/T_{NI} = 0.82$), it does not exhibit the XY critical behavior which is expected in such systems.[17] It also does not exhibit the anisotropic critical exponents which have been observed in nonpolar materials (for example, $\bar{8}S5$ and 4O.7) with large nematic ranges. For pure DB_6, the measured critical exponents γ and $\nu_{\parallel}$ agree fairly well with the XY values. However, combining the measured exponents via the anisotropic scaling relation[10] $\nu_{\parallel} + 2\nu_{\perp} = 2 - \alpha$ predicts a specific-heat exponent $\alpha = 0.29 \pm 0.09$, in clear disagreement with the XY value of -0.02. Although no heat-capacity measurements on DB_6 have been reported yet, this transition should definitely be studied to determine whether anisotropic scaling works in this system. This is a particularly important test since all the nonpolar materials which have been studied satisfy anisotropic scaling. The behavior observed in DB_6 clearly represents a departure from that of nonpolar materials with large nematic ranges, which have measured specific-heat exponents α close to the XY value. These discrepancies may be due to the additional degree of freedom introduced by the dipolar order parameter Ψ_2.[5,6]

Measurements on systems with a wider A_d temperature range would help to clarify this issue, as would measurements searching for possible crossover phenomena where the N-A_d line meets the N-A_2 line. Clearly, further experimental and theoretical work is needed on the role of the extra order parameter in these transitions.

The measured exponents for the N-A_1 transition are nonuniversal and anisotropy in the correlation-length exponents persists along the entire N-A_1 line. For the mixture closest to the triple point (11.9 mol% TBBA), $\gamma = 1.09 \pm 0.05$, $\nu_{\parallel} = 0.57 \pm 0.03$, and $\nu_{\perp} = 0.43 \pm 0.03$. These values are similar to the results obtained for systems near the tricritical point[21] where the N-A transition turns first order due to the coupling between the smectic and nematic order parameters. Specifically, our susceptibility exponent γ tends toward the mean-field prediction of 1.0. In fact, it actually becomes equal to ~ 1.1, the value observed for the nonpolar tricritical N-A materials $\overline{10}S5$ and $9CB$. In addition, the average correlation-length exponent, $\overline{\nu} \equiv \frac{1}{2}(\nu_{\parallel} + \nu_{\perp})$, is equal to the mean-field prediction for an isotropic transition $\nu = 0.50$; just as the measured exponents for the tricritical nonpolar materials seem to approach this value ($\overline{10}S5$, $\overline{\nu} = 0.56$; $9CB$, $\overline{\nu} = 0.47$).

The measured critical exponents for the A_1-A_2 transition are $\gamma = 1.46 \pm 0.05$ and $\nu_{\parallel} = \nu_{\perp} = 0.74 \pm 0.03$. The correlation-length exponents are isotropic within the measurement errors. This is an important result because it provides the first experimental support for the hypothesis that the divergent phase fluctuations of the smectic order parameter are responsible for the anisotropy in the N-A transition.[27] Presumably, the A_1-A_2 transition is isotropic because the Ψ_1 phase fluctuations are quenched due to lockin with the established Ψ_2 field, even though both the A_1 and A_2 phases have QLRO.

The measured critical exponents are also consistent with the measured specific-heat exponent, $\alpha = -0.14 \pm 0.06$[28] via the scaling law $3\nu + \alpha = 2$ (experimental value 0.28 ± 0.15). Although the measured correlation-length and susceptibility exponents disagree with the predictions for a simple Ising transition and for an Ising-like transition on a QLRO lattice,[26] they are in agreement with "Fisher renormalized" values of the simple Ising values.[30] However, the Fisher renormalized values are the asymptotic values near the transition and the theory

predicts a crossover from bare to renormalized values with decreasing reduced temperture. Since crossover effects are difficult to measure, the data often satisfy a simple power law with measured exponents that are somewhere between the bare and renormalized values. Although it is possible that the agreement between the measured and renormalized values is fortuitous, the fact that the renormalized heat-capacity exponent α_x has the opposite sign of the bare exponent α_I [i.e., $\alpha_x = -\alpha_I/(1 - \alpha_I)$] supports the renormalization hypothesis. Using the theoretical Ising value, $\alpha_I = 0.11$, the predicted value, $\alpha_x = -0.12$, is in reasonable agreement with the measured value $\alpha = -0.14$.

The pretransitional smectic-A_2 fluctuations in the smectic-A_1 phase were observed to be well described by a Lorentzian in both the longitudinal and the transverse directions in sharp contrast to the non-Lorentzian character of the transverse fluctuations for all known nematic—to—smectic-A transitions. This suggests that the inherent anisotropy in the N-A transition is related to the anomalous q^4 transverse tails and should provide a useful clue for future theoretical efforts. Since both γ and ν are normalized in the same way, the fact that the scaling law $\gamma = 2(\nu - \eta)$ predicts an η very close to zero ($\eta = 0.01 \pm 0.05$) implies that η is essentially zero. This contrasts with the usual anisotropic nematic—to—smectic-A transitions for which $\eta_{\perp} < 0 < \eta_{\parallel}$.

In view of the fact that this is the only example of the A_1-A_2 transition that has been studied in detail there is great need for further experiments. Although the Fisher-renormalization approach appears plausible, it may not provide the correct interpretation of the DB_6-TBBA data. For example, the A_2 fluctuation scattering was also observed to have a temperature-dependent incommensuration relative to the smectic-A_1 peak. This suggests that the coupling between the phases of Ψ_1 and Ψ_2 may not be in a "strong field" limit, and it is possible that this might be responsible for the differences between the observed exponents and the bare Ising values. Further experiments on other systems with an A_1-A_2 transition are needed to test whether the behavior observed in this system is universal or whether the A_1-A_2 transitions will exhibit the nonuniversal behavior so characteristic of the nematic—to—smectic-A transition.

The unusually rich behavior observed in the DB_6-TBBA system clearly demonstrates the intrinsic interest of the polar nematic and smectic-A phases. In addition, the observed similarities and differences between this polar system and the previously studied nonpolar systems shows the relevance of continued studies of the polar materials to the eventual resolution of the nematic—to—smectic-A enigma.

ACKNOWLEDGMENTS

This work was supported in part by grants from the National Science Foundation through Grants No. NSF-DMR-82-12189 and No. NSF-DMR-83-16979.

*Present address: National Synchrotron Light Source, Brookhaven National Laboratory, Upton, NY 11973-5000.

†Present address: Department of Physics, University of Washington, Seattle, WA 98195.

[1]F. Hardouin, A. M. Levelut, M. F. Achard, and G. Sigaud, J. Chim. Phys. **80**, 53 (1983).

[2]G. Sigaud, F. Hardouin, M. F. Achard, and H. Gasparoux, J. Phys. (Paris) Colloq. **40**, C3-356 (1979).

[3]J. Prost, J. Phys. **40**, 581 (1979).

[4]J. Prost, in *Proceedings of the Conference on Liquid Crystals of One- and Two-Dimensional Order, Garmisch Partenkirchen* (Springer Verlag, Berlin, 1980), p. 125.

[5]J. Prost and P. Barois, J. Chim. Phys. **80**, 65 (1983).

[6]P. Barois, J. Prost, and T. C. Lubensky, J. Phys. **46**, 391 (1985).

[7]K. Kobayashi, Phys. Lett. A **31**, 125 (1970).

[8]K. Kobayashi, J. Phys. Soc. Jpn. **29**, 101 (1970).

[9]W. L. McMillan, Phys. Rev. A **4**, 1238 (1971).

[10]T. C. Lubensky, J. Chim. Phys. **80**, 31 (1983).

[11]C. Dasgupta, Phys. Rev. Lett. **55**, 1771 (1985).

[12]P. G. de Gennes, Solid State Commun. **10**, 753 (1972).

[13]P. G. de Gennes, Mol. Cryst. Liq. Cryst. **21**, 49 (1973).

[14]J. C. LeGuillou and J. Zinn-Justin, Phys. Rev. B **21**, 3976 (1980).

[15]T. C. Lubensky, S. G. Dunn, and J. Isaacson, Phys. Rev. Lett. **22**, 1609 (1981).

[16]J. Toner, Phys. Rev. B **26**, 462 (1982).

[17]P. Brisbin, R. De Hoff, T. E. Lockhart, and D. L. Johnson, Phys. Rev. Lett. **43**, 1171 (1979).

[18]D. L. Johnson, J. Chim. Phys. **80**, 45 (1983).

[19]C. W. Garland, M. Meichle, B. M. Ocko, A. R. Kortan, C. R. Safinya, L. J. Yu, J. D. Litster, and R. J. Birgeneau, Phys. Rev. A **27**, 3234 (1983).

[20]C. R. Safinya, R. J. Birgeneau, J. D. Litster, and M. E. Neubert, Phys. Rev. Lett. **47**, 668 (1981).

[21]B. M. Ocko, R. J. Birgeneau, J. D. Litster, and M. E. Neubert, Phys. Rev. Lett. **52**, 208 (1984).

[22]J. Thoen, H. Marynissen, and W. Van Dael, Phys. Rev. Lett. **52**, 204 (1984).

[23]E. K. Riedel and F. J. Wegner, Phys. Rev. Lett. **29**, 349 (1972).

[24]K. K. Chan, M. Deutsch, B. M. Ocko, P. S. Pershan, and L. B. Sorensen, Phys. Rev. Lett. **54**, 920 (1985).

[25]M. E. Fischer and A. Aharony, Phys. Rev. Lett. **31**, 1238 (1973).

[26]J. Wang and T. C. Lubensky, Phys. Rev. A **29**, 2210 (1984).

[27]K. K. Chan, P. S. Pershan, L. B. Sorensen, and F. Hardouin, Phys. Rev. Lett. **54**, 1694 (1985).

[28]C. W. Garland, C. Chiang, and F. Hardouin, Phys. Rev. A (to be published).

[29]D. A. Huse, Phys. Rev. Lett. **55**, 2228 (1985).

[30]M. E. Fisher, Phys. Rev. **176**, 257 (1968).

[31]G. S. Pawley, R. H. Swendson, D. J. Wallace, and K. G. Wilson, Phys. Rev. B **29**, 4030 (1984), and references therein.

[32]J. Stamatoff, P. E. Cladis, D. Guillon, M. C. Cross, T. Bilash, and P. Finn, Phys. Rev. Lett. **44**, 1509 (1980).

[33]L. Gunther, Y. Imry, and J. Lajzerowicz, Phys. Rev. A **22**, 1733 (1980).

[34]D. W. Marquardt, J. Soc. Indust. Appl. Math. **11**, 431 (1963).

[35]P. R. Bevington, *Data Reduction and Error Analysis for the Physical Sciences* (McGraw-Hill, New York, 1969).

[36]W. L. McMillan, Phys. Rev. A **7**, 1419 (1973).

[37]D. Davidov, C. R. Safinya, M. Kaplan, S. S. Dana, R. Schaetzing, R. J. Birgeneau, and J. D. Litster, Phys. Rev. B **19**, 1657 (1979).

5. IN PLANE ORDER-HEXATIC PHASES-THEORY

5.1. Halperin, B. I. and Nelson, D. R. (1978). Theory of two
      dimensional melting. *Phys. Rev. Lett.* **41**, 121–124. [Erratum:
      (1978) *Phys. Rev. Lett.* **41**, 519]   247

5.2. Birgeneau, R. J. and Litster, J. D. (1978). Bond orientational
      order model for smectic B liquid crystals. *J. Phys. (Paris) Lett.*
      39, L399–L402   251

5.3. Nelson, D. R. and Halperin, B. I. (1980). Solid and fluid phases in
      smectic layers with tilted molecules. *Phys. Rev.* **B21**, 5312–5329   255

Theory of Two-Dimensional Melting

B. I. Halperin and David R. Nelson

Department of Physics, Harvard University, Cambridge, Massachusetts 02138
(Received 17 May 1978)

The consequences of a theory of dislocation-mediated two-dimensional melting are worked out for triangular lattices. Dissociation of dislocation pairs first drives a transition into a "liquid crystal" phase with exponential decay of translational order, but power-law decay of sixfold orientational order. A subsequent dissociation of *disclination* pairs at a higher temperature then produces an isotropic fluid. The critical behavior, as well as the effect of a periodic substrate, is discussed.

Kosterlitz and Thouless[1] have proposed a model of two-dimensional melting, in which the "topological order" of a solid phase is destroyed by the dissociation of dislocation pairs. Similar ideas,[1] with vortices taking the place of dislocations, have led to a rather detailed theory[2] of the superfluid transition in two dimensions. In this Letter, we summarize the consequences of dislocation-mediated melting of triangular lattices, on both smooth and periodic substrates. A more detailed derivation will be given elsewhere.[3]

Consider the properties of a two-dimensional triangular solid on a smooth substrate. By definition, the solid has nonzero long-wavelength elastic constants. The structure factor exhibits[4] power-law singularities, $S(\vec{q}) \sim |\vec{q} - \vec{G}|^{-2+\eta_{\vec{G}}}$, near a set of reciprocal lattice vectors $\{\vec{G}\}$, with exponents $\eta_{\vec{G}}$ related to the Lamé elastic constants $\mu_R(T)$ and $\lambda_R(T)$ by $\eta_{\vec{G}} = k_B T |\vec{G}|^2 (3\mu_R + \lambda_R) / 4\pi\mu_R(2\mu_R + \lambda_R)$. These singularities, which replace the δ-function Bragg peaks found in three-dimensional solids, reflect power-law decay at large distances of the correlation function $\langle \exp\{i\vec{G} \cdot [\vec{u}(\vec{r}) - \vec{u}(\vec{0})]\}\rangle$, where $\vec{u}(\vec{r})$ is the lattice displacement at point $\vec{r}$. One can also define an order parameter (analogous to $e^{i\vec{G}\cdot\vec{u}}$) for bond orientations, namely $\psi \equiv e^{6i\theta}$, where $\theta(\vec{r})$ is the orientation, relative to the x axis, of a bond between two nearest-neighbor atoms at $\vec{r}$. In a solid, θ is given in terms of the displacement field, $\theta = \frac{1}{2}(\partial_y u_x - \partial_x u_y)$. The solid phase exhibits long-range orientational order, since $\langle \psi^*(\vec{r})\psi(\vec{0})\rangle$ approaches a nonzero constant at large $\vec{r}$.[5]

If melting is indeed characterized by an unbinding of dislocation pairs at a temperature T_m, one expects that a density $n_f(T)$ of free dislocations above T_m will lead to exponential decay of the

translational order parameter $e^{i\vec{G}\cdot\vec{u}}$, with a correlation length $\xi_+(T) \approx n_f^{-2}$. This length diverges as $T \to T_m^+$ [see (6) below]. The structure factor $S(\vec{q})$ is now finite at all Bragg points, and the Lamé coefficients vanish at long wavelengths. We shall see, however, that orientational order persists, in the sense that bond-angle correlations now decay algebraically, $\langle \psi^*(\vec{r})\psi(\vec{0})\rangle \sim 1/r^{\eta_6(T)}$. This phase can be described as a liquid crystal, similar to a two-dimensional nematic, but with a sixfold rather than twofold anisotropy. The exponent $\eta_6(T)$ is related to the Franck constant $K_A(T)$, which is the coefficient of $\frac{1}{2}|\nabla\theta|^2$ in the free-energy density: $\eta_6(T) = 18 k_B T/\pi K_A(T)$. We find that K_A is infinite just above T_m, but decreases with increasing temperatures, until a temperature T_i, where dissociation of *disclination* pairs drives a transition into an isotropic phase in which both the translational and orientational order decay exponentially.

The liquid-crystal phase is isomorphic to a two-dimensional superfluid, except that $\pm 60°$ disclinations play the role of vortices. The transition at T_i should belong to the same universality class as the superfluid transition, and we expect, in particular, that[2] $\eta_6(T_i) = \frac{1}{4}$. Although disclination pairs are very tightly bound in the solid phase, screening by a gas of free dislocations produces a weaker logarithmic binding for $T_m < T < T_i$. It is interesting to note that an isolated dislocation can itself be regarded as a tightly bound disclination pair,[6] separated by one lattice constant.

To see the origin of these results, let us decompose the displacement field of a solid into a smoothly varying phonon field $\vec{\varphi}(\vec{r})$, and a part due to dislocations.[1] The Hamiltonian $\mathcal{K}_E$ for the solid, within continuum elasticity theory,[6] then breaks into two parts, $\mathcal{K}_E = \mathcal{K}_0 + \mathcal{K}_D$, with

$$\frac{\mathcal{K}_0}{k_B T} = \frac{1}{2} \int \frac{d^2 r}{a_0^2} [2\bar{\mu}\,\varphi_{ij} + \bar{\lambda}\varphi_{ii}^2], \tag{1}$$

$$\frac{\mathcal{K}_D}{k_B T} = -\frac{1}{8\pi} \sum_{\vec{R} \neq \vec{R}'} \left[K_1 \vec{b}(\vec{R}) \cdot \vec{b}(\vec{R}') \ln\left(\frac{|\vec{R} - \vec{R}'|}{a}\right) + K_2 \frac{\vec{b}(\vec{R}) \cdot (\vec{R} - \vec{R}')\vec{b}(\vec{R}') \cdot (\vec{R} - \vec{R}')}{|\vec{R} - \vec{R}'|^2} \right] + \frac{E_c}{k_B T} \sum_{\vec{R}} |\vec{b}(\vec{R})|^2. \tag{2}$$

In (1), φ_{ij} is related to the smooth part of the displacement field, $\varphi_{ij} = \frac{1}{2}(\partial_i \varphi_j + \partial_j \varphi_i)$, and $\bar{\mu}$ and $\bar{\lambda}$ are "reduced" elastic constants, given by the usual Lamé coefficients μ and λ multiplied by the square of the lattice spacing, a_0^2, and divided by $k_B T$. In (2), $\vec{b}(\vec{R})$ is a dimensionless dislocation Burgers vector of the form $\vec{b}(\vec{R}) = m(\vec{R})\vec{e}_1 + n(\vec{R})\vec{e}_2$, where $m(\vec{R})$ and $n(\vec{R})$ are integers, and $\vec{e}_1$ and $\vec{e}_2$ are unit vectors spanning the underlying Bravais lattice. The sums in (2) are over, say, a square mesh with spacing a of sites in physical space, and the $\vec{b}(\vec{R})$ must satisfy a vector charge neutrality condition, $\sum_{\vec{R}} \vec{b}(\vec{R}) = 0$. The quantities K_1 and K_2 are equal, $K_1 = K_2 \equiv K = 4\bar{\mu}(\bar{\mu}+\bar{\lambda})/(2\bar{\mu}+\bar{\lambda})$, and E_c is the core energy of a dislocation.

If dislocations only exist in bound pairs at low temperatures, one expects that they can be ignored, and that the long-wavelength properties of the solid will simply be given by (1), with suitably renormalized elastic constants. The properties of the solid phase quoted above follow directly from this observation.

One of us[7] has studied the properties of (2) in the absence of the dot product or angular terms ($K_2 = 0$). It was found that dislocations are indeed unimportant at low temperatures (large K_1), and that a dislocation-unbinding transition was controlled by the terminus of a fixed *surface*, parametrized by K_1 and $\vec{e}_1 \cdot \vec{e}_2$. Here, we restrict ourselves to the triangular lattice ($\vec{e}_1 \cdot \vec{e}_2 = \frac{1}{2}$), and extend this treatment to the full dislocation Hamiltonian $\mathcal{K}_D$, taking into account the neglected angular terms. Recursion relations for K and $y \equiv \exp(-E_c/k_B T)$ can in fact be obtained rather straightforwardly, by considering the renormalization of elastic constants due to dislocation pairs, in analogy to calculations of the effect of vortices on the superfluid density in a ^{4}He film.[8] Integrating over mesh sizes between a and ae^l, we obtain partially dressed parameters $\bar{\mu}(l)$, $\bar{\lambda}(l)$, $y(l)$, and $K(l)$, which satisfy, to $O(y^2(l))$,

$$\frac{d\bar{\mu}^{-1}}{dl} = 3\pi y^2 e^{-K/8\pi} I_0\left(\frac{K}{8\pi}\right), \tag{3}$$

$$\frac{d[\bar{\mu}+\bar{\lambda}]^{-1}}{dl} = 3\pi y^2 e^{-K/8\pi}\left[I_0\left(\frac{K}{8\pi}\right) + I_1\left(\frac{K}{8\pi}\right)\right], \tag{4}$$

$$\frac{dy}{dl} = \left(2 - \frac{K}{8\pi}\right)y + 2\pi y^2 e^{-K/16\pi} I_0\left(\frac{K}{16\pi}\right), \tag{5}$$

where $I_0(x)$ and $I_1(x)$ are modified Bessel functions of the first and second kind. We find that

$$K^{-1}(l) = \frac{1}{4}\left\{\bar{\mu}^{-1}(l) + [\bar{\mu}(l) + \bar{\lambda}(l)]^{-1}\right\}$$

for all l, so that its recursion relation can be obtained trivially from (3) and (4).

As in Ref. 7, $y(l)$ is driven to zero at large l, for all temperatures below a critical value T_m. Above T_m, $y(l)$ is ultimately driven toward large values and $K(l)$ is driven towards zero, an instability we associate with dislocation-pair unbinding.

Following Kosterlitz[2] and Ref. 7, we determine the behavior near T_m by studying Eqs. (3)–(5) near the critical value $K_c = 16\pi$. We identify the correlation length $\xi_+(T)$ with ae^{l*}, with $l*$ chosen such that $K(l*) \approx \frac{1}{2}K_c$. In this way, we find that

$$\xi_+(T) \approx a \exp[b/(T/T_m - 1)^{0.44817\cdots}], \tag{6}$$

as $T \to T_m^+$, where b is a constant, and $0.44817\ldots$ can be expressed in terms of the roots of a quadratic equation with Bessel-function coefficients. The specific heat exhibits only an essential singularity, $C_p \sim \xi_+^{-2}$, while the structure factor at the Bragg points is given by $S(\vec{G}) \sim \xi_+^{2-\eta_G}$. Taking over the discussion for the superfluid density in Ref. 8, we find that the reduced shear modulus in the solid phase is

$$\bar{\mu}_R(T) = \lim_{l\to\infty} \bar{\mu}(l).$$

It follows from Eqs. (3)–(5) that $\mu_R(T)$ approaches a finite limiting value as $T \to T_m^-$. Just below T_m we find $\mu_R(T) = \mu_R(T_m)[1 + \text{const}(T_m - T)^{0.44817\cdots}]$, with a similar result for $\lambda_R(T)$. There is a universal relationship involving the elastic constants at T_m,

$$\lim_{T\to T_m^-}\left\{\frac{1}{\mu_R(T)} + \frac{1}{\mu_R(T) + \lambda_R(T)}\right\} = \frac{a_0^2}{4\pi k_B T_m}. \tag{7}$$

This corresponds to the critical value $K_c = 16\pi$, and is also suggested by the "entropy argument" of Kosterlitz and Thouless.[1]

The results for orientational correlations above T_m follow from a calculation of the Franck constant K_A:

$$\frac{k_B T}{K_A} = \lim_{q\to 0} q^2 \langle \theta(q)\theta(-q)\rangle$$

$$= \lim_{q\to 0} \frac{q_i q_j}{q^2}\langle b_i(q) b_j(-g)\rangle, \tag{8}$$

where $\theta(q)$ and $b_i(q)$ are the Fourier-transformed orientational and Burgers-vector fields, respectively. The second line of (8) follows because the contribution of $\vec{\varphi}(\vec{r})$ to K_A^{-1} is zero and because the dislocation part of $\theta(q)$ is just[3,6] $\theta(q) = -iq_i b_i(q)/q^2$. To estimate K_A just above T_m, we use its transformation properties under the renormal-

VOLUME 41, NUMBER 2   PHYSICAL REVIEW LETTERS   10 JULY 1978

ization group, $K_A(K(0), y(0)) = e^{2l}K_A(K(l), y(l))$. Choosing $l = l^* = \ln(\xi_+/a)$, we can evaluate K_A using Debye-Hückel theory, which amounts to treating $\vec{b}(\vec{R})$ as a continuous vector field, rather than restricting it to discrete points on a Bravais lattice. Upon Fourier transformation, $\mathcal{K}_D$ becomes

$$\frac{\mathcal{K}_D}{k_B T} = \frac{1}{2V} \sum_{\vec{q}} \left[\frac{K}{2q^2}\left(\delta_{ij} - \frac{q_i q_j}{q^2}\right) + \frac{2E_c a^2}{k_B T} \delta_{ij} \right] b_i(\vec{q}) b_j(-\vec{q}), \tag{9}$$

where V is the volume. Since the term proportional to the transverse projection operator in (9) does not contribute to (8), one obtains $K_A(K(l^*), y(l^*)) \approx 2E_c(l^*)a^2 = O(k_B T_m)$. It follows that the physical Franck constant is $K_A \sim \xi_+^2(T)$. The algebraic decay of orientational order above T_m and the relationship between $\eta_b(T)$ and $K_A(T)$ are straightforward consequences of this result.

Many experimental investigations of two-dimensional melting are carried out on films adsorbed onto a regular substrate,[9] and so it is important to determine the effect of a periodic potential on our results. One must now distinguish between a "floating solid," characterized by power-law Bragg singularities at reciprocal lattice vectors $\{\vec{G}\}$ which vary continuously with coverage and temperature, and an "epitaxial solid," having δ-function Bragg peaks at a lattice of vectors including the substrate reciprocal lattice $\{\vec{K}\}$ as a proper subset. The floating solid should be rather similar to the solid on a smooth substrate discussed in this paper. Figure 1 shows a schematic phase diagram with fluid, floating solid, and epitaxial phases. A region of two-phase coexistence is also shown, separating epitaxial phase I from a dilute fluid or "vapor."[8,10] We expect an increasing multiplicity and complexity of epitaxial phases with decreasing temperatures.

To understand Fig. 1, consider first the effect of a weak substrate potential commensurate with the lattice of the adsorbed film. Let $\{\vec{M}\}$ be the set of vectors common to $\{\vec{G}\}$ and $\{\vec{K}\}$, and let M_0 be the minimum length of nonzero vectors in $\{\vec{M}\}$. Let us expand the potential on the reciprocal lattice $\{\vec{K}\}$, and focus our attention on

$$\mathcal{K} = \mathcal{K}_E + \sum_{|\vec{M}| = M_0} h_{\vec{M}} \sum_{\vec{r}} e^{i\vec{M} \cdot \vec{u}(\vec{r})}, \tag{10}$$

where $h_{\vec{M}}$ is the potential strength at $\vec{M}$; the terms displayed in (10) are the most important ones for weak potentials. The renormalization-group eigenvalue for $h_{\vec{M}}$ is easily shown to be $\lambda_{\vec{M}} = 2 - \frac{1}{2}\eta_{\vec{G}}|_{\vec{G}=\vec{M}}$, so that any commensurate perturbation becomes relevant at sufficiently low temperatures. If M_0 is sufficiently small ($M_0 \lesssim 8\pi/a_0$), λ_{M_0} remains relevant out to quite high temperatures and a floating solid can never exist; there is then a transition directly from a low-tempera-

ture expitaxial phase into a fluid, as shown for epitaxial phases I and III. For large enough M_0 ($M_0 \gtrsim 8\pi/a_0$), however, there is a temperature window where $\lambda_{M_0} < 0$ and $K_R > 16\pi$, indicating that a floating solid is stable to both substrate perturbations and dislocation unbinding. The dotted line in Fig. 1 shows a locus of such points, where the floating solid has the same periodicity as epitaxial phase II, which exists at lower temperatures.

It can be shown[3] that the transitions from floating solid to fluid and from floating solid to epitaxial phase II (marked A and B in Fig. 1) are both describable at long wavelengths by a Hamiltonian of the form (2) with $K_1 \neq K_2$. Indeed, these transitions are essentially dual to each other.[3] The situation is very similar to the "$\cos p\theta$" perturbations discussed by José et al.[11] The transition from epitaxial phase II to a floating solid at points other than B may connect two phases with different periodicities; its nature is not yet known. We expect that the transition from floating solid to fluid will be everywhere qualitatively similar to our discussion of dislocation unbinding on a smooth substrate. The orientational bias imposed by the substrate, however, should alter or eliminate the liquid-crystal isotropic transition discussed above.

We wish to emphasize in closing that we have only explored consequences of the dislocation model of melting perturbatively in $y = \exp(-E_c/$

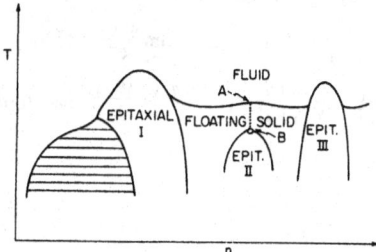

FIG. 1. Hypothetical phase diagram for a submonolayer adsorbed film on a periodic substrate as a function of density n and T.

$k_B T$). Although the theory is stable and self-consistent, we cannot rule out other mechanisms for melting, perhaps leading to a first-order transition. A "premature" unbinding of disclinations (before dislocations dissociate) might constitute such a mechanism.

We have benefitted from discussions with S. Aubry, A. N. Berker, M. Kléman, and R. Peierls. This work was supported by the National Science Foundation, under Grant No. DMR 77-10210, and by the Harvard Materials Research Laboratory program. One of us (D.R.N.) received a Junior Fellowship from the Harvard Society of Fellows.

[1]J. M. Kosterlitz and D. J. Thouless, J. Phys. C 6, 118 (1973), and to be published.

[2]J. M. Kosterlitz, J. Phys. C 7, 1046 (1974); see also J. José, L. P. Kadanoff, S. Kirkpatrick, and D. R. Nelson, Phys. Rev. B 16, 1217 (1977).

[3]D. R. Nelson and B. I. Halperin, to be published.

[4]See, e.g., Y. Imry and L. Gunther, Phys. Rev. B 3, 3939 (1971).

[5]N. D. Mermin, Phys. Rev. 176, 250 (1968).

[6]F. R. N. Nabarro, Theory of Dislocations (Clarendon, New York, 1967).

[7]D. R. Nelson, Phys. Rev. B (to be published).

[8]D. R. Nelson and J. M. Kosterlitz, Phys. Rev. Lett. 39, 1201 (1977).

[9]J. G. Dash, Films on Solid Surface (Academic, New York, 1975).

[10]See, e.g., A. N. Berker, S. Ostlund, and F. A. Putnam, Phys. Rev. B 17, 3650 (1978).

[11]José et al., Ref. 2.

ERRATUM: (1978) *Phys. Rev. Lett.* 41, 519

THEORY OF TWO-DIMENSIONAL MELTING. B. I. Halperin and David R. Nelson [Phys. Rev. Lett. 41, 121 (1978)].

Dr. A. P. Young has kindly called our attention to a sign error in the second term on the right-hand side of Eq. (2), which led to incorrect coefficients in the recursion relations (3)–(5). Consequently, the numerical value of the exponent in Eq. (6), for the correlation length, and in the singular part of the elastic constants is wrong. The correct value of this exponent (quoted as 0.448 17 ... in our Letter) is 0.369 63 .... Dr. Young had independently obtained the correct behavior of the correlation length.

The corrected forms of Eqs. (3)–(5) are

$$\frac{d\bar{\mu}^{-1}}{dl} = 3\pi y^2 e^{K/8\pi} I_0\left(\frac{K}{8\pi}\right), \tag{3}$$

$$\frac{d[\bar{\mu}+\bar{\lambda}]^{-1}}{dl} = 3\pi y^2 e^{K/8\pi}\left[I_0\left(\frac{K}{8\pi}\right) - I_1\left(\frac{K}{8\pi}\right)\right]. \tag{4}$$

$$\frac{dy}{dl} = \left(2 - \frac{K}{8\pi}\right)y + 2\pi y^2 e^{K/16\pi} I_0\left(\frac{K}{8\pi}\right). \tag{5}$$

Two additional misprints should be corrected: Equations (8) and (9) should read

$$\frac{k_B T}{K_A} = \lim_{q\to 0} q^2 \langle \hat{\theta}(\vec{q})\hat{\theta}(-\vec{q})\rangle = \lim_{q\to 0} \frac{q_i q_j}{q^2}\langle \hat{\delta}_i(\vec{q})\hat{\delta}_j(-\vec{q})\rangle a_0^2, \tag{8}$$

$$\frac{\mathcal{K}_D}{k_B T} = \frac{1}{2V}\sum_{\vec{q}}\left[\frac{K}{q^2}\left(\delta_{ij} - \frac{q_i q_j}{q^2}\right) + \frac{2E_c a_0^2}{k_B T}\delta_{ij}\right]\hat{\delta}_i(\vec{q})\hat{\delta}_j(-\vec{q}). \tag{9}$$

LE JOURNAL DE PHYSIQUE — LETTRES

TOME **39**, 1ᵉʳ NOVEMBRE **1978**, PAGE L-399

Classification
Physics Abstracts
61.30

BOND ORIENTATIONAL ORDER MODEL
FOR SMECTIC B LIQUID CRYSTALS

R. J. BIRGENEAU (*) and J. D. LITSTER (**)

Department of Physics, Massachusetts Institute of Technology,
Cambridge, Massachusetts, 02139, U.S.A.

(*Reçu le 27 juin 1978, révisé le 8 septembre 1978, accepté le 11 septembre 1978*)

Résumé. — Nous utilisons les concepts pris dans les théories récentes sur la fusion des solides à 2 dimensions pour définir les paramètres d'ordre des phases solide, smectiques A et B des cristaux liquides. Dans les phases smectiques bien ordonnées, de type B par exemple, nous introduisons, un ordre à longue distance entre les orientations des liaisons des molécules proches voisines en plus de l'ordre de position à courte distance dans le plan des couches. Les implications de ce modèle pour la diffusion des rayons X sont discutées en détail.

Abstract. — We show how recent theories of two dimensional melting can be carried over to provide natural order parameters for solid, smectic B, and smectic A phases of liquid crystals. Smectic B and other well-ordered smectic phases correspond to systems with bond orientational long range order and positional short range order in the plane of the smectic layers. The X-ray scattering predicted by this model is discussed in detail.

Recently smectic liquid crystals, which have long range orientational order of the molecules combined with varying degrees of positional order, have been the subject of extensive experimental and theoretical investigation. As a result we have rather good theoretical models for the smectic A (SmA) and smectic C (SmC) phases characterized by one dimensional density waves whose wave vector is along (SmA) or at an angle (SmC) to the nematic director [1]. Also, as a result of combined light and high resolution X-ray scattering, a rather good quantitative experimental description of the SmA phase is now emerging [2]. One of the more interesting features of the SmA phase is that the lower marginal dimensionality d^0 is three [3]; this is the spatial dimensionality at which fluctuations prevent the establishment of true long range order [4, 5, 6]. Between the SmA or SmC phases and the crystalline solid of many materials are one or more intermediate phases which are characterized by a considerable degree of order within and between the smectic layers. Although beautiful experimental

work has been done on the SmB and similar phases, especially by the Orsay group [7, 8], little progress has been made on the theoretical front. In particular no convincing microscopic definition has been given for the order parameter(s) of these phases to distinguish them from the SmA or SmC phases and the crystalline solid.

Marginal dimensionality in ordinary solids occurs at $d^0 = 2$ and considerable theoretical work has been carried out for two dimensional (2D) solids [9, 10]. In this paper we suggest how the concepts introduced by Halperin and Nelson (HN) [10] for the 2D melting problem can be carried over to 3D liquid crystals. We emphasize that our discussion is essentially qualitative in nature. However, this 2D melting analogy appears to be very successful in accounting for the overall features of smectic liquid crystals. We hope therefore that our observations will serve to inspire more rigorous theory for liquid crystals of the sort carried out by HN for 2D melting.

We begin with a summary of the experimentally known facts for the various smectic phases. Most of these, especially with respect to macroscopic properties, have been discussed elsewhere [1, 11], and we limit our attention to X-ray scattering patterns. In the SmA and SmC phases the distribution of molecules

(*) Supported by Joint Services Electronics Program, contract n⁰ DAAG-29-78-C-0020.
(**) Supported by the National Science Foundation, grant n⁰ DMR-76-18035.

in the planes normal to the density wave is random as in a conventional liquid ; the X-ray powder pattern [12] shows a sharp ring at $q_0 = 2 \pi/d$, where d is the density wave period, and a diffuse outer ring corresponding to the first peak in the in-plane liquid structure factor. In the phases between SmA or SmC and the crystalline solid the diffuse outer ring becomes a series of sharp rings, reminiscent of the rings in a crystalline powder pattern [7, 8, 11], but with several essential differences. First of all, only a small number of primary reflections are observed. In multidomain SmB phases often only the primary (110) in plane reflections are clearly visible, while in SmH or SmE phases one can typically resolve (110), (111), (200), (201), (210), and (211) reflections. The important observation is that only reflections with small Miller indices are clearly visible, and this has led most workers to describe these SmB like phases as ones with considerable local order or quasi-long range order, but without a precise definition of the latter concept. A second feature, especially for SmB's, is that the observable reflections are accompanied by very pronounced diffuse scattering whose intensity varies as $\Delta q^{-2} = | q - G |^{-2}$, where q is the momentum transfer and G is a reciprocal lattice position. The third feature, and most important from our point of view, is that one can prepare *single domain* samples in which one observes, for example, six well defined (200) or equivalent reflections rather than a ring of scattering. This feature precludes any model based on short range order (S.R.O.) alone.

A possible association between SmB liquid crystals and 2D melting was first noticed by de Gennes and Sarma [13]. Explicitly they considered a set of idealized 2D harmonic systems stacked to yield a 3D structure. Since de Gennes and Sarma proposed their model considerable progress has been made in the theory of 2D melting. In particular it is now believed that the harmonic model does not provide a realistic picture of melting and that for an isotropic 2D model melting occurs as a result of the disassociation of bound pairs of point dislocations — this is the Kosterlitz-Thouless mechanism [9]. However, in a crystal the orientational anisotropy, such as the six-fold symmetry of a triangular or hexagonal lattice, must be explicitly considered in the theory. Such a theory has recently been given by Halperin and Nelson (HN) ; they point out that one must consider both positional and orientational correlation functions. The positional order may be quantitatively described by the correlation function

$$P(G, r) = \langle e^{iG.[u(r) - u(0)]} \rangle \tag{1}$$

where $u(r)$ is the displacement of the lattice at r. The orientational order parameter [14] for six-fold symmetry is defined by $\psi(r) \equiv e^{i6\theta(r)}$ where θ specifies the orientation of a bond (i.e. the line between the centres of mass) between two nearest neighbour molecules at r (see Fig. 1). Then the orientational

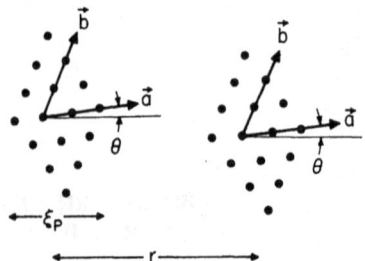

Orientational L.R.O. , Positional S.R.O.

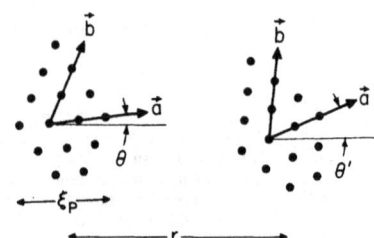

Orientational S.R.O. , Positional S.R.O.

FIG. 1. — Correlated triangular droplets in state (ii), upper, and state (iii), lower, of the Halperin Nelson model for two dimensional melting. Here d is the separation between the droplets and ξ_p is the positional correlation length within the droplets.

correlation function is

$$O(r) = \langle \psi^*(r) \, \psi(0) \rangle \tag{2}$$

Mermin has shown rigorously [14] for an ideal 2D system that the positional order parameter $\langle e^{iG.u} \rangle$ is zero at all finite temperatures, but that a 2D *solid* may have bond orientational long range order (L.R.O.) since $O(r)$ does not decay to zero at large r.

Halperin and Nelson have shown by explicit calculation with an elastic continuum Hamiltonian that a 2D crystal has three distinct phases. These are phase (i), at low temperatures with orientational L.R.O. of near neighbour bonds and algebraic decay of position order, so that

$$P(G, r) \sim r^{-\eta_G(T)} ; \qquad O(r) \sim \langle \psi \rangle^2 .$$

At intermediate temperatures is phase (ii) with positional short range order and algebraic decay of orientational correlations, thus

$$P(G, r) \sim \frac{e^{-r/\xi_p}}{r^{1/2}} ; \qquad O(r) \sim r^{-\eta_6(T)} .$$

In phases (i) or (ii) the algebraic decay of correlation functions is accompanied by an infinite corresponding susceptibility. The high temperature phase (iii) has only short range order with

$$P(\mathbf{G}, \mathbf{r}) \sim \frac{e^{-r/\xi_p}}{r^{1/2}}; \qquad O(\mathbf{r}) \sim \frac{e^{-r/\xi_O}}{r^{1/2}}.$$

We suggest the following heuristic application of these results to SmB and similar phases. Analogous to de Gennes and Sarma, we identify the solid phase with stacked layers of HN's phase (i). The infinite positional susceptibility means that an infinitesimal interaction between layers will convert the algebraic decay of positional correlation functions into true 3D long range order. For this same reason it seems highly unlikely that the SmB phase can be formed of stacked Kosterlitz-Thouless quasi-ordered 2D solids, as was recently proposed by Huberman, Lublin, and Doniach [15]. In the 2D harmonic system the positional susceptibility is finite above a certain temperature so that layers may be stacked to form a SmB phase in the de Gennes Sarma model.

We propose that the SmB and higher ordered smectic phases can be described as stacked layers of HN phase (ii). We expect that interactions between layers will cause true in-plane bond orientational L.R.O. rather than the algebraic decay of $O(\mathbf{r})$ in HN phase (ii). Since the SmA phase has algebraic decay of positional correlations [16] along the 1D density wave and by analogy with 2D crystals has bond orientational L.R.O. in the same direction, the SmB phase should have true 3D bond orientational long range order. Since the positional in-plane correlations of phase (ii) decay exponentially, the corresponding susceptibility will be finite. Thus we expect stacked layers of phase (ii) can exist as well-ordered smectic phases with layers free to slide on one another provided the between layer interactions are sufficiently weak. The melting of the aliphatic chains [17] would provide a mechanism to reduce the interaction, which should be weaker for longer chains; this probably explains why only the longer chain members of a homologous series exhibit SmB like phases.

Finally, we identify stacked layers of HN phase (iii) with SmA or SmC phases. As mentioned above, the layers are stacked with algebraic decay of the positional correlations along the 1D density wave directions. The *bond orientational* L.R.O. which is present by analogy with 2D solids manifests itself as the fixed angle between the molecular axes and the smectic layers. The states of our model are summarized in table I.

TABLE I

Correlations in liquid crystal phases

Phase	In-plane bond orientational correlations	In-plane positional correlations
Solid	L.R.O.	L.R.O.
SmB	L.R.O.	Exponential decay (S.R.O.)
SmA	Exponential decay (S.R.O.)	S.R.O.

There are further subdivisions to be made to describe the various well-ordered smectic phases; these include tilt angles, rotational degrees of freedom about molecular long axes, and the type of in-plane lattice. In the tilted phases, the molecular tilt means twofold orientational in-plane L.R.O. even in the SmC phase [18] which will induce some six-fold order. Thus the well-ordered tilted phases differ from SmC in that the bond orientational L.R.O. is a primary order parameter rather than induced. Most of the discussion in the literature [e.g. 8] can be incorporated directly into our bond orientational L.R.O. model.

We now discuss some of the consequences of our model. Clearly the primary consequence is the identification of a microscopic order parameter which enables a quantitative differentiation between SmA, SmC, and the well-ordered SmB type phases.

The model also suggests a physical explanation for the strange temperature dependence of the SmA phase elastic constants B and D [19]. These constants give, respectively, the restoring force for compression of the smectic layers and for tilt of the molecules away from perpendicular to the layers. Several experiments have established [19, 20, 21] that B vanishes at the SmA-nematic transition with a power law quite different from that predicted by analogy with 3D superfluid helium; we believe this to be the result of the divergent phase fluctuations of the SmA order parameter and the absence of true positional L.R.O. In addition, D goes to zero with a different power law than B [19, 2], a result quite unexpected from the ⁴He analogue. In our view, since D is directly associated with the angle between the molecules and the smectic layers, which has true L.R.O., we might expect different critical behaviour for D compared with B.

To begin a discussion of the implications of our model for X-ray diffraction patterns, we remind the reader that the results of an X-ray experiment are sensitive to the positions of the molecules, and that one normally observes peaks in the scattered intensity at momentum transfers corresponding to reciprocal lattice vectors. Our model for the SmA and SmC phases is identical to the conventional one, a 1D density wave with algebraic decay of positional correlations together with a weakly correlated liquid transverse to the density wave direction. Hence we predict the usual X-ray pattern.

For SmB type phases important differences occur. The in-plane orientational L.R.O. makes it possible to prepare single domain samples with well defined bond orientations across the entire sample. The effect is illustrated in figure 1. Instead of the ring one would observe with orientational S.R.O., for a single domain sample with orientational L.R.O. we predict resolved peaks at the reciprocal lattice positions of the in-plane lattice (hexagonal for SmB). We also discuss the shape of these peaks. Halperin and Nelson compare their phase (ii) to a 2D nematic; this would have very short positional correlation lengths even in the presence of orientational L.R.O. It seems clear however that in order to have six-fold in-plane orientational L.R.O. combined with algebraic decay of positional correlations perpendicular to the planes, there must be positional correlations that extend over quite large distances. Thus we expect the correlation length $\xi_p \gg$ in-plane nearest neighbour separation; in HN phase (ii) ζ_p ranges from ~ 6 lattice parameters to ∞ as the melting transition is approached. We therefore predict an X-ray pattern with well defined Lorentzian peaks (with Δq^{-2} tails) centred about the lower order reciprocal lattice positions in the plane. The width of the peaks provides a measure of ξ_p and as the in-plane order increases more reflections will become observable.

These predictions are consistent with the available information in the literature [7, 8, 11, 12]. Studies so far have been carried out with relatively low resolution; ultrahigh resolution studies, similar to those of Als-Nielsen et al. [3], are required to provide a definitive test of our model. The de Gennes Sarma model is probably incorrect to the extent it is based on a 2D harmonic system, but an experimental distinction between their model and ours is possible. Our model predicts an analytic structure factor

corresponding to vectors of the in-plane reciprocal lattice of the SmB phase while the de Gennes Sarma picture predicts a cusp. Finally, the de Gennes Sarma model does not provide a natural transition between the SmB and SmA phases, whereas this transition, HN phase (ii) to phase (iii), is an essential feature of our model.

In conclusion, we note that if our adaptation of the Halperin-Nelson 2D melting picture to smectic liquid crystals is correct, then smectics in turn provide a rich laboratory for investigation of the ideas developed for systems whose marginal dimensionality d^0 is two. Primarily because of substrate difficulties it is difficult to imagine, for example, a real physical 2D fluid in state (ii). However, smectics may be studied in bulk without serious substrate problems and may be probed with a wide variety of experimental tools. Thus smectic phases may provide important insights into the problem of melting in two dimensions.

Acknowledgments. — We should like to thank B. Halperin, M. Kaplan and D. Nelson for stimulating conversations about this work, and P. G. de Gennes for important comments on the manuscript.

References

[1] DE GENNES, P. G., *The Physics of Liquid Crystals* (Oxford University Press) 1974.
[2] LITSTER, J. D., BIRGENEAU, R. J., ALS-NIELSEN, J., DAVIDOV, D., DANA, S. S., GARCIA-GOLDING, F., KAPLAN, M., SAFINYA, C. R. and SCHAETZING, R., paper presented at International Conference on Liquid Crystals, Bordeaux, 1978, to be published in *J. Physique Colloq.*
[3] ALS-NIELSEN, J., BIRGENEAU, R. J., KAPLAN, M., LITSTER, J. D. and SAFINYA, C. R., *Phys. Rev. Lett.* **39** (1977) 1668.
[4] PEIERLS, R. E., *Helv. Phys. Acta* **7** (1934) 81.
[5] LANDAU, L. D. and LIFSHITZ, E. M., *Statistical Physics*, (Addison Wesley, Reading, Mass.) 1969, p. 402.
[6] DE GENNES, P. G., *J. Physique Colloq.* **30** (1969) C4-65.
[7] DOUCET, J., LEVELUT, A. M., LAMBERT, M., *Phys. Rev. Lett.* **32** (1974) 301.
[8] LEVELUT, A. M., DOUCET, J. and LAMBERT, M., *J. Physique* **35** (1974) 773 ; DOUCET, J., LEVELUT, A. M., LAMBERT, M. and STRZELECKI, L., *ibid* **36** (1975) C1-13 ; DOUCET, J., KELLER, P., LEVELUT, A. M. and PORQUET, P., *ibid* **39** (1978) 548.

[9] KOSTERLITZ, J. M. and THOULESS, D. J., *J. Phys.* **C 6** (1973) 118 ; *Prog. Low Temp. Phys.* (to be published).
[10] HALPERIN, B. I. and NELSON, D. R., to be published.
[11] DE VRIES, A. and FISHEL, D. L., *Mol. Cryst. Liq. Cryst.* **16** (1972) 311.
[12] DE VRIES, A., *J. Physique Colloq.* **36** (1975) C1-1 gives a review of the literature up to 1975.
[13] DE GENNES, P. G. and SARMA, G., *Phys. Lett.* **38A** (1972) 219.
[14] MERMIN, N. D., *Phys. Rev.* **176** (1968) 250.
[15] HUBERMAN, B. A., LUBLIN, D. M. and DONIACH, S., *Solid State Commun.* **17** (1975) 485.
[16] CAILLÉ, A., *C.R. Hebd. Séan. Acad. Sci.* **274B** (1972) 891 ; eq. (9) is in error, see [1], p. 298.
[17] POLDY, F., DVOLAITZKY, M. and TAUPIN, C., *J. Physique Colloq.* **36** (1975) C1-27.
[18] HALPERIN, B. I., private communication.
[19] BIRECKI, H., SCHAETZING, R., RONDELEZ, R. and LITSTER, J. D., *Phys. Rev. Lett.* **36** (1976) 1376.
[20] RIBOTTA, R., *C.R. Hebd. Séan. Acad. Sci.* **279B** (1974) 295.
[21] CLARK, N. A., *Phys. Rev.* **14A** (1976) 1551.

PHYSICAL REVIEW B   VOLUME 21, NUMBER 11   1 JUNE 1980

Solid and fluid phases in smectic layers with tilted molecules

David R. Nelson and B. I. Halperin

Department of Physics, Harvard University, Cambridge, Massachusetts 02138

(Received 10 September 1979)

A recent theory of two-dimensional melting is applied to freely suspended liquid-crystal films with a tilt degree of freedom. As many as seven distinct phases are possible, including those we identify with smectic-A, -B, -C, and -H liquid crystals. Some of these phases may survive when stacked to form bulk smectics.

I. INTRODUCTION

A. Purpose

There has been considerable progress recently in understanding the mechanism of dislocation-mediated melting in two dimensions proposed by Kosterlitz and Thouless.[1,2] Although dislocations have long been suggested as a mechanism for *three*-dimensional melting,[3] only in two dimensions has substantial analytical progress been made. A study of a simplified model of interacting dislocations[4] was followed by detailed analyses[5,6] of dislocation-unbinding transitions in more realistic situations. In Ref. 5 it was found that dislocations drive a transition into a kind of liquid-crystal phase, with persistent correlations in the orientation of "bond angles." Since triangular lattices melt into a phase with persistent sixfold orientational order, this new phase might be called a "hexatic liquid crystal." A second *disclination* unbinding transition is necessary to complete the transition to an isotropic liquid.

Recent molecular-dynamics simulations by Frenkel and McTague[7] support this picture of two-dimensional melting as a two-stage process. Some evidence that dislocations were important had been obtained previously by Cotterill and Pedersen.[8] A computer simulation by Morf[9] suggests that the two-dimensional electron crystal observed by Grimes and Adams[10] melts at a temperature consistent with the dislocation-unbinding picture, provided the renormalization of the shear modulus is taken into account.

Free-standing smectic liquid-crystal films[11,12] may provide particularly interesting experimental tests of two-dimensional melting theories. One might plausibly associate a thin smectic-B film with a two-dimensional solid, and smectic-A films with two-dimensional hexatic or liquid phases. Birgeneau and Litster,[13] on the other hand, have suggested that the *bulk* smectic-B phases may be understood as three-dimensional analogues of the hexatic phase, with long-range orientational order and short-range translational order. If this conjecture were correct, smectic-B films would be examples of hexatic liquid

crystals rather than two-dimensional solids. X-ray diffraction measurements by Moncton and Pindak,[14] and related work by Pershan *et al.*,[15] however, indicate that the bulk smectic-B phase of butoxybenzylidene-octylaniline (BBOA) possesses very long-range translational correlations. Infinite-range translational order would contradict the Birgeneau-Litster hypothesis.

Whichever interpretation is correct, it seems important to extend the theory of two-dimensional

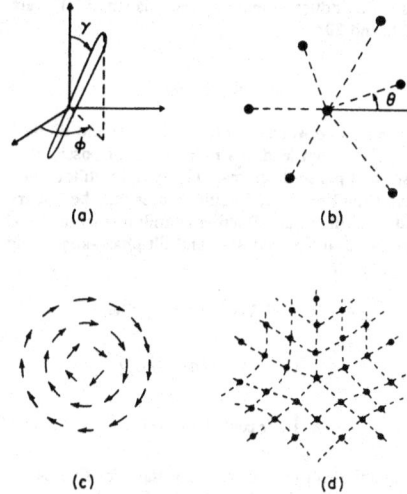

FIG. 1. Degrees of freedom and singularities necessary to describe fluid phases in smectic layers. (a) Tilted liquid-crystal molecule with polar angle γ and orientation angle ϕ. (b) "Bonds" (dashed lines) joining a central atom to its six nearest neighbors. Each such bond makes a bond angle θ with the x axis. (c) Vortex in the orientation angle field ϕ. The vectors are given by $\sin\gamma(r)[\cos\phi(r), \sin\phi(r)]$. (d) Fivefold disclination imbedded in a square lattice with bond angles shown as dashed lines.

melting to a variety of experimentally relevant situations. Here we generalize the theory to consider systems in which there is a two-dimensional vector order parameter, coupled to the order parameters that characterize melting of a triangular solid. Our work should apply in particular to thin films of materials which exhibit (in bulk) a smectic-C and/or a smectic-H phase in which there is a molecular axis tilted relative to smectic layers. Here, one must introduce an order parameter $\Phi(\vec{r}) = (\sin\gamma) \times \exp[i\phi(\vec{r})]$, which describes the projection of the tilt axis in the xy plane [see Fig. 1(a)]. Interesting effects arise from the coupling between $\Phi(\vec{r})$ and the parameters $\psi(\vec{r}) = \exp[6i\theta(\vec{r})]$ and $\rho_{\vec{G}}(\vec{r}) = \exp[i\vec{G}\cdot\vec{u}(\vec{r})]$ describing bond-angle orientations and translational order in the material [see Fig. 1(b) for a definition of the bond-orientation field].

Although we shall concentrate on applications to freely suspended films, parts of the theory may also have relevance to magnetic or orientational transitions in adsorbed films. We do not consider any electric dipolar forces associated with the molecular tilts; this is correct in a suspended film with nonchiral molecules, and may be a valid approximation in other cases if the induced dipole moment is small (cf. Refs. 11, 12, and 20).

B. Fluid phases

Couplings between orientational order and tilt degrees of freedom lead to a rich variety of possible phases and phase diagrams. Tilted and untilted versions of the hexatic and liquid phases may be understood in terms of an effective Hamiltonian functional of the bond-angle field $\theta(r)$ and tilt-phase-angle field $\phi(r)$, namely,

$$\frac{H}{k_B T} = \frac{1}{2}\int d^2 r\,[\,K_6|\vec{\nabla}\theta|^2 + K_1|\vec{\nabla}\phi|^2$$
$$+ 2g(\vec{\nabla}\theta)\cdot(\vec{\nabla}\phi)]$$
$$- h\int d^2 r\cos[6(\theta - \phi)]\;. \tag{1.1}$$

The quantity K_1 is a stiffness constant for fluctuations in the tilt orientations, while K_6 is the Frank constant for fluctuations in the bond orientation.[16] The term proportional to h occurs because both $\theta(\vec{r})$ and $\phi(\vec{r})$ feel a sixfold symmetric potential when rotated with the other field held fixed. The gradient cross-coupling, proportional to g, is generated by the renormalization group discussed in Sec. II even if it is initially absent. "Vortices" in the tilt-angle field and "disclinations" in the bond-orientation-angle fluctuations [see Figs. 1(c) and 1(d)] will also be taken into account. These excitations renormalize the elastic constants at large distances, and can also drive phase

transitions by unbinding from a state containing bound pairs only. The "bare" constants in Eq. (1.1) will themselves have an analytic dependence on temperature due to the effects of fluctuations on the atomic length scale.

The constant K_6 in Eq. (1.11) is related to the Frank constant K_A used in previous papers,[5] by $K_6 = K_A/k_B T$. The subscripts 6 and 1, which we use in the present paper to indicate quantities referring to the angles θ and ϕ, respectively, were chosen because the bond orientation θ is defined modulo $\frac{2}{6}\pi$, while the tilt orientation ϕ is defined on the entire range from 0 to 2π.

A variety of possible phases follow from this model, which may be distinguished by the large-distance behavior of the correlation functions

$$C_6(\vec{r}) = \langle\exp\{6i[\theta(\vec{r})-\theta(\vec{0})]\}\rangle\;, \tag{1.2a}$$

$$C_1(\vec{r}) = \langle\exp\{i[\phi(\vec{r})-\phi(\vec{0})]\}\rangle\;. \tag{1.2b}$$

We summarize here the results of our analysis, whose details are given in Secs. II and III below. One possible phase diagram is shown in Fig. 2, as a function of the inverse "bare" Frank constants K_1^{-1} and K_6^{-1}, with g and h small and fixed. The quantities K_1^{-1} and K_6^{-1} should both be monotonically increasing functions of temperature, so that a given material will trace a path from lower left to upper right in the figure, as temperature is increased. The solid phases shown in this diagram, in which $K_6 = \infty$, will be discussed later.

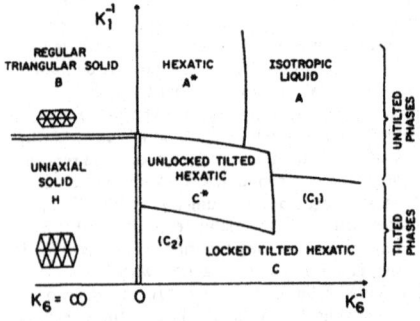

FIG. 2. Phase diagram for smectic liquid-crystal layers, as a function of the inverse temperature-dependent Frank coefficients $K_1^{-1}(T)$ and $K_6^{-1}(T)$. Both solid and fluid phases are shown, and these can be either tilted or untilted. The Frank constant K is infinite in the solid phases. Experiments with varying temperature might trace a path from the lower-left to the upper-right portion of the figure, with increasing temperatures.

Four *fluid* phases are indicated in the diagram, labeled A, A^*, C, and C^*. Phase A^* is identical to the hexatic phase of Ref. 5, with algebraic decay of $C_6(\bar{r})$ and exponential decay of $C_1(\bar{r})$,

$$C_6(\bar{r}) \simeq r^{-\eta_6(T)} \ , \qquad C_1(\bar{r}) \simeq e^{-r/\xi_1(T)} \ . \qquad (1.3)$$

[We shall use the term "quasi-long-range order" for correlations such as $C_6(\bar{r})$ in Eq. (1.3) which decay to zero as power laws.]

Phase A, in Fig. 2, is an isotropic liquid phase, where both $C_1(\bar{r})$ and $C_6(\bar{r})$ decay exponentially at large r. The remaining fluid phases, C and C^*, have quasi-long-range order for both tilt and bond orientations; i.e., for large r we have

$$C_6(\bar{r}) \simeq r^{-\eta_6(T)} \ , \qquad (1.4)$$

$$C_1(\bar{r}) \simeq r^{-\eta_1(T)} \ . \qquad (1.5)$$

Phase C is a "locked" tilted hexatic phase in which long-wave-length fluctuations in ϕ are tied to fluctuations in θ. The phase is characterized by a single renormalized (i.e., macroscopic) Frank constant K_+^R which describes the increase in energy caused by equal gradients in θ and ϕ. (Roughly one has $K_+^R \simeq K_1 + K_6 + 2g$). In this phase, the exponents η_1 and η_6 are related by

$$\eta_6 = 36\eta_1 = \frac{18}{\pi K_+^R} \ . \qquad (1.6)$$

Phase C^* is an "unlocked" tilted hexatic phase, in which long-wavelength fluctuations in ϕ and θ are independent. This phase has three renormalized Frank constants, K_1^R, K_6^R, and g_R, corresponding to the constants in Eq. (1.1), and there is no simple relation between η_1 and η_6

$$\eta_6 = \frac{18 K_1^R}{\pi (K_6^R K_1^R - g_R^2)} \ , \qquad (1.7a)$$

$$\eta_1 = \frac{K_6^R}{2\pi (K_6^R K_1^R - g_R^2)} \ . \qquad (1.7b)$$

Further insight into the difference between the phases C and C^* may be obtained by examining the correlation function

$$C_\Delta(\bar{r}) = \langle \Delta(\bar{r}) \Delta^*(\bar{0}) \rangle \ , \qquad (1.8)$$

where $\Delta(\bar{r})$ is a variable which measures the relative orientation of the bond and tilt axes,

$$\Delta(\bar{r}) = \exp\{6i[\theta(\bar{r}) - \phi(\bar{r})]\} \ . \qquad (1.9)$$

In the unlocked state C^*, we find the large-distance behavior

$$C_\Delta(\bar{r}) \sim (\text{const}) + r^{-\eta_\Delta} \ , \qquad (1.10)$$

with

$$\eta_\Delta = \frac{18(K_6^R + K_1^R + 2g_R)}{\pi(K_6^R K_1^R - g_R^2)} \ . \qquad (1.11)$$

In phase C, the locked tilted state, the correlation function $C_\Delta(\bar{r})$ tends *exponentially* towards a constant value at large r.

The unlocked phase C^* is found to be present in the phase diagram when the bare coupling constant h is small, but the state may *not* exist when h is large. Some necessary conditions for stability of the C^* phase are

$$\eta_\Delta > 4 \ , \qquad (1.12)$$

$$K_1^R > 2/\pi \ , \qquad (1.13)$$

$$K_6^R > 72/\pi \ . \qquad (1.14)$$

It should be noted that quasi-long-range order in tilt orientation always induces algebraically decaying correlations of bond orientation. Thus, phases with short-range bond order but quasi-long-range tilt order are impossible. We expect that short-range bond orientational order will also be incompatible with long-range tilt-angle order in *bulk* smectic liquid crystals.

In the right-hand portion of the C phase, labeled C_1 in Fig. 2, one is in a region where there would be no bond order if the molecules were not tilted. The correlation function $C_6(\bar{r})$ has algebraic decay at long distances only because of the coupling h between tilt and bond angles, and the amplitude of the correlations should be proportional to h^2. In the left-hand portion of the C phase (labeled C_2), the bond angles would tend to order, (forming a hexatic phase) even in the absence of tilt. The amplitude of $C_6(\bar{r})$ will be independent of h in this region, and hence much larger than in the region C_1. Since there is no change in symmetry, there is no necessity for a sharp phase transition between the regions C_1 and C_2. However, there may be a first-order transition in some cases.[17,18]

C. Solid phases

In addition to the fluid phases described above, there are two solid phases (B and H) indicated in Fig. 2. These phases have true long-range order in the bond orientation,[19]

$$\langle e^{6i\theta} \rangle = (\text{const}) e^{6i\theta_0} \neq 0 \ , \qquad (1.15)$$

where θ_0 is the orientation of the crystal axis in the xy plane. The Frank constant K_6 entering Eq. (1.1) should be considered infinite in the solid phases; however, we must now take into account coupling of the tilt orientation to the strain field of the crystal. Possible solid phases may then be understood in terms of the effective Hamiltonian

$$\frac{H}{k_B T} = \frac{1}{2} \int d^2r \, [2\mu u_{ij}^2 + \lambda u_{kk}^2$$
$$+ 2w(u_{ij} - \frac{1}{2}\delta_{ij} u_{kk}) s_i s_j]$$
$$- h \int d^2r \cos[6(\phi(r) - \theta_0)]$$
$$+ \frac{1}{2} K_1 \int d^2r \, (\vec{\nabla}\phi)^2 \, , \qquad (1.16)$$

where the strain tensor $u_{ij}(\vec{r})$ is the symmetric derivative of the displacement field $\vec{u}(\vec{r})$

$$u_{ij} = \frac{1}{2}\left[\frac{\partial u_i}{\partial x_j} + \frac{\partial u_j}{\partial x_i}\right] \, , \qquad (1.17)$$

μ and λ are the (bare) Lamé elastic constants, and

$$\vec{s} = \begin{bmatrix} \cos(\phi - \theta_0) \\ \sin(\phi - \theta_0) \end{bmatrix} \, . \qquad (1.18)$$

If the coupling proportional to w in Eq. (1.16) were neglected, one could apply the analysis of José et al.[20] to the tilt degrees of freedom and find three solid phases: (i) a regular triangular solid with short-range order in tilt angles; (ii) a regular triangular solid with quasi-long-range tilt-angle order; and (iii) an anisotropic solid with genuine long-range order in $\phi(\vec{r})$, i.e.,

$$\langle e^{i\phi(\vec{r})} \rangle \neq 0 \, . \qquad (1.19)$$

We find, however, that the anharmonic coupling between phonons and $\vec{s}$ destabilizes the intermediate phase (ii) above. Presumably, this instability leads to a tilted anisotropic solid identical to (iii). One would then expect the line of phase transitions directly from a regular triangular untilted solid (labeled B) to an anisotropic solid with tilt (labeled H), as shown in Fig. 2.

D. Discussion

Phase boundaries shown as light solid lines in Fig. 2 are "Kosterlitz-Thouless"-type phase transitions, with unobservable essential singularities in the specific heat, but with jumps in appropriate stiffness constants. The double lines represent transitions whose character has not been analyzed.

On the line connecting locked tilted hexatic phase C to the isotropic liquid (this may be considered as a line of smectic-A to smectic-C phase transitions), one has $\eta_1 = \frac{1}{4}$, as discussed by Nelson and Kosterlitz[21] and Pelcovits and Halperin.[22] It follows from Eq. (1.6) that $\eta_6 = 9$ on this line. On the line joining the hexatic and isotropic liquid phases, one finds $\eta_6 = \frac{1}{4}$, as discussed in Ref. 5. In contrast, η_1, η_6, and η_Δ are nonuniversal and subject to only mild restrictions on the boundaries of the unlocked tilted hexatic phase (see Sec. II B).

The complicated array of phases and phase transitions described above is interesting for a number of reasons. First, the phases with quasi-long-range or long-range order in the tilt angle should be optically active and accessible to light-scattering studies. Orientational order in the hexagonal bond fields, which cannot be directly probed by optical methods, can thus be studied indirectly through its coupling to the tilt angle. Second, one might hope our results have some relevance to bulk smectic liquids crystals. The labels A and C in Fig. 2 were chosen because the corresponding phases have the properties of an isolated layer of the bulk phases known as smectic A and smectic C, respectively. Similarly, phases B and H correspond to the most commonly accepted description of the bulk smectic-B and -H phases, in which the smectic layers are believed to be two-dimensional solids. A stack of two-dimensional solids with any finite coupling between the layers would be expected to form (in thermal equilibrium) a three-dimensional solid, with conventional long-range translational order in all directions. Recent x-ray measurements on the smectic-B phase of the compound BBOA support this description.[14,15]

As pointed out by Birgeneau and Litster,[13] a stack of weakly coupled hexatic layers (the A^* phase) would form a bulk liquid-crystal phase, with short-range translational order parallel to the layers, but long-range order in the bond-angle field, $\langle\psi\rangle \neq 0$. Experimental identification of such a phase has not been reported, however.

The unlocked tilted hexatic C^* should not have any analog in three dimensions. If the tilt orientation has true long-range order, it will always be locked to θ.

In Sec. II, below, we shall discuss the fluid phases A, A^*, C, and C^*, in the limits of both weak and strong coupling between bond and tilt orientations. The effect of tilt on two-dimensional solids is described in Sec. III. Derivations of renormalization-group recursion relations are contained in an Appendix A, and the behavior of a correlation function is worked out in Appendix B.

II. HEXATIC AND LIQUID PHASES WITH TILT

A. Weak-coupling Hamiltonian

As discussed in the Introduction, a reduced effective Hamiltonian describing hexatic and liquid phases with tilt is[16]

$$-\vec{H} = \frac{H}{k_B T} = \frac{1}{2} \int d^2r \, [K_6(\vec{\nabla}\theta)^2 + K_1(\vec{\nabla}\phi)^2$$
$$+ 2g(\vec{\nabla}\theta)\cdot(\vec{\nabla}\phi)]$$
$$- h \int d^2r \cos[p(\theta - \phi)] \, , \qquad (2.1)$$

where we have generalized Eq. (1.1) to allow for a p-fold symmetric coupling between $\theta(\vec{r})$ and $\phi(\vec{r})$. The probability of a given configuration of bond and tilt orientation angles is proportional to $\exp \overline{H}$, and the partition function corresponding to Eq. (2.1) is given by a functional integral over θ and ϕ,

$$Z = \int D\theta \int D\phi \exp \overline{H} \ . \tag{2.2}$$

Although we shall focus primarily on the case $p=6$, other values of p are possible. Square lattices should melt into "tetratic" liquid-crystal phases,[5] in which $p=4$ would be appropriate. One can use the model with $p=3$ to describe symmetric molecules tilted completely into the plane of the film $[\gamma = \frac{1}{2}\pi$ in Fig. 1(a)], provided one redefines θ and ϕ to be twice the angle between the x axis and the bond or molecular orientation. (If θ and ϕ are not redefined in this way, one has $p=6$, but then half-integer vortices in the θ field must be considered.[23]) Here we consider only integral vortices in ϕ, but allow for a p-fold symmetric θ field. Our analysis can easily be extended to completely general situations.

In the limit of infinite K_6, the θ field is locked to some constant value, and Eq. (2.1) reduces to a model where rotational invariance in ϕ is broken by a p-fold symmetric "crystal field." With vortices taken into account, this is the "$\cos p\phi$" model of xy magnetism studied extensively by José et al.[20] Our conclusions should reduce to the known results for this system in the limit $K_6 \rightarrow \infty$ (see below). In the limit $K_1 \rightarrow \infty$, the ϕ field is locked, and one is left with a p-fold-symmetric field θ subject to a $\cos p\theta$ perturbation. Since this situation is like an xy model in a magnetic field, one expects no phase transition with varying K_6 in this limit. Note that one must have

$$g^2 < K_6 K_1 \tag{2.3}$$

in Eq. (2.1) for stability.

Strictly speaking, all terms in Eq. (2.1) should be periodic under the transformation

$$\theta(\vec{r}) \rightarrow \theta(\vec{r}) + \frac{2\pi}{p} m(\vec{r}) \ , \quad m(\vec{r}) = 0, \pm 1, \ldots \ , \tag{2.4a}$$

$$\phi(\vec{r}) \rightarrow \phi(\vec{r}) + 2\pi n(\vec{r}) \ , \quad n(\vec{r}) = 0, \pm 1, \ldots \ , \tag{2.4b}$$

and the functional integrals in Eq. (2.2) need only be carried out over the range

$$-\frac{\pi}{p} \leq \theta(\vec{r}) \leq \frac{\pi}{p} \ , \tag{2.5a}$$

$$-\pi \leq \phi(\vec{r}) \leq \pi \ . \tag{2.5b}$$

This periodicity reflects the invariance of the systems to discrete local rotations of θ and ϕ, with angles in other regions of space held fixed. In practice, however, studies of the two-dimensional xy model[20,24] sug-gest that one can integrate freely provided the disclination and vortex singularities allowed by Eq. (2.4) are taken into account explicitly.

These textural singularities can be included in Eq. (2.1) for small h as follows. At every point in space, the exponentiated periodic potential can be expanded in a Fourier series,

$$\exp \left(h \cos \{ p[\theta(\vec{r}) - \phi(\vec{r})] \} \right)$$
$$= \sum_{s(\vec{r})=-\infty}^{+\infty} A_s(h) \exp \{ips(\vec{r})[\theta(\vec{r})-\phi(\vec{r})]\} \ . \tag{2.6}$$

It can readily be shown that, for small h,

$$A_s(h) \simeq (\tfrac{1}{2}h)^{s^2}, \quad s = 0, \pm 1 \tag{2.7}$$

and $A_s(h)$ is negligible for large s. More generally, one can take

$$A_s(h) = A_0 \exp[(\ln y_h)s^2] \ , \tag{2.8}$$

where Eq. (2.7) is recovered if we take

$$y_h = \tfrac{1}{2}h, \quad A_0 = 1 \ . \tag{2.9}$$

Inserting the decomposition (2.6) into Eq. (2.2), one finds that the partition sum acquires a sum over an integer-valued field $\{s(\vec{r})\}$,

$$Z = \sum_{\{s(\vec{r})\}} \int D\theta \int D\phi \exp[\overline{H}(\{\theta\}, \{\phi\}, \{s\})] \ , \tag{2.10}$$

with

$$\overline{H} = -\frac{1}{2} \int d^4r \ [K_6(\vec{\nabla}\theta)^2 + K_1(\vec{\nabla}\phi)^2 + 2g(\vec{\nabla}\theta)\cdot(\vec{\nabla}\phi)]$$
$$+ ip \int d^2r \ s(\vec{r})[\theta(\vec{r}) - \phi(\vec{r})]$$
$$+ (\ln y_h) \int d^2r \ s^2(\vec{r}) \ . \tag{2.11}$$

As h tends to zero, $\ln y_h$ becomes large and nega-tive, and excitations with $s(r)$ nonzero at some point become very unlikely. "Vortices" and "disclinations" can then be defined as solutions of

$$K_6\nabla^2\theta + g\nabla^2\phi = 0 \ , \tag{2.12a}$$

$$K_1\nabla^2\phi + g\nabla^2\theta = 0 \ , \tag{2.12b}$$

subject to conditions on contour integrals taken around isolated points $\vec{r}$,

$$\oint \vec{\nabla}\theta \cdot d\vec{l} = \frac{2\pi}{p} m(\vec{r}) \ , \quad m(\vec{r}) = 0, \pm 1, \ldots \ , \tag{2.13a}$$

$$\oint \vec{\nabla}\phi \cdot d\vec{l} = 2\pi n(\vec{r}) \ , \quad n(\vec{r}) = 0, \pm 1, \ldots \ . \tag{2.13b}$$

Expanding $\theta(r)$ and $\phi(r)$ about disclination and vortex complexions described by $\{m(\bar{r})\}$ and $\{n(\bar{r})\}$ amounts to making the substitutions

$$\theta(\bar{r}) \rightarrow \delta\theta(\bar{r}) + \frac{1}{6}\sum_{\bar{r}' \neq \bar{r}} m(\bar{r}')\tilde{G}(\bar{r},\bar{r}') \ , \qquad (2.14a)$$

$$\phi(\bar{r}) \rightarrow \delta\phi(\bar{r}) + \sum_{\bar{r}' \neq \bar{r}} n(\bar{r}')\tilde{G}(\bar{r},\bar{r}') \qquad (2.14b)$$

and then integrating freely over $\delta\phi$ and $\delta\phi$. The summations in Eq. (2.14) cover a lattice of possible sites for disclinations and vortices. This same lattice

can be used to provide an ultraviolet cutoff of order the inverse core diameter a^{-1} for the functional integrations in Eq. (2.2). It has no further physical significance. For large separations $(\bar{r} - \bar{r}')$ and far from boundaries, the Green function $\tilde{G}(\bar{r},\bar{r}')$ is

$$\tilde{G}(\bar{r},\bar{r}') \simeq \tan^{-1}\left|\frac{y-y'}{x-x'}\right| \ , \qquad (2.15)$$

where $\bar{r} = (x,y)$ and $\bar{r}' = (x',y')$.

Upon making the replacements (2.14) in Eq. (2.11), one easily carries out the functional integrations over $\delta\theta$ and $\delta\phi$ to find

$$Z = (\text{const})\sum_{\{s(\bar{r})\}}' \sum_{\{m(r)\}}' \sum_{\{n(r)\}}' \exp[\bar{H}(\{s\},\{m\},\{n\})] \ , \qquad (2.16)$$

where $\bar{H}$ is expressible in terms of three coupled Coulomb-gas Hamiltonians

$$\bar{H} = \bar{H}_c(K_h, y_h, \{s\}) + \bar{H}_c\left(\frac{K_6}{p^2}, y_6, \{m\}\right) + \bar{H}_c(K_1, y_1, \{n\}) + i\sum_{\bar{r} \neq \bar{r}'} s(\bar{r})m(\bar{r}')\tan^{-1}\left|\frac{y-y'}{x-x'}\right|$$

$$- ip\sum_{\bar{r} \neq \bar{r}'} s(\bar{r})n(\bar{r}')\tan^{-1}\left|\frac{y-y'}{x-x'}\right| + \frac{2\pi g}{p}\sum_{r \neq r'} m(\bar{r})n(\bar{r}')\ln\left|\frac{|\bar{r}-\bar{r}'|}{a}\right| \ . \qquad (2.17)$$

The Hamiltonian $\bar{H}_c$ is the usual[24] scalar Coulomb gas,

$$\bar{H}_c(K, y, \{m\}) = \pi K \sum_{\bar{r} \neq \bar{r}'} m(\bar{r})m(\bar{r}')\ln\left|\frac{|\bar{r}-\bar{r}'|}{a}\right| + \ln y \sum_r m^2(r) \ . \qquad (2.18)$$

The disclination and vortex fugacities y_6 and y_1 are

$$y_6 = \exp(-CK_6/p^2), \quad y_1 = \exp(-C'K_1) \ , \qquad (2.19)$$

where C and C' are cutoff-dependent constants. The strength K_h of the logarithmic interaction between the $\{s(\bar{r})\}$ generated by integrating over θ and ϕ is

$$K_h = \frac{p^2(K_6 + K_1 + 2g)}{4\pi^2(K_6 K_1 - g^2)} \ , \qquad (2.20)$$

and the corresponding fugacity is given by Eq. (2.9). The primes on the summations over the three integer fields in Eq. (2.16) mean they are subject to the constraints

$$\sum_{\bar{r}} s(\bar{r}) = \sum_{\bar{r}} m(\bar{r}) = \sum_{\bar{r}} m(\bar{r}) = 0 \ . \qquad (2.21)$$

In deriving Eq. (2.17), it is helpful to use the identity

$$[\vec{\nabla}\tilde{G}(\bar{r},\bar{r}')]^2 = [\vec{\nabla}G(\bar{r},\bar{r}')]^2 \ , \qquad (2.22)$$

where $G(r,r')$ is the harmonic conjugate of $\tilde{G}(r,r')$ and satisfies

$$\nabla^2 G(\bar{r},\bar{r}') = -2\pi\delta(\bar{r}-\bar{r}') \ . \qquad (2.23)$$

Far from boundaries and for large $(\bar{r} - \bar{r}')$, one has

$$G(\bar{r},\bar{r}') \simeq \ln(|\bar{r}-\bar{r}'|/a) + 2\pi C \ . \qquad (2.24)$$

If we take the limit $K_6 \rightarrow \infty$ or $K_1 \rightarrow \infty$ in Eq. (2.17), so that either disclination or vortex pair excitations are unlikely, then one can safely set either all $m(\bar{r})$ or all $n(\bar{r})$ to zero. In this case, the Hamiltonian (2.17) reduces to an expression derived by Kadanoff[25] for the xy models with $\cos p\theta$ perturbations using a different method.

B. Weak-coupling phase diagram

The statistical mechanics of the interacting disclinations, vortices, and "periodic excitations" $\{s(\bar{r})\}$ described by Eq. (2.17) can be studied by a triple expansion in the fugacities y_6, y_1, and y_h. Experience with the xy model[20,24] suggests that such perturbation series will either diverge, or instead give only small corrections to the behavior obtained in the absence of disclinations, vortices, or periodic couplings. The corresponding fugacities will be "relevant" or "irrelevant" variables depending on the magnitudes of K_6, K_1, and g. If it is known that vortices and the y_h

coupling can be neglected, for example, then the crucial integral which determines the importance of disclinations is

$$2\pi \int_a^\infty r^3 \langle m(\vec{r}) m(\vec{0}) \rangle \, dr$$

$$\simeq -4\pi y_6^2 \int_a^\infty dr \, r^{3-2\pi K_6/p^2} \qquad (2.25)$$

The corresponding renormalization-group eigenvalue for the fugacity y_6 is

$$\lambda_6 = 2 - \pi K_6/p^2 \quad , \qquad (2.26)$$

where the sign of λ_6 determines the convergence of the integral in Eq. (2.25). Similar considerations lead to the eigenvalues for y_1 and y_h, namely,

$$\lambda_1 = 2 - \pi K_1 \quad , \qquad (2.27)$$

$$\lambda_h = 2 - \pi K_h \quad , \qquad (2.28)$$

where K_h is given by Eq. (2.20). [More precisely, for large length scales we must use the renormalized elastic constants in Eqs. (2.26)−(2.28).]

A more detailed derivation of these results, together with more complete renormalization-group recursion relations (see Appendix A), will be given in Sec. II C. A crude, but qualitatively correct phase diagram, however, follows from considering regions with different stability properties determined by the eigenvalues quoted above. For example, for $p = 6$, there is a region in the space of K_1^{-1}, K_6^{-1}, and g in which all three fugacity eigenvalues are negative. In Fig. 3, which illustrates the plane $g = 0$, the region is the trapezoid $ABCD$. Although the figure is drawn for $g = 0$, it is qualitatively similar for finite g. The trapezoid becomes a triangle for $p \leq 4$, and vanishes altogether for slightly smaller p's.

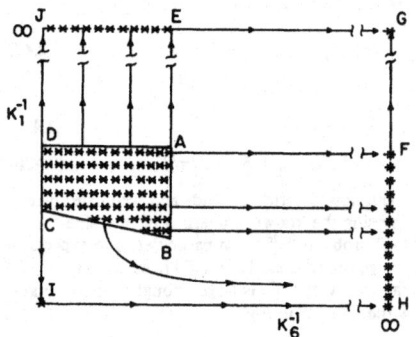

FIG. 3. Schematic renormalization-group flows describing different smectic fluid phases. Asterisks indicate fixed points. In addition to isolated fixed points, there are fixed lines JE and FH, and fixed surface $ABCD$.

Disclinations, vortices, and the h coupling term are irrelevant inside the trapezoid, and the long-wavelength properties of the resulting state (C^* phase) should be describable by an effective Hamiltonian

$$\overline{H} = \frac{1}{2} \int d^2r \, [K_6^R (\nabla \theta)^2 + K_1^R (\nabla \phi)^2$$

$$+ 2g_R (\nabla \theta) \cdot (\nabla \phi)] \quad , \qquad (2.29)$$

where K_6^R, K_1^R, and g^R are renormalized versions of the couplings appearing in Eqs. (1.1) and (2.1). The inequalities (1.12)−(1.14) follow directly from the conditions that λ_1, λ_6, and λ_h be negative.

Since correlations can be determined by integrating freely over θ and ϕ in this phase, it is easy to verify the results (1.7a) and (1.7b) quoted in the introduction. All three stiffness constants in Eq. (2.29) could, in principle, be determined by measuring the algebraic falloff of correlations indicated in Eq. (1.4), together with the decay of the four-point function C_Δ discussed in the introduction.

The cross correlation function, $C_x(r)$ $= \langle \exp[6i\theta(r) - 6i\phi(0)] \rangle$, is more complicated. If C_x is evaluated at the fixed point, when $h = 0$, we simply find $C_x = 0$, since the Hamiltonian then contains no term which depends on the overall phase difference between ϕ and θ. Nevertheless, at any finite length scale, the renormalized value of h is finite, and we expect to find a nonzero result. The evaluation of C_x in this phase is carried out in Appendix B.

In analogy to the universal jump prediction for the stiffness in the xy model,[21] we expect that $K_6^R = 2/\pi$ just inside the line AB of Fig. 3, and that $K_1^R = 2/\pi$ just inside BC. These stiffnesses should approach their universal values with square-root cusp singularities.[24] Such conditions are not enough to completely determine η_1 and η_6, if g_R is unspecified, so these exponents should be nonuniversal on these lines. The exponent η_Δ equals $\frac{1}{4}$ on BC. The stiffness K_6^R is infinite on CD, while K_1^R varies smoothly from $9/2\pi$ at C to $2/\pi$ at D on this line.

The arrows in Fig. 3 show schematically our expectations for renormalization-group flows in K_6^{-1} and K_1^{-1} outside the stable trapezoid. Flows immediately above AD, which is a locus of vortex-unbinding transitions, should tend toward a fixed line JE at $K_1 = 0$. The physics on this fixed line is just that discussed for the hexatic phase in Ref. 5; correlations in the $\theta(\vec{r})$ should be describable by a Hamiltonian of the form

$$\overline{H}_0 = \frac{1}{2} \int d^2r \, K_6 (\nabla \theta)^2 \quad . \qquad (2.30)$$

Disclinations must still be included in the θ field.[5] On the line segment JE, disclinations are bound in pairs and may be neglected at large length scales. At point E, however, disclination unbind, driving an instability into a liquid phase described by the fixed point G. The domain of attraction of the fixed line JE determines an untilted hexatic phase, with the properties (1.3) quoted in the Introduction.

Unstable flows immediately to the right of AB are triggered by a disclination unbinding, and should tend toward the fixed line FH at $K_A = 0$. This fixed line describes smectic-C liquid-crystal films, with power-law decay in tilt-angle correlations, and a vortex-unbinding transition at F. In an analogy to Eq. (2.30), we expect that correlations in the $\theta(\vec{r})$ are given by a Hamiltonian

$$\bar{H}_0 - \tfrac{1}{2}\int d^2r\, K_1(\nabla\phi)^2 \qquad (2.31)$$

provided vortices are also taken into account.

There is, however, *induced* quasi-long-range order in bond-angle correlations in the C phase. To describe bond-angle order, we must add the periodic coupling between the θ and ϕ fields to Eq. (2.31)

$$\bar{H}_0 \to \bar{H} - \bar{H}_0 + h\int d^2r \cos[6\theta(\vec{r}) - 6\phi(\vec{r})] \ . \quad (2.32)$$

This addition has no effect on tilt-angle correlations, since the periodic term disappears when integrated over $\theta(\vec{r})$. In the language of renormalization theory, h is a "redundant" coupling. Consider, though,

$$C_6(\vec{r}) - \langle\exp\{6i[\theta(\vec{r}) - \theta(\vec{0})]\}\rangle_{\bar{H}}\ , \qquad (2.33)$$

where $\langle\ \rangle_{\bar{H}}$ means an average evaluated in the ensemble specified by $\bar{H}$. One finds immediately after a simple change of variables that

$$C_6(\vec{r}) - A^2(h)\,\langle\exp\{6i[\phi(\vec{r}) - \phi(\vec{0})]\}\rangle_{\bar{H}_0}\ , \quad (2.34)$$

where

$$A(h) - \frac{\int_{-\pi/6}^{\pi/6} d\theta\,\exp(6i\theta)\exp(h\cos6\theta)}{\int_{-\pi/6}^{\pi/6} d\theta\,\exp(h\cos6\theta)} \qquad (2.35)$$

Below F, where vortices can be neglected, one finds immediately from Eqs. (2.34) and (2.31) that

$$C_6(\vec{r}) \simeq A^2(h)\,r^{-36\eta_1}\ , \qquad (2.36)$$

where η_1 is defined by Eq. (1.4). Similar manipulations suffice to show that

$$C_\times(\vec{r}) - \langle\exp\{6i[\theta(r) - \phi(0)]\}\rangle \simeq A(h)\,r^{-36\eta_1} \qquad (2.37)$$

so one has $\eta_6 - \eta_\times - 36\eta_1$, as claimed in the Introduction. This property of induced order in bond angle correlations should be a quite general feature of smetic-C liquid crystals in two and three dimensions.

In principle, the Hamiltonian (2.30) appropriate to the hexatic phase should also be replaced with some-

thing analogous to Eq. (2.32). There are *no* induced tilt-angle correlations, however, since one now finds a result like Eq. (2.34) for C_1, with Eq. (2.35) replaced by a quantity which vanishes identically,

$$A'(h) - \frac{\int_{-\pi/6}^{\pi/6} d\phi\,\exp(i\phi)\exp(h\cos6\phi)}{\int_{-\pi/6}^{\pi/6} d\phi\,\exp(h\cos6\phi)} - 0\ . \quad (2.38)$$

This result is consistent with the exponential decay of $C_1(\vec{r})$ quoted in the Introduction for the hexatic phase.

Finally, we discuss the unstable flows just below CB in Fig. 3, which are triggered by an unbinding of the periodic excitations in Eq. (2.17). As discussed in Sec. II A, the line IH at $K_1 - \infty$ is like the xy model in a magnetic field, with fixed points at $K_6 - 0$ and $K_6 - \infty$. Indeed, the flows immediately below CB may ultimately tend toward the fixed line FH, and the phase below this line could just be part of the smectic-C phase discussed above.

To study this point further, consider the Hamiltonian (2.1) in the absence of disclinations and vortices. This should be a good approximation near CB. Equation (2.1) then simplifies upon defining fields $\theta_\pm(\vec{r})$,

$$\theta_+(\vec{r}) - \alpha\theta(\vec{r}) + \beta\phi(\vec{r})\ , \qquad (2.39a)$$

$$\theta_-(\vec{r}) - \theta(\vec{r}) - \phi(\vec{r})\ , \qquad (2.39b)$$

where

$$\alpha - (1-\beta) - \frac{K_6 + g}{K_6 + K_1 + 2g}\ . \qquad (2.40)$$

In terms of the $\theta_\pm(\vec{r})$, the Hamiltonian breaks into a sum of a Gaussian in $\vec{\nabla}\theta_+$ and "sine-Gordon" Hamiltonian in θ_-,

$$\bar{H} - \tfrac{1}{2}\int d^2r\,(K_+|\vec{\nabla}\theta_+|^2 + K_-|\vec{\nabla}\theta_-|^2)$$
$$+ h\int d^2r \cos(p\theta_-)\ , \qquad (2.41)$$

with

$$K_+ - K_6 + K_1 + 2g\ , \qquad (2.42a)$$

$$K_- - (K_1 K_6 - g^2)/K_+ - 9/\pi^2 K_h\ . \qquad (2.42b)$$

We can now determine bond- and tilt-angle correlations using the known properties of the "sine-Gordon" problem.[20,26,27] In particular, one expects long-range order in $\exp[iq\theta_-(\vec{r})]$, where q is arbitrary, provided K_- is large enough so that we are below the line CB in Fig. 3,

$$\lim_{r\to\infty}\langle\exp[iq\theta_-(\vec{r})]\exp[-iq\theta_-(\vec{0})]\rangle - \mathrm{const} \neq 0\ .$$
$$(2.43)$$

Correlations in $\exp[iq\theta_+(\vec{r})]$, on the other hand, are always algebraic in these regions. Since Eqs. (2.39) can be inverted to read

$$\phi(\vec{r}) = \theta_+(\vec{r}) - \alpha\theta_-(\vec{r}) \quad , \tag{2.44a}$$

$$\theta(\vec{r}) = \theta_+(\vec{r}) + \beta\theta_-(\vec{r}) \quad , \tag{2.44b}$$

it is straightforward to determine $C_6(\vec{r})$, $C_1(\vec{r})$, and $C_x(\vec{r})$ at large distances using the properties quoted above. Below the line CB all these correlations decay as power laws, with

$$\eta_6 - \eta_x - 36\eta_1 - 18/\pi K_+ \quad . \tag{2.45}$$

Since this is also the behavior we associated with the fixed line FH, this phase may be indentical with the C phase described above. On the other hand, we cannot rule out a first-order transition, separating C_1 and C_2 phases with the same symmetry.

C. Weak-coupling recursion relations

We now study the neighborhood of the special points A and B in Fig. 3, using renormalization equations constructed as expansions in the fugacities y_6, y_1, and y_h. Since only two of these fugacities become relevant near A and B (for $p > 4$), we shall supress one "irrelevant" set of excitations. One is left with two coupled scalar Coulomb gases. (Away from A and B, and in particular near the lines AD, AB, and BC, at most one set of excitations is important, and one can then take over the results of Kosterlitz[24] for a single scalar Coulomb gas.)

Consider first recursion relations valid near the point A of Fig. 3. The coupling y_h may be set equal to zero in this region, and the Hamiltonian (2.17) simplifies to

$$\bar{H} = \bar{H}_c\left[\frac{K_6}{p^2}, y_6, \{m\}\right] + \bar{H}_c(K_1, y_1, \{n\})$$

$$+ \frac{2\pi g}{p} \sum_{\vec{r} \neq \vec{r}'} m(\vec{r})n(\vec{r}') \ln\left[\frac{|\vec{r} - \vec{r}'|}{a}\right] \quad . \tag{2.46a}$$

The corresponding partition sum is

$$Z = (\text{const}) \sum_{\{m(\vec{r})\}}' \sum_{\{n(\vec{r})\}}' e^{\bar{H}} \quad . \tag{2.46b}$$

The couplings K_6, y_6, K_1, y_1, and g are altered slightly from their values in Eq. (2.17) by the suppressed $\{s(\vec{r})\}$ excitations. In particular, the $\{s(\vec{r})\}$ generate a nonzero value of g even if this coupling is initially zero.

Recursion formulas appropriate to Eq. (2.46a) are constructed in Appendix A, using the method of Kosterlitz[24] and of Anderson and Yuval.[28] Upon scaling up the core diameters of disclinations and vortices from a to ae^l in Eq. (2.46), we find an effective

Hamiltonian like Eq. (2.46a), but with renormalized couplings $K_6(l)$, $y_6(l)$, $K_1(l)$, $y_1(l)$, and $g(l)$. These couplings are solutions of differential recursion relations, which, to leading order in $y_6(l)$ and $y_1(l)$, read

$$\frac{dK_6}{dl} = \frac{-4\pi^3}{p^2} K_6^2 y_6^2 - 4\pi^3 g^2 y_1^2 \quad , \tag{2.47a}$$

$$\frac{dy_6}{dl} = \left[2 - \frac{\pi K_6}{p^2}\right]y_6 \quad , \tag{2.47b}$$

$$\frac{dK_1}{dl} = -4\pi^3 K_1^2 y_1^2 - \frac{4\pi^3}{p^2} g^2 y_6^2 \quad , \tag{2.47c}$$

$$\frac{dy_1}{dl} = (2 - \pi K_1)y_1 \quad , \tag{2.47d}$$

$$\frac{dg}{dl} = 0 \quad . \tag{2.47e}$$

Note that, with y_h set to zero, $g(l)$ remains fixed at its initial value to this order.

To study the system (2.47) near $K_A \simeq 2p^2/\pi$ and $K_1 \simeq 2/\pi$, it is helpful to study deviations $x_a(l)$ and $x_b(l)$ defined by

$$K_6^{-1}(l) = \frac{\pi}{2p^2}[1 + x_a(l)] \quad , \tag{2.48a}$$

$$K_1^{-1}(l) = \frac{1}{2}\pi[1 + x_b(l)] \quad , \tag{2.48b}$$

as well as rescaled fugacities

$$y_a^2(l) = 8\pi^2 y_6^2(l) \quad , \tag{2.48c}$$

$$y_b^2(l) = 8\pi^2 y_1^2(l) \quad .$$

To lowest order in these variables, the recursions relations (2.47a)–(2.47d) become

$$\frac{dx_a}{dl} = y_a^2 + \gamma y_b^2 \quad , \tag{2.49a}$$

$$\frac{dy_a}{dl} = 2x_a y_a \quad , \tag{2.49b}$$

$$\frac{dx_b}{dl} = y_b^2 + \gamma y_a^2 \quad , \tag{2.49c}$$

$$\frac{dy_b}{dl} = 2x_b y_b \quad , \tag{2.49d}$$

with

$$\gamma = \pi^2 g^2/2p^2 \quad . \tag{2.49e}$$

Although the flows generated by the system (2.49) are complicated, it is easily checked that the quantity

$$C = 2x_a^2(l) + 2x_b^2(l) - 4\gamma x_a(l)x_b(l)$$
$$- (1 - \gamma^2)[y_a^2(l) + y_b^2(l)] \tag{2.50}$$

is invariant along the trajectories,

$$\frac{dC}{dl} = 0 \quad . \tag{2.51}$$

Since C is entirely determined by Eq. (2.50) evaluated at $l = 0$, it is an analytic function of the initial conditions.

The detailed phase diagram near A in Fig. 3 suggested by these recursion relations is shown in Fig. 4. Both $y_a(l)$ and $y_b(l)$ ultimately tend to zero in the shaded unlocked-tilted-hexatic phase. A line AD of vortex unbindings is determined by a locus of initial conditions such that $y_a(l) \rightarrow 0$, $y_b(l) \rightarrow 0$, and $x_b(l) \rightarrow 0$ as l tends to infinity. In this case, we have from Eq. (2.50) the requirement

$$C = 2x_a^2(\infty) \ , \tag{2.52a}$$

where $x_a(\infty) = \lim_{l \to \infty} x_a(l)$. Similarly, the requirement

$$C = 2x_b^2(\infty) \tag{2.52b}$$

determines a line of disclination unbindings AB. These two lines intersect at A, where

$$x_a(\infty) = x_b(\infty) = 0 \ . \tag{2.52c}$$

According to the recursion formulas (2.47a) and (2.47c), both $K_6(l)$ and $K_1(l)$ are destabilized and pushed toward smaller values when $y_6(l)$ starts to grow. This happens whenever $K_6 \lesssim 2p^2/\pi$. Since this double instability also occurs when $y_1(l)$ becomes unstable, we expect a portion A_1AA_2 of the unlocked-tilted-hexatic phase boundary when disclinations and vortices unbind simultaneously. In the limit $g = 0$, this region shrinks to a point.

The hexatic-liquid phase boundary A_1E can be determined as follows. Assume initial conditions in the hexatic phase just above A_1D, so that vortices have unbound. The recursion relations can then be integrated until one is far enough from A_1D, to treat the unbound vortices using a "Debye-Hückel" approximation[29]: the vortex degrees of freedom must be sufficiently excited to allow one to integrate rather than sum over the integers $\{n\}$ in Eq. (2.46b).

Let us assume that this condition is satisfied when $l = l^*$ such that $K_1(l^*)$ equals, say, π^{-1}, twice its crit-

ical value. In terms of the Fourier-transformed integer fields

$$\hat{m}(\vec{q}) = \sum_{\vec{r}} \exp(i\,\vec{q} \cdot \vec{r})m(\vec{r}) \ , \tag{2.53a}$$

$$\hat{n}(\vec{q}) = \sum_{\vec{r}} \exp(i\,\vec{q} \cdot \vec{r})m(\vec{r}) \ , \tag{2.53b}$$

the Hamiltonian (2.46a) becomes

$$\overline{H} = -\frac{1}{2} \int_q \left[\frac{4\pi^2 K_6}{q^2 p^2} + B_6 \right] \hat{m}(\vec{q})\hat{m}(-\vec{q})$$

$$-\frac{1}{2} \int_q \left[\frac{4\pi^2 K_1}{q^2} + B_1 \right] \hat{n}(\vec{q})\hat{n}(-\vec{q})$$

$$-\frac{1}{2} \int_q \frac{4\pi^2 g}{q^2 p} [\hat{m}(\vec{q})\hat{n}(-\vec{q}) + \hat{m}(-\vec{q})\hat{n}(-\vec{q})] \ , \tag{2.54}$$

where B_6 and B_1 are constants depending on the fugacities y_6 and y_1, and on the cutoff. Integrating freely over the vortex field $\hat{n}(\vec{q})$ when $l = l^*$, we find an effective Hamiltonian for disclinations of the form

$$\overline{H}_{\rm eff} = -\frac{1}{2} \int_q \left[\frac{4\pi^2 K_6^{\rm eff}}{q^2 p^2} + B_6^{\rm eff} + O(q^2) \right] \hat{m}(\vec{q})\hat{m}(-\vec{q}) \ , \tag{2.55a}$$

with

$$K_6^{\rm eff} = K_6(l^*) - g^2/K_1(l^*) \tag{2.55b}$$

and a similar expression for $B_6^{\rm eff}$. Rewriting Eq. (2.55) in terms of the $\{m(\vec{r})\}$, we find a Coulomb gas like Eq. (2.18) with $K = K_6^{\rm eff}/p^2$ and an effective fugacity $y_6^{\rm eff}$. The disclinations will unbind when $K_6^{\rm eff} \simeq 2/\pi p^2$, a requirement which determines the line A_1E.

Evidently, the effect of the cross coupling g is to *depress* the disclination unbinding temperature. Physically, we can imagine oppositely charged "red" and "blue" vortices coexisting with bound disclination pairs. Because of the logarithmic interaction proportional to g in Eq. (2.46a), plus disclinations will be screened by a cloud of, say, red vortices, and minus disclinations screened by a cloud of blue ones. This reduces the strength of the logarithmic interaction between disclination pairs, and lowers T_c.

A very similar treatment applies to the unstable flows to the right of the line AB in Fig. 4. Integrating now until $K_6(l^*)$ equals, say, twice its critical value, and treating disclinations in the Debye-Hückel approximation, one obtains an effective Coulomb gas for vortex excitations with

$$K_1^{\rm eff} = K_1(l^*) - g^2/p^2 K_6(l^*) \ . \tag{2.56}$$

Vortices will unbind when $K_6^{\rm eff} \simeq 2/\pi$ (this deter-

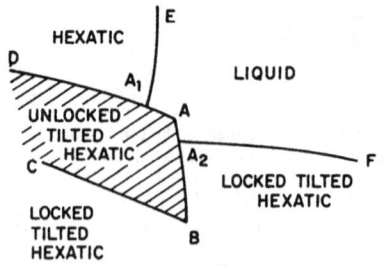

FIG. 4. More detailed version of the region surrounding the point A in Fig. 3. Four possible fluid phases are indicated.

mines the line A_2F), and the transition temperatures are again reduced due to vortex-disclination interactions.

It is interesting to consider the behavior of the susceptibilities

$$\chi_6 = \int d^2r \, C_6(\vec{r}) \ , \tag{2.57a}$$

$$\chi_1 = \int d^2r \, C_1(\vec{r}) \ , \tag{2.57b}$$

when approaching the different transition lines in Fig. 4. Results can be extracted in a straightforward manner from our recursion relations using the method of Kosterlitz.[24] Although χ_6 may be difficult to measure, χ_1 can be probed directly in light-scattering experiments. When the lines AD or AA_2 are approached in the hexatic or liquid phases, one finds

$$\chi_1 \simeq \xi_+^{2-\eta_1^*} \ , \tag{2.58}$$

where η_1^* is the exponent defined by Eq. (1.4) at the point being crossed. The correlation length $\xi_+(T)$ has the temperature dependence found by Kosterlitz,[24]

$$\xi_+(T) \simeq \exp(b/t^{1/2}) \ , \tag{2.59}$$

where t is the deviation from the critical temperature, $t = (T - T_c)/T_c$. The susceptibility χ_6 is infinite in the hexatic phase, but upon crossing AA_1 or AB from the liquid or unlocked-tilted-hexatic phases, one has for the singular part of χ_6,

$$\chi_6 \simeq \xi_+^{2-\eta_6^*} \ , \tag{2.60}$$

where η_6^* is the exponent of the point being crossed. Even though C_6 decays algebraically to zero in the C phase, χ_6 is finite then, provided $\eta_6 = 36\eta_1 \geq 2$. Both susceptibilities are infinite in the unlocked-tilted-hexatic phase. The quantity χ_6 diverges like $\xi_+^{7/4}$ upon crossing A_1E from the liquid side, while χ_1 remains finite. Similarly, χ_1 behaves like $\xi_+^{7/4}$ upon crossing A_2F, while χ_6 remains finite. As usual,[24] there are only experimentally undetectable essential singularities in the specific heat on all the above transition lines.

Finally, we consider briefly recursion relations near the point B in Fig. 3. Here, we can neglect vortices, and Eq. (2.17) can be replaced by

$$\overline{H} = \overline{H}_c(K_h, y_h, \{s\}) + \overline{H}_c\left(\frac{K_6}{p^2}, y_6, \{m\}\right)$$

$$+ i \sum_{\vec{r} \neq \vec{r}'} s(\vec{r}) m(\vec{r}') \tan^{-1}\left(\frac{y-y'}{x-x'}\right) \ , \tag{2.61}$$

with slightly altered parameters K_h, y_h, K_6, and y_6. According to Appendix A, the recursion relations for this problem are

$$\frac{dK_6}{dl} = -\frac{4\pi^3}{p^2} K_6^2 y_6^2 + p^2 y_h^2 \ , \tag{2.62a}$$

$$\frac{dy_6}{dl} = 2 - \left(\frac{\pi K_6}{p^2}\right) y_6 \ , \tag{2.62b}$$

$$\frac{dK_h}{dl} = -4\pi^3 K_h^2 y_h^2 + \pi y_6^2 \ , \tag{2.62c}$$

$$\frac{dy_h}{dl} = (2 - \pi K_h) y_h \ . \tag{2.62d}$$

These formulas may be analyzed near B just as was done for the recursion near A. One recovers the transition lines BC and BA, and finds that χ_6 diverges as in Eq. (2.60). The quantity χ_1 is infinite in both tilted hexatic phases.

The Debye-Hückel approximation is more difficult to carry out here, because of subtleties associated with the invariance of the inverse tangent part of $e^{\overline{H}}$ to transformations like $s(\vec{r}) \rightarrow s(\vec{r}) + 2\pi$ and $m(\vec{r}) \rightarrow m(\vec{r}) + 2\pi$.

D. Strong coupling

There is one remaining limit in which the Hamiltonian (2.1) simplifies: the limit of infinite h. To study this situation, we rewrite Eq. (2.1) in a form which makes manifest its periodicity as a function of θ and ϕ,

$$\overline{H} = \frac{K_6}{p^2} \sum_{\langle \vec{r}, \vec{r}' \rangle} \cos[p\theta(\vec{r}) - p\theta(\vec{r}')] + K_1 \sum_{\langle \vec{r}, \vec{r}' \rangle} \cos[\phi(\vec{r}) - \phi(\vec{r}')]$$

$$+ \frac{g}{p} \sum_{\langle \vec{r}, \vec{r}' \rangle} \sin[p\theta(\vec{r}) - p\theta(\vec{r}')] \sin[\phi(\vec{r}) - \phi(\vec{r}')] + h \sum_{\vec{r}} \cos[p\theta(\vec{r}) - p\phi(\vec{r})] \ , \tag{2.63}$$

where $\sum_{\langle \vec{r}, \vec{r}' \rangle}$ denotes a sum over nearest-neighbor sites on, say, a square lattice. This expression reduces to Eq. (2.1) in the continuum limit, and displays particularly simple periodic interactions between adjacent lattice sites.

When h is infinite the functional integrals in Eq. (2.?) become restricted to complexions such that

$$\theta(\bar{r}) = \phi(\bar{r}) + 2\pi s(\bar{r})/p \quad , \tag{2.64}$$

where $s(\bar{r})$ is an integer which can vary from site to site. The partition sum is then

$$Z = (\text{const}) \int D\phi \, e^{\bar{H}_\infty} \quad , \tag{2.65a}$$

where

$$\bar{H}_\infty = \sum_{\langle \tau, \tau' \rangle} \left\{ \frac{K_6}{p^2} \cos[p\phi(\bar{r}) - p\phi(\bar{r}')] + K_1 \cos[\phi(\bar{r}) - \phi(\bar{r}')] + \frac{g}{p} \sin[p\phi(\bar{r}) - p\phi(\bar{r}')] \sin[\phi(\bar{r}) - \phi(\bar{r}')] \right\} \tag{2.65b}$$

and the dependence on the $\{s(\bar{r})\}$ has dropped out. This Hamiltonian can be thought of as representing an xy model of magnetism, with a particularly complicated periodic interaction between nearest neighbors.

Although we have not studied Eq. (2.65b) in detail (this could be done using the Migdal approximate recursion formula,[30,31] for example), a qualitatively correct phase diagram follows from considering different limiting cases. For sufficiently small K_6 and K_1, both $C_6(\bar{r})$ and $C_1(\bar{r})$ should decay exponentially in an isotropic fluid phase. In the limit $K_1 \to 0$, g must also tend to zero if, for simplicity, we impose the restriction (2.3). Equation (2.65b) then exhibits an unbinding of $\frac{1}{6}$ - integer vortices in the θ field, by virtue of Eq. (2.64). For small K_6^{-1}, we expect a hexatic phase, with power-law decay of $C_6(\bar{r})$, and exponential decay of $C_1(\bar{r})$. If $K_6 = 0$, however, integer vortices unbind for large enough K_1^{-1}. For K_1^{-1} small, it is easily shown that correlations exhibit the locked algebraic decay (with $\eta_6 = \eta_x = 36\eta_1$) we identified earlier with a locked-tilted-hexatic (smectic-C) phase.

Finally, consider the limit $K_6 \to \infty$. One then has

$$\phi(\bar{r}) = \phi_0 + \frac{2\pi}{p} t(\bar{r}) \quad , \tag{2.66}$$

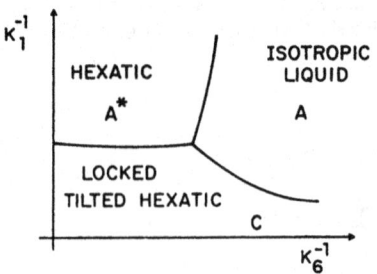

FIG. 5. Possible fluid phase diagram in the strong-coupling ($h = \infty$) limit. The unlocked tilted phase, which is not shown, may not exist in this limit.

where $t(r)$ is an integer field, and the partition sum becomes

$$Z = (\text{const}) \sum_{\{t(\bar{r})\}} \exp \left[K_1 \sum_{\langle \tau, \tau' \rangle} \cos \left(\frac{2\pi t(\bar{r})}{p} - \frac{2\pi t(\bar{r}')}{p} \right) \right] \tag{2.67}$$

This is the p-state clock model considered by José et al.[20] The Migdal recursion formula and perturbative renormalization-group analysis[20] suggest that these models have two phase transitions for p greater than a critical value p_c. Although $p_c = 4$ in a weak-coupling approximation,[20] this critical value is not known precisely for the Hamiltonian (2.67). For $p = 6$, there could be a single phase transition from a phase with exponential decay of $C_1(\bar{r})$ to a phase with long-range order $\langle \exp[i\theta(\bar{r})] \rangle \neq 0$, and locked long-range order in $\exp[6i\theta(\bar{r})]$. For any finite K_6, long-range order in $\exp[i\phi(\bar{r})]$ and $\exp[6i\theta(\bar{r})]$ is impossible.

A tentative phase diagram consistent with these observations is shown in Fig. 5. Hexatic, liquid, and locked-tilted-hexatic phases are shown. Although the unlocked-tilted-hexatic phase is absent, we cannot be positive that this is indeed the case in the strong-coupling limit.

III. SOLID PHASES

The starting Hamiltonian for investigation of the solid phases is Eq. (1.16) above. We shall assume that the elastic constants are sufficiently large so that dislocations are bound, and shall investigate the possibilities for the tilt orientation parameter, $\Phi(\bar{r}) = (\sin\gamma)e^{i\phi(\bar{r})}$. For a rigid lattice, $u_{ij} = 0$, and we need only consider the coupling h of ϕ to the hexagonal asymmetry, plus the effects of vortices in the ϕ field, which have not been written explicitly in Eq. (1.16). As discussed in Ref. 20, three phases are possible for this system, at least if the bare value of h is not too large. At intermediate temperatures, an xy-like phase occurs, with quasi-long-range order in the correlation function $C_1(\vec{r})$. For the phase to be stable, the exponent η_1 must lie in the range

$$\frac{1}{9} < \eta_1 < \frac{1}{4} \ . \tag{3.1}$$

At high temperatures, where η_1 exceeds $\frac{1}{4}$, the system is unstable to vortex formation, and a paramagnetic phase results, with exponential decay of $C_1(\vec{r})$. For low temperatures, where η_1 is less that $\frac{1}{9}$, the coupling h becomes a "relevant perturbation", and the order parameter is locked to one of the six easy axes, $\langle e^{i\phi} \rangle \neq 0$. The three phases of the tilted molecules on a rigid lattice are indicated in Fig. 6.

We now show that for a nonrigid lattice this intermediate phase is destabilized by the "magnetoelastic" term in Eq. (1.16). To see this, consider the renormalized elastic constants determined from the Hamiltonian appropriate to the intermediate phase,

$$\overline{H}_I = \frac{1}{2}\int d^2r(2\mu u_{ij}^2 + \lambda\, u_{kk}^2) + w\int d^2r\,(u_{ij} - \tfrac{1}{2}\delta_{ij}u_{kk})s_i s_j$$
$$+ \frac{1}{2}K_1\int d^2r(\vec{\nabla}\phi)^2 \ . \tag{3.2}$$

The inverse tensor of renormalized elastic constant may be written

$$(C_R^{-1})_{ijkl} = (\langle U_{ij}U_{kl}\rangle - \langle U_{ij}\rangle\langle U_{kl}\rangle)/A \ , \tag{3.3}$$

where

$$U_{ij} = \int d^2r\, u_{ij}'(\vec{r}) \tag{3.4}$$

and A is the area of the system. Terms linear in u_{ij} in $\overline{H}_I$ may be eliminated upon defining a new strain field,

$$\tilde{u}_{ij} = u_{ij} + \frac{w}{2\mu}(s_i s_j - \tfrac{1}{2}\delta_{ij}) \tag{3.5}$$

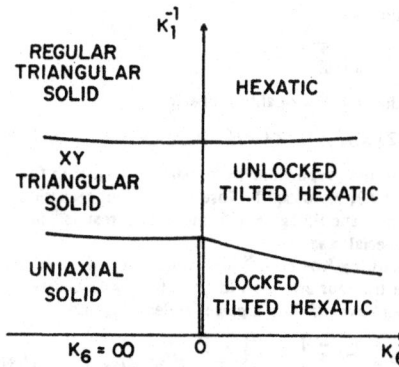

FIG. 6. Solid and fluid phases when the coupling w of the orientational order parameter to the elastic degrees of freedom is negligible. The xy triangular solid phase is unstable when this coupling in nonzero.

in terms of which $\overline{H}_I$ becomes

$$\overline{H}_I = (\text{const}) + \frac{1}{2}\int d^2r\,(2\mu\tilde{u}_{ij}^2 + \lambda\tilde{u}_{kl}^2)$$
$$+ \frac{1}{2}K_1\int d^2r\,(\vec{\nabla}\phi)^2 \ . \tag{3.6}$$

The elastic tensor is then

$$(C_R^{-1})_{ijkl} = \frac{\langle \tilde{U}_{ij}\tilde{U}_{kl}\rangle}{A} + \frac{w^2}{4\mu^2 A}(\langle S_{ij}S_{kl}\rangle - \langle S_{ij}\rangle\langle S_{kl}\rangle) \ , \tag{3.7}$$

where

$$\tilde{U}_{ij} = \int d^2r\,\tilde{u}_{ij}(\vec{r}) \ , \quad S_{ij} = \int d^2r\, s_i(\vec{r})s_j(\vec{r}) \ . \tag{3.8}$$

Substituting Eq. (3.8) into Eq. (3.7), it is now straightforward to evaluate the averages in Eq. (3.13) and find

$$(C_R^{-1})_{ijkl} = \frac{1}{4\mu}(\delta_{ik}\delta_{jl} + \delta_{il}\delta_{jk}) - \frac{\lambda}{4\mu(\mu+\lambda)}\delta_{ij}\delta_{kl}$$
$$+ \frac{w^2}{4\mu^2}I(K_1)(\delta_{ik}\delta_{jl} + \delta_{il}\delta_{jk} - \delta_{ij}\delta_{kl}) \ , \tag{3.9}$$

where

$$I(K_1) = \int d^2r\,\langle\exp[2i\phi(\vec{r}) - 2i\phi(\vec{0})]\rangle \ . \tag{3.10}$$

Although the elastic part of the bulk modulus is unaffected,

$$(\mu_R + \lambda_R)^{-1} = (C_R^{-1})_{iijj} = (\mu+\lambda)^{-1} \ , \tag{3.11}$$

the magnetoelastic coupling does alter the shear modulus,

$$\mu_R^{-1} = (C_R^{-1})_{ijij} - \frac{1}{2}(C_R^{-1})_{iijj} = \frac{1}{\mu}[1 + w^2 I(K_1)] \ . \tag{3.12}$$

Since one readily finds that

$$I(K_1) \simeq 2\pi\int_a^\infty dr\, r^{1-\eta_2} \ , \tag{3.13}$$

where

$$\eta_2 = 2/\pi K_1^R = 4\eta_1 < 1 \ , \tag{3.14}$$

the renormalization of μ_R^{-1} is infinite.

Thus, μ_R vanishes, and the intermediate xy-like phase in Fig. 6 is unstable to shear distortions. In the language of the renormalization group the eigenvalue of w is

$$\lambda_w = 1 - \frac{1}{2}\eta_2 \ , \tag{3.15}$$

which is relevant (positive) in the intermediate phase.

The shear instability of the xy-like phase, leads, presumably, to a distorted lattice, in which there is a nonzero value of $\langle e^{i\phi}\rangle$, as well as a uniform shear strain

$$\langle u_{ij}\rangle \approx \frac{-w}{2\mu}(\langle s_i\rangle\langle s_j\rangle - \frac{1}{2}\delta_{ij}|\langle\vec{s}\rangle|^2) \ . \tag{3.16}$$

Of course, the same phase would have resulted if we had added the coupling to lattice distortions directly to the low-temperature ordered phase of Fig. 6. Thus, on the nonrigid lattice, the two lower phases of Fig. 6 will be replaced by a single uniaxial phase, labeled H in Fig. 2. Although this is the simplest possible situation, we cannot, of course, rule out a first-order phase transition between uniaxial phases with the same symmetry.

To investigate the stability of the long-range tilt-angle order in the H phase against fluctuations in ϕ, we return to the full solid-phase free energy (1.16), and expand u_{ij} about its minimum (3.16). We assume for convenience that $\theta_0 = 0$, that h is positive, and the tilt axis has aligned itself along the x axis, so that $\langle \phi \rangle = 0$. We work, for simplicity, at sufficiently low temperatures so that $|\langle \vec{s} \rangle|$ can be replaced by unity. Substituting Eq. (3.16) into Eq. (1.16), and expanding in ϕ, we find, to quadratic order in $\bar{u}_{ij}$ and ϕ,

$$\frac{H}{k_B T} = \frac{1}{2} \int d^2 r \left[2\mu \bar{u}_{ij}^2 + \lambda \bar{u}_{kk}^2 + \frac{1}{2} K_1 (\vec{\nabla} \phi)^2 + 36h\phi^2 + \frac{w^2}{\mu} \phi^2 + 2w \bar{u}_{xy} \phi \right] . \tag{3.17}$$

Upon Fourier transforming and using Eq. (1.17), this becomes

$$\frac{H}{k_B T} = \frac{1}{2} \int_q \left[(2\mu + \lambda) |q_i \hat{u}_i(\vec{q})|^2 + \mu(q^2 \delta_{ij} - q_i q_j) \hat{u}_i(\vec{q}) \hat{u}_j(-\vec{q}) \right.$$
$$\left. + \left[K_1 q^2 + 36h + \frac{w^2}{\mu} \right] \hat{\phi}(\vec{q}) \hat{\phi}(-\vec{q}) + 2iw[q_x u_y(\vec{q}) + q_y u_x(\vec{q})] \hat{\phi}(-\vec{q}) \right] . \tag{3.18}$$

It follows that

$$\langle |\phi(\vec{q})|^2 \rangle = \left(K_1 q^2 + 36h + \frac{4(\mu + \lambda)w^2}{\mu(2\mu + \lambda)} \frac{q_x^2 q_y^2}{q^4} \right)^{-1} . \tag{3.19}$$

Fluctuations in $\phi(r)$ may be calculated by taking the Fourier transform of Eq. (3.19)

$$\langle |\phi(\vec{r})|^2 \rangle = \int \frac{d^2 q}{(2\pi)^2} \langle |\phi(\vec{q})|^2 \rangle . \tag{3.20}$$

Note that the integral converges, even if $h = 0$. Thus, if the renormalized value of w^2/μ is of order unity or larger, fluctuations in $\phi(r)$ will be of order unity or smaller, and we expect that $\langle e^{i\phi} \rangle \neq 0$. The resulting phase is a uniaxial solid, labeled H in Fig. 2.

The nature of the transitions between the uniaxial solid phase H and the various adjacent phases in Fig. 2 (B, C, or C^*) has not been determined.

ACKNOWLEDGMENTS

The authors are grateful for discussions with R. J. Birgeneau, S. Hikami, J. D. Litster, P. Pershan, and R. Pindak. Research has been supported by the NSF through the Harvard Material Research Laboratory and through Grant No. DMR77-10210.

APPENDIX A: DERIVATION OF SCALING EQUATIONS

In this appendix, we outline a derivation of the renormalization-group or "scaling" recursion equations used in the text. Consider first recursion relations valid near the point B in Fig. 4. We shall discuss a generalization of the Hamiltonian (2.61), namely,

$$\bar{H} = \bar{H}_c(K_a, y_a, \{m\}) + \bar{H}_c(K_b, y_b, \{n\})$$
$$+ iq \sum_{\vec{r} \neq \vec{r}'} m(\vec{r}) n(\vec{r}) \tan^{-1} \left[\frac{y - y'}{x - x'} \right] , \tag{A1}$$

where q is an integer. With $q = 1$, $K_a = K_6/p^2$, $y_a = y_6$, $K_b = K_h$, and $y_b = y_h$, this reduces to Eq. (2.61). For $K_a = q^2/4\pi^2 K_b = K$, however, it reduces to a representation of an xy model with vortices and a $\cos(p\theta)$ crystal field obtained by Kadanoff.[25] The partition sum

$$Z = \sum_{\{m\}}' \sum_{\{s\}}' e^{\bar{H}} \tag{A2}$$

then has the self-duality property

$$Z(K, y_a, y_b) \propto Z(q^2/4\pi^2 K, y_b, y_a) \tag{A3}$$

(where the proportionality constant is a smooth function of K), discussed by Jose et al.[20] Our recursion formulas should agree with the known results[20] in this special case.

Following Kosterlitz[24] and Anderson and Yuval,[28] we restrict our attention to excitations with charges ± 1, and expand Z in a power series in y_a and y_b

$$Z = \sum_{M=0}^{\infty} \sum_{N=0}^{\infty} \left(\frac{1}{M!} \right)^2 \left(\frac{1}{N!} \right)^2 y_a^{2M} y_b^{2N} Z_{M,N} . \tag{A4}$$

It will be convenient in what follows to call the $\{m\}$ and $\{n\}$ excitations green and white "vortices," respectively. The quantity $Z_{M,N}$ is then the configurational partition sum for $2M$ green and $2N$ white vortices,

$$Z_{M,N} = \left(\prod_{i=1}^{2M} \int' d^2 r_i\right)\left(\prod_{j=1}^{2N} \int' d^2 R_j\right) e^{\bar{H}_{M,N}} , \quad (A5a)$$

with

$$\bar{H}_{M,N} = 2\pi K_a \sum_{i \neq j} G(\bar{r}_i - \bar{r}_j) m(\bar{r}_i) m(\bar{r}_j)$$

$$+ 2\pi K_b \sum_{i \neq j} G(\bar{R}_i - \bar{R}_j) n(\bar{R}_i) n(\bar{R}_j)$$

$$+ 2iq \sum_{i \neq j} \tilde{G}(\bar{r}_i - \bar{R}_j) m(\bar{r}_i) n(\bar{R}_j) , \quad (A5b)$$

and where the $\{\bar{r}\}$ denote $2M$ green vortices, and the $\{\bar{R}_j\}$ signify $2N$ white ones. The green and white vortices each obey charge neutrality. Each integration in Eq. (A5) is excluded from circles of radius a about all others vortices, and the functions $G(\bar{r})$ and $\tilde{G}(\bar{r})$ are harmonic conjugates,

$$G(\bar{r}) = \ln(|\bar{r}|/a) , \quad (A6)$$

$$\tilde{G}(\bar{r}) = \tan^{-1}\left|\frac{y-y'}{x-x'}\right| . \quad (A7)$$

A renormalization transformation can now be constructed by integrating over those configurations where two oppositely charged vortices of the same color approach each other with separations between a and ae^δ, with δ small. Breaking up in integrations in Eq. (A5) in this way, we can write, to leading order in δ

$$\left(\prod_{i=1}^{2M}\int' d^2 r_i\right)\left(\prod_{j=1}^{2N}\int' d^2 R_j\right) e^{\bar{H}_{M,N}} = \left(\prod_{i=1}^{2M}\int^> d^2 r_i\right)\left(\prod_{j=1}^{2N}\int^> d^2 R_j\right) e^{\bar{H}_{M,N}}$$

$$+ \sum_{k=1}^{M}\sum_{l=1}^{M}\left(\prod_{\substack{i=1\\i\neq k,l}}^{2M}\int^> d^2 r_i\right)\left(\prod_{j=1}^{2N}\int^> d^2 R_j\right)\int d^2 x \int^\delta d^2 y \, e^{\bar{H}_{M,N}}$$

$$+ \sum_{r=1}^{N}\sum_{t=1}^{N}\left(\prod_{i=1}^{2M}\int^> d^2 r_i\right)\left(\prod_{\substack{j=1\\j\neq t,s}}^{2N}\int^> d^2 R_j\right)\int d^2 X \int^\delta d^2 Y \, e^{\bar{H}_{M,N}} + O(\delta^2) , \quad (A8)$$

where the integrals $\int^>$ are excluded from circles of radii ae^δ around all other vortices, and

$$\bar{x} = \bar{r}_k - \bar{r}_l , \quad (A9)$$

$$\bar{y} = \tfrac{1}{2}(\bar{r}_k + \bar{r}_l) , \quad (A10)$$

$$\bar{X} = \bar{R}_s - \bar{R}_t , \quad (A11)$$

$$\bar{Y} = \tfrac{1}{2}(\bar{R}_s + \bar{R}_t) . \quad (A12)$$

The second term of Eq. (A8) sums over pairs of oppositely charged green vortices at $\bar{r}_k$ and $\bar{r}_l$, while the third term sums over white charge-neutral pairs at $\bar{R}_s$ and $\bar{R}_t$. These pairs must be within ae^δ of each other. The integrations over $\bar{y}$ and $\bar{Y}$ are restricted to annuli of width δ, while the $\bar{x}$ and $\bar{X}$ integrations are essentially unrestricted. Other possible pairings of vortices, neglected in Eq. (A8), generate ± 2 charged green and white vortices, and hybrid green-white ones. The corresponding fugacities turn out to be irrelevant variables in range of couplings we are interested in, so we can neglect these effects.

To implement the renormalization procedure, we want to carry out explicitly only the integrals over $\bar{x}$, $\bar{y}$, $\bar{X}$, and $\bar{Y}$, and leave the remaining integrations intact. Separating out the parts of $\bar{H}_{M,N}$ which depend on $\bar{r}_k$ and $\bar{r}_l$, or $\bar{R}_s$ and $\bar{R}_t$, we must consider

$$I_{kl} = \int d^2 \bar{x} \int^\delta d^2 \bar{y} \exp\left(2\pi K_a \sum_{\substack{i=1\\i\neq k,l}}^{M} m(\bar{r}_i)[G(\bar{r}_i - \bar{r}_k) - G(\bar{r}_i - \bar{r}_l)]\right)\exp\left(2iq \sum_{j=1}^{N} n(\bar{R}_j)[\tilde{G}(\bar{R}_j - \bar{r}_k) - \tilde{G}(\bar{R}_j - \bar{r}_l)]\right)$$

$$(A13)$$

and

$$I_{st} = \int d^2 \bar{X} \int^\delta d^2 \bar{Y} \exp\left(2\pi K_b \sum_{\substack{j=1\\j\neq s,t}}^{N} n(\bar{R}_j)[G(\bar{R}_j - \bar{R}_s) - G(\bar{R}_j - \bar{R}_t)]\right)\exp\left(2iq \sum_{i=1}^{M} m(\bar{r}_i)[\tilde{G}(\bar{r}_i - \bar{R}_s) - \tilde{G}(\bar{r}_i - \bar{R}_t)]\right)$$

$$(A14)$$

In writing Eqs. (A13) and (A14), we have set $m(\bar{r}_k)$ $=-m(\bar{r}_l)=+1$, and $n(\bar{R}_s)=-n(\bar{R}_t)=1$. These expressions simplify upon using Eqs. (A9) – (A12) to eliminate $\bar{r}_k$, $\bar{r}_l$, $\bar{R}_s$, and $\bar{R}_t$, and then expanding in y and Y

$$G(\bar{r}_j-\bar{r}_k)-G(\bar{r}_i-\bar{r}_l) \simeq (\bar{y}\cdot\vec{\nabla}_x)G(\bar{r}_i-\bar{x}) ,$$
(A15)

$$\hat{G}(\bar{R}_j-\bar{r}_k)-\hat{G}(\bar{R}_j-\bar{r}_l) \simeq (\bar{y}\cdot\vec{\nabla}_x)\hat{G}(\bar{R}_j-\bar{x}) ,$$
(A16)

$$G(\bar{R}_j-\bar{R}_s)-G(\bar{R}_j-\bar{R}_t) \simeq (\bar{Y}\cdot\vec{\nabla}_X)G(\bar{R}_j-\bar{X}) ,$$
(A17)

$$\hat{G}(\bar{r}_i-\bar{R}_s)-\hat{G}(\bar{r}_i-\bar{R}_t) \simeq (\bar{Y}\cdot\vec{\nabla}_X)\hat{G}(\bar{r}_i-\bar{X}) .$$
(A18)

Upon expanding the exponentials to second order in $\bar{y}$ and $\bar{Y}$, and averaging over orientations of these variable, we can eliminate $\hat{G}$ in favor of G by exploiting the identity

$$(\vec{\nabla}G)^2=(\vec{\nabla}\hat{G})^2 .$$
(A19)

An integration by parts then allows us to use the relation

$$\nabla^2 G(\bar{r})=2\pi\delta(\bar{r})$$
(A20)

to eliminate the integrations over $\bar{x}$ and $\bar{X}$. After manipulations of this kind, and taking the limit of small δ, we find results which are independent of k, l, s, and t, namely,

$$I_{kl}=2\pi\delta a\left[A-2\pi^3 K_a^2 a^2 \sum_{\bar{r}\neq\bar{r}'} m(\bar{r})m(\bar{r}')G(\bar{r}-\bar{r}')\right.$$
$$\left.+\tfrac{1}{2}\pi q^2 a^2 \sum_{\bar{r}\neq\bar{r}'} n(\bar{r})n(\bar{r}')G(\bar{r}-\bar{r}')\right] ,$$
(A21)

$$I_{st}=2\pi\delta a\left[A-2\pi^3 K_b^2 a^2 \sum_{\bar{r}\neq\bar{r}'} n(\bar{r})n(\bar{r}')G(\bar{r}-\bar{r}')\right.$$
$$\left.+\tfrac{1}{2}\pi q^2 a^2 \sum_{\bar{r}\neq\bar{r}'} m(\bar{r})m(\bar{r}')G(\bar{r}-\bar{r}')\right] ,$$
(A22)

where A is the area of the system. The summations are over pairs of the remaining $(M-2)$ green and $(N-2)$ white vortices.

Inserting Eqs. (A21) and (A22) into Eqs. (A8) and (A5b), we see that terms proportional to the area renormalize the charge-independent part of $\bar{H}$. Since there are now two fewer vortices, the remaining terms in Eqs. (A21) and (A22) give $O(y_a^2)$ and $O(y_b^2)$ renormalizations of the logarithmic interaction in $\bar{H}$. Proceeding exactly as in the Appendix of the paper by Kosterlitz,[24] we can rescale y_a and y_b to pro-

duce a Hamiltonian of the same form as Eq. (A1). Upon iterating this procedure until the effective core diameter is increased from a to ae^l, we find renormalized couplings $K_a(l)$, $y_a(l)$, $K_b(l)$, and $y_b(l)$ which are solutions of

$$\frac{dK_a}{dl}=-4\pi^3 K_a^2 y_a^2+\pi q^2 y_b^2 ,$$
(A23)

$$\frac{dy_a}{dl}=(2-\pi K_a)y_a ,$$
(A24)

$$\frac{dK_b}{dl}=-4\pi^3 K_b^2 y_b^2+\pi q^2 y_a^2 ,$$
(A25)

$$\frac{dy_b}{dl}=(2-\pi K_b)y_b .$$
(A26)

As a check on these results, we set $K_a=q^2/4\pi^2 K_b$ $=K$ initially, and find that this self-duality condition is preserved under our renormalization transformation

$$\frac{d}{dl}\left[K_a(l)-\frac{q^2}{4\pi^2 K_b(l)}\right]=0$$
(A27)

as it should be. Indeed, in this limit we recover the results of José et al.[20] Recursion relations for this special self-dual problem have also been derived using the Kosterlitz method by Elitzur et al.[32] Upon setting $q=1$, K_6/p^2, $y_b=y_6$, $K_b=K_h$, and $y_b=y_h$, we find the results (2.62) quoted in the text.

Recursion formulas for the Hamiltonian (2.46a) appropriate near A in Fig. 4 can be found in the same way provided we make the replacement

$$iq\hat{G}(\bar{r}-\bar{r}')\to\frac{2\pi g}{p}G(\bar{r}-\bar{r}')$$
(A28)

in Eq. (A1), and in subsequent equations. Since there is now a logarithmic interaction between green and white vortices, one expects that configurations of, say $(+)$ green and $(-)$ white pairs will be favored in addition to $(\pm)$ green and $(\pm)$ white pairings. These new complexions generate hybrid vortex-disclinations like the one shown in Fig. 7. It is easily checked

FIG. 7. Hybrid "vortex disclination" on a square lattice. Excitations of this kind are shown to be irrelevant in the text.

however, that the fugacity of such hybrids has the eigenvalue

$$\lambda_{\text{hybrid}} = 2 - \pi K_a - \pi K_b , \qquad (A29)$$

which is negative (irrelevant) for $K_a \geq 2/\pi$. Ignoring these irrelevant hybrids, and setting $q = -i\pi g/p$, $K_a = K_6/p^2$, $y_a = y_6$, $K_b = K_1$, and $y_b = y_1$, we recover the recursion relations (2.47) quoted in Sec. II C.

APPENDIX B: EVALUATION OF $C_x(\vec{r})$

We now evaluate the cross-correlation function $C_x(r)$, defined in Sec. II B, in the unlocked phase C^*. For this purpose we may ignore vortices and disclinations; however, we cannot set $h = 0$, even though h is formally an "irrelevant variable." As we shall see, $C_x(r)$ is itself of order h. To first order in h, we have

$$C_x(r) = \langle \exp[6i\theta(\vec{r})] \exp[-6i\phi(\vec{0})]$$
$$\times \exp(h \int d^2r' \cos[6\theta_-(\vec{r}')]) \rangle_{h=0}$$
$$\approx h \int d^2r' \langle \exp[6i\theta(\vec{r})] \exp[-6i\phi(\vec{0})]$$
$$\times \cos[6\theta(\vec{r}')] \rangle_{h=0} , \qquad (B1)$$

where the expectation values on the right-hand side of Eq. (B1) are taken with $h = 0$, as indicated. After expressing θ and ϕ in terms of the variables θ_+ and θ_-, defined in Sec. II B, we find

$$C_x(\vec{r}) = h \langle \exp\{6i[\theta_+(\vec{r}) - \theta_+(\vec{0})]\} \rangle_{h=0}$$
$$\times \int d^2r' \langle \exp\{6i[\alpha\theta_-(\vec{0}) + \beta\theta_-(\vec{r})$$
$$- \theta_-(\vec{r}')]\} \rangle_{h=0} . \qquad (B2)$$

The expectation values may be evaluated using Eq. (2.41) with $h = 0$. Making use of the relation $\alpha + \beta = 1$, which implies that

$$[\alpha\theta_-(\vec{0}) + \beta\theta_-(\vec{r}) - \theta_-(\vec{r}')]^2$$
$$= \alpha[\theta_-(\vec{r}') - \theta_-(\vec{0})]^2 + \beta[\theta_-(\vec{r}') - \theta_-(\vec{r})]^2$$
$$- \alpha\beta[\theta_-(\vec{r}) - \theta_-(\vec{0})]^2 \qquad (B3)$$

we find

$$C_x(r) = h \left(\frac{1}{r}\right)^{\eta_+ - \alpha\beta\eta_\Delta} \int d^2r' \left(\frac{1}{r'}\right)^{\alpha\eta_\Delta} \left(\frac{1}{|\vec{r} - \vec{r}'|}\right)^{\beta\eta_\Delta} , \qquad (B4)$$

where $\eta_+ = 18/\pi K_+$ and η_Δ is given by Eq. (1.11).

In the C^* phase, we necessarily have $\eta_\Delta > 4$, so that the integral (B4) is convergent at large distances. If the gradient-coupling constant g is not too large, then the inequalities (1.13) and (1.14) imply that $\beta\eta_\Delta < 2$ and hence $\alpha\eta_\Delta > 2$. Thus, the integral is formally divergent at $\vec{r}' = \vec{0}$, and the major contribution to the integral comes from the region where r' is close to the short-distance cutoff.

We therefore find

$$C_x(r) = (\text{const}) hr^{-\eta_x} , \qquad (B5)$$

where

$$\eta_x = \eta_+ + \beta(1 - \alpha)\eta_\Delta = \eta_6 . \qquad (B6)$$

The value of the constant in Eq. (B5) depends on the nature of the short-distance cutoff, and of course, this will be modified if we include the effects of dislocations and vortices. However, the dependence on r is correct, for sufficiently large r, provided we use the renormalized value of η_6.

In principle, if g is large, there can be a region of the C^* phase where $\beta > \alpha$, or equivalently a region where $\eta_6 > 36\eta_1$. In this case, the integral (B4) will be dominated by the region $|\vec{r}' - \vec{r}| \approx 0$, and the exponent η_x will be equal to $36\eta_1$. It should be noted, however, that the analysis used here to calculate the cross-correlation function $C_x(r)$ will also lead to a term in $C_6(r)$ proportional to $h^2(1/r)^{36\eta_1}$. This term will actually dominate at large distances if we have $\eta_6 > 36\eta_1$. In this case, the leading terms in the long-distance behavior of the correlation function would have the same form as in the locked tilted phase C, and the C^* phase would be distinguished only by the behavior of corrections to the leading power law.

In the limit $r \to 0$, the correlation function $C_x(r)$ becomes the expectation value $\langle \Delta(\vec{0}) \rangle$, where Δ is defined by Eq. (1.9). Clearly $\langle \Delta \rangle$ is nonzero, and is proportional to h for small h, in the C^* phase. The constant term in Eq. (1.10), for the large distance behavior of $C_\Delta(\vec{r})$ is of course equal to $|\langle \Delta \rangle|^2$. The second term in Eq. (1.10) can be obtained by the methods of Sec. II B, with $h = 0$ for convenience.

[1] J. M. Kosterlitz and D. J. Thouless, J. Phys. C **6**, 1181 (1973).

[2] See also, V. L. Berezinskii, Zh. Eksp. Teor. Fiz. **59**, 907 (1970) [Sov. Phys. JETP **32**, 493 (1971)].

[3] See, e.g., W. Shockley, in *L'Etat Solide*, Proceedings of the Neuvieme Couseil de Physique, edited by R. Stoops (Institut International de Physique Solvay, Brussels, 1952).

[4] D. R. Nelson, Phys. Rev. B **18**, 2318 (1978).

[5] B. I. Halperin and D. R. Nelson, Phys. Rev. Lett. **41**, 121 (1978); **41**, 519(E) (1978); D. R. Nelson and B. I Halperin, Phys. Rev. B **19**, 2457 (1979).

[6] A. P. Young, Phys. Rev. B **19**, 1855 (1979).

[7] D. Frenkel and J. P. McTague, Phys. Rev. Lett. **42**, 1632 (1979).

[8] R. M. J. Cotterill and L. B. Pedersen, Solid State Commun. **10**, 439 (1972).

[9] R. Morf, Phys. Rev. Lett. **43**, 931 (1979); see also, D. J. Thouless, J. Phys. C **11**, L189 (1978).

[10] C. C. Grimes and G. Adams, Phys. Rev. Lett. **42**, 795 (1979).

[11] C. Y. Young, R. Pindak, N. A. Clark, and R. B. Meyer, Phys. Rev. Lett. **40**, 773 (1978).

[12] C. Rosenblatt, R. Pindak, N. A. Clark, and R. B. Meyer, Phys. Rev. Lett. **42**, 1220 (1979).

[13] R. J. Birgeneau and J. D. Litster, J. Phys. Lett. (Paris) **39**, L399 (1978).

[14] D. E. Moncton and R. Pindak, Phys. Rev. Lett. **43**, 70 (1979).

[15] P. Pershan, G. Aeppli, R. Birgeneau, and J. Litster (private communication); see also, J. D. Litster, R. J. Birgeneau, M. Kaplan, C. R. Safinya, and J. Als-Nielsen, Lecture notes for NATO Advanced Study Institute, Geilo, Norway, 1979 (Plenum, New York, to be published).

[16] The quantities K_1, K_6, and g are ordinary stiffness constants divided by $k_B T$; thus, they are dimensionless. On a microscopic level, the constants K_1, K_6, and g are actually *tensors* with different eigenvalues referring to gradients parallel to and perpendicular to the local tilt orientation. We ignore this complication in the present paper. In any case, the tensors become isotropic in the limit of very long wavelengths. See Refs. 11, 12, and 21; and D. R. Nelson and R. Pelcovits, Phys. Rev. B **16**, 2191 (1977).

[17] It is intriguing to speculate that the C_1 and C_2 phases, if they exist as distinct phases in two dimensions, may have three-dimensional analogs in bulk smectic-C and smectic-F liquid crystals. See. Ref. 18. We are grateful to R. Pindak for bringing the smectic-F phase to our attention.

[18] For x-ray-diffraction studies of smectic-H, -F, and -C phases, see A. J. Leadbetter, J. P. Gaughan, B. Kelley, G. W. Gray, and J. Goodby, J. Phys. (Paris) **40**, C3-178 (1979). See also J. J. Benattar, J. Doucet, M. Lambert, and A. M. Levelut (unpublished).

[19] See Ref. 5 and N. D. Mermin, Phys. Rev. **176**, 250 (1968).

[20] J. José, L. P. Kadanoff, S. Kirkpatrick, and D. R. Nelson, Phys. Rev. B **16** 1217 (1977).

[21] D. R. Nelson and J. M. Kosterlitz, Phys. Rev. Lett. **39**, 1201 (1977).

[22] R. A. Pelcovits and B. I. Halperin, Phys. Rev. B **19**, 4614 (1979).

[23] See D. Stein, Phys. Rev. B **18**, 2397 (1978).

[24] J. M. Kosterlitz, J. Phys. C **7**, 1046 (1974).

[25] L. P. Kadanoff, J. Phys. C **11**, 1399 (1978).

[26] A. Luther and I. Peschel, Phys. Rev. B **9**, 2911 (1974).

[27] S. Coleman, Phys. Rev. D **11**, 2088 (1972).

[28] P. W. Anderson and G. Yuval, J. Phys. C **4**, 607 (1971).

[29] See, e.g., Appendix A of A. N. Berker and D. R. Nelson, Phys. Rev. B **19**, (1979), as well as Appendix B of the second part of Ref. 5.

[30] A. A. Migdal, Zh. Eksp. Teor. Fiz. **69**, 810 (1975) [Sov. Phys. JETP **42**, 413 (1976)]; Zh. Eksp. Teor. Fiz. **69**, 1457 (1975) [Sov. Phys. JETP **43**, 743 (1976)].

[31] L. P. Kadanoff, Ann. Phys. (N.Y.) **100**, 359 (1976).

[32] S. Elitzur, R. Pearson, and J. Shigemitzu, Phys. Rev. D **19**, 3698 (1979).

6. IN PLANE ORDER-EXPERIMENT

6.1. Levelut, A. M. (1976). Étude de l' ordre local lié a la rotation des molécules dans la phase smectique B. *J. Phys. (Paris)* **37**, C3-51-C3-54 275

6.2. Leadbetter, A. J., Frost, J. C. and Mazid, M. A. (1979). Interlayer correlations in smectic B phases. *J. Phys. (Paris) Lett.* **40**, L325–L329 279

6.3. Moncton, D. E. and Pindak, R. (1979). Long-range order in two- and three-dimensional smectic-B liquid-crystal films. *Phys. Rev. Lett.* **43**, 701–704 284

6.4. Aeppli, G., Litster, J. D., Birgeneau, R. J. and Pershan, P. S. (1981). High resolution x-ray study of the smectic A-smectic B phase transition and the smectic B phase in butyloxybenzylidene octylaniline. *Mol. Cryst. Liq. Cryst.* **67**, 205–214 288

6.5. Pindak, R., Moncton, D. E., Davey, S. C. and Goodby, J. W. (1981). X-ray observation of a stacked hexatic liquid-crystal B phase. *Phys. Rev. Lett.* **46**, 1135–1138 298

6.6. Collett, J., Sorensen, L. B., Pershan, P. S., Litster, J., Birgeneau, R. J. and Als-Nielsen, J. (1982). Synchrotron x-ray study of novel crystalline-B phases in heptyloxybenzylidene-heptylaniline (70.7). *Phys. Rev. Lett.* **49**, 553–556 302

6.7. Hirth, J. P., Pershan, P. S., Collett, J., Sirota, E. and Sorensen, L. B. (1984). Dislocation model for restacking phase transitions in crystalline-B liquid crystals. *Phys. Rev. Lett.* **53**, 473–476 306

6.8. Collett, J., Sorensen, L. B., Pershan, P. S. and Als-Nielsen, J. (1985). X-ray scattering study of restacking transitions in the crystalline-B phases of heptyloxybenzylidene heptylaniline 70.7. *Phys Rev.* **A32**, 1036–1043 310

6.9. Sirota, E. B., Pershan, P. S., Sorensen, L. B. and Collett, J. (1987). X-ray and optical studies of the thickness dependence of the phase diagram of liquid crystal films. *Phys. Rev.* **A36**, 2890–2901 318

JOURNAL DE PHYSIQUE

Colloque C3, *supplément au n⁰ 6, Tome* 37, *Juin* 1976, *page* C3-51

ÉTUDE DE L'ORDRE LOCAL LIÉ A LA ROTATION DES MOLÉCULES DANS LA PHASE SMECTIQUE B (*)

A. M. LEVELUT

Laboratoire de Physique des Solides (**) Université Paris-Sud, 91405 Orsay, France

Résumé. — Il a déjà été prouvé en particulier par des mesures de diffusion inélastique de neutrons que, dans la phase smectique B, les molécules tournent autour de leur grand axe. La répartition de l'intensité des rayons X diffusés dans le plan réciproque équatorial (perpendiculaire aux molécules) montre que cette rotation se fait par sauts orientationnels de $\pi/3$, les rotations de molécules voisines étant corrélées. On a ainsi des domaines de fluctuations dont la taille croît lorsque la température s'abaisse ; mais, si dans la phase smectique B_A toutes les positions occupées sont équivalentes, au contraire, dans la phase smectique B_C du TBBA, lorsqu'on abaisse la température, les molécules occupent de plus en plus souvent les positions qui sont celles de la phase smectique E_C de basse température.

Abstract. — From inelastic neutron scattering experiments, it has been proved that there is a rotational motion of molecules about their long axis in the smectic B phase. From the repartition of the X ray diffuse scattering intensity in the equatorial reciprocal plane (perpendicular to the molecules), it is shown that in fact orientational jumps of $\pi/3$ take place with correlations between neighbouring molecules. So there are fluctuations domains the size of which is increasing with decreasing temperature. In the smectic B_A phase all occupied positions are equivalent ; on the contrary in the smectic B_C phase of TBBA when the temperature is decreasing, the molecules occupied more often the positions corresponding to the smectic E_C lower temperature phase.

Introduction. — La phase smectique B est une mésophase dans laquelle les molécules sont disposées en couches, aux nœuds d'un réseau hexagonal plan à l'intérieur de chaque couche [1]. Ce réseau est rigoureusement hexagonal, lorsque le grand axe des molécules est perpendiculaire au plan des couches (Sm B_A) ; il est rectangle centré lorsque les molécules sont inclinées sur le plan smectique (Sm B_C), et, dans ce cas, les axes des molécules sont équidistants à 1 % près. L'existence d'une telle structure implique que les molécules peuvent tourner autour de leur grand axe. Ce mouvement de rotation a été observé par l'étude de la diffusion inélastique incohérente des neutrons [2]. En fait, les paramètres du réseau dans les couches sont tels que la rotation ne peut être libre ; les molécules voisines tournent ensemble comme le feraient les roues d'un engrenage, leur rotation pouvant se faire par sauts orientationnels.

L'examen de l'intensité diffusée des rayons X en dehors des taches de Bragg nous permet de décrire ce mécanisme de *rotation engrenée* ; cet effet est présent dans toutes les phases smectiques B (uniaxes et biaxes) que nous avons pu étudier sous forme de monodomaines.

1. Méthodes expérimentales. — Les monodomaines de la phase Sm B sont obtenus soit en chauffant un

(*) Cet article ne recouvre qu'une partie de la conférence *structure des phases smectiques*, le reste ayant déjà été publié dans les articles cités en références.
(**) Laboratoire associé au C. N. R. S.

monocristal de la substance étudiée, soit en refroidissant l'échantillon lentement depuis la phase nématique, si celle-ci suit la phase smectique B. Les diagrammes de rayons X ont été obtenus en éclairant l'échantillon fixe par la radiation monochromatique Cu Kα, le rayonnement diffusé est recueilli sur une plaque photographique plane. C'est l'intensité diffusée et diffractée dans le plan réciproque (h, k, o), perpendiculaire aux molécules, qui traduit l'ordre des molécules d'une même couche. Etant donné l'extension non négligeable de part et d'autre de ce plan, il est possible d'obtenir simultanément l'image d'une portion importante de ce plan avec un faisceau de rayons X peu incliné par rapport à la direction d'allongement des molécules.

Deux substances ont été étudiées en détail :

1.1. — Le Terephtal bis butylaniline, TBBA

$$C_4H_9 - \langle O \rangle - N = CH - \langle O \rangle - CH =$$
$$= N - \langle O \rangle - C_4H_9$$

dont les phases mésomorphes se succèdent suivant le schéma ci-dessous :

$$Cr \xrightarrow{114°} Sm\ B_C \xleftrightarrow{139°} Sm\ C \quad etc.$$
$$VII \xleftrightarrow[68°]{84°} Sm\ E_C(VI) \qquad [3]$$

1.2. — Le Paraphényl benzylidène 4-n-butyl aniline, PBBA

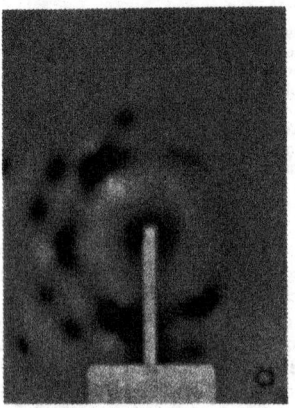

qui possède 2 phases mésomorphes :

$$Cr \overset{83°}{\longleftrightarrow} Sm\, B_A \overset{120°}{\longleftrightarrow} N .$$

Dans le premier cas, les molécules sont inclinées sur les couches tant dans la phase smectique **B** que dans la phase smectique E (Sm E_C). Elles sont perpendiculaires aux plans smectiques dans la phase smectique du second composé.

2. **Description de la répartition de l'intensité diffusée dans le plan [h, k, o].** — Pour toutes les substances Sm B que nous avons étudiées, la répartition de l'inten-

sité diffusée par un monodomaine [5] dans le plan h, k, o est analogue (Fig. 1) et nous l'avons représentée schématiquement sur la figure 2.

Nous avons déjà étudié [6] l'intensité diffusée localisée autour des taches de diffraction qui traduisent l'agitation thermique des molécules (écart autour de leur position moyenne). Il existe un second type de taches diffuses, au nombre de 12, repérées par (a) sur la figure 2. Si l'on représente le réseau hexagonal ou pseudo-hexagonal plan des couches smectiques par une maille multiple rectangle centrée (Fig. 3a). Ces douze taches ont pour coordonnées dans le réseau réciproque correspondant (Fig. 3b) : $\pm$ 2, $\pm$ 1, 0 ; $\pm \frac{1}{2}$, $\pm \frac{1}{2}$, 0 et $\pm \frac{1}{2}$, $\pm \frac{3}{2}$, 0. Remarquons que dans un réseau hexagonal vrai, les 8 dernières taches se déduisent des quatre premières par des rotations de $\pi/3$ autour de l'axe sénaire, mais la même symétrie pseudo sénaire apparaît dans le cas des smectiques B_C.

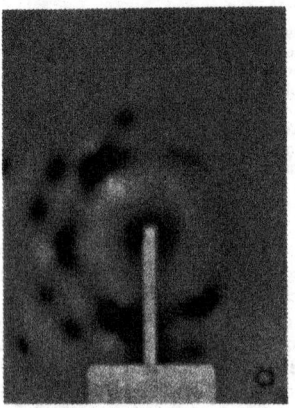
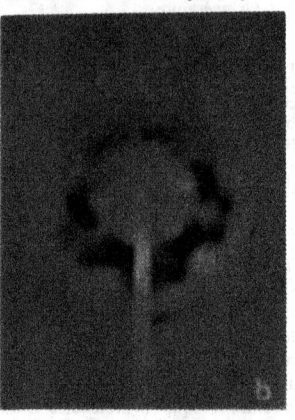
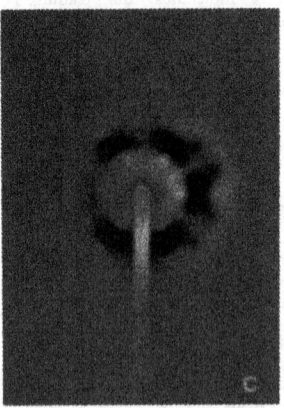
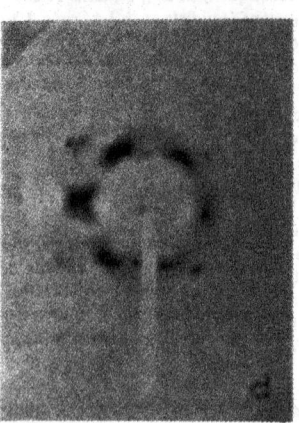

FIG. 1. — Clichés montrant une portion du plan réciproque h, k, o perpendiculaire aux molécules : a) PBBA phase smectique B_A (90 °C) (distance film-échantillon 40 mm) ; b) Para butyloxybenzylidène para éthyl aniline phase smectique B_C (40 °C) (distance film-échantillon 30 mm) ; c) TBBA phase smectique B_C (118 °C) (distance film-échantillon 27 mm) ; d) TBBA phase smectique E_C (80 °C) (distance film-échantillon 30 mm).

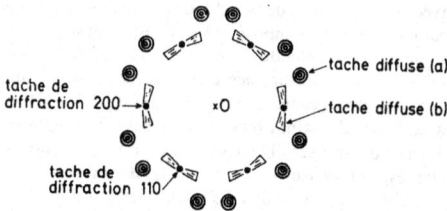

Fig. 2. — Représentation schématique du plan réciproque équatorial h, k, o de la phase smectique B.

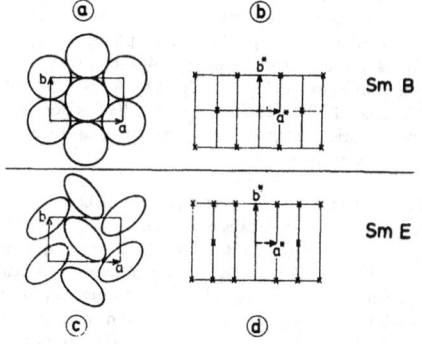

Fig. 3. — Représentation de la maille élémentaire dans le plan des couches smectiques pour les phases Sm B (*a*) et Sm E (*c*) et mailles réciproques correspondantes Sm B (*b*) et Sm E (*d*).

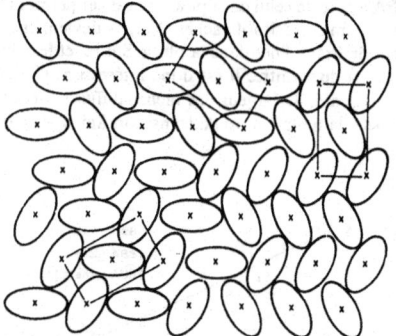

Fig. 4. — Représentation schématique de l'arrangement des molécules à l'intérieur d'une couche dans la phase smectique B. Les rectangles indiquent les trois mailles locales de type Sm E possibles.

3. Cas du smectique B_A.

— La présence de taches diffuses supplémentaires indique que la symétrie locale est plus basse que la symétrie moyenne. Les quatre points ± 2, ± 1, 0 appartiennent au réseau réciproque d'une maille rectangulaire non centrée à deux molécules par maille identique à celle décrite pour la phase smectique E [6] (Fig. 3c et 3d). Dans une telle phase, les molécules sont disposées en chevrons dans chaque couche et il ne peut plus y avoir, dans une telle phase, qu'une rotation de Π des molécules autour de leur grand axe. La présence sur le diagramme de la phase smectique B de quatre taches diffuses situées en ± 2, ± 1, 0 et l'absence de taches aux points ± 1, 0, 0 et 0, ± 1, 0 montre qu'il existe de petits domaines de fluctuation corrélés où les molécules sont disposées comme dans la phase smectique E. Les 8 autres taches observées sont liées à d'autres domaines dont le réseau rectangulaire serait tourné de $\pm \frac{2}{3} \Pi$ autour de l'axe sénaire, respectant ainsi la symétrie hexagonale moyenne de la phase smectique B. La figure 4 schématise la disposition des molécules de la phase smectique B à un instant donné. La rotation des molécules a lieu par sauts orientationnels de façon à conserver ces domaines corrélés de molécules en chevrons. Pour chaque type de domaine, la molécule peut prendre quatre orientations différentes, compte tenu des possibilités de retournement face pour face ; il y a donc

douze orientations possibles par molécule, se reduisant à six si les directions principales de la molécule dans le plan des couches font des angles de $\pi/3$ et $\pi/6$ avec les axes du réseau.

4. Cas des phases smectiques B_C.

— En dépit de la symétrie monoclinique de ces phases, il existe encore douze taches diffuses et dans certains cas il est difficile de distinguer les diagrammes des smectiques B_C de ceux des smectiques B_A, si l'on se limite à l'observation du plan (h, k, o). Un arrangement analogue à celui de la phase smectique E peut exister lorsque les molécules sont inclinées sur les plans smectiques et il a été décrit dans le cas du TBBA [3] : un monodomaine de la phase Sm B_C donne alors un seul domaine de la phase Sm E_C par refroidissement en dessous de 84 ºC. La maille rectangulaire du réseau dans la couche conserve sensiblement les mêmes paramètres, mais il apparaît des taches supplémentaires aux nœuds ± 2, ± 1, 0 et 0, ± 1, 0. Puisque dans la phase Sm B_C on observe des taches diffuses, non seulement aux points (± 2, ± 1, 0), mais aussi dans des positions correspondant à des rotations de $\pm \pi/3$ autour d'un axe perpendiculaire au plan (h, k, o), il y a encore trois types de domaines en chevrons d'orientation différente, et l'arrangement local des molécules est identique à celui de la phase B_A représenté figure 4 : les molécules peuvent tourner autour de leur axe en prenant, six ou douze positions sensiblement équivalentes, la symétrie monoclinique de cette phase étant essentiellement caractérisée par l'inclinaison des molécules sur les couches dans un plan perpendiculaire à l'axe binaire.

En fait, la symétrie hexagonale du plan réciproque (h, k, o) des phases smectiques B_C n'est qu'approchée. En effet, les 4 taches diffuses du type 2, 1, 0 sont souvent plus fines que les 8 autres. De plus, on peut aussi observer une tache diffuse en 0, 1, 0, point où se trouve une tache de Bragg de la phase Sm E_C. La figure 1 montre le diagramme de la phase Sm E_C du

A. M. LEVELUT

TBBA à côté de celui de la phase Sm B_C et permet ainsi la comparaison : les 4 taches 2, 1, 0 deviennent des taches de Bragg fines dans la phase Sm E_C et les autres tendent à disparaître. Il est donc intéressant d'étudier la variation de la largeur des taches diffuses lorsqu'on approche la température de transition Sm B_C-Sm E_C.

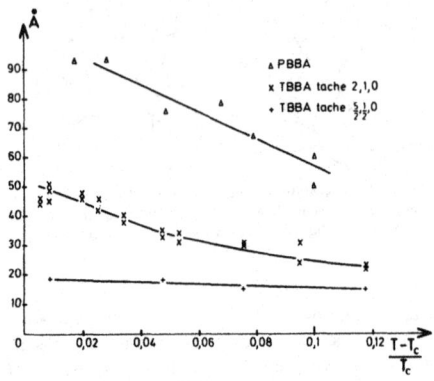

FIG. 5. — Variation de la taille des domaines de fluctuation avec la température, T_C est la température de transition Sm B_A → Cristal pour le PBBA et Sm B → Sm E pour le TBBA.

5. Dépendance des longueurs de corrélations avec la température.

La largeur des taches diffuses est inversement proportionnelle à la taille des domaines *engrenés* qui sont à l'origine de ces taches. Nous avons étudié la variation de la dimension de ces domaines (dans le plan des couches) pour tout le domaine d'existence de la phase smectique B_A ou B_C : dans le cas du TBBA, on oriente le cristal en phase Sm B de façon à observer simultanément, dans les mêmes conditions, une tache 2, 1, 0 et une tache $\frac{1}{2}$, $\frac{1}{2}$, 0, puis on fait varier la température depuis la transition Sm E → Sm B jusqu'à la transition Sm B-Sm C ; pour le PBBA, on peut obtenir un monodomaine Sm B_A en refroidissant la phase nématique et suivre ce monodomaine jusqu'à la cristallisation. La largeur des taches diffuses dans le plan (h, k, o) est ensuite mesurée au moyen d'un enregistrement microdensitométrique (ces taches sont alors sensiblement isotropes). La figure 4 donne la variation

avec la température de la taille des domaines engrenés pour les deux composés. Le PBBA présente des domaines assez étendus (ils varient entre 50 et 90 Å lorsque la température décroît), mais ce n'est pas une caractéristique générale de la phase Sm B_A. Pour TBBA la taille des domaines reliés aux taches d'indice demi-entiers ne varie sensiblement pas avec la température (elle est inférieure à 20 Å), celle des autres croît lorsqu'on s'approche de la phase smectique E et varie entre 20 et 50 Å. Bien que la transition Sm B Sm E soit du 1er ordre et que la largeur des taches 2, 1, 0 diminue alors brutalement, l'ordre local de la phase smectique B se rapproche de celui de la phase Sm E lorsqu'on abaisse la température. Il existe d'autres substances où la dissymétrie de la phase Sm B_C est moins accentuée, par exemple le 4-4'-butyloxybenzylidène éthyl aniline [5] dont le diagramme est porté sur la figure 1. Dans tous les cas, la taille des domaines corrélés varie lentement avec la température et ne dépasse guère 20 mailles.

6. Conclusion.

La présence de taches diffuses dans la phase Sm B, localisées sur des nœuds du réseau réciproque correspondant à la phase smectique E, nous a permis d'établir que les molécules y étaient localement disposées en chevrons comme dans la phase Sm E, mais pouvaient prendre en moyenne six orientations (différant entre elles de $\pi/3$) autour de leur grand axe. Ces six positions sont strictement équivalentes dans le cas d'une phase Sm B_A (hexagonale) et elles le sont presque à haute température dans la phase Sm B_C (monoclinique). Dans le cas du TBBA qui présente une transition Sm B_C → Sm E_C, les domaines de fluctuations ayant la structure de la phase Sm E_C s'étendent de plus en plus à basse température. Comme de plus la symétrie de la phase E n'exclut pas la possibilité d'un mouvement de retournement face pour face de la molécule, on conçoit aisément que la transition Sm B_C → Sm E_C du TBBA n'induise pas forcément une discontinuité dans les propriétés liées à la dynamique des mouvements moléculaires, tel par exemple la diffusion inélastique incohérente des neutrons mesurée sur une poudre [7]. Une étude du caractère inélastique des taches diffuses observées sur les diagrammes de rayons X permettrait d'apporter des informations plus complètes sur la dynamique des molécules dans la phase smectique B.

Bibliographie

[1] LEVELUT, A. M., LAMBERT, M., *C. R. Hebd. Séan. Acad. Sci.* **272** (1972) 1018.

[2] HERVET, H., VOLINO, F., DIANOUX, A., LECHNER, R. E., *J. Physique Lett.* **35** (1974) 151.

[3] DOUCET, J., LEVELUT, A. M., LAMBERT, M., *Phys. Rev. Lett.* **32** (1974) 301.

[4] Ce composé a été synthétisé par LIEBERT, L. et STRZELECKI, L. Les propriétés des phases mésomorphes de cette

substance et d'autres homologues seront publiées ultérieurement.

[5] LEVELUT, A. M., DOUCET, J., LAMBERT, M., *J. Physique* **35** (1974) 773.

[6] DOUCET, J., LEVELUT, A. M., LAMBERT, M., LIEBERT, L., STRZELECKI, L., *J. Physique Colloq.* **36** (1975) C 1-13.

[7] HERVET, H., DIANOUX, A. J., VOLINO, F., *J. Physique Colloq.* **37** (1976) C 3.

LE JOURNAL DE PHYSIQUE — LETTRES

TOME **40**, 15 JUILLET 1979, PAGE L-325

Classification
Physics Abstracts
61.30

Interlayer correlations in smectic B phases

A. J. Leadbetter, J. C. Frost and M. A. Mazid

Chemistry Department, University of Exeter, Exeter EX4 4QD, U.K.

(*Reçu le 2 avril 1979, accepté le 29 mai 1979*)

Résumé. — Les diagrammes de diffraction de rayons X sur des échantillons très bien orientés de plusieurs phases smectiques B mettent en évidence l'existence de plusieurs types de corrélations entre les couches smectiques. Les couches hexagonales peuvent s'organiser suivant l'ordre ABA... ou ABCA... (A, B et C décrivent la position relative des couches) avec des couches désordonnées de type ABC, ou être très peu corrélées entre elles. On n'a pas trouvé de bon exemple d'une structure simple monocouche de type AA...

Abstract. — X-ray diffraction results are presented on very well aligned samples of several smectic B phases which show the existence of several kinds of interlayer correlations. The hexagonal closepacked layers may be stacked with ordered ABA... or ABCA... arrangements (where A, B and C denote the relative position of the layers), with a random array of ABC-type planes, or with almost no correlation between layers. No clear examples of the simple monolayer packing AA... have been found.

This note is concerned principally with the uniaxial ordered smectic phase in which the average arrangement of molecules in the layers is hexagonal close packing and their long axes are approximately orthogonal to the layers. There is no ambiguity about the nomenclature S_B for such phases which have hexagonal symmetry [1] (e.g. the S-3 phases of $C_nH_{2n+1}OPhCHNPh\ C_mH_{2m+1}\ [nO.m]$ compounds) [2]. However, the same name has often been used for the phase of average monoclinic symmetry in which the (pseudo) hexagonal packing is significantly tilted (e.g. $\sim 25^\circ$) relative to the layer normal [1] (e.g. the S-V phase of TBBA [3]). Following the recent statement of the nomenclature problem for tilted smectics by Goodby and Gray [4] and the recognition that S_B and S_H are indeed distinct phases, we follow their recommendation of calling the tilted phase S_H.

From X-ray diffraction measurements on specimens prepared by careful melting of single crystals it has been established that for both S_B and S_H phases single domain specimens may be prepared in which the hexagonal or monoclinic symmetry extends throughout the bulk sample [1]. This implies a finite yield stress for the sliding of layers relative to each other and also some long range interlayer correlation. The nature of this correlation has been established for a number of S_H phases (notably 4O.2 [1, 2, 5] and other $nO.m$ compounds [2]) by X-ray measurements both on monodomain samples and on powders.

It is clear from the observed hkl reflections that there exists an average monoclinic unit cell with $c \approx l$ where l is the molecular length. Adjacent layers are thus related by a translation of c in the direction of the long molecular axis and there exists definite three dimensional order. The same is also true of S_G phases. What remains to be determined for particular compounds is a *quantitative* description of the extend of the order and the nature of the disorder.

For the S_B structures no such positive information has been established about the interlayer correlation. The possibility of two extreme cases of extensive and of no interlayer correlations has long been recognized [6, 7] but little direct evidence exists. The proven existence in several cases of a hexagonal lattice extending throughout the bulk sample implies at least a long range correlation of the *orientation* of the layers.

We report here the results of X-ray diffraction experiments on a variety of S_B compounds in which extremely well aligned samples were obtained mostly by slow cooling from the isotropic liquid phase in à magnetic field of $\gtrsim 2$ T. Such samples have been shown to be almost randomly disordered about the field direction — which for these phases is identical with the molecular long axis (**m**) and the layer normal ($\langle 00l \rangle$, **c** or z). This means that all of the reciprocal lattice (for l not too large) is observed in a simple

Table 1. — *Type of diffraction pattern shown by some* S_B *compounds.*

Compound	Code	Phase sequence	Diffraction profile along $\langle 00l \rangle$, see figure 2
$C_nH_{2n+1}OPhCHNPhC_mH_{2m+1}$	$nO.m$ (40.8 ; 50.7 ; 70.5 ; 60.4 ; 70.7 ; 50.6)	I-N-S_A-S_C-S_B-(S_H)-C (see refs. [2, 10])	2(e) (sometimes 2(c), once (60.4) 2(f))
$PhPhCHNPhCHCHCOOCH_2CH(CH_3)_2$	IBPBAC	I-N-S_A-S_B-S_E-C	2(d) (+ 2(e)) ; (2(c))
$PhPhCHNPhCHCHCOOC_4H_9$	BPBAC	I-S_A-S_B-S_E-C	2(c)
$C_6H_{13}OPhCHNPhPh$	60.Ph	I-N-S_A-S_B-C	2(d) (+ 2(e))
$C_8H_{17}OPhPhCOOPhOC_5H_{11}$	80.O5	I-N-S_A-S_C-S_B-C	2(b)
$C_8H_{17}PhPhCOOPhC_8H_{17}$	8.8	I-S_A-S_B-C	2(b)
$C_8H_{17}OPhPhCOOPhOC_8H_{17}$	80.O8	I-N-S_A-S_C-S_B-C	2(c) ; 2(b)
$C_5H_{11}OPhCHNPhPh$	50.Ph	I-N-S_B-C	Failed to prepare aligned sample
$C_{16}H_{33}OPhCHNPhPh$	160.Ph	I-S_A-S_B-C	2(b) but sample not well aligned

flat plate experiment with fixed sample. Measurements were made photographically using graphite-mono-chromated CuK$_\alpha$ radiation. An experiment was performed on a sample of 40.8 aligned by surface treatment using a rubbed film of polyvinyl alcohol which gave identical results to magnetically aligned samples. Some experiments were also made on specimens obtained by careful melting of single crystals ; these were generally less well ordered (except for the uniaxial disorder) than the magneti-cally-aligned specimens but enabled us to confirm the hexagonal symmetry of the phase. The liquid crystals investigated together with the type of dif-fraction pattern observed are summarized in table I.

In all cases the gross features of the diffraction pattern with the X-ray beam parallel to the layers were as shown in the photographs of figure 1. Two or more orders of the 00l reflections are observed and within the experimental resolution these are delta functions, implying a crystal-like structure [7]. There is also a strong bar of scattering associated with the reciprocal lattice points 100, $\bar{1}$00, 010, 0$\bar{1}$0, $\bar{1}$10 and $1\bar{1}$0 (hexagonal indexing) and of extent $\sim \pm c^*$. Very weak reflections of type 110, $\bar{1}$20, etc. are also found but the second order reflections (200, etc.) have not been observed (Fig. 1). Strong diffuse scattering is also observed. In many cases, however, with shorter exposure times and *provided that the sample is very well aligned* the bars of scattering associated with the first hexagonal ring of the reci-procal lattice showed pronounced structure. An example of such a photograph is shown in figure 1 and intensity profiles along the $\langle 00l \rangle$ rows for a variety of specimens obtained from photographs like figure 1 using a low resolution home-made microdensitometer are shown in figure 2. Profile (d)

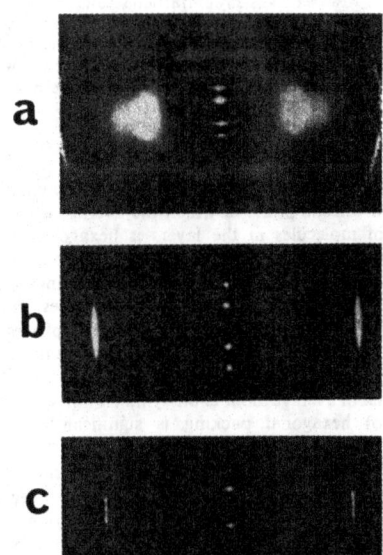

Fig. 1. — X-ray diffraction photographs of S_B phases with the beam parallel to the smectic layers. (a) IBPBAC ; long exposure which shows the extensive diffuse scattering, 00l reflections up to $l = 3$ and the 110 reflections. The strong outer arcs are container scattering. (b) IBPBAC ; shorter exposure, to show the bars of scattering associated with the first (hk) ring of the reciprocal lattice (100, $1\bar{1}$0, etc.). (c) 40.8 which gives the clearest example of the structure of the bars of scattering in the first (hk) ring. The micro-densitometer profile along such an $\langle 00l \rangle$ row is shown in figure 2(e).

for IBPBAC was the first observed [8] and an expla-nation was suggested in which the structure arose

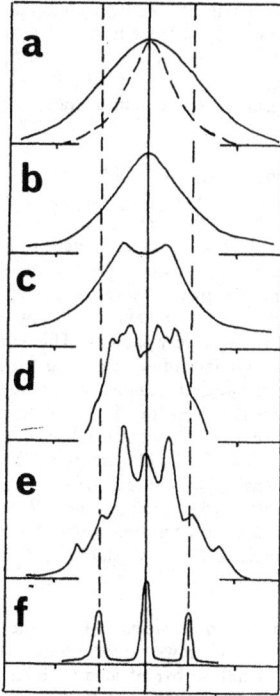

Fig. 2. — Intensity profiles along $\langle 00l \rangle$ for the first (hk) ring of the reciprocal lattice for S_B phases. (a) The Debye-Waller factor $\exp - Q^2 \langle z^2 \rangle$ for $\langle z^2 \rangle^{1/2}/d = 0.1$ (full line) and $\langle z^2 \rangle^{1/2}/d = 0.2$ (dashed line). (b) Experimental result for 8.8 showing uncorrelated layers. (c) Experimental result for IBPBAC sample prepared from a single crystal showing random ABC type layers. (d) Experimental result for IBPBAC sample prepared by cooling in a 2 T field showing predominantly ABCA... packing plus some ABA... packing. (e) Experimental result for 4O.8 showing ABA... packing. (f) Schematic intensity profile expected for simple AA... packing.

from a local tilt of $\sim 5°$ of the hexagonally packed molecules about an axis joining opposite corners of a hexagon, which results in the diffraction spots for the uniaxially disordered hexagonal reciprocal lattice appearing at two levels, both displaced from the zero $(l = 0)$ layer. The average uniaxial properties were accounted for by a short range interlayer correlation for the tilt direction. A similar type of diffraction pattern and explanation has been reported by Benattar et al. [9] on some compounds of the type $PhPhCH\!=\!NPhC_nH_{2n+1}$. Subsequently, however, the results shown in figures 1 and 2 have been accumulated and it has been shown elsewhere [10] that the profile type 2 e observed for essentially all of the $nO.m$ series studied by us so far which could be well aligned (> 5 examples) are consistent both in symmetry and intensity with an S_B phase which is a bilayer structure with an ABAB type packing of the molecules in which A is at (0) and B at

$(2\,\mathbf{a}/3 + \mathbf{b}/3)$, where $\mathbf{a}$ and $\mathbf{b}$ are the hexagonal unit cell vectors of the A layer. This leads to a simple consistent explanation for all of the results so far obtained in terms of the possible stacking arrangements of the hexagonal layers and we believe that our earlier description of the S_B structure of IBPBAC (although still internally consistent in this case) is not the correct one. Furthermore for the $nO.m$ compounds 7 spots are observed along the $\langle 00l \rangle$ row for the first ring of the reciprocal lattice (i.e. at $l = \pm 3/2, \pm 1, \pm 1/2$ and 0) and the position of these spots cannot be explained in terms of a slightly tilted monolayer structure. It is of course not contradictory to our conclusions if the compound studied by Benattar et al. was a truly tilted phase with small tilt angle but it should then be called a S_H phase.

The intensity of scattering from the S_B phases may be written quite generally as

$$I(hkl) \sim F_c^2(hkl) \langle F_m(hkl) \rangle^2$$

where $F_c(hkl)$ denotes the structure factor for the average molecular centres (the lattice points) including any interlayer disorder while $\langle F_m(hkl) \rangle$ is the structure factor for the average molecule associated with each point.

If we separate out the important contribution of disorder to $\langle F_m(hkl) \rangle$, this may be written

$$\langle F_m(hkl) \rangle \sim \langle F_m^1(hkl) \rangle \times$$

$$\times \exp - \left[\left(\frac{2\pi}{d_{00l}} \right)^2 \frac{\langle z \rangle^2}{2} \right] f(hx, ky)$$

where the various components of disorder are assumed to be independent and z is the component of molecular displacement perpendicular to the layers (assumed Gaussian) and $f(hx, ky)$ is a disorder function arising from rigid molecule displacement within the layer, and everything else (i.e. the head to tail disorder and any non-uniform atomic displacements) is included in $F_m^1(hkl)$. Calculation of $F_m^1(hkl)$ requires a detailed model of the molecular conformation and packing but it will generally be a relatively slowly varying function of Q at least in comparison with the very rapid decrease with Q arising from the disorder and to a first approximation we assume it to be constant over the restricted Q-range in question (i.e. along the $\langle 00l \rangle$ rows for $|l| \gtrsim 3$) $\langle z^2 \rangle$ may then be obtained from the intensities of the $(00l)$ reflections and for S_B phases relatively few orders are observed ($l_{max} \sim 2$ to 5) so that generally $\langle z^2 \rangle^{1/2}/d \lesssim 0.1$. $f(hx, ky)$ will include the six fold orientational disorder of molecular orientation about m together with the large librational amplitude and lateral displacements.

The envelope of the Debye-Waller factor $\exp[- Q^2 \langle z^2 \rangle]$ for typical values of $\langle z^2 \rangle$ for S_B compounds is shown in figure 2. The structure factors $F_c^2(hkl)$ may readily be determined for various types of packing of the hexagonal layers, both ordered and disordered. This leads to a straightforward interpretation of the main features of results like those shown in figure 2. There are five simple cases, all of which give sharp $00l$ reflections but are distinguished by their scattering profiles along $\langle 00l \rangle$ for $h - k \neq 3\,n$ [11]. For convenience of comparison the coefficient l will be defined in relation to the layer spacing d rather than the unit cell length c so that the observed $00l$ reflections in all cases (monolayer, bilayer, etc.) are $l = 1, 2, \ldots$ The five cases are as follows :

1. A simple monolayer (AAA...) packing with adjacent layers related by a simple translation of d along c. The resultant diffraction profile expected along $\langle 00l \rangle$ rows for h and/or $k \neq 0$ is shown schematically in figure 2(f) ; except for one poorly defined example observed for 6O.4, we have never seen such a profile, which is not surprising as such a stacking configuration must be unstable relative to those below.

2. A simple bilayer (ABAB) packing with adjacent layers displaced alternately by $+$ and $- (2\,\mathbf{a}/3 + \mathbf{b}/3)$ where $\mathbf{a}$ and $\mathbf{b}$ are the vectors defining the hexagonal unit cell of a layer. This will give diffraction peaks along $\langle 00l \rangle$ for $h - k \neq 3\,n$ with relative intensities 3 at $l = p + 1/2$ and 1 at $l = p$ where p is an integer. A finite correlation length along $c*$ will result in broadening of the peaks and this may arise for example from a finite domain size due to the disorder about c, or from stacking faults. Several of the $n\mathrm{O}.m$ compounds (Table I) have been shown to possess such a structure and a typical diffraction profile is shown for 4O.8 in figure 2(e). Within the rather large uncertainties caused by overlapping peaks and/or a diffuse background arising perhaps from regions of random incommensurate stacking at domain boundaries, the intensities as well as the positions of the peaks are consistent with this packing and with $\langle z^2 \rangle^{1/2} \lesssim 4$ Å. The correlation length of the ordered regions is $\lesssim 5$ molecules ($\lesssim 100$ Å).

3. A trilayer ABC... packing with B at $(2\,\mathbf{a}/3 + \mathbf{b}/3)$ and C at $(\mathbf{a}/3 + 2\,\mathbf{b}/3)$ relative to A. This will give diffraction peaks along $\langle 00l \rangle$ of equal intensity at $l = p + 1/3$ for $h - k = 3\,n + 1$ and at $l = p - 1/3$ for $h - k = 3\,n - 1$ and for a sample which is disordered about c this will result in diffraction rings of equal intensity at $l = 1/3, 2/3\ldots$ etc. Disorder will again cause the peaks to broaden and in addition stacking and distortion faults cause the peaks to be shifted from their ideal positions.
We have observed two examples where this is the predominant packing arrangement (Table I) and the diffraction profile for IBPBAC is shown in figure 2(d). The intensities of the peaks at $l = \pm 1/3$ and $\pm 2/3$ are also approximately consistent with this packing and $\langle z^2 \rangle^{1/2} \sim 2.5$ Å although the decrease in inten-

sity at higher l indicates some additional disorder. In neither of the two examples, however, is the packing purely of ABC... type and the profile of figure 2(d) shows evidence of regions of ABAB packing and possibly considerable incommensurate disorder as well.

4. Random packing of ABC-type layers : this will give a diffuse scattering profile along $\langle 00l \rangle$ at $h - k \neq 3\,n$ with maxima at $l = n + 1/2$ of intensity 9 relative to minima of intensity 1 at $l = n$. A number of compounds have been found to show a diffraction profile like that in figure 2(c) which approximates to this behaviour except that the minimum at $l = 0$ is very shallow. This is probably a result of additional diffuse scattering from incommensurately disordered regions (see below). Only BPBAC so far has been found to show this pattern exclusively (see Table I) but it has been found additionally for IBPBAC samples prepared from single crystals *via* the S_E phase and for several $n\mathrm{O}.m$ compounds prepared by cooling in a field. In both cases the preparation conditions presumably did not allow formation of the more stable ABC or ABAB packings.

5. No correlation between layers : the intensity along $\langle 00l \rangle$ for all rows except $h = k = 0$ will be a structureless diffuse bar of length about $2/d$. The profile shown in figure 2(b) is typical of those observed for a number of compounds. It is roughly Gaussian in shape with a full width at half height of $2/d$ which implies a correlation length of only about one layer thickness. This type of diffraction profile might also be considered as arising from, say, an ABAB packing with a sufficiently low correlation length to broaden the profile of 2(e) to give that of 2(b). This requires a correlation length of less than about two layers so that in any case a profile like that of 2(b) must mean that $\mathbf{a}$, $\mathbf{b}$ correlations between the smectic layers do not extend beyond adjacent layers.
This type of pattern is the one observed most frequently (Table I) but it does not necessarily imply that a more ordered packing cannot be produced — for example by long annealing.
All samples which are completely disordered around c must have *some* incommensurately disordered layers at the boundaries between domains of differing orientation and hence their scattering along $\langle 00l \rangle$ except for $h = k = 0$ must have a diffuse component of type 2(b).
The above results make it clear that many S_B phases have a regular ordered packing arrangement of the hexagonal layers giving a three dimensional crystalline structure. However the energy minima for the relative location of the layers are in general shallow so that disorder is very easily introduced and may be enough to give essentially uncorrelated layers. The relationship between the extent and type

of interlayer correlation and the chemical constitution remains to be established.

A feature of all S_B phases studied is the small number of diffraction peaks observed. The rapid decrease in intensity with increasing l means that $\langle z^2 \rangle^{1/2}/d_{001} \lesssim 0.1$ while the decrease with increasing h, k requires a similar value for the reduced rms centre of mass displacement in the plane $[\langle X^2 \rangle^{1/2}/d_{100}]$ in addition to the 6 fold orientational disorder. This cannot account however for the much smaller intensity of (200) compared with (110) reflections (the former are in fact not observed). This is associated with disorder in the stacking of the hexagonal layers since faults in the regular packing cause the reciprocal lattice nodes for $h - k \neq 3n$ (e.g. 200 but *not* 110) to broaden and consequently to appear weaker.

We have established for the first time the nature of the interlayer correlations in a number of S_B phases. It would now be useful to extend these measurements to a wider variety of compounds in order to develop an understanding of the causes underlying the occurrence of particular structures. To this end it will also be necessary to develop further the art of preparing very well aligned (and preferably true monodomain) samples and to study them by accurate intensity measurements with higher resolution than has generally been used previously. We are now beginning this work.

References

[1] LEVELUT, A. M., DOUCET, J. and LAMBERT, M., *J. Physique* **35** (1974) 773.

[2] DOUCET, J. and LEVELUT, A. M., *J. Physique* **38** (1977) 1163.

[3] DOUCET, J., LEVELUT, A. M. and LAMBERT, M., *Phys. Rev. Letts.* **32** (1974) 201.

[4] GOODBY, J. W. and GRAY, G. W., *Mol. Cryst. Liq. Cryst. Lett.*, in press.

[5] DE VRIES, A. and FISCHEL, D. L., *Mol. Cryst. Liq. Cryst.* **16** (1972) 311.

[6] DE VRIES, A., *Chem. Phys. Letts.* **28** (1974) 252.

[7] DE GENNES, P. G. and SARMA, G., *Phys. Lett.* **38A** (1972) 219 ; DE GENNES, P. G., *Mol. Cryst. Liq. Cryst.* **21** (1973) 49.

[8] RICHARDSON, R. M., LEADBETTER, A. J. and FROST, J. C., *Ann. Phys.* **3** (1978) 177.

[9] BENATTAR, J. J., LEVELUT, A. M. and STRZELECKI, L., *J. Physique* **39** (1978) 1233.

[10] LEADBETTER, A. J., KELLY, B. A., MAZID, M. A., GOODBY, J. W. and GRAY, G. W., Submitted to *Phys. Rev. Letts.*

[11] GUINIER, A., *X-ray Diffraction* (W. H. Freeman) 1963.

Volume 43, Number 10   PHYSICAL REVIEW LETTERS   3 September 1979

Long-Range Order in Two- and Three-Dimensional Smectic-B Liquid-Crystal Films

D. E. Moncton and R. Pindak
Bell Laboratories, Murray Hill, New Jersey 07974
(Received 21 May 1979)

This Letter presents x-ray studies of freely suspended smectic liquid-crystal films from 2 to > 100 layers thick. Smectic-B thick films of butyloxybenzylidene octylaniline (4O.8) have three-dimensional long-range order while thin films are two-dimensional crystals. Large diffuse scattering is observed from soft shear waves, propagating perpendicular to the layers, and three new low-temperature phases with different interlayer order in thick films are found.

Freely suspended films of smectic liquid crystals are attractive systems for the study of the effects of reduced structural dimensionality. Success in optical experiments[1] has led to attempt x-ray studies of intermolecular correlations in the various smectic phases and at phase transitions. The film technique is unique for two reasons. First, uniform, stable, large-area (1 cm²) films can be made which consist of two molecular layers ($N = 2$). These films are the first truly two-dimensional (2D) systems which allow the study of structural correlations. Second, the film thickness may be increased to study the evolution to 3D. For thick films ($N > 100$), this technique eliminates layer pinning at sample-cell walls and extraneous forces of alignment fields. In the smectic-B phase, the films are single crystals with negligible mosaic spread.

The principal motivation for the present work was to investigate the 2D melting process (i.e., the smectic $B \rightarrow A$ transition) and the interlayer correlations with increasing thickness. A 2D lattice possesses both conventional long-range bond-orientational order and power-law decay of positional correlations.[2] The δ-function Bragg scattering of a 3D solid is replaced by the structure factor

$$S(\vec{q}) \sim (\vec{q} - \vec{G})^{-2 + \eta_{\vec{G}}(T)} , \qquad (1)$$

where $\vec{G}$ is a 2D reciprocal-lattice vector and $\eta_{\vec{G}}(T)$ is wave-vector and temperature dependent. During our experiments Halperin and Nelson,[3] building upon ideas of Kosterlitz and Thouless,[4] showed that melting could occur in two steps. With increasing temperature, one would first observe a transition to an intermediate "hexatic" phase with power-law decay of bond order and short-range positional order. This state would have a Lorentzian structure factor

$$S(\vec{q}) \sim [\xi_{\parallel}^2(\vec{q} - \vec{G})^2 + 1]^{-1}, \qquad (2)$$

where $\xi_{\parallel}$ is the in-plane positional correlation length. Then in an additional transition at higher

temperatures, this phase would transform to a 2D liquid. Later Birgeneau and Litster[5] proposed that a 3D system of stacked hexatic layers would have long-range orientational order and exponential decay of both in-plane and interlayer positional correlations. They offered this explanation for previous x-ray observations[6,7] suggesting finite interlayer correlations in smectic-B phases.

Here we present our progress in the study of films of butyloxybenzylidene octylaniline (4O.8), which has a bulk smectic $B \rightarrow A$ transition at T_{AB} = 49°C. This material was chosen because previous work[6] on the B phase suggested the behavior described above. However our results prove that the bulk smectic-B phase of 4O.8 does *not* have a short $\xi_{\perp}$ and the Birgeneau-Litster model does not apply. Thin films in the smectic-B temperature range are 2D crystals and not the intermediate hexatic phase discussed by Halperin and Nelson, although this phase may occur at temperatures above 49°C.

Our experiments used a spectrometer based on a 50-kW rotating–Cu-anode x-ray generator. A vertically bent pyrolytic graphite(002) crystal focused Cu $K\alpha$ x rays to a 1×3-mm² spot on the film and scattered radiation was analyzed with a flat pyrolyite graphite(002) crystal and scintillation detector. Our resolution function is Gaussian with full widths at half maxima given by

$$\Delta q_x = 0.046 \cos\theta \text{ Å}^{-1},$$
$$\Delta q_y = 0.046 \sin\theta \text{ Å}^{-1},$$
$$\Delta q_z \approx 0.1 \text{ Å}^{-1},$$

where $\hat{x}$, $\hat{y}$ are in the scattering plane, parallel and perpendicular, respectively, to the nominal momentum transfer, $\vec{Q}$, and $\hat{z}$ is perpendicular to the scattering plane.

Films were drawn inside an oven with 0.05°C temperature control, evacuated to 500 mTorr to minimize the background. Since the scattering from a 100-Å thick film of hydrocarbons is small, it was crucial to have no windows within the scat-

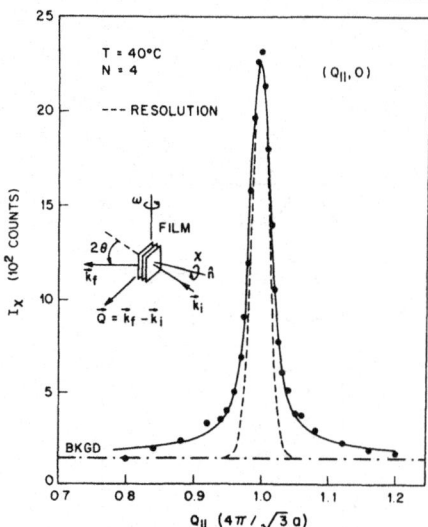

FIG. 1. Intensity integrated over a 60° χ segment for a four-layer film, vs the in-plane momentum transfer $Q_{\parallel}$. The solid line represents the structure factor described in the text. The inset describes the experimental geometry, with $\hat{n}$ being the film normal.

tering volume. Typically the beam-on, no-film background was less than 0.025 count per second.

Our data are obtained as scans in the hexagonal reciprocal lattice where $a^* [4\pi/(3^{1/2}a)]$ varies linearly from 1.4457 Å^{-1} at 25°C to 1.4360 Å^{-1} at 49°C and $C^* (2\pi/c) = 0.217$ Å^{-1} essentially independent of temperature. Momentum transfers $\vec{Q} = (Q_{\parallel}, Q_{\perp})$ are defined by their components parallel and perpendicular to the smectic layer plane (see inset to Fig. 1). In thin films we find well-defined lattice orientational order as evidenced by narrow peaks in a χ scan. However, there is often more than one such peak in a 60° χ segment. Since the beam area is large, ~3 mm², we conclude that the film consists of a few well-ordered domains. In thin films the orientation of these domains varies slowly with time. At temperatures near 48°C (43°C) the peak rotates 2° in χ in about 20 minutes (1.5 hours). This domain structure and dynamics provide evidence that the 2D film can support mobile dislocations.

To establish the extent of in-plane positional correlations with moving domains, we employed a rotational averaging technique to obtain the data in Fig. 1. At each value of momentum trans-

fer, the intensity was integrated over a 60° χ segment. Successive scans were accumulated until fluctuations due to counting statistics dominated the random errors due to the averaging procedure. To compare with experiment, structure factors (1) and (2) have been integrated over the angle χ and convoluted with the measured resolution. Equally good fits can be obtained by either form provided $\xi_{\parallel} > 800$ Å. The fact that this length is so large argues strongly against the applicability of the hexatic description in favor of the 2D crystal interpretation. Such a model for 2D films predicts that thick films would have conventional 3D order.

To test this behavior we studied interlayer correlations as a function of film thickness. As shown in Fig. 2 the slowly varying layer interference function characteristic of a four-layer film evolves continuously into a spectrum of sharp Bragg peaks riding on a large diffuse background. We note that peak widths are fully accounted for

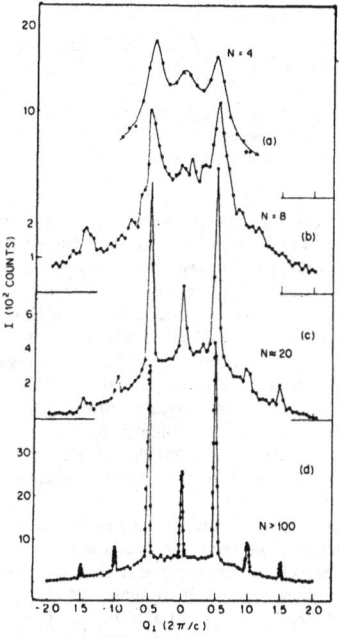

FIG. 2. Scans taken at $\vec{Q} = (1, Q_{\perp})$ for four different film thicknesses show the development of interlayer correlations. The data for the $N = 4$ film was integrated over χ.

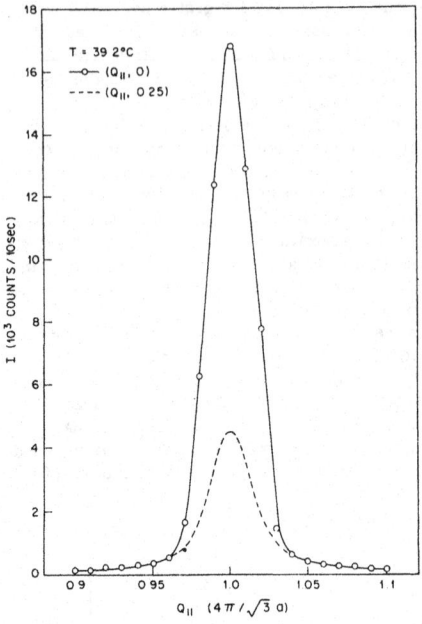

FIG. 3. Longitudinal scans through the Bragg peak at
$\vec{Q} = (1, 0)$ (solid line) and through the diffuse scattering
at $\vec{Q} = (1, 0.25)$ (dashed line) for a thick film with $N > 100$.

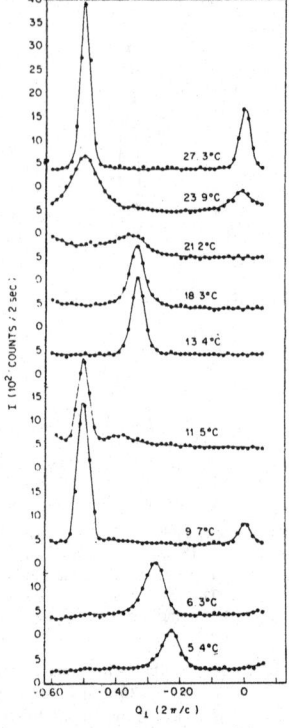

FIG. 4. Temperature dependence of the interlayer
ordering in a thick film ($N > 100$) is monitored by transverse scans with $\vec{Q} = (1, Q_\perp)$. Note phase transitions at
about 22°C, 12°C, and 7°C.

by a combination of resolution and the finite number of layers for $N = 4$, 8, and 20. For the case
$N > 100$ the peaks are limited solely by resolution
and have profiles which are identical within experimental error to those produced by a perfect
Ge single crystal. There is *no doubt* that this
smectic-B phase has conventional 3D long-range
order. The spectrum of peaks indexes on half-
integer units of the single-layer reciprocal lattice constant. Intensity arguments and systematic extinction rules imply that the interlayer
ordering consists of 3D hexagonal-close-packed
($ABAB$) stacking.

The large diffuse scattering seen in Fig. 2 is
clearly distinguished from the Bragg scattering
as the number of layers increases. For a thick
film, Fig. 3 shows longitudinal scans both
through a Bragg point ($Q_\perp = 0$) and through the
diffuse scattering only ($Q_\perp = 0.25$). Subtraction
of the two scans shows that the coherent Bragg
scattering is identical to the Gaussian resolution
function within experimental error (5% in FWHM).

The diffuse scattering is sharply peaked in $Q_\parallel$
but is neither Gaussian nor of resolution width.
It has long wings characteristic of scattering
from phonons but its intensity is extraordinary
compared with that from a normal solid. We believe that this scattering derives from soft shear
waves propagating perpendicular to the layers.
Neutron scattering experiments[8] on terephthal-
bis-butylaniline (TBBA) have shown the zone-
boundary frequency for these TA [00ξ] modes to
be less than 1 MeV.

Having established that the smectic-B phase
has 3D long-range order but anomalously soft
TA [00ξ] excitations, we believe that interlayer
ordering, although present, is energetically weak.
Unexpected evidence for weak interlayer forces
is provided by a remarkable series of three first-

order phase transitions which occurs between 27°C and 5°C. In Fig. 4, transverse scans show that these transitions involve changes in the interlayer periodicity from two layers to three layers near 22°C and back to two layers again near 12°C. Finally below 7°C the lattice becomes monoclinic with the $\bar{c}*$ axis tilting as a function of temperature.

We conclude by emphasizing that our results on thick films clearly establish that the smectic-B phase of 4O.8 has 3D long-range order. Although it is apparently a solid, unusually low-frequency layer shear modes exist. Furthermore, the energy of interlayer ordering must be weak to account for the observed phase transitions below 27°C. In thin-film experiments we have shown that it is possible to use x-ray diffraction techniques which have thus far demonstrated that the smectic-B phase is a 2D crystal. From this basis it is now possible to approach the problem of 2D melting. Improved synchrotron-based x-ray experiments are now in progress.

We have had a number of stimulating discussions with R. J. Birgeneau, W. F. Brinkman, D. S. Fisher, B. I. Halperin. J. M. Kosterlitz, and D. R. Nelson which have been important to

the development of these studies. Also we thank S. C. Davey for technical assistance.

[1]C. Y. Young, R. Pindak, N. A. Clark, and R. B. Meyer, Phys. Rev. Lett. 40, 773 (1978); C. Rosenblatt, R. Pindak, N. A. Clark, and R. B. Meyer, Phys. Rev. Lett. 42, 1220 (1979).

[2]B. Jancovici, Phys. Rev. Lett. 19, 20 (1967); N. D. Mermin, Phys. Rev. 176, 250 (1968).

[3]B. I. Halperin and D. R. Nelson, Phys. Rev. Lett. 41, 121 (1978); D. R. Nelson and B. I. Halperin, Phys. Rev. B 19, 2457 (1979).

[4]J. M. Kosterlitz and D. J. Thouless, J. Phys. C 6, 1181 (1973).

[5]R. J. Birgeneau and J. D. Litster, J. Phys. Lett. (Paris) 39, 399 (1978).

[6]A.-M. Levelut, J. Doucet, and M. Lambert, J. Phys. (Paris) 35, 773 (1974).

[7]A. deVries, A. Ekachai, and N. Spielberg, J. Phys. (Paris), Colloq. 40, C3-147 (1979); A. J. Leadbetter, J. Frost, J. P. Gaughan, and M. A. Mazid, J. Phys. (Paris), Colloq. 40, C3-185 (1979); J. Doucet, A.-M. Levelut, and M. Lambert, Ann. Phys. (N.Y.) 3, 157 (1978).

[8]J. J. Benattar, A.-M. Levelut, L. Liebert, and F. Moussa, J. Phys. (Paris), Colloq. 40, C3-115 (1979).

Mol. Cryst. Liq. Cryst., 1981, Vol. 67, pp. 205-214

High-Resolution X-Ray Study of the Smectic A-Smectic B Phase Transition and the Smectic B Phase in Butyloxybenzylidene Octylaniline

P. S. Pershan,[†] G. Aeppli, J. D. Litster and R. J. Birgeneau

Center for Materials Science and Engineering,
Massachusetts Institute of Technology,
Cambridge, MA, U.S.A.

Received August, 5, 1980

Abstract We have carried out a high resolution X-ray study of smectic phases of Butyloxybenzylidene Octylaniline. We find that the phase previously identified as Smectic-B in this material is crystalline with in-plane order extending over at least $1.4\mu m$. The in-plane Bragg peaks are accompanied by anomalously strong diffuse scattering that can be described by a form $1/(q_\perp^2 + \gamma^2 q_z^2)$. Unless the elastic constant C_{44} is more than an order of magnitude smaller than previously reported values of $\sim 10^8$ ergs/cm^3 the diffuse scattering can not be due to acoustic phonons. The crystalline-B to Smectic-A melting transition is strongly first order with no observable pre-transition effects on either side of the transition.

In spite of extensive research, the well-ordered smectic phases such as B, E, F and H are still only poorly understood. A variety of theoretical models have been proposed for these phases, beginning with the stacked two dimensional harmonic crystal model of de Gennes and Sarma.[1] Very recently, important progress has occurred in the theory of two dimensional melting. The relevance of two dimensional (2d) dislocation-mediated melting, proposed by Kosterlitz and Thouless,[2] to liquid crystals was first realized by Huberman *et al.*[3] Recently the theory of 2d melting has been advanced by Halperin and Nelson and Young.[4] Halperin and Nelson have suggested that a 2d solid may first melt into a

---

[†]Permanent address: Division of Applied Sciences, Harvard University, Cambridge, MA, U.S.A.

Paper presented at 8th International Liquid Crystal Conference, Kyoto, Japan, June 30-July 4, 1980.

Halperin and Nelson have suggested that a 2d solid may first melt into a 2d "hexatic phase" characterized by very long range bond orientational order but short range positional order and this hexatic phase will then melt into an isotropic fluid. Two of us (RJB and JDL) have proposed a model for smectic B's which is, in essence, an amalgam of the de Gennes-Sarma and Halperin-Nelson ideas.[5] In this model a SmA liquid crystal may transform on cooling either into a 3d solid with positional long range order or into a 3d "stacked hexatic phase" with 3d bond orientational long range order and positional short range order. The former transition would be strongly first order whereas the latter could be second order. Only the stacked hexatic phase would be a true liquid crystal.

Previous low resolution X-ray experiments[6-11] have suggested that Butyloxybenzylidene Octylaniline (40.8) has a B-phase exhibiting the characteristics expected for the stacked hexatic model. Therefore, we have carried out a high resolution X-ray study of the B-phase and the B–A transition in this material. Concurrent with this work Moncton and Pindak[12] have carried out X-ray studies of freely suspended films of the B-phase of 40.8. Other workers have searched for pre-transition effects in the viscosity,[13] linear optical properties,[14] and heat capacity of 40.8 at the smectic A to smectic B transition.[15] We will show below that this transition is strongly first order, and that the phase, previously identified as smectic B, has 3d long range crystalline order in this material. Henceforth we will refer to this phase as *B– type lamellar solid* to distinguish it from a true smectic B.[5] Finally we will describe the results of our study on the anomalously strong diffuse scattering that accompanies the Bragg peaks.

The transition temperatures of our 40.8 as received from CPAC-Organics were slightly lower than accepted values for pure 40.8. However, after pumping on the material for eight hours at 75°C the following transition temperatures were obtained: isotropic → nematic 78.8°C, nematic→ smectic A 63.6°C, and smectic A→ B-solid 49.6°C. The sample cell consisted of a 2 mm×1cm × 1cm beryllium can sealed by a teflon plug held in the mouth of the can by a brass fixture. We filled the cell with 40.8 and placed it onto a heated (~ 80°C) vacuum stage for approximately eight hours. After that the cell was brought down to room temperature under vacuum and sealed in dry nitrogen. With the sample prepared in this way and held below 49°C, the smectic A to B-solid transition temperature was virtually unchanged for approximately four weeks.

An oriented B-solid sample was prepared by heating the sample into the isotropic phase and then slowly cooling it down to 45°C in the presence of a magnetic field of approximately 5000 Gauss. Cooling rates varied from $\sim$ 1°C per hour to as low as $\sim$ 0.01°C per hour near the smectic A to B transition. Since the magnetic field was not necessary for maintaining orientational order in the smectic phases, the samples could be grown in a large electromagnet and then moved to the X-ray spectrometer. Our samples were oriented in the sense that all of the smectic layers were parallel to one another with cooling-rate dependent angular mosaicities that varied from ±0.3°C to ±2°C. Within the layers, however, the single crystal grain size was much less than the overall cell dimension and the sample appeared as a two-dimensional powder.

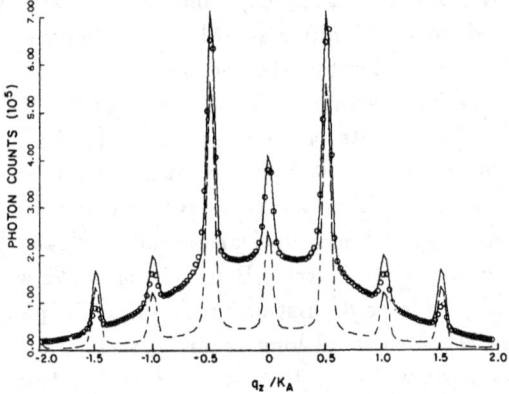

Figure 1. $(1, 0, q_z/K_A)$ scan showing the anomalously intense diffuse scattering and the $(1, 0, n/2)$ Bragg peaks. The circles represent experimental data, the full line the 2d powder average of the theoretical $S_{th}(1, 0, q_z/K_A)$, and dashed line the phonon and Bragg contributions to $S_{th}(1, 0, q_z/K_A)$, assuming $C_{44} \sim 5 \times 10^7$ ergs/cm^3. S_{th} accounts for the acoustic phonons, Bragg peaks, additional q^{-2} singularities at peaks, and form- and Debye-Waller factors. $K_A = 2\pi/L$ where $L = 28.7$Å is the layer spacing in the solid-B phase.

The X-ray spectrometer was similar to the one described by Als-Nielsen *et al.*[16] We discuss first the results for the B-solid phase below 49°C. In agreement with others[11,12] we observe a series of Bragg peaks superposed on a very strong diffuse background. Figure 1 illustrates the results for one $(1, 0, q_z)$ scan through the Bragg peaks. For CuKα radiation $(\lambda = 1.542$Å$)$ the $(1, 0, 0)$ peak occurs at a $2\theta = 20.35°$ corresponding to a lattice spacing of 4.36Å. Successive peaks imply a

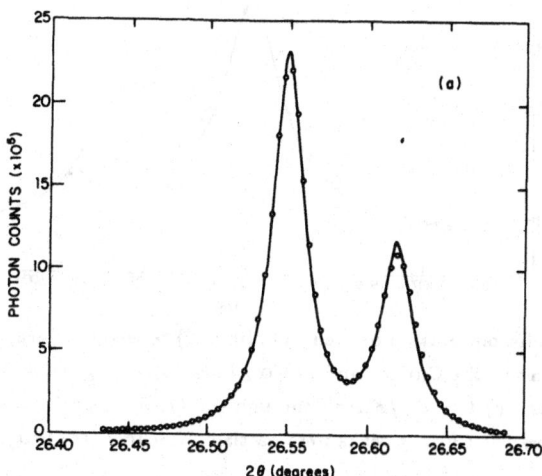

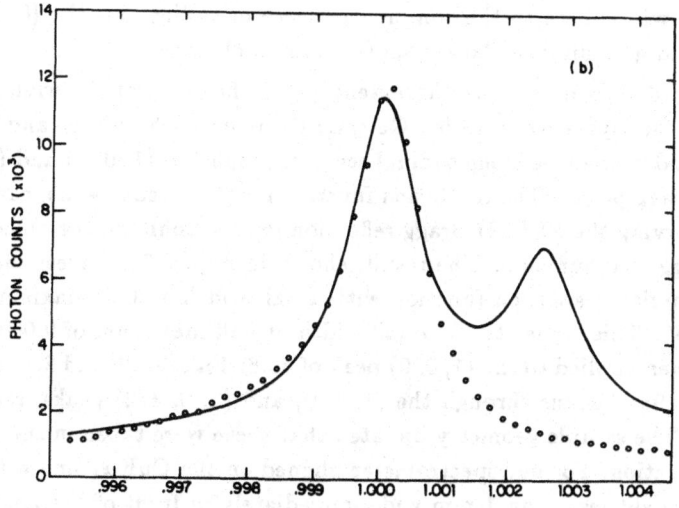

periodicity normal to the layer of 57.4Å, two times longer than the layer spacing as determined from the Bragg peak indexed as (0, 0, 1). This is consistent with a hexagonal close packed, ABAB, structure[11,12] in which the lowest order (0, 0, 1/2) reflection is missing by virtue of the structure

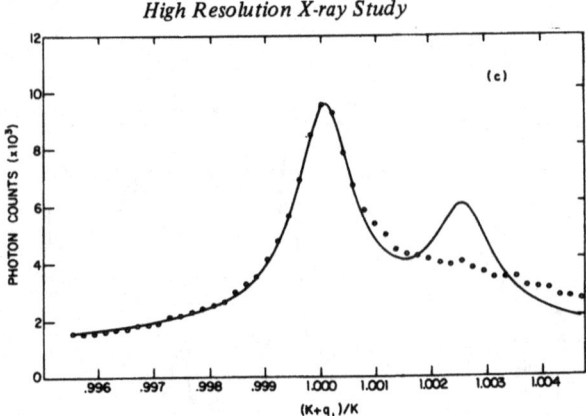

Figure 2a. Resolution function derived from (0, 0, 2) reflection of graphite. The two peaks correspond to the $CuK\alpha_1$ and $CuK\alpha_2$ lines. b) $1 - q_\perp/K$ scan through the (1, 0, 1/2) peak. c) $1 - q_\perp/K$ scan through the (1, 0, 0) peak. As the result of geometric factors the $CuK\alpha_2$ peaks were partially obscured (see text).

factor. That is, the unit cell is actually two layers thick and correct crystallographic notation would assign the small angle ($2\theta = 3.09°$) peak corresponding to the lowest observed order of Bragg reflection to (0, 0, 2). We will adhere to the common practice of calling this the (0, 0, 1) peak and assigning the larger angle peaks to (1, 0, $m/2$).

In order to determine the extent ξ_x of the positional correlations within the layers we installed Ge crystals as monochromator and analyzer and carried out longitudinal scans through the (1, 0, 0) and (1, 0, 1/2) Bragg peaks. The resolution function for these scans was measured by observing the (0, 0, 2) Bragg reflection from graphite in the reflection or Bragg configuration. The result, shown in Figure 2a, corresponds to a Lorentzian resolution function with a half width at half maximum of 0.01155°. This translates to a half width at half maximum of 0.0008165 $Å^{-1}$ when applied to the (1, 0, 0) peak of 40.8. Figures 2b and 2c display the results of scans through the (1, 0, 0) and (1, 0, 1/2) peaks, respectively. The sample geometry dictated that these were taken in the Laue configuration. For our spectrometer aligned on the $CuK\alpha_1$ line with an entrance slit less than 1 mm wide immediately in front of the analyzer crystal, and approximately 1.5 m between analyzer and monochromator, the $CuK\alpha_2$ line is not passed in the transmission configuration. The solid line is the sum of the resolution function shown in Figure 2a and a diffuse scattering signal that is calculated as the convolution of the resolution function with a form $[q_x^2 + q_y^2 + \gamma^2 q_z^2]^{-1}$. This is then averaged over the $x - y$ plane as is appropriate to a two-dimensional powder. Both Bragg and diffuse terms were averaged over the measured mosaic distribution for this sample. We will discuss the suitability of the above form of the diffuse background signal below. The principal result that the $(1, 0, 0)$ and $(1, 0, 1/2)$ Bragg peaks are resolution-limited does not depend on how we represent the diffuse background. A non-linear least square fit of the data to the above theoretical form, allowing the intensities of the peak, of the background, and the width of the resolution function to be free parameters yields a half width at half maximum for both the $(1, 0, 0)$ and $(1, 0, 1/2)$ scans of $\Delta q_\perp = 0.79 \pm 0.05 \times 10^{-3}$ Å^{-1}; the measured width of the resolution function is $0.82 \pm 0.014 \times 10^{-3}$Å^{-1}. By taking the extreme limits of both error bars we conclude that the in-plane correlation length ξ_x is greater than 1.4μm. A straightforward geometrical analysis applied to the $(1, 0, 1/2)$ scan demonstrates that the between-plane correlation length ξ_z is greater than 0.14μm. Thus, our bulk samples of (40.8) in the B-solid phase have positional correlations at 45.5°C extending at least 1.4μm within the layer plane and 0.14μm normal to it. We next discuss the first order character of the smectic A to B-solid transition and the absence of any observed pretransition phenomena.

The $(0, 0, 1)$ and $(1, 0, m/2)$ peaks and the broad background intensities of an oriented sample of 40.8 were monitored as the sample temperature was increased from 45.5°C over a period of approximately two weeks. Aside from the subjective observation that certain features of the mosaic distribution function smoothed out slightly between 47°C and 48°C there was no observable change in *any detail* of the X-ray spectra until the temperature reached 49.1°C. Between 49.1°C and 49.3°C the intensities of the $(1, 0, m/2)$ peaks decreased to zero linearly with temperature. Beyond 49.3°, the only observable feature in $(\theta, 2\theta)$ scans around the position of the $(1, 0, m/2)$ peaks was a broad maximum characteristic of a liquid with a temperature-independent width corresponding to an in-plane correlation length of $\sim$ 20Å. There was absolutely no indication of anything interpretable as a pretransition increase in the positional correlations.

The above data were taken with a moderate resolution X-ray spectrometer. At a later stage (by then the sample had become less pure), we monitored the A, B coexistence by measuring the $(0, 0, 1)$ peaks

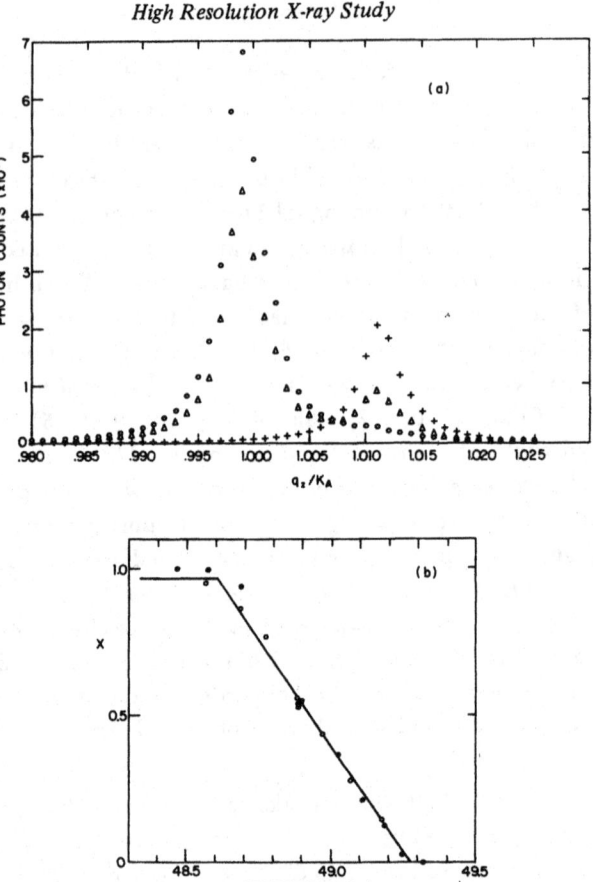

Figure 3a. $(0, 0, 1 - q_z/K_A$ scans at temperatures $T = 49.3°C$ (Smectic A phase) $(+)$, $T = 48.9°C$ (coexistence region) Δ, and $T = 48.7°C$ (solid-B phase) 0. b) Relative concentration $\chi = I_B(I_B + 4I_A)$ of A to B phases are plotted against the temperature T. Filled circles are for heating and open circles for cooling.

of the A and B phases. The planar spacings in the A and B phases differ by 1.3% so that the A and B peaks may be readily separated using the high resolution configuration described previously. We show in Figure 3a a set of scans just above, in the middle, and just below the coexistence region. The results are just those expected from a strongly first order transition that is broadened by a small impurity concentration. If I_B and I_A are the intensities of the $(1, 0, 0)$ peaks in the B and A phases respectively the relative concentration of B to A is given by $\chi =$

$I_B (I_B + 4I_A)$ where the factor 4 arises from the larger intensity of the (0, 0, 1) peak in the B phase. Figure 3b is a plot of χ vs. T on heating and then on cooling through the transition; the linear behavior is evident. These data are consistent with the low resolution results on the purer sample which had a correspondingly narrower coexistence regime. Although it is conceivable that impurities can convert an otherwise second order transition into a first order transition there is absolutely no reason to believe that this effect is occurring here. For example, the complete absence of pretransition phenomena expected for second or weakly first order transitions weighs heavily against any such hypothesis.

It is evident, therefore, that 40.8 exhibits the less interesting of the two possibilities suggested by Birgeneau and Litster,[5] viz., a first order transition from the A phase into a 3d positionally ordered solid. The label "smectic B" is incorrect since in 40.8 that phase is not liquid crystalline at all; we suggest instead *B— type lamellar solid.* It remains to be proven whether or not the "stacked hexatic phase" does indeed exist in Nature.

The B-solid is unusual since the in-plane Bragg peaks are accompanied by extraordinarily strong diffuse scattering. We have carried out a detailed analysis of this scattering which will be the subject of a future publication. The chief results are as follows.

When we represented the diffuse background for the high resolution data near the Bragg peaks as a two dimensional powder average of the sum of anisotropic Lorentzians.

$$S(\mathbf{q}) \sim (\xi_\perp^2 q_\perp^2 + \xi_z^2 q_z^2 + 1)^{-1} \qquad (1)$$

with $q_\perp^2 = q_x^2 + q_y^2$, each centered at the Bragg positions, a least squares analysis yielded 3000 Å$< \xi_\perp < \infty$. This implies that the diffuse scattering is dominated by q^{-2} singularities as could arise for example from acoustic phonons. We therefore attempted to analyze the low resolution data, shown in Figures 1 and 4, with the background signal represented by a form that approximates the contributions of transverse acoustic phonons with anisotropies and polarization factors comparable to previously reported values.[17,18] Although this could describe the data very well for $|q_\perp - K| \gtrsim 0.05$, where K is the reciprocal lattice vector for the (1, 0, 0) peak, it seriously underestimated the intensities close to the Bragg peaks. The solid lines in Figures 1 and 4 are the results of fitting the data with a background signal that is the superposition of the

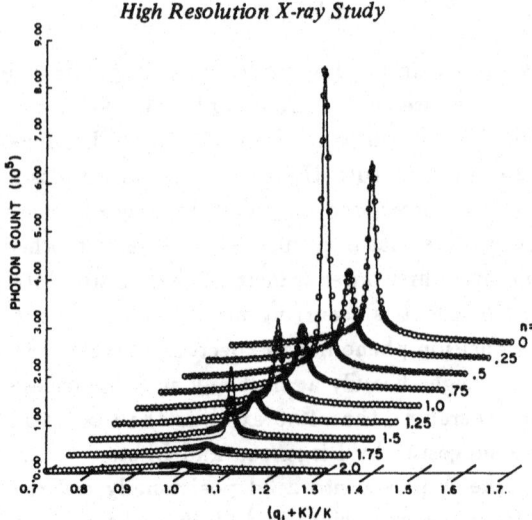

Figure 4. Circles represent the experimental structure factor $S_{expt}(q_\perp, 0, q_z)$, sólid lines the two dimensional powder average of the theoretical structure factor $S_{th}(q_\perp, 0, q_z)$, for $q_z = K_A n$, n = 0, 1/4, 12,..., 2. Bragg peaks correspond to n = 0, 1/2, 1, 1/2, 2.

polarized transverse phonons and an additional sharp ridge that is approximately of the form $1/q_\perp^2$, independent of q_z. The difference between the solid and dashed lines in Figure 1 represents the contribution from this term. On the other hand, in contrast with previously reported values of $C_{44} \sim 10^8$ ergs/cm^3 values as low as 10^6 ergs/cm^3 were reported at the 8th International Liquid Crystal Conference.[19,20] This is nearly two orders of magnitude smaller than previously reported values and would result in anisotropies that allow the diffuse background to be interpreted as orginating in acoustic phonons. If the original, larger values of C_{44} were to be correct the origin of the diffuse scattering would have to be some other long range, nearly two dimensional fluctuations including, for example, either static or dynamic correlations of the aliphatic tails. The explicit origin of this strong 2d diffuse scattering would then remain an open question.

Acknowledgements

We would like to acknowledge helpful conversations with B. I. Halperin, C. W. Garland, D. E. Moncton, D. R. Nelson and R. Pindak.

This work was principally supported by the National Science Foundation Grant No. DMR-78-23555. However, one of us (PSP) received supplemental support from Grant No. DMR-76-22452.

P. S. Pershan, et al.   297

References

1. P. G. De Gennes and G. Sarma, *Phys. Letter*, **A38**, 219 (1972).
2. J. M. Kosterlitz and D. G. Thouless, *J. Phys.*, **C6**, 118 (1973).
3. B. A. Huberman, D. M. Lublin, and S. Doniach, *Solid State Commun.*, **17**, 485 (1975).
4. B. I. Halperin and D. R. Nelson, *Phys. Rev. Letter*, **41**, 121 (1978); A. P. Young, *Phys. Rev.*, **B19**, 1855 (1979); D. R. Nelson and B. I. Halperin, *Phys. Rev.* **B19**, 2457 (1979).
5. R. J. Birgeneau and J. D. Litster, *J. Phys. Lett.* (*Paris*), **39**, 399 (1978).
6. A. M. Levelut, J. Doucet, and M. Lambert, *J. Physique (Paris)* **35**, 773 (1974).
7. A. de Vries, A. Ekachai, and N. Spielberg, *J. Physique (Paris) Colloq.*, **40**, C3-147 (1979).
8. A. Leadbetter, J. Frost, J. P. Gauglan, and M. A. Mazid, *J. Physique (Paris) Colloq.*, **40**, C3-185(1975).
9. J. Doucet, A. M. Levelut, and M. Lambert, *Ann. Phys.* (N.Y.), **3**, 157 (1978).
10. J. Doucet and A. M. Levelut, *J. Physique (Paris)*, **38**, 1163 (1977).
11. A. J. Leadbetter, M. A. Mazid, B. A. Kelley, J. Goodby, and G. W. Gray, *Phys. Rev. Letter*, **43**, 630 (1979).
12. D. E. Moncton and R. Pindak, *Phys. Rev. Letter*, **43**, 701 (1979).
13. S. Bhattacharya and S. V. Letcher, *Phys. Rev. Letter*, **42**, 485 (1979).
14. C. L. Khoon and J. T. Ho, *Phys. Rev. Letter*, **43**, 1167 (1979).
15. K. J. Lushington, G. B. Kasting, and C. W. Garland, *Calorimetric Study of Phase Transitions in the Liquid Crystal Butozybenzylidene Octylaniline* (preprint).
16. J. Als-Neilsen, R. J. Birgeneau, M. Kaplan, J. D. Litster, and C. R. Safinya, *Phys. Rev. Letter*, **39**, 1668 (1977).
17. L. York, N. A. Clark, and P. S. Pershan, *Phys. Rev. Letters* **30**, 639 (1973).
18. S. Bhattacharya and S. V. Letcher, (Private communication).
19. P. Martinoty and Y. Thiriet, *Shear Wave Attenuation Near a Smectic-A to Smectic-B Phase Transition*. 8th International Liquid Crystal Conference, Kyoto, Japan (1980) (unpublished).
20. M. Cagnon and G. Durand, *Mechanical Shear of Layers in Smectic-A and Smectic-B Liquid Crystals*. 8th International Liquid Crystal Conference, Kyoto, Japan (1980) (unpublished).

VOLUME 46, NUMBER 17   PHYSICAL REVIEW LETTERS   27 APRIL 1981

X-Ray Observation of a Stacked Hexatic Liquid-Crystal B Phase

R. Pindak, D. E. Moncton, S. C. Davey, and J. W. Goodby

Bell Laboratories, Murray Hill, New Jersey 07974

(Received 28 January 1981)

X-ray studies have been performed on a new liquid-crystal material which exhibits a noncrystalline B phase. Using free-standing liquid-crystal film techniques, we find that this B phase has short-range in-plane positional correlations but long-range, three-dimensional, sixfold bond-orientational order. We interpret our results in terms of a system of interacting two-dimensional hexatic layers.

PACS numbers: 64.70.Ew, 61.30.Eb

The liquid-crystal B phase is a layered phase with the molecules oriented perpendicular to the layer planes and hexagonally ordered within each layer. X-ray structural studies[1,2] recently demonstrated that this hexagonal order in the prototypical B-phase liquid crystal N-(4-n-butyloxybenzylidene)-4-n-octylaniline (4O.8) involves positional correlations which are three dimensional (3D) and long range. This B phase is, therefore, crystalline. It supports a shear both within and between its layers[3,4] and melts into the smectic-A phase, a higher-temperature phase with fluidlike layers, by a first-order transition.[5] Crystalline B phases have also been observed by us in N-(4-n-butyloxybenzylidene)-4-n-butylaniline (4O.4) and $trans$-1,4-cyclohexane-di-4-n-octyloxybenzoate (TCOB).[6] Although many B phases appear to be crystals, Leadbetter, Frost, and Mazid[7] claimed that some materials exhibited B phases which lacked interlayer correlations. In this paper we report x-ray studies on a similar material using free-standing liquid-crystal film techniques. This approach enables us to ascertain the novel structural nature of this new B phase. Unlike previous crystalline B phases, this new phase has short-range in-plane positional correlations, but it differs from the A phase in having long-range, sixfold bond-orientational order. Halperin and Nelson first proposed the possibility of bond-orientational ordering in their treatment of two-dimensional (2D) melting.[8] They found that a 2D phase having algebraically decaying bond-orientational order (a hexatic phase) would occur between the 2D solid and liquid phases if 2D melting was a dislocation-mediated second-order phase transition. Subsequently, Birgeneau and Litster[9] suggested a 3D liquid-crystal phase consisting of 2D hexatic layers which interact to produce long-range, 3D bond-orientational order. As we will discuss below, this stacked hexatic phase has the structural properties which we have observed in the B phase of a new liquid-crystal material. We refer to this phase as a hexatic B. The discovery of this phase has already motivated experiments which show that this hexatic B phase does not support an in-plane shear[10] and that it melts by a second-order transition.[5]

In the present experiments we have used the same rotating-anode x-ray techniques as in our previous study of the crystalline B phase in 4O.8.[1] Samples were free-standing films at least 100 molecular layers thick. Our work on 4O.8 demonstrated that these films are single-crystal–quality samples with in-plane domains ~ 1 mm^2 and with layer alignment better than 0.01°. These qualities offer considerable technical advantages over the conventional field-aligned bulk samples. For this experiment, field-aligned samples would not permit a study of the bond-orientation structure within the layers since such samples are completely disordered with respect to rotation about the layer normal.

We chose to examine the liquid-crystal material n-hexyl-4'-n-pentyloxybiphenyl-4-carboxylate (65OBC),[11] which exhibits the following sequence of phase transitions on cooling:

$$\text{Isotropic} \xrightarrow{86°} \text{smectic-}A \xrightarrow{68°} B \xrightarrow{60°} E \,(\text{crystal}).$$

Three different types of scans are necessary to establish the structural properties of 65OBC (see inset to Fig. 1). These scans are referred to in hexagonal reciprocal-lattice coordinates with $a^* = 4\pi/(3^{1/2}a) = 1.420$ Å^{-1} and $c^* = 2\pi/c = 0.242$ Å^{-1}. Scans with $Q = Q_\parallel$ (momentum transfer in the plane of the layers) probe the extent of in-plane positional correlations. Scans which rotate the film about the layer normal (χ scans) probe bond-orientational order. Finally, scans along $Q_\perp$ at $Q_\parallel = 1$ probe the extent of interlayer correlations.

Scans in the B phase of 65OBC are shown in Fig. 1. There is no evidence for any Bragg peaks such as seen in the crystalline B phase of 4O.8. Rather we find a rod of scattering along $Q_\perp$ which

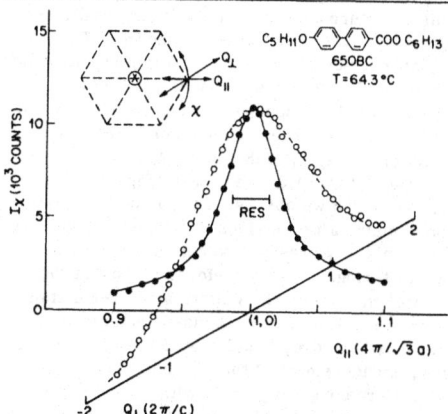

FIG. 1. χ-averaged intensity for a $Q_{\parallel}$ scan (closed circles) and a $Q_{\perp}$ scan (open circles) in the B phase of 65OBC. Note that the $Q_{\parallel}$ scale has been expanded relative to the $Q_{\perp}$ scale. The scattering along $Q_{\perp}$ is a diffuse rod. The resolution width for the $Q_{\perp}$ scan ($\Delta Q_{\perp} = 0.006$ Å^{-1} full width at half maximum) was too small to illustrate. The inset describes the three scan directions.

indicates the complete absence of interlayer correlations. Furthermore, the peak is not resolution limited in $Q_{\parallel}$. The solid line through the data points is the result of a least-squares fit of the χ-averaged 2D Lorentzian structure factor $S(Q_{\parallel}) = 1/[(Q_{\parallel} - Q_0)^2 + \kappa^2]$ convoluted with the Gaussian instrumental resolution function of width $\Delta Q_{\parallel} = 0.038$ Å^{-1} full width at half maximum. This scan demonstrates that the in-plane positional correlations decay exponentially with a correlation length $\xi_{\parallel} = 1/\kappa \simeq 100$ Å.

Although the positional order is short range in the B phase of 65OBC, the scattering differs distinctly from the smectic-A phase in a χ scan. A 60° segment of a χ scan is shown in Fig. 2 for several temperatures. From heat-capacity measurements[5] the $A \rightarrow B$ transition temperature is $T_{AB} = 67.9$ °C. Above T_{AB} the scattering shown in Fig. 2(a) is a constant, independent of χ. For temperatures below T_{AB} the χ scan develops substantial sine-wave modulation [Fig. 2(b)] indicative of a sixfold periodicity. This scattering is direct evidence for hexagonal bond-orientational correlations which are three-dimensionally ordered over distances of the order of the area illuminated by the x-ray beam (~ 2 mm²).

Structure factor calculations[12] for the stacked

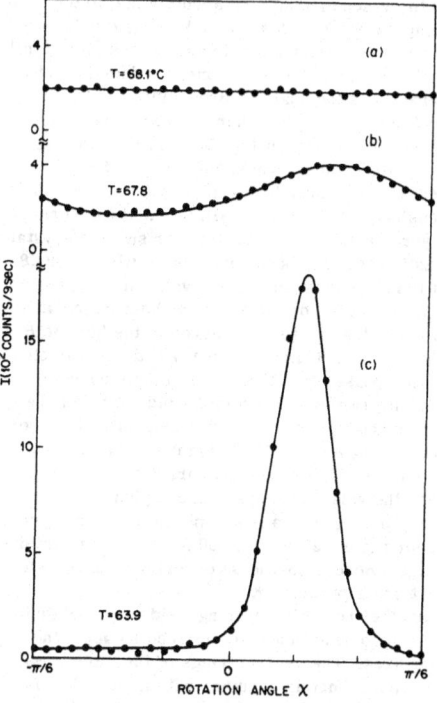

FIG. 2. χ scans for three different temperatures: (a) In the smectic-A phase; (b) just below the smectic-$A \rightarrow B$ transition; (c) well into the B phase (below T_0).

hexatic phase demonstrate that, in principle, a measurement of the amplitude of the χ-scan modulation provides a method for measuring the temperature dependence of the bond-orientational order parameter. In practice, the situation is complicated by the presence of more than one orientational domain within the scattering volume. If different domains produce sine-modulated scattering with different χ phase shifts, the resulting scattering will still be a single sine wave of reduced amplitude. It is not possible to interpret unambiguously the amplitude of the modulation unless one knows that the film contains only a single domain. Since the amplitude that we measure is not a reproducible function of temperature in the vicinity of T_{AB}, we believe that a small but variable number of domains are probably present within the area probed.

As the hexatic B phase is cooled, the amplitude

of the χ-scan modulation abruptly stabilizes at a temperature $T_0 = 66.4\,°C$. Below T_0, the $60°$ χ scans can develop a single sharp peak [Fig. 2(c)] indicating a single, well-oriented domain. This behavior is thermally reproducible and there is no anomaly at T_0 in either the positional correlation length (arrow in Fig. 3) or in the heat capacity.[5] Mechanical measurements[10] on 65OCB suggest this enhanced bond-orientational order is possibly due to the crystallization of the surface layers of the film at T_0. Similar surface crystallization has also been observed in films of 4O.8 at temperatures above its crystalline B phase.[3]

We next consider whether the bond-orientational order that we have observed in the hexatic B phase of 65OBC is the primary order parameter of this phase or whether it is being induced by coupling to another ordering field. As we have shown above, long-range lattice positional order does not develop at T_{AB}. Figure 3 shows measurements of the positional correlation length over the relevant temperature region. In the vicinity of T_{AB}, there is a rapid increase in $\xi_\parallel$ on cooling (from about 20 to 60 Å) but it does not diverge. Hence, positional ordering is not a relevant order parameter.

Another possible ordering field is associated with the orientation of the molecules about their long axes. The benzene rings of the molecules are approximately coplanar with a lateral van der Waals size $6.7 \times 3.7\,Å^2$. It is evident that such molecules cannot pack with a spacing $a = 5.11\,Å$ and still maintain complete rotational freedom about their long axes. A local herringbone packing structure was proposed by Levelut.[13]

This structure explains the additional scattering observed at $Q_\parallel = 1.32$, which is shown in Fig. 4. The absence of a peak at $Q_\parallel = 0.5$ implies the presence of a glide-plane symmetry which is contained in the herringbone packing model. The local herringbone order can assume three distinct directions $\vec{x}_1, \vec{x}_2, \vec{x}_3$ each contributing four diffraction spots at $Q_\parallel = 1.32$ (see inset, Fig. 4). Below T_0, even when we observe a single bond-orientation domain, all directions of the herringbone are simultaneously present and contribute equally to the scattering. It is, therefore, clear that the herringbone structure is correlated over distances much smaller than the bond-orientational order. Furthermore, it is known from neutron scattering measurements[14] that the molecular-orientational ordering is dynamic with a molecular orientation time $\sim 10^{-11}$ sec. For these reasons we rule out the possibility that the molecular-orientational order is driving the bond-orientational order, although it is certainly possible that the molecular orientational order enhances the bond-orientational order.

In conclusion, our x-ray study has established the existence of a 3D stacked hexatic B phase with sixfold bond-orientation as the order parameter. Since the existence of a 2D hexatic phase

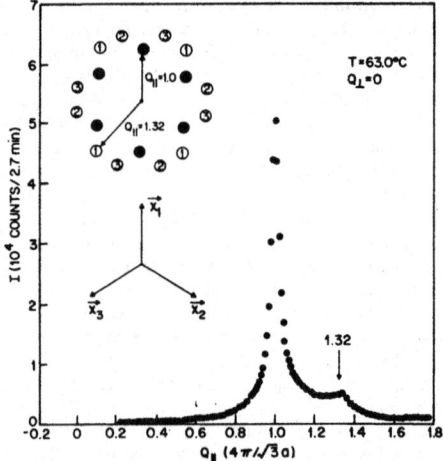

FIG. 4. An expanded $Q_\parallel$ scan showing the additional scattering at $Q_\parallel = 1.32$. The inset describes the scattering that results from a local herringbone packing: A herringbone oriented in the direction $\vec{x}_1$ $(\vec{x}_2, \vec{x}_3)$ gives rise to four diffraction spots labeled 1 (2, 3).

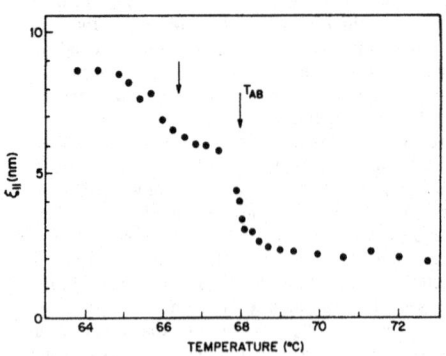

FIG. 3. Temperature dependence of the in-plane positional correlation length $\xi_\parallel$.

has not yet been experimentally confirmed, our study gives important experimental support to the existence of hexatic ordering. Furthermore, there exists the possibility that films of 65OBC will exhibit a 2D hexatic phase when studied in the thin film limit (2 molecular layers). These experiments are currently in progress. We note that the crossover from 3D to 2D crystalline order has already been observed in thin films of 4O.8 in its crystalline B phase.[1] Finally, we have also used the free-standing film technique to study smectic phases in which the molecules are tilted with respect to the layer planes. Our results[15] are in agreement with previous studies[16] which indicated a new phase without long-range positional order at temperatures below the smectic-C phase (tilted analogue of the smectic-A phase). In this case of tilted molecules, the coupling of the tilt direction to sixfold bond-orientational order adds a new and potentially interesting aspect to the physics.

We are grateful for helpful discussions with W. F. Brinkman, D. S. Fisher, B. I. Halperin, and D. R. Nelson.

[1]D. E. Moncton and R. Pindak, Phys. Rev. Lett. 43, 701 (1979).

[2]P. S. Pershan, G. Aeppli, J. D. Litster, and R. J. Birgeneau, in Proceedings of the Eighth International Liquid Crystal Conference, Kyoto, Japan, June–July 1980 (to be published); A. J. Leadbetter, M. A. Mazid, B. A. Kelley, J. Goodby, and G. W. Gray, Phys. Rev. Lett. 43, 630 (1979); J. Doucet and A. M. Levelut, J. Phys. (Paris) 38, 1163 (1977).

[3]R. Pindak, D. J. Bishop, and W. O. Sprenger, Phys. Rev. Lett. 44, 1461 (1980).

[4]M. Cagnon and G. Durand, Phys. Rev. Lett. 45, 1418 (1980).

[5]C. C. Huang, J. M. Viner, R. Pindak, and J. W. Goodby, to be published.

[6]R. Pindak, D. E. Moncton, M. E. Neubert, and M. E. Stahl, unpublished.

[7]A. J. Leadbetter, J. C. Frost, and M. A. Mazid, J. Phys. (Paris), Lett. 40, 325 (1979).

[8]B. I. Halperin and D. R. Nelson, Phys. Rev. Lett. 41, 121 (1978).

[9]R. J. Birgeneau and J. D. Litster, J. Phys. (Paris), Lett. 39, 399 (1978).

[10]R. Pindak, D. J. Bishop, W. O. Sprenger, and D. D. Osheroff, to be published.

[11]J. W. Goodby and R. Pindak, to be published.

[12]R. Bruinsma and D. R. Nelson, Phys. Rev. B 23, 402 (1981).

[13]A. M. Levelut, J. Phys. (Paris), Colloq. 37, C3-51 (1976).

[14]H. Hervet, F. Volino, A. Dianoux, and R. E. Lechner, J. Phys. (Paris), Lett. 35, 151 (1974).

[15]D. E. Moncton, R. Pindak, and J. W. Goodby, Bull. Am. Phys. Soc. 25, 213 (1980).

[16]J. J. Benattar, F. Moussa, and M. Lambert, J. Phys. (Paris), Lett. 41, 1371 (1980); P. A. C. Gane, A. J. Leadbetter, and P. G. Wrighton, in Proceedings of the Eighth International Liquid Crystal Conference, Kyoto, Japan, June–July 1980 (to be published).

VOLUME 49, NUMBER 8   PHYSICAL REVIEW LETTERS   23 AUGUST 1982

Synchrotron X-Ray Study of Novel Crystalline-B Phases in Heptyloxybenzylidene-Heptylaniline (7O.7)

J. Collett, L. B. Sorensen, and P. S. Pershan

Division of Applied Sciences, Harvard University, Cambridge, Massachusetts 02138

and

J. D. Litster and R. J. Birgeneau

Department of Physics, Massachusetts Institute of Technology, Cambridge, Massachusetts 02139

and

J. Als-Nielsen

Risø National Laboratory, DK-4000 Roskilde, Denmark

(Received 10 May 1982)

This paper reports an x-ray diffraction study of structures and restacking transitions within the B phases of heptyloxybenzylidene-heptylaniline. The system evolves from a hexagonal close-packed structure, through intermediate orthorhombic and monoclinic phases, to a simple hexagonal structure. The monoclinic phase has a temperature-dependent shear which transforms the system from orthorhombic to hexagonal. The latter three phases exhibit a single-$\bar{q}$ sinusoidal modulation of the molecular layers.

PACS numbers: 61.30.Eb, 64.70.Ew, 81.30.Hd

The realization that an infinitesimally weak interlayer coupling will induce three-dimensional (3D) long-range order in a stack of two-dimensional (2D) crystals has altered our view of the phases previously identified as ordered smectics.[1] Many phases classified as smectic B actually are layered molecular crystals (crystalline B) with unusual features which are as yet both poorly characterized and barely understood. X-ray diffraction experiments show strong diffuse scattering due to modes polarized in the plane of the layers. This scattering has intralayer correlation lengths greater than 3000 Å but has little interlayer correlation.[2,3] This indicates that layers slide rigidly over one another with relative ease. A second feature is the presence of restacking transitions in which there are dra-

matic shifts in the relative positions of adjacent layers.[2,4-7] No latent heat or other thermal anomaly has been associated with these transitions, leaving their detailed nature a mystery. It is also not known whether the restacking transition is driven by the interlayer coupling or if it arises from a transition occurring within the individual 2D layers. These crystalline-B solids thus represent prototypes of 3D systems with unusually large directional anisotropy. We expect that elucidation of their properties will enhance generally our understanding of crossover from 2D to 3D collective behavior.

High-resolution x-ray diffraction studies of freely suspended thick films of heptyloxybenzylidene-heptylaniline (7O.7) were carried out by using synchrotron radiation and the triple-axis

VOLUME 49, NUMBER 8   **PHYSICAL REVIEW LETTERS**   23 AUGUST 1982

spectrometer installed by one of us (J. A.-N.) at HASYLAB in Hamburg, Germany. Single crystals of Ge(111) were used as monochromator and analyzer. Lower-resolution studies were performed on similar spectrometers at Harvard and the Massachusetts Institute of Technology, using pyrolytic graphite crystals and Cu $K\alpha$ radiation from rotating-anode sources. Freely suspended films 600 to 3000 layers thick were prepared in a manner similar to others.[2] We were thus able to examine samples with a few single-crystal domains. The samples were oriented with the normal to the molecular layers always in the scattering plane. Scans along $\vec{q}_{x,y}$, in the plane of the layers, or along $\vec{q}_z$, normal to the layers, were done. A third rotation about the normal to the layers allowed us to choose the direction of $\vec{q}_{x,y}$ within the layer.

We find that 7O.7 exhibits rich structural behavior between 63° and 59 °C with a series of three transitions involving restacking and subtle changes in the intralayer packing.[8] In order of decreasing temperature, the structures are hexagonal close-packed (69–63 °C), orthorhombic F (63–60.1 °C), monoclinic C (60.1–59.75 °C), and hexagonal (59.75–55 °C). In the last three phases there is also a long-wavelength modulation in which the smectic layers are displaced along their normals. The relative stacking of adjacent layers is shown in Fig. 1. The real and reciprocal lattices for the three lower-temperature phases are also given in the figure.

In the temperature range 69 °C > T > 63 °C, 7O.7 exhibits the hcp structure with $ABAB$ stacking common in crystalline-B materials.[2-4] Scans along $\vec{q}_z$ passing through any of the six Bragg peaks with $\vec{q}_z = 0$ show peaks at intervals of $2\pi/c$ (c is the spacing between smectic layers) as shown in Fig. 2(a). At $T = 63$ °C there is an abrupt first-order transition to a phase which can be described either by an orthorhombic-F structure or by a triclinic primitive cell with one 7O.7 molecule per unit cell. Here $\vec{q}_z$ scans through $\vec{b}_1$ and $\vec{b}_2$ are not equivalent [Figs. 2(b) and 2(d)]. High-resolution measurements show that the lattice is compressed along $\hat{y}$ and extended along $\hat{x}$ [see Fig. 3(a)]. The relative changes are parametrized by ϵ_x and ϵ_y in Fig. 1. The distortion is very near to being area conserving with $\epsilon_x = \epsilon_y$ = 0.0010 ± 0.0001. The choice $s = \tfrac{1}{2}$ in Fig. 1 yields the lattice vector of the triclinic primitive cell.

At 60.1 °C the orthorhombic-F structure shears into a monoclinic-C structure. Within the monoclinic-C phase the displacement between adjacent layers is $s\vec{a}_1$ (Fig. 1) where s varies approximately linearly with decreasing temperature from 0.45 to 0.05 [Fig. 3(b)]. The lattice vectors given in Fig. 1 are again those of the primitive cell. There are small, weakly first-order jumps in s at both the upper and lower transitions. Scans along $\vec{q}_z$ through $\vec{b}_1$ [Fig. 2(c)] and through $\vec{b}_2$ [Fig. 2(d)]

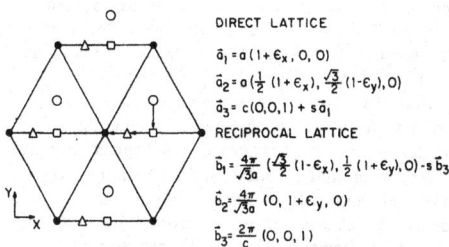

DIRECT LATTICE

$\vec{a}_1 = a(1+\epsilon_x, 0, 0)$

$\vec{a}_2 = a(\tfrac{1}{2}(1+\epsilon_x), \tfrac{\sqrt{3}}{2}(1-\epsilon_y), 0)$

$\vec{a}_3 = c(0,0,1) + s\vec{a}_1$

RECIPROCAL LATTICE

$\vec{b}_1 = \tfrac{4\pi}{\sqrt{3}a}(\tfrac{\sqrt{3}}{2}(1-\epsilon_x), \tfrac{1}{2}(1+\epsilon_y), 0) - s\vec{b}_3$

$\vec{b}_2 = \tfrac{4\pi}{\sqrt{3}a}(0, 1+\epsilon_y, 0)$

$\vec{b}_3 = \tfrac{2\pi}{c}(0, 0, 1)$

FIG. 1. Relative displacements of adjacent layers in the various phases: closed circles, reference layer position; open circles, hexagonal closed packed; squares, orthorhombic F; and triangles, monoclinic C. In the hexagonal AAA phase the adjacent layer positions coincide with the reference layer. At the right are the primitive unit cells for the real and reciprocal lattices of the orthorhombic-F, monoclinic-C, and hexagonal AAA structures.

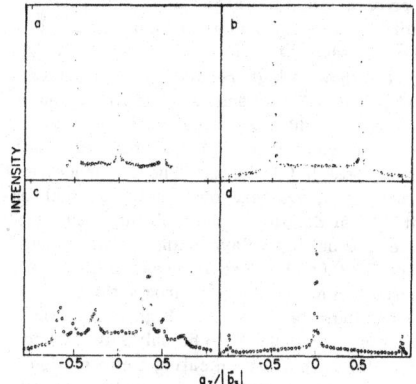

INTENSITY

$q_z/|\vec{b}_3|$

FIG. 2. (a) q_z scan through an in-plane peak for $T > 63$ °C ($T = 65.43$ °C). (b) $\vec{q}_z$ scan through $\vec{b}_1$ in orthorhombic-F phase at $T = 60.88$ °C. (c) $\vec{q}_z$ scan through $\vec{b}_1$ in monoclinic-C phase at $T = 60.02$ °C. (d) $\vec{q}_z$ scan through $\vec{b}_2$ in hexagonal AAA phase at $T = 59.35$ °C. Identical scans through $\vec{b}_2$ are found in both the orthorhombic and monoclinic phases.

304

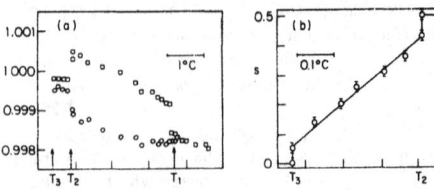

FIG. 3. (a) Measurement of distortion in intralayer packing: Circles, $|\vec{b}_1 \times \hat{z}|$; squares, $|\vec{b}_2|$. Data are plotted in units of 1.434 Å^{-1}. Transition temperatures are marked; T_1, hexagonal close packed to orthorhombic F; T_2, orthorhombic to monoclinic; T_3, monoclinic to hexagonal AAA. (b) Adjacent layer displacement vs temperature in the monoclinic-C phase.

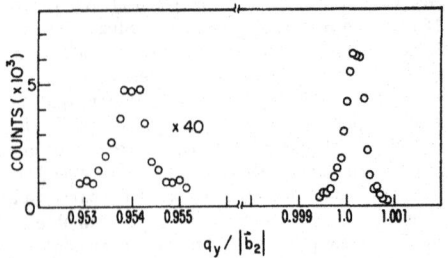

FIG. 4. Synchrotron high-resolution $\vec{q}_y$ scans through $\vec{b}_2 + \vec{b}_3$ and through the satellite peak at $0.954\vec{b}_2 + \vec{b}_3$ in hexagonal AAA phase.

show that the layers shear along the $\vec{a}_1$ direction. The presence of peaks at both $\pm 0.3|\vec{b}_3|$ at $T = 60$ °C indicates that although the shear is along a single direction, there are both $\pm s\vec{a}_1$ displacements. Although the relative intensities of the $\pm s\vec{a}_1$ peaks are sample dependent, there is no evidence of hysteresis. There are small residual peaks at $\pm 0.5\vec{b}_3$ that could arise from strain-induced coexistence. The distortion from hexagonal intralayer packing relaxes in this phase to $\epsilon_x \simeq \epsilon_y \simeq 0.0002 \pm 0.0001$.

At $T = 59.75$ °C the monoclinic-C phase transforms into a simple hexagonal phase with AAA stacking. Molecules in adjacent layers occupy the reference sites in Fig. 1 and the intralayer distortion vanishes. Scans along $\vec{q}_z$ through $\vec{b}_1$ and $\vec{b}_2$ are identical and are shown in Fig. 2(d). Figure 4 shows a high-resolution $\vec{q}_y$ scan taken at HASYLAB. The intensity of the Bragg peak relative to the thermal diffuse background is about 500. The full width at half maximum of the primary peak is $0.0006|\vec{b}_2|$. The orthorhombic-F, monoclinic-C, and hexagonal AAA phases all exhibit a 1D modulation of the molecular layers polarized normal to the layers and having a wave vector $\vec{Q} \simeq 0.046 \vec{b}_2$. These appear as pairs of sidebands in scans along $\vec{q}_y$ through the $(0,1,\pm 1)$ peaks as indicated in Fig. 4. High-resolution data show the sidebands to be only slightly broader than the resolution indicating a correlation length greater than 4000 Å. Leadbetter et al.[4] have reported the existence of modulations with this polarization in crystalline-B materials that have G phases at temperatures immediately below the B. The high-quality samples we were able to produce allowed us to show that in 70.7 the modulation $\vec{q}$ vector is in the direction of the

molecular displacements that occur at the hcp-to-orthorhombic-F restacking transition.

In discussing the above structures in 70.7 we should note that an AAA simple hexagonal structure with one high-symmetry unit per primitive cell does not appear to have been observed previously in nature. However, the very sharp $\vec{b}_2$ satellite peaks which we observe in these novel crystalline-B phases suggest that the underlying structures must be more complicated. From the intensity of the satellite peak shown in Fig. 4 we infer that the amplitude of the $\vec{b}_2$ transverse sinusoidal modulation wave is about $0.05c$ in the direction perpendicular to the planes. Since the period is about 20 molecules this corresponds to a nearest-neighbor vertical displacement of only 1% of the molecular length. In the lower-temperature G phase of 70.7 the corresponding displacement[5] is about 8%. The driving mechanism for the sinusoidal modulation is almost certainly the tilt degree of freedom. From Fig. 1 it is evident that a tilt of the molecule in the $\vec{b}_2$ direction will facilitate the packing of adjacent layers. We have not, however, succeeded in producing a microscopic model that can account quantitatively for the sinusoidal modulation. Similar modulations observed in lipid-water systems have been modeled as a martensitic transformation by Chan and Webb.[9] However, their inferred displacements are much too large to account for our observed intensities in 70.7. It is possible, however, that their model is relevant to systems such as 50.7 where the B-G transition is more weakly first order.[4] Indeed the temperature variation of the observed intensities in 50.7 can be simply understood on the basis of an increasing amplitude of the density wave.

Finally, we should emphasize that the new features which our high-resolution single-domain studies have revealed are by no means unique to 7O.7. Doucet and Levelut[5] have reported lower-resolution x-ray studies of a number of nOm compounds. In 7O.7, 7O.5, and 5O.7 they report a B phase with an extra "ring" of low-angle scattering in powder patterns; this may now be identified as originating from the sinusoidal modulation peaks around $(0,0,1)$. Moncton and Pindak[2] reported some single-domain data on the B-phase restacking transitions in 4O.8. Although their investigation was not extensive enough to make unambiguous identifications, it seems likely that the lowest two phases which they observe correspond to our orthorhombic-F and monoclinic-C phase.

It is evident that these quasi-2D solids exhibit highly unusual structures and structural phase transitions. The continuous shear transformation in the monoclinic phase of 7O.7 is particularly interesting since adjacent layers slide over one another with only minute distortions in the intralayer packing. The sinusoidal corrugation of the layers throughout the B phases is a novel feature which remains unexplained. In order to disentangle intralayer versus interlayer effects we plan to study these structures as a function of film thickness. Finally, it is clear that there is need for a significant development of theories for the structures of quasi-2D molecular solids.

The research reported here was supported in part by the National Science Foundation under Grants No. DMR-79-19479, No. DRM78-24185, and No. DMR78-23555, in part by the Joint Ser-

vices Electronics Program under Grant No. DAAG-29-80-C-0104, and in part by a NATO research grant. The triple-axis spectrometer at HASYLAB was funded by the Danish National Science Foundation. We also wish to express our thanks to the staff of HASYLAB, especially Dr. G. Materlik and Dr. V. Saile, for their assistance.

---

[1]This has been known for a long time in magnetic systems, and was implicit in the de Gennes–Sarma model of smectic B [P. G. de Gennes and G. Sarma, Phys. Lett. 38A, 219 (1972)]. It has been reemphasized for liquid crystals by R. J. Birgeneau and J. D. Litster, J. Phys. (Paris), Lett. 39, 399 (1978).

[2]D. E. Moncton and R. Pindak, Phys. Rev. Lett. 43, 701 (1979).

[3]P. S. Pershan, G. Aeppli, J. D. Litster, and R. J. Birgeneau, Mol. Cryst. Liq. Cryst. 67, 205 (1981).

[4]A. J. Leadbetter, M. A. Mazid, B. A. Kelley, J. Goodby, and G. W. Gray, Phys. Rev. Lett. 43, 630 (1979).

[5]J. Doucet and A. M. Levelut, J. Phys. (Paris) 38, 1163 (1977).

[6]A. J. Leadbetter, J. C. Frost, and M. A. Mazid, J. Phys. Lett. 40, 325 (1979).

[7]A. J. Leadbetter, M. A. Mazid, and R. M. Richardson, in Liquid Crystals, edited by S. Chandrasekhar (Cambridge Univ. Press, London, 1980), pp. 65–79.

[8]The original phase diagram for 7O.7 [K 33 G 55 B 69.0 C 72.0 A 83.7 N 84.0 I, Verbit notation] was obtained by G. W. Smith, Z. G. Garland, and R. J. Curtis, Mol. Cryst. Liq. Cryst. 19, 327 (1973); G. W. Smith and Z. G. Garland, J. Chem. Phys. 59, 3214 (1973). The G phase was labeled H in Ref. 4.

[9]W. K. Chan and W. W. Webb, Phys. Rev. Lett. 46, 39 (1981).

306

Dislocation Model for Restacking Phase Transitions in Crystalline-B Liquid Crystals

J. P. Hirth

Department of Metallurgical Engineering, The Ohio State University, Columbus, Ohio 43210

and

P. S. Pershan, J. Collett,[a] E. Sirota, and L. B. Sorensen[b]

Division of Applied Sciences, Harvard University, Cambridge, Massachusetts 02138

(Received 20 March 1984; revised manuscript received 14 June 1984)

A dislocation-mediated model is presented for restacking phase transitions that have been observed in a variety of lamellar (liquid crystalline) systems. The model explains the existence of nonhexagonal crystalline (smectic)-B phases in terms of dislocation-induced tilting of hexagonally packed layers. Ordered dislocation arrays explain both the symmetry and the amplitude of observed modulations. It is likely that the model will also be applicable to modulated lipid-water phases.

PACS numbers: 64.70.Ew, 61.30.−v, 81.30.Kf

Current ideas concerning the types of order and phase transitions in two-dimensional systems[1,2] have stimulated renewed theoretical interest in dislocation-mediated phase transitions in both two and three dimensions.[3,4] Unfortunately, there are not many experimentally observed phase transitions that can be unambiguously interpreted in terms of dislocations. One of the more interesting properties of many liquid crystal materials is that on cooling they form hexagonal, or nearly hexagonal, layered crystals that exhibit subtle structural phase transitions within a small temperature range.[5-7] In this Letter we demonstrate a dislocation model that explains most of the features of the structural phase transitions observed in three-dimensional samples of heptyloxybenzylidene-heptylaniline, 7O.7.[7] Nearly all of the liquid crystalline systems that have phases with long-range in-plane positional order (i.e., smectic B,G,H, etc.)[8,9] exhibit similar phase transitions. In most cases where there is a low-temperature phase with tilted molecules, they are accompanied by an unexplained long-wavelength modulation that is similar to the modulation accompanying one of the principal phase transitions commonly observed in lipid-water mixtures.[10] We believe the proposed model is generally applicable to all of the liquid crystal systems and may also be applicable to lipid-water systems. In the latter case, Chan and Webb interpreted the appearance of long-wavelength modulation in terms of a martensitic transformation. Although dislocation models for martensitic transformations have been discussed previously, the present specific model has new features that might well explain the lipid-water system and also be applicable to metals.

Between 69 and 63 °C thick films of 7O.7 form an hexagonal close packed (hcp) crystal with $ABAB$

stacking. Figure 1(a) displays the A, B, and C sites of a two-dimensional hexagonal lattice. At 63 °C 7O.7 undergoes a first-order transition in which the macroscopic symmetry changes from hcp to a modulated orthorhombic-F phase. The x-ray structure of the latter has been explained by assuming that the molecules at the B sites move to the position marked by X one layer below (or above) the A

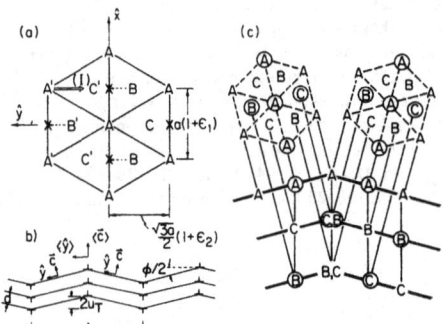

FIG. 1. (a) Hexagonal lattice sites A, B, and C. The hexagonal lattice with alternate layers in the A and B positions transform to orthorhombic-F by displacing the B positions to the X position. Partial dislocations with Burgers vector (I) in every second layer transforms the hexagonal $ABAB$ to a hexagonal $ABCABC$. The notation for in-plane strains ϵ_1 and ϵ_2 is shown. (b) A regular array of edge dislocations of opposite sign produces tilt boundaries as shown. (c) Projection of lattices that are locally hexagonal ABC and ACB onto tilted layers to obtain a modulated orthorhombic-F lattice. Circle letters indicate equivalent x coordinates. Viewed from opposite sides of the tilt boundary, single molecular sites in the interface are either B or C.

473

sites.[7] Accompanying this is the appearance of side-bands on the $(0,0,1)$, $(1,0,l)$ Bragg peaks with $l \neq 0$ due to a long-wavelength modulation, $\lambda \simeq 18.8a$, with wave vector parallel to the BX direction and displacements normal to the layers. For the $(1,0,1)$ Bragg peak the only observed sideband is the lowest order, with the intensity $\simeq 2\%$ of the main peak. Shifts in the positions of the $(1,0,l)$, $(0,1,l)$, and $(1,1,l)$ Bragg peaks imply strains of $\epsilon_1 = -\epsilon_2 = 0.001$.

The literature abounds with successful phase transformation models involving localized defects of either the shear type, the shuffle type, or a mixture of the two.[11,12] The common feature is the assumption of spontaneously nucleated dislocation loops of small Burgers vector that grow and interact to produce a macroscopic transformation. In the present case, consider a partial dislocation in one layer of A sites that transforms the crystal from $ABABABABA$ to $ABCACACAC$ stacking. The Burgers vector (I) is shown in Fig. 1(a).[13] Consider further that if a similar dislocation is repeated every second layer throughout the crystal the original lattice is converted to $ABCABCABC$. This is an obvious possibility for the molecule pentyloxybenzylidine-hexylaniline (5O.6) that is only slightly different from 7O.7. In the case of (5O.6) there is a hexagonal phase with ABC stacking between a hexagonal phase with $ABAB$ stacking and the orthorhombic-F phase.[14]

If, however, we assume that in 7O.7 the Burgers vector of each of the successive dislocations were parallel to one another, and that the dislocations interacted to be above one another the result would be a tilt boundary. Figure 1(b) illustrates a regular array of tilt boundaries of opposite sign. For 7O.7 the layer thickness $d \simeq 6a$. According to the Frank formula[15] the angle $\phi/2 = \sin^{-1}(a\sqrt{3}/6d) = 0.048$ is exactly the correct tilt to put the sites A', B', and C' [Fig. 1(a)] in a common vertical plane as required for the orthorhombic-F.

Figure 1(c) illustrates the molecular positions. On the two sides of the tilt boundary the local packing is either hexagonal ABC or ACB. To the right of the tilt boundary the transformation is as described above. To the left, however, the dislocations occur in the alternate layers and the transformation is from $ABABAB$ to $ACBACB$. However, as viewed along the average $\langle \bar{c} \rangle$ axis the molecules form the face-centered rectangle of the orthorhombic-F phase.

A model to explain the formation of a regular dislocation array is illustrated in Figs. 2(a) and 2(b). With the assumption of spontaneous nucleation of a

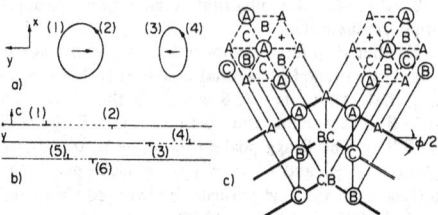

FIG. 2. (a) Top view of two dislocation loops (1)-(2) and (3)-(4). The arrows inside the loops indicate Burgers vectors and those on the loops the senses. (b) Side view of dislocations loops whose elastic interactions favor mutual formation. (c) Projection of a lattice that is hexagonal ABC onto tilted layers to obtain a modulated hexagonal AAA lattice. The plus indicates a missing molecule due to the interposition of dislocations.

single dislocation loop (1)-(2) of the type discussed above, the elastic-strain field provides a loop-loop elastic interaction favoring nucleation of the loop (5)-(6) two planes directly above or below.[16] These two will then produce a field favoring nucleation of a (3)-(4), one plane removed from each, displaced laterally, and of opposite sign. Successive loop nucleations of this kind produce the edge dislocation tilt walls of Fig. 1(b). The screw portions of the loops, with lines parallel to the y direction could also meet and interact to form an array. Such an array, however, would not be a low-energy one (a crossed-grid of screw dislocations is required for a stable, low-angle twist boundary).[12] The attendant repulsive interactions among screw segments would tend to cause them to slip out of the crystal, leaving the edge array of Fig. 1(b).

To evaluate the quantitative predictions of the model note that the values quoted above for the strains ϵ_1 and ϵ_2 were obtained from shifts in the positions of the Bragg peaks. If we define $(\epsilon_1^l, \epsilon_2^l)$ to be the strains measured along the local $\hat{x}, \hat{y}$ axis, then one can show that the strains (ϵ_2, ϵ_2) measured relative to the average $\hat{x}, \langle \hat{y} \rangle$ axis are $\epsilon_1 = \epsilon_1^l$, $\epsilon_2 = \epsilon_2^l - \phi^2/8$. Under the assumption that the transformation is accompanied by a simple area dilation in the local $\hat{x}$-$\hat{y}$ smectic layers, one obtains $\epsilon_1^l = \epsilon_2^l = 0.001$ and $\phi/2 \simeq (0.004)^{1/2} = 0.063$. This should be compared with the value of 0.048 obtained from the Frank formula under the implicit assumption that $(\epsilon_1^l, \epsilon_2^l) = 0$. With the assumption of the triangular modulation shown in Fig. 1, the predicted ratio of sideband to $(1,0,1)$ peak intensity $(8u_T/\pi d)^2 \simeq (\phi\lambda/\pi d)^2 \approx 0.016$. The measured ratio was $\simeq 0.02$. Although we are not able to predict the value of λ this quantitative agreement strongly supports the model.

Further qualitative support for the model is the above-mentioned hexagonal phase with ABC stacking that occurs in 5O.6 between the hexagonal $ABAB$ phase and the orthorhombic-F (48 to 43.8 °C). The Bragg peaks that occur in this phase at $(1, 0, \pm \frac{2}{3})$ and $(1, 0, \pm \frac{1}{3})$ are accompanied by diffuse sidebands cylindrically distributed about the peak.[14] The transition at 48 °C produces layers with ABC stacking; however, according to the model the accompanying dislocations have not condensed into the regular array. This happens in 5O.6 at 43.8 °C where one observes a modulated othorhombic-F phase identical to that in 7O.7.

At 60.1 °C thick films of 7O.7 undergo a second transition from the orthorhombic-F to a monoclinic phase. The x-ray data for this phase are consistent with the assumption that the molecules located at the X position in Fig. 1(a) moved along the line XA. Following a small sudden displacement at 60.1 °C the molecules move continuously with decreasing temperature until at 59.75 °C there is a second small displacement that places this layer directly under the A position in the adjacent layer. These effects can also be explained by an extension of the previous model. For example, Fig. 2(c) displays a structure similar to that of Fig. 1(c) except that it has a modulated AAA structure. The tilt angle $\phi/2$ required to produce this structure, $\tan^{-1}(\sqrt{3}a/3d)$, is exactly twice the tilt required to produce the orthorhombic-F structure. One way to produce this starting from the orthorhombic phase depicted in Fig. 1(c) is illustrated in Fig. 3. Starting from a structure that is locally $ABCABCA$ (orthorhombic-

F) a partial dislocation in the second layer (of the same type as invoked previously) with Burgers vector (II) converts the structure to the $ACABCAB$ structure when in the center panel. A second dislocation, either the same type (II) or of the type (III) in the third layer converts this to $ACBCABC$. If this sequence is repeated one obtains an $ACBAC-BACB$ structure with an increase in the Frank tilt by $\Delta(\phi/2) = \tan^{-1}(\sqrt{3}a/6d)$. This is exactly the correct value to be added to the orthorhombic-F phase to obtain the structure in Fig. 2(c).

The monoclinic phase that separates the orthorhombic-F and the hexagonal AAA would then be an intermediate phase for which there were some statistical distribution in the number of type II and III dislocations to produce a domain with a component of tilt along the x direction. Obviously the sequence could have begun with a type III, rather than a type II, dislocation. According to this model the maximum component of tilt due to a coherent stacking of only type (II) dislocations would be $\phi_y/2 \simeq \tan^{-1}(a/2d) \simeq 0.083$ about the $\hat{y}$ axis and $(\phi_x/2) \simeq \tan^{-1}(\sqrt{3}a/3d) = 0.096$, about the $\hat{x}$ axis. Assuming the intralayer strain remains isotropic (e.g., $\epsilon_1^x = \epsilon_2^y$) the predicted anisotropy in the reciprocal lattice vectors (see Fig. 3 of Ref. 7) is $\frac{1}{4}[(\phi_x/2)^2 - (\phi_y/2)^2] \simeq 0.00086$. Although this is approximately three times larger than the observed splitting in the monoclinic phase the actual splitting is a very delicate function of not only the distribution of (II) and (III) dislocations but also of their physical distribution within the domains. The observed anisotropies could also be reduced by allowing $\epsilon_1^x \neq \epsilon_2^y$. The more significant feature is that it provides a rational explanation for the observed reduction in the splitting.

In conclusion, we have presented a dislocation model for restacking transitions observed in bulk phases of crystalline-B liquid crystals. Features of these transitions appear similar to phase transitions in other systems, such as lipid-water mixtures, which have been widely studied and it is likely that the model here is relevant to those systems. The model also predicts that the orthorhombic-F to hexagonal-ABC transition in 5O.6 should be an example of the crystallization of dislocations. Further experimental studies of this transition should be carried out with this in mind. The model is sufficiently specific that one might hope to develop a molecular model for the forces that generate the dislocations. For example, if these transitions are driven by a mechanism that causes a shearing of the original unit cell, the model illustrates how a set of minimum molecular displacements can accommo-

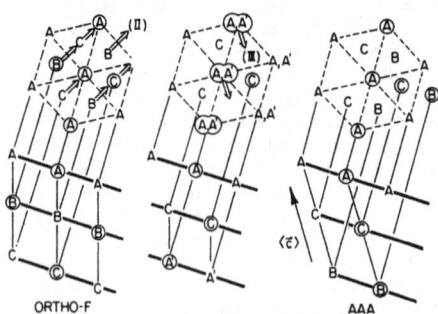

FIG. 3. Intermediate steps in the dislocation-induced transition from the orthorhombic-F to the hexagonal-AAA phase.

date both the shear and the local molecular packing.

Finally, we *do not* believe there is a direct relation between the present model and the tilted hexatic phases that appear in thin films of 7O.7.[17] The large molecular tilt ($\simeq 24°$) is indicative of a separate thermodynamic phase that is probably stabilized by surface effects. Similarly, since the dislocations in the present model run parallel to the smectic layers they are not the same as dislocations normal to the layers that are commonly invoked to discuss two-dimensional melting.[1,2]

The research reported here was supported in part by the National Science Foundation under Grants No. DMR-82-12189 and No. DMR-80-20247. In addition one of us (J.P.H.) would like to acknowledge support from both Ohio State University and the Harvard Materials Research Laboratory while on sabbatical leave. Conversations with both D. Nelson and B. Halperin are also gratefully acknowledged.

(a)Present address: IBM Research Laboratory, Yorktown Heights, N.Y. 10598.

(b)Present address: Department of Physics, FM-15, University of Washington, Seattle, Wash. 98195.

[1]B. I. Halperin and D. R. Nelson, Phys. Rev. Lett. 41, 121 (1978). D. R. Nelson and B. I. Halperin, Phys. Rev. B 19, 2456 (1979).

[2]A. P. Young, Phys. Rev. B 19, 1855 (1979).

[3]C. Deutsch and S. Doniach, Phys. Rev. B 29, 2724 (1984).

[4]H. Kleinert, Phys. Lett. 96A, 302 (1983).

[5]A. J. Leadbetter, M. A. Mazid, B. A. Kelley, J. Goodby, and G. W. Gray, Phys. Rev. Lett. 43, 630 (1979).

[6]J. Doucet and A. M. Levelut, J. Phys. (Paris) 38, 1163 (1977).

[7]J. Collett, L. B. Sorensen, P. S. Pershan, J. D. Litster, R. J. Birgeneau, and J. Als-Nielsen, Phys. Rev. Lett. 49, 553 (1982).

[8]A. J. Leadbetter, M. A. Mazid, and R. M. Richardson, in *Liquid Crystals,* edited by S. Chandrasekhar (Cambridge Univ. Press, London, 1980), pp. 65–79.

[9]P. A. C. Gane, A. J. Leadbetter, P. A. Tucker, G. W. Gray, and A. R. Tajbaksh, J. Chem. Phys. 77, 6215 (1982).

[10]W. K. Chan and W. W. Webb, Phys. Rev. Lett. 46, 39 (1981). See also Refs. 6–9, and J. W. Christian and A. G. Crocker, in *Dislocations of Solids,* edited by F. R. N. Nabarro (North-Holland, Amsterdam, 1980), Vol. 3, p. 165, and J. P. Hirth and J. Lothe, *Theory of Dislocations* (Wiley, New York, 1982), p. 707.

[11]Christian and Crocker, Ref. 10.

[12]Hirth and Lothe, Ref. 10.

[13]Similar dislocations appear in computer simulations of dislocation mediated melting carried out by R. M. J. Cotterill, W. Damgaard Kristensen, and E. J. Jensen, Philos. Mag. 30, 245 (1974).

[14]Jeffrey Allen Collett, "X-Ray Scattering Study of Liquid Crystal Thin Films," Ph.D. thesis, Harvard University, June 1983 (unpublished).

[15]F. C. Frank, Discuss. Faraday Soc. 25, 19 (1958).

[16]Hirth and Lothe, Ref. 10, p. 733.

[17]Jeffrey Collett, P. S. Pershan, Eric B. Sirota, and L. B. Sorensen, Phys. Rev. Lett. 52, 356 (1984).

310

PHYSICAL REVIEW A   VOLUME 32, NUMBER 2   AUGUST 1985

X-ray scattering study of restacking transitions in the crystalline-B phases of heptyloxybenzylidene heptylaniline (7O.7)

Jeffrey Collett,* L. B. Sorensen,† and P. S. Pershan

Division of Applied Sciences, Harvard University, Cambridge, Massachusetts 02138

J. Als-Nielsen

Risø National Laboratory, DK-4000 Roskilde, Denmark

(Received 28 January 1985)

This paper reports a comprehensive x-ray diffraction study of the structure and the restacking transitions which occur within the crystalline-B phases of N-[4-(n-heptyl)oxybenzylidene]-4'-(n-heptyl)aniline (7O.7). Between 63 °C and 59 °C, this system exhibits a rich structural behavior with a series of three transitions involving restacking and subtle changes in the intralayer packing. In order of decreasing temperature, the structures are hexagonal close packed (69—63 °C), orthorhombic F (63—60.1 °C), monoclinic C (60.1—59.75 °C), and simple hexagonal (59.75—55 °C). In the last three phases, there is also a long-wavelength modulation ($\lambda \sim 100$ Å) in which the smectic layers are displaced along their normals. The wave vector of the modulation lies in the plane of the smectic layers directed along a reciprocal-lattice vector and exhibits a weak ($\sim 20\%$) temperature variation. Both low-resolution (rotating-anode) and high-resolution (synchrotron) measurements of the structures and of the diffuse scattering are reported.

I. INTRODUCTION

Interest in the properties of smectic-B phases has been greatly stimulated by recent theoretical and experimental developments. Since the smectic phases have extremely weak interlayer coupling and a variety of two-dimensional (2D) intralayer structures, they have become particularly valuable models for the study of weakly interacting 2D systems. Recent advances in the theory of melting in two dimensions[1] have led to theoretical models of the smectic-B phase based on weakly coupled two-dimensional systems.[2−4] One of these models predicted two possible smectic-B phases: a crystalline-B phase with true long-range positional and orientational order and a three-dimensional (3D) hexatic-B phase consisting of coupled 2D hexatic layers.[4] The 3D hexatic phase does not have long-range positional order, but does possess long-range order in bond-angle orientation.

Early low-resolution x-ray studies of several smectic-B materials indicated that the smectic-B phase had a structure corresponding to the hexatic-B phase.[5] However, later experiments showed that although the original materials actually had crystalline-B order, both kinds of order occur, with the crystalline-B order occurring much more often than the hexatic-B order.[6−10] The resolution of the original controversy was due to both high-resolution x-ray diffraction measurements and to the use of freely-suspended liquid-crystal films.[6,8]

The measurements made on the crystalline-B materials showed a variety of unusual features due to the weak coupling between layers which makes these quasi-2D phases extremely interesting in their own right. X-ray diffraction experiments have revealed strong diffuse-scattering modes which can be interpreted as low-energy phonons with polarizations in the plane of the smectic layers and wave

vectors normal to the smectic layers.[7,8] This strong diffuse scattering and the poor sample mosaic found in the earlier experiments hindered the identification of these materials as true 3D crystals. The second feature observed in these systems is the preponderance of restacking transitions involving dramatic shifts in the relative positions of adjacent layers.[6,9,11−13] Prior to the present study, no latent heat or other thermal anomaly had been observed in conjunction with the restacking transitions, leaving their detailed nature unknown. High-resolution ac calorimetry measurements,[14] motivated by our initial publication of the restacking-transition sequence in 7O.7,[15] verified the presence of the weak first-order transitions observed in this study. The third feature of these phases which merits closer examination is the presence of modulations polarized normal to the layers with wave vectors in the plane of the layers.[9,11,13]

In this study we carefully examined freely-suspended crystalline-B films of N-[4-(n-heptyl)oxybenzylidene]-4'-(n-heptyl)aniline (7O.7). The crystalline-B phases were originally thought to consist of just the two hexagonal-close-packed structures with $ABAB$ and $ABCABC$ stacking; subsequently, phases with AAA stacking were suggested by Gane, Leadbetter and Wrighton.[9] Prior to the present study, the crystalline-B phase of 7O.7 has been identified as a hexagonal-close-packed structure with $ABAB$ stacking and with layer modulations observed in the low-temperature region of the phase.[11−13] We found that the crystalline-B phase of 7O.7 actually has four distinct crystal structures separated by three restacking transitions;[15,16] these phases (in order of decreasing temperature) are the following: a hexagonal-close-packed structure (hcp) with $ABAB$ stacking, an orthorhombic-face-centered structure (ortho-F), a monoclinic-C-centered structure (mono-C), and a simple hexagonal structure.

We also found that the last three of these structures possess a one-dimensional modulation of the smectic layers with a single (temperature-dependent) wave vector in the direction of the molecular motion at the transition from the hcp to the ortho-F structure.

II. EXPERIMENTAL TECHNIQUES

Two techniques were crucial in developing the present understanding of the microscopic structure of smectic-B phases; both freely suspended liquid-crystal film techniques and high-resolution x-ray diffraction techniques allowed the unambiguous determination of these structures. High-resolution diffraction clearly distinguishes between crystalline phases with true long-range order and disordered phases with abnormally long correlation lengths. This was crucial in the experiments of Pershan et al.[7] on bulk samples, and of Moncton and Pindak[6,8] on freely-suspended films, in which the smectic-B phase of N-[4-(n-butyl)oxybenzylidene]-4'-(n-octyl)aniline (4O.8) was unambiguously shown to be crystalline. The freely-suspended-film technique allows the preparation of single-crystal domains with lateral dimensions of 1 mm or more and allows a study of the interlayer correlations uncomplicated by the problem of sample mosaic.[6] This was crucial for the present study of the rich crystallography of the crystalline-B phases of 7O.7. In this section the details of the x-ray spectrometers used in the present measurements and a description of the free-film x-ray oven are presented.

High-resolution x-ray measurements were carried out using synchrotron radiation and the triple-axis spectrometer at the Hamburger Synchrotronstrahlungslabor, Deutschis Elektronen-Synchrotron (HASYLAB) in Hamburg, Germany. Single-crystal Ge(111) reflections were used in the monochromator and analyzer. The wavelength was chosen to be 1.5 Å. For the in-plane scattering angle of 20° appropriate for 7O.7, this spectrometer configuration had a longitudinal resolution of 6×10^{-4} Å^{-1} and a transverse resolution of $\sim 1 \times 10^{-4}$ Å^{-1} [full width at half maximum (FWHM)]. This resolution allowed the measurement of correlation lengths up to 1 μm and the determination of lattice-constant differences to better than one part in 10^4. Absolute lattice-constant measurements were not as good due to systematic drifts in the x-ray wavelength with time. The vertical resolution of the spectrometer was determined by the vertical acceptance of the detector and was varied from 8×10^{-4} to 4×10^{-2} Å^{-1}, with a slit at the detector position.

Lower-resolution data were taken on similar spectrometers at Harvard University and the Massachusetts Institute of Technology (MIT) using Cu $K\alpha$ radiation from rotating-anode sources and highly oriented pyrolytic-graphite (002) reflections in the monochromator and analyzer. For these spectrometers, the longitudinal and transverse resolutions were 3.5×10^{-2} Å^{-1} and 6.3×10^{-3} Å^{-1} (FWHM), respectively. The vertical resolution was determined by slits and was varied from 7×10^{-2} to 3×10^{-1} Å^{-1}. Some low-resolution measurements were also made at MIT using a pyrolytic-graphite monochromator and a TEC model-210 linear position-sensitive proportional counter. The longitudinal resolution in this case

was about a factor of 4 better than that obtained with graphite monochromator and analyzer crystals.

All of these spectrometers were two-circle (planar) instruments. In order to observe the diffraction from the single-crystal samples, it was necessary to introduce a rotation stage inside the liquid-crystal oven to allow the desired Bragg vector of the crystal to be brought into the scattering plane of the spectrometer. This rotation stage has its axis of rotation normal to the smectic layers and in the scattering plane of the two-circle diffractometer. The two-circle diffractometer allowed us to perform scans along q_z (normal to the smectic layers) and along q_{xy} (parallel to the smectic-layer planes). The internal rotation stage controlled the angle χ which determined the direction of q_{xy} within the smectic layers.

The oven used in these experiments had to satisfy a number of design criteria.[16] The four major requirements were: millikelvin temperature stability, extremely low-background scattering, the ability to make films and measure their thickness while in place, and the incorporation of the necessary remotely controlled χ rotational degree of freedom described above. In addition to these features, provisions were made to mount samarium-cobalt permanent magnets inside the rotation circle to allow magnetic orientation of the director for smectic phases in which the long axis of the molecules is tilted with respect to the layer normal.

Since the samples used in these experiments were thin films, it was necessary to reduce the background scattering as much as possible. To accomplish this, the region around the sample from which a photon from the direct beam can be scattered into the detector should contain no scatterers except the sample. This region is referred to as the scattering volume and should not contain any windows or air. For this reason, the oven was designed for vacuum operation. Pressures of 10^{-1} Torr obtained with a mechanical roughing pump were adequate for all of the present measurements. In practice, the presence of air (at 1 atm) did not cause a significant background problem in the measurement of films several hundred layers thick. However, the background had to be reduced to perform measurements on thin films.

The general layout of the oven is shown in Fig. 1. The oven consists of two concentric cylindrical brass cans with the inner can heated and insulated from the outer can by vacuum and by Delrin support posts. Heaters on the top and bottom plates of the inner can could be independently adjusted to minimize temperature gradients. The outer can is sealed using O-ring seals. An Ardel kinematics rotation stage is rigidly suspended from the inner cover by a brass plate [Fig. 1(a)]. A cross section of the rotation stage in the plane normal to the rotation axis of the stage is shown in Fig. 1(b). The drive shaft from this stage is coupled to the rotary-vacuum feedthrough in the outer cover by a stainless-steel bellows which takes up mechanical stresses and provides thermal isolation from the outside environment. The rotation stage is driven by a computer-controlled stepping motor mounted on top of the oven. A cylindrical beryllium window in the inner can and a corresponding Kapton window in the outer can provide 300° x-ray access in the scattering plane.

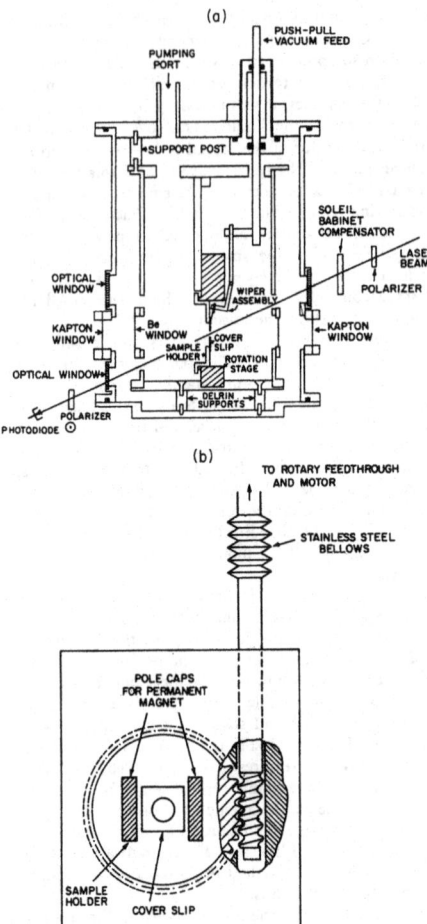

FIG. 1. (a) Schematic cross section of the free-film oven. (b) Cross section of the rotation stage normal to its axis of rotation. (The soft-iron return path for the permanent magnet is not shown.)

Films were produced on a sample holder mounted in the 2-in.-diam aperture in the rotation stage. As in the other free-film experiments,[6,17] the films were produced by pulling a microscope cover slip (attached to the wiper assembly) across a polished hole in a second cover slip mounted in the rotation stage. For these experiments, we used a 7-mm-diam round hole in a 0.2-mm-thick glass cover slip. Permanent magnets mounted in a soft-iron sample holder provide a magnetic field in the plane of the cover slip [Fig. 1(b)]. Before closing the oven, a small amount of the liquid-crystal compound was melted

around the edge of the hole. The oven was then sealed, evacuated, and heated to bring the sample into the smectic-*A* phase. The first cover slip was then drawn across the second via a push-pull vacuum feedthrough; the liquid crystal then spontaneously produced a free film across the open hole (like a soap bubble). Films were produced repeatedly until one of the desired thickness was obtained.

A birefringence measurement was used to determine the film thickness; the simple reflectivity technique used in previous studies of thin films is only useful up to thicknesses of about 25 layers where the reflectivity reaches its first maximum. The birefringence method was better for thick films because the optical phase retardation is a single-valued function over the entire range of thicknesses of interest (50—1000 layers) whereas the reflectivity goes through many maxima and minima. Optical access to the film was provided by three fused-quartz windows adjacent to the x-ray window in the outer can. A He-Ne laser beam was incident as shown in Fig. 1(a) with its polarization at a 45° angle to the optic axes of the liquid-crystal film. A Soleil-Babinet compensator was placed in the beam with its axes parallel to those of the film. By adjusting the compensator to minimize the transmission of the beam through the set of crossed polarizers, the optical phase retardation introduced by the film could be determined. Using estimated values for the refractive indices of 7O.7,[18] the film thickness could then be calculated. The film thicknesses obtained using this method were uncertain by about 20% due to the uncertainty in the refractive-index values. Thin-film thicknesses were determined using the standard reflectivity method.

In order to accurately determine the temperature of thin films, it is necessary to surround the sample with a very uniform temperature distribution. Since the heat flow through the film to the supporting cover slip and rotation stage is proportional to the thickness of the film, while the radiative and conductive heat flow from the film surface to the oven walls scales with the surface area, the surface terms will eventually dominate the thermal transport for thin enough films. In our evacuated oven, the heat flow was dominated by line-of-sight conduction and radiation to the oven walls nearest the film. For this reason we used beryllium (a good thermal conductor) windows near the sample and carefully balanced the power to the heaters to minimize the gradient between the rotation stage and the surrounding walls. In addition, the optical windows were designed to subtend the smallest solid angle which would still allow the necessary thickness measurements.

The overall temperature stability of the oven was very good. Short-term stability (20 min) of better than 1 mK was typical. The behavior over longer periods depended on the temperature stability of the laboratory. For typical variations in the room temperature of less than 3 °C, the oven was stable to better than 5 mK for many hours. One of the reasons for the excellent short-term stability was the massiveness of the inner-oven assembly. This resulted in good stability but slow response; the full equilibration time (to ±5 mK) was about 2 h under vacuum conditions.

III. EXPERIMENTAL RESULTS

These studies have revealed an unusually rich structural behavior within the crystalline-B range ($69\,°C > T > 55\,°C$) with a series of three first-order transitions between $63\,°C$ and $59\,°C$ that involve restacking of the layers and subtle changes in the intralayer packing. In order of decreasing temperature, the system goes from a hexagonal-close-packed structure (hcp) with $ABAB$ stacking (hcp, $69\,°C > T > 63\,°C$) to an orthorhombic-face-centered structure (ortho-F, $63\,°C > T > 60.1\,°C$), a monoclinic-C-centered phase (mono-C, $60.1\,°C > T > 59.75\,°C$), and a simple hexagonal structure (hex-AA, $59.75\,°C > T > 55\,°C$). At $55\,°C$, 7O.7 transforms into a smectic-G phase, a crystalline, mono-C-centered structure with the long axes of the molecules tilted by $23°$ with respect to the normals to the smectic layers.[9,13] The ortho-F, mono-C, and hex-AA phases all exhibit a one-dimensional modulation of the smectic layers in which the displacement is normal to the layers and the modulation wave vector is in the plane of the layers. (See Note added in proof below.)

The relative stacking of the adjacent layers in these phases is shown in Fig. 2. The primitive real and reciprocal lattices for the three lowest-temperature phases are also given. The modulation wave vector is directed along the y direction. In order to present the data in a unified form, all the peaks have been indexed by the reciprocal lattice which uses the in-plane (x,y) components of the $\mathbf{b}_1$ and $\mathbf{b}_2$ to form a lattice which is the reciprocal of the lattice of the individual layers. The third reciprocal-lattice vector was chosen along the z direction with a length equal to $2\pi/(\text{molecular length})$. Although the peaks no longer occur only at integer values, this description clearly separates the changes in the intralayer packing (apparent in $\mathbf{b}_{1xy}$ and $\mathbf{b}_{2xy}$) from the changes in the stacking (reflected by changes in the period and displacements along the z direction).

At temperatures between $63\,°C$ and $69\,°C$, 7O.7 exhibited the $ABAB$-stacked hcp structure common in crystalline-B materials. Scans along q_z through the six lowest-order in-plane peaks showed peaks at integral and half-integral multiples of $2\pi/c$ (where c is the smectic layer spacing). The observed peaks at the half-integral positions were three times as intense as the peaks at the integral positions as expected from the calculated structure factor (after correcting for the form factor and the Debye-Waller effects). Figure 3(a) shows a low-resolution q_z scan in this phase.

At $63\,°C$ there was an abrupt first-order transition to an ortho-F phase. This phase can be described by the primitive triclinic unit cell of the lattice in Fig. 2, or by a larger orthorhombic cell. The choice $s = \frac{1}{2}$ for the representation in Fig. 2 produces the orthorhombic unit cell. When freely-suspended films in the ortho-F phase were cooled slowly through the transition into the hex-AA phase and were then heated slowly back into the ortho-F phase, large single domains were formed. Figures 3(b) and 3(d) show low-resolution q_z scans through $\mathbf{b}_1$ and $\mathbf{b}_2$. The high-resolution synchrotron measurements showed that the direct lattice was compressed along y and was extended along x. Because of this distortion, the magnitudes of $\mathbf{b}_1$ and $\mathbf{b}_2$ were different; the synchrotron measurements of the distortion as a function of temperature are shown in Fig. 4(a). This distortion is parametrized by ϵ_x and ϵ_y in the representation in Fig. 2; the measured distortion was consistent with an area-conserving change with $\epsilon_x = \epsilon_y = 10^{-3}$. This structure has not previously been associated with crystalline-B phases because the distortion is extremely small. The use of freely-suspended films together with the high resolution provided by the HASYLAB spectrometer were crucial for the identification of this phase. The free-film technique allowed the preparation of large single-domain samples which were not available using standard sample-preparation techniques. In a normal sample which has many domains, the peak locations along q_z are the same for the ortho-F and

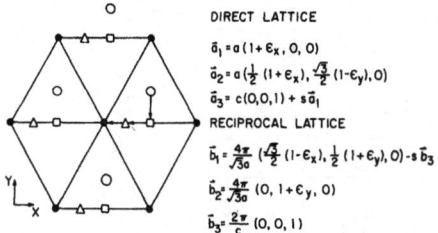

DIRECT LATTICE

$$\vec{a}_1 = a(1+\epsilon_x, 0, 0)$$
$$\vec{a}_2 = a\left(\tfrac{1}{2}(1+\epsilon_x), \tfrac{\sqrt{3}}{2}(1-\epsilon_y), 0\right)$$
$$\vec{a}_3 = c(0,0,1) + s\vec{a}_1$$

RECIPROCAL LATTICE

$$\vec{b}_1 = \tfrac{4\pi}{\sqrt{3}a}\left(\tfrac{\sqrt{3}}{2}(1-\epsilon_x), \tfrac{1}{2}(1+\epsilon_y), 0\right) - s\vec{b}_3$$
$$\vec{b}_2 = \tfrac{4\pi}{\sqrt{3}a}(0, 1+\epsilon_y, 0)$$
$$\vec{b}_3 = \tfrac{2\pi}{c}(0, 0, 1)$$

FIG. 2. Relative positions of the adjacent layers in the various phases: solid circles, reference layer position; open circles, hexagonal close packed; squares, ortho-F; and triangles, mono-C. In the hex-AA phase the adjacent-layer position coincides with the reference layer. The primitive unit cells for the real and reciprocal lattices of the ortho-F, mono-C, and hex-AA structures are shown on the right.

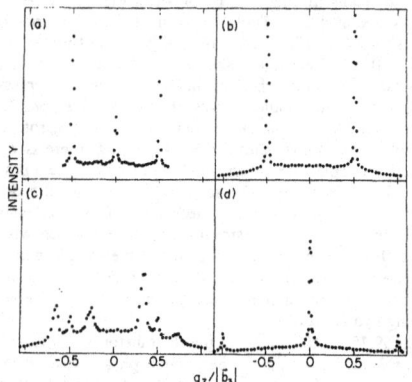

FIG. 3. Low-resolution q_z scans: (a) through an in-plane peak for $T > 63\,°C$ ($T = 65.43\,°C$); (b) through $\mathbf{b}_1$ in the ortho-F phase at $T = 60.88\,°C$; (c) through $\mathbf{b}_1$ in the mono-C phase at $T = 60.02\,°C$; (d) through $\mathbf{b}_2$ in the hex-AA phase at $T = 59.35\,°C$; identical scans through $\mathbf{b}_2$ are found in both the ortho-F and mono-C phases.

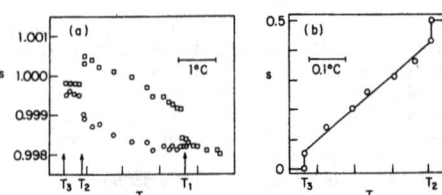

FIG. 4. (a) Measurement of distortion in intralayer packing: circles, $|b_1|$; squares, $|b_2|$. Data are plotted in units for s of $4\pi/\sqrt{3}a = 1.434$ Å^{-1}. Transition temperatures are marked: T_1, hcp to ortho-F; T_2, ortho-F to mono-C; T_3, mono-C to hex-AA. (b) Adjacent layer displacement versus temperature in the mono-C phase.

the hcp structures. Consequently, although the relative intensities are quite different (the half-integer peaks are twice as intense as the integer-order peaks in the ortho-F phase instead of the threefold ratio found in the hcp phase), it is still extremely difficult to distinguish these two phases in a low-resolution measurement. At high resolution, the phases can be distinguished by the distortion of the intralayer packing mentioned above; this is how the phase was originally identified using polydomain samples at HASYLAB before the technique for producing monodomain samples was discovered.

At $T = 60.1$°C, the ortho-F structure transforms into the mono-C structure and the distortion in the intralayer packing relaxes to $\epsilon_x = \epsilon_y = 2 \times 10^{-4}$. This transformation can be described by a local shear along x as shown in Fig. 2. Within the mono-C phase the displacement between adjacent layers is sa_1, where s varies approximately linearly with temperature from 0.45 to 0.05 as shown in Fig. 4(b). The temperature dependence of s was determined from q_z scans through b_1, as shown in Fig. 2; the peaks are located at $q_z = s$. Typical q_z scans through b_1 and b_2 are shown in Figs. 3(c) and 3(d), respectively. At $T = 60.02$°C, the main peaks in Fig. 3(c) are located at $q_z = \pm 0.3$, so that $s = \pm 0.3$ at this temperature. The presence of $\pm s$ values indicates that the peaks correspond to crystallographic twins. Both twins must be present to prevent macroscopic shears from occurring; there is no net macroscopic shear in an equal mixture of the two possible shear directions. The small peaks at $q_z = \pm 0.5$ decrease in intensity as $|s|$ decreases. These remnant peaks might arise from strain-induced coexistence with the ortho-F phase or from regions of the sample where the shears go in opposite direction from one layer to the next; such an alternation would produce a unit cell containing two molecules.

At 59.75°C, the mono-C phase transformed into the hex-AA phase, a simple hexagonal phase with AAA stacking, and the intralayer distortion vanished (within our resolution of 1×10^{-4}). This transition was indicated by a small first-order jump in s. In this phase the molecules in adjacent layers occupy sites directly above those in the layer below. Scans along q_z for the six low-order in-plane peaks are now identical to the scan in Fig. 3(d). This is the clearest evidence for the existence of an AAA-

stacked hexagonal phase in the crystalline-B materials, although, Gane, Leadbetter, and Wrighton[9] previously inferred the presence of a hex-AA phase in 6O.4. This is an exceedingly rare structure in nature because it is normally unstable against a transition to one of the close-packed structures.

The ortho-F, mono-C, and hex-AA structure all have a one-dimensional modulation. As the temperature is lowered from the hcp phase, the modulation first appears at the hcp—ortho-F phase transition with a wave vector in the direction of the molecular motion (along y). As the temperature is lowered further, the modulation persists until the smectic-G phase occurs; there is a slow increase in the intensity and a slow decrease in the wave vector as the temperature is reduced. The shearing which occurs in the monoclinic phase does not appear to affect the modulation; the shearing is in the direction perpendicular to the modulation wave vector.

A careful characterization of the satellite peaks was done in the hex-AA phase of 7O.7. We found that we could produce samples in which there was a single domain of the modulated phase. Figure 5(b) shows the observed location of the peaks corresponding to the simple hexagonal lattice and of the superlattice satellite peaks produced by the modulation. The region around the (001) peak is inaccessible in our scattering geometry. High-resolution scans were done in the q_x, q_y, and q_z directions through the satellites of the (011) Bragg peak. In the x and z

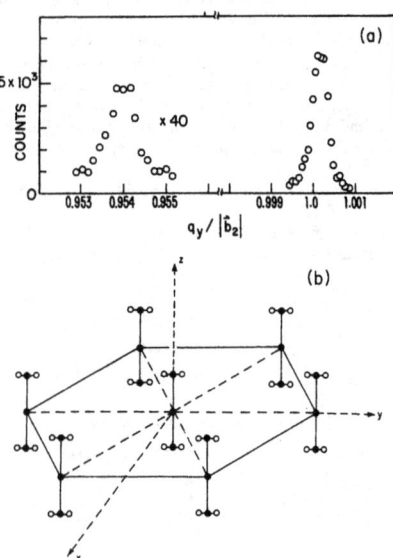

FIG. 5. (a) Synchrotron high-resolution q_y scans through the (011) peak and through its satellite in the hex-AA phase. (b) Locations of the satellite peaks. Solid circles represent the locations of the Bragg peaks corresponding to the simple hexagonal structure and the open circles indicate the locations of the satellite peaks.

directions the scans through the satellites were indistinguishable from scans through the (011) Bragg peak. The q_y scan is shown in Fig. 5(a). The FWHM of the primary peak is $4.6 \times 10^{-4} | \mathbf{b}_2 |$; the satellite peak is slightly wider, indicating a correlation length for the modulation of ~ 4000 Å. At this temperature the wave vector of the modulation was $\mathbf{Q}_m = 0.046 \mathbf{b}_2$, corresponding to a modulation wavelength of about 95 Å. Although this was not a commensurate wave vector, it was very close to $\mathbf{b}_2 / 22$. As the temperature was lowered, the modulation wave vector gradually decreased to $0.038 \mathbf{b}_2$ just above the transition to the smectic-G phase, corresponding to a gradual modulation-wavelength increase to about 115 Å. The resolution of the modulation-wave-vector measurements was not sufficient to determine whether it decreases continuously or had discrete jumps.

As in the previous measurements,[11-13] no satellites were observed in the $q_z = 0$ plane indicating that the polarization of the modulation is perpendicular to the plane of the smectic layers. The amplitude of the modulation was determined from the intensity relative to the primary peak of the simple hexagonal lattice and the polarization was determined from the systematic extinction of the $q_z = 0$ satellites. The relative intensity for a weak-modulation satellite peak is given by

$$I_{\text{sat}} = I_{\text{Bragg}} (\tfrac{1}{2} \mathbf{q} \cdot \mathbf{u})^2 , \qquad (1)$$

where $\mathbf{q}$ is the scattering vector and $\mathbf{u}$ is the amplitude of the sinusoidal modulation; the modulation must be very nearly sinusoidal because of the absence of any detectable higher-order satellites. In the $q_z = 2\pi/c$ plane the satellite intensities are given by

$$I_{\text{sat}} = I_{\text{Bragg}} \left[\frac{\pi u_0}{c} \right]^2 , \qquad (2)$$

where u_0 is the amplitude of the modulation and c is the layer spacing. The measured intensity ratio $I_{\text{sat}} / I_{\text{Bragg}} \sim 2\%$ at $T = 59$°C indicated an amplitude $u_0 = 0.05c$. Near the transition to the smectic-G phase, the intensity ratio increased to about 8% indicating that the amplitude doubled, so $u_0 = 0.1c$.

Another unusual feature of crystalline-B materials is the intense diffuse scattering that is observed. Both hydrodynamical theory[19] and light-scattering measurements[20] indicate that there are two sets of modes which have much lower frequencies than the rest of the modes. The first set of low-frequency modes have their wave vectors normal to the smectic layers and their polarizations in the plane of the layers. Scattering from these modes produces the intense diffuse scattering observed in a q_z scan through $\mathbf{b}_1$ or $\mathbf{b}_2$. Our measurements of this scattering for the hcp, ortho-F, and hex-AA phases are shown in Fig. 6. The data in the hex-AA phase was taken with much better vertical resolution; the factor of 20 increase in the vertical resolution is responsible for the observed decrease in the diffuse intensity for the hex-AA phase. The peaks at half-integral q_z positions in Fig. 6(b) and the peaks at integral q_z positions in Fig. 6(c) are present because we did not have a single-domain sample. Because $| \mathbf{b}_1 | < | \mathbf{b}_2 |$, we could separate the two ridges quite well

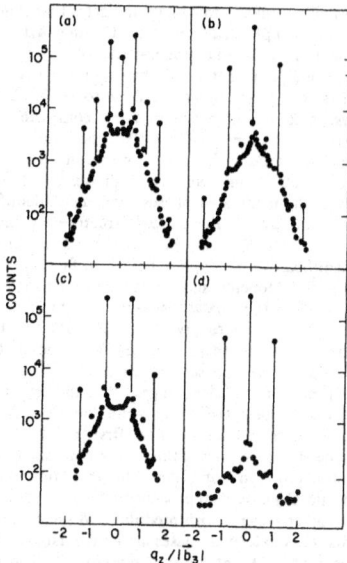

FIG. 6. High-resolution q_z scans illustrating the diffuse scattering characteristic of crystalline-B materials. Vertical lines through the Bragg points are guides for the eye. (a) hcp structure at $T = 66$°C. (b) ortho-F structure through $\mathbf{b}_2$ at $T = 62$°C. (c) ortho-F structure through $\mathbf{b}_1$ at $T = 62$°C. (d) hex-AA phase at $T = 57$°C.

with high resolution; however, the tails of the resolution function extended far enough along the ridge to pick up some scattering from the nearby Bragg peak. The envelope of the Bragg intensities and of the diffuse scattering are given by the product of the Debye-Waller factor and the molecular form factor. The observed envelopes for each phase (corrected for the resolution) superimpose nicely, indicating that these modes are largely unaffected by the phase transitions.

The second set of low-frequency modes are layer-undulation modes. These modes are polarized normal to the smectic layers and propagate in the plane of the layers. In the hex-AA phase, these modes produce a scattering intensity in the region between the Bragg peak and the satellite peak in the $q_z = 2\pi/c$ plane which is about $\frac{1}{500}$ of the Bragg intensity (measured at higher resolution). This intensity is comparable to the intensity of the diffuse ridges produced by the first set of low-frequency modes.

IV. DISCUSSION

Several of the features of this system merit further discussion. The simple hexagonal structure with one molecule per unit cell is an exceedingly rare structure since it is usually unstable and transforms to one of the more stable close-packed structures. However, the presence of the sharp satellite peaks in this phase indicates that it actually has a more complicated structure. The hex-AA stacking is presumably stabilized by the modulation.

The second point that should be emphasized is that the structures found here are not unique to 7O.7. Evidence for layer modulations exists in every compound in the homologous $nO.m$ N-(4-n-alkoxybenzylidene)-4'-(n-alkyl)anilines series which has a crystalline-B phase just above a smectic-G or a smectic-F phase.[12,13] It is very likely that the modulated phases previously identified as hcp structures are actually ortho-F structure; we have also studied 5O.6 and have shown that the phase which had been identified as a modulated hcp is, in fact, an ortho-F structure.[16,21] Moncton and Pindak[6] reported some single domain data on the crystalline-B-phase restacking transitions in 4O.8 which strongly suggest that the two lowest-temperature phases they observed correspond to the ortho-F and mono-C phases identified here.

Two pieces of evidence suggest that molecular tilt is one of the important driving factors in the restacking transitions observed here. The first is that all the crystalline-B phases, except the hcp, are modulated. The modulation grows in amplitude as the transition to the tilted smectic-G phase is approached. This trend has been noted in other systems with modulated B phases and suggests that the modulated phases are a precursor to the tilted phases which lie at lower temperatures. The second point is that the displacements of the molecules within a given layer with respect to their neighbors is similar to the displacements which occur when the molecules tilt with respect to the layer normals. The amplitude and wavelength of the modulation correspond to a nearest-neighbor displacement of about 2% of the molecular length at the low-temperature end of the hex-AA phase. This should be compared with an 8% displacement in the lower-temperature smectic-G phase.

The fact that all the structures which are not close packed have the associated modulation suggests that the modulation plays a fundamental role in stabilizing these structures. This stabilization has been explained by a dislocation-mediated model[22] for the restacking transitions which occur in 7O.7. This model accounts for the non-close-packed structures by a dislocation-induced tilting of the layers. The spontaneous ordering of the dislocations into an ordered array explains both the symmetry of the phases and the amplitude of the observed modulations. The only free parameter in the model is the modulation wave vector which is set equal to the measured wave vector; the in-plane lattice spacing of the dislocation array is equal to the modulation wavelength, but the lattice spacing is a very delicate function of the dislocation interactions and cannot be calculated easily. This model also provides an excellent description of the observed restacking transitions in 5O.6

This study has shown that the behavior of bulk samples (thick films) of these weakly coupled 2D crystalline systems is extremely rich. Recent studies of thin films of 7O.7 have shown that there are radical changes in the phase diagram as a function of the film thickness.[21,23] Two tilted hexatic phases, the smectic-F and smectic-I phases, are introduced between the crystalline-B and the smectic-G phases in very thin films. The smectic-F phase first appears over a narrow temperature range when the film thickness is reduced to about 180 layers and then gradually widens its temperature range as the thickness is reduced further. The smectic-I phase first appears at a lower film thickness (about 25 layers) and also gradually widens as the thickness is reduced. The surface interactions evidently destabilize the untilted crystalline phases and stabilize the tilted hexatic phases; this is consistent with the known tendency of the molecules in these materials to tilt at the film surface[24] and reemphasizes the importance of the tilt degree of freedom in determining the phase transitions of these materials. We are currently studying the phase transitions of these weakly coupled 2D hexatics and the crossover to truly 2D systems (two-layer films). The phase transition behavior of these remarkable hexatics appear to be as rich as the behavior already observed in the crystalline systems. Finally, it is clear that much more theoretical and experimental work is needed on these unusual quasi-2D systems.

Note added in Proof. Recent experiment results by Eric Sirota and P.S.P. indicate that on cooling the ortho-F phase the one-dimensional modulation breaks up into a two-dimensional structure. In addition, the modulation in the hex-AA and mono-C phases is shown to be two dimensional.

ACKNOWLEDGMENTS

This work was supported by the National Science Foundation under Grant No. NSF-DMR-82-12189 and No. DMR-80-20247 and from the Danish National Science Foundation. Support for foreign travel for J.C., P.S.P., and J.A.-N. was provided by a NATO Research Grant. In addition, Dave Litster participated in the HASYLAB measurements displayed in Figs. 4(a), 5(a), and 6. We gratefully acknowledge both his assistance and his permission to publish the results in the present format. Participation of Bob Birgeneau in an early form of this work is also gratefully acknowledged.

---

*Present address: IBM Corporation, Department 242/040-2, Rochester, MN 55901.

†Present address: Department of Physics, University of Washington, Seattle, WA 98195.

[1]D. R. Nelson and B. I. Halperin, Phys. Rev. B **19**, 2457 (1979).

[2]P. G. de Gennes and G. Sarma, Phys. Lett. **38A**, 219 (1972).

[3]B. A. Huberman, D. M. Lublin, and S. Doniach, Solid State Commun. **17**, 485 (1975).

[4]R. J. Birgeneau and J. D. Litster, J. Phys. (Paris) Lett. **39**, L399 (1978).

[5]A. M. Levelut, J. Doucet, M. Lambert, J. Phys. (Paris) **35**, 773 (1974). A. de Vries, A. Ekachai, and N. Spielberg, J. Phys. (Paris) Colloq. **40**, C3-147 (1975); A. Leadbetter, J. Frost, J. P. Gauglan, and M. A. Mazid, J. Phys. (Paris) Colloq. **40**, C3-185 (1975); J. Doucet, A. M. Levelut, and M. Lambert, Ann. Phys. (N.Y.) **3**, 157 (1978); J. Doucet and A. M. Levelut, J.

Phys. (Paris) 38, 1133 (1977).

[6]D. E. Moncton and R. Pindak, Phys. Rev. Lett. 43, 701 (1979).

[7]P. S. Pershan, G. Aeppli, J. D. Litster, and R. J. Birgeneau, Mol. Cryst. Liq. Cryst. 67, 205 (1981).

[8]D. E. Moncton and R. Pindak, in Ordering in Two Dimensions: Proceedings of an International Conference Held at Lake Geneva, U.S.A., edited by S. K. Sinha (North-Holland, New York, 1980), pp. 83—90.

[9]P. A. C. Gane, A. J. Leadbetter, and P. G. Wrighton, Mol. Cryst. Liq. Cryst. 66, 247 (1981).

[10]R. Pindak, D. E. Moncton, J. W. Goodby, and S. C. Davey, Phys. Rev. Lett. 46, 1135 (1981).

[11]A. J. Leadbetter, J. C. Frost, and M. A. Mazid, J. Phys. (Paris) 40, 325 (1979).

[12]A. J. Leadbetter, M. A. Mazid, and B. A. Kelly, Phys. Rev. Lett. 43, 630 (1979).

[13]A. J. Leadbetter, M. A. Mazid, and R. M. Richardson, in Liquid Crystals, edited by S. Chandrasekhar (Cambridge University Press, London, 1980), pp. 65—79.

[14]J. Thoen and G. Seynhaeve (unpublished).

[15]J. Collet, L. B. Sorensen, P. S. Pershan, R. J. Birgeneau, J. D.

Litster, and J. Als-Nielsen, Phys. Rev. Lett. 49, 553 (1982).

[16]Jeffrey Allen Collett, Ph.D. thesis, Harvard University, 1983.

[17]C. Y. Young, R. Pindak, N. A. Clark, and R. B. Meyer, Phys. Rev. Lett. 40, 773 (1978).

[18]Index of refraction values were assumed to be approximately the same as those published for 40.8 by P. S. Pershan, in The Molecular Physics of Liquid Crystals, edited by G. R. Luckhurst and G. W. Gray (Academic, New York, 1979), p. 365.

[19]P. C. Martin, O. Parodi, and P. S. Pershan, Phys. Rev. A 6, 2401 (1972).

[20]Y. Liao, N. A. Clark, and P. S. Pershan, Phys. Rev. Lett. 30, 639 (1973).

[21]Jeffrey Collett, P. S. Pershan, E. B. Sirota, and L. B. Sorensen (unpublished).

[22]John Hirth, P. S. Pershan, J. Collett, E. B. Sirota, and L. B. Sorensen, Phys. Rev. Lett. 53, 473 (1984).

[23]Jeffrey Collett, P. S. Pershan, E. B. Sirota, and L. B. Sorensen, Phys. Rev. Lett. 52, 356 (1984).

[24]C. Rosenblatt, R. Pindak, N. A. Clark, and R. B. Meyer, Phys. Rev. Lett. 42, 1220 (1979).

PHYSICAL REVIEW A   VOLUME 36, NUMBER 6   SEPTEMBER 15, 1987

X-ray and optical studies of the thickness dependence of the phase diagram of liquid-crystal films

E. B. Sirota* and P. S. Pershan

Department of Physics, Harvard University, Cambridge, Massachusetts 02138

L. B. Sorensen

Department of Physics, University of Washington, Seattle, Washington 98195

J. Collett†

*Division of Applied Sciences, Harvard University, Cambridge, Massachusetts 20138
and IBM Thomas J. Watson Research Center, Yorktown Heights, New York 10598*
(Received 12 January 1987)

A comprehensive study of the thickness dependence of the phase diagram of freely suspended films of the liquid crystal 4-n-heptyloxybenzylidene-4-n-heptylaniline (7O.7) between 50 and 69 °C is reported. In thick films (thicker than about 300 layers and characteristic of bulk samples) there is a low-temperature crystalline-G phase followed by five crystalline-B phases with different stacking arrangements at higher temperatures. In thinner films there are two additional crystalline-B phases and two tilted hexatic phases, smectic-F and smectic-I, which do not appear in bulk samples. The in-plane and interlayer correlations in the tilted hexatic phases are anisotropic with a clear dependence on the molecular tilt direction; the in-plane correlations are more developed (longer range) perpendicular to the molecular tilt direction and the interlayer correlations are more developed parallel to the tilt direction.

I. INTRODUCTION

The liquid-crystalline phases of matter, which have order intermediate between that of three-dimensional crystals and fluids, have been extremely fertile systems for studies of the influence of symmetry and dimensionality on phase transitions. Because of their relatively weak interlayer coupling and strong in-plane interactions, these materials can be successfully modeled as a collection of weakly coupled two-dimensional systems. The weak interlayer coupling allows these materials to form three-dimensional stacked hexatic phases without spontaneously converting to a three-dimensional crystal.[1] The freely suspended film technique allows the thickness of the unusual materials to be varied from two molecular layers to thousands of molecular layers, allowing direct experimental studies of the two-dimensional to three-dimensional crossover. In addition to probing the effects due to the dimensionality, thin films also provide important information about the influence of the surface field caused by the termination of the material at the surface.

This paper reports our comprehensive studies of the thickness dependence of the phase diagram of freely suspended films of 4-n-heptyloxybenzylidene-4-n-heptylaniline (7O.7). The chemical structure of 7O.7 is shown in Fig. 1. The phase diagram is extremely rich—there are 11 distinct bulk phases between 30 and 85 °C and there are four additional phases that only appear in thin films.[2-4]

The phase sequence of bulk 7O.7 was first investigated by Smith and co-workers[5,6] who found the following

sequence: crystal $\rightarrow$ 33 °C $\rightarrow$ crystalline-G (CrG) $\rightarrow$ 55° $\rightarrow$ crystalline-B (CrB) $\rightarrow$ 69° $\rightarrow$ smectic-C(SmC) $\rightarrow$ 72° $\rightarrow$ smectic-A(SmA) $\rightarrow$ 83.7° $\rightarrow$ nematic $\rightarrow$ 84° $\rightarrow$ isotropic. Subsequent x-ray studies of 7O.7 and of homologous[7-13] compounds with similar phase sequences showed that the CrB phase in these materials consisted of a series of crystalline phases with different stacking sequences. Motivated by these studies we performed high-resolution x-ray scattering measurements on freely suspended films of 7O.7 and found that the crystalline-B phase actually consisted of five distinct phases with different stacking sequences:[14,15]   55 °C $\rightarrow$ hexagonal-AAA $\rightarrow$ 61° $\rightarrow$ monoclinic-C $\rightarrow$ 61.5° $\rightarrow$ orthorhombic-F $\rightarrow$ 64° $\rightarrow$ hexagonal-close-packed-$ABAB$ $\rightarrow$ 69°. We also found that all of these crystalline-B phases, except the hexagonal-close-packed phase, also have a one-dimensional modulation of the smectic layers with the modulation wave vector in the plane of the layers and the polarization normal to the layers.[14,15] Recently, two of us found that there are also two-dimensional modulations of the smectic layers; these results are reported in the companion paper to this paper.[16]

This paper describes the first comprehensive study of the thickness dependence of phase sequence in freely suspended liquid-crystal films. Our initial expectation that the system would simply cross over from three-dimensional to two-dimensional behavior at some small thickness (we guessed two or three layers) greatly underestimated the richness of the behavior we subsequently found—for example, the phase diagram starts to differ from the bulk diagram for films thinner than about

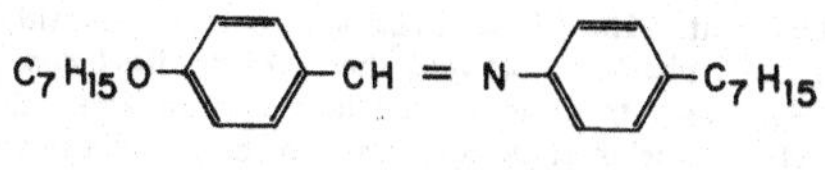

$$C_7H_{15}O-\bigcirc-CH=N-\bigcirc-C_7H_{15}$$

FIG. 1. Chemical structure of 7O.7.

300 layers. We now believe that there is a competition between the effects of the surface field (generated at the film-vacuum interface) and the effects of reduced dimensionality. Some of the observed phase boundaries move to lower temperature as the film thickness is decreased (as expected for reduced dimensionality) and some of the transitions move to higher temperature (as expected for a surface field which favors the ordered low-temperature phase).

II. EXPERIMENTAL TECHNIQUES

The thickness dependence of the phase diagram for freely suspended 7O.7 films was determined by combining optical microscopy observations with laboratory-based low-resolution x-ray diffraction measurements and with synchrotron-based high-resolution measurements. By combining these techniques, we have been able to study the unusual evolution of the phase diagram for film thicknesses between two molecular layers and about 300 layers where the behavior approaches that of bulk samples.

The films were drawn across a 7-mm-diam. hole in a 0.2-mm-thick glass cover slip. The oven has been described in detail previously.[15] Two modifications were made to improve the performance. (1) The original kapton-x-ray windows were replaced by 0.25-mm bent beryllium windows. This modification produced a substantial reduction in the background scattering from the windows and greatly facilitated the high-resolution x-ray studies of very thin films. (2) The original optical windows were mounted using epoxy. To reduce the strain-induced birefringence in the windows, the epoxy seals were replaced by O-ring seals.

Temperature control was provided by a computer-interfaced controller.[17] Under typical operating conditions (± 3 K room-temperature variations) the present single-stage oven provided a temperature stability better than ± 5 mK over many hours.

The oven was evacuated and then backfilled with approximately 1 Torr of helium gas. This greatly improved the temperature equilibrium times of the films and produced only a small increase in the x-ray scattering background. Permanent magnets did not align the tilt direction of thin hexatic or crystalline phases of 7O.7 and therefore were not used.

A. Film-thickness determination

For very thin films the thickness can be determined very easily by measuring the reflected intensity of a low-power laser.[18] For films thinner than about 12 layers, the reflected intensity scales as N^2, where N is the number of layers. Since N is quantized, a reflected-intensity-

versus-film-thickness calibration can be easily obtained by measuring many films. This calibration can be extended up to about 20 layers. For thicker films, the quantized intensity variation gets quite small and it becomes difficult to determine the exact thickness. For thicker films, the thickness can be determined by measuring the optical phase retardation of a transmitted laser beam[15] or by measuring the apparent reflected color produced for incident white light. The reflected intensity for light of wavelength λ and incidence angle θ from an optically isotropic film with index of refraction n consisting of N layers each of thickness d is given by[19]

$$I(\lambda,N)=I_0(\lambda)\frac{F\sin^2\phi}{1+F\sin^2\phi}, \tag{1}$$

where $I_0(\lambda)$ is the incident intensity, $F=4R/(1-R)^2$, and $\phi=(2\pi/\lambda)Nd(n^2-\sin^2\theta)^{1/2}$. Although the general expression for the reflectivity from uniaxially anisotropic liquid-crystal films in an arbitrary direction is quite complicated, the special case of s-polarized light reflected from a smectic-A film (where the optic axis is perpendicular to the surface of the film) is given by Eq. (1) with n replaced by the ordinary index n_0 and

$$R=\left|\frac{\cos\theta-(n_0^2-\sin^2\theta)^{1/2}}{\cos\theta+(n_0^2-\sin^2\theta)^{1/2}}\right|^2. \tag{2}$$

For 7O.7 films in the smectic-A phase, $n_0=1.60$,[20] and $d=29.8$ Å;[12] the angle of incidence for our x-ray film oven is $\theta=25°$. To compute the expected colors versus thickness, the intensity distribution $I(\lambda,N)$ was calculated by assuming that the Bausch and Lomb 31-33-05 illuminator produced an incident blackbody spectrum $I_0(\lambda)$ with a color temperature of $T=2655$ K. The corresponding tristimulus values were computed using the 1931 Commission Internationale de l'Eclairage (CIE) tables[21] for the tristimulus responses $\bar{x}_\lambda$, $\bar{y}_\lambda$, and $\bar{z}_\lambda$

$$X=\int I(\lambda,N)\bar{x}_\lambda d\lambda,$$
$$Y=\int I(\lambda,N)\bar{y}_\lambda d\lambda, \tag{3}$$
$$Z=\int I(\lambda,N)\bar{z}_\lambda d\lambda.$$

The resulting tristimulus values were then transformed to chromaticity coordinates (x and y) and a relative brightness value (b)

$$x=\frac{X}{X+Y+Z},$$
$$y=\frac{Y}{X+Y+Z}, \tag{4}$$
$$b=X+Y+Z.$$

The loci of the calculated chromaticity coordinates versus film thickness are shown on the standard CIE chromaticity diagram in Fig. 2. For films thinner than about 20 layers, this calculation shows that the films will appear gray, becoming whiter (brighter) as the thickness is increased. The color sequence then is orange ($N\simeq30$), red ($N\simeq53$), blue ($N\simeq62$), green ($N\simeq71$), yellow ($N\simeq85$), red ($N\simeq113$), blue ($N\simeq127$), green

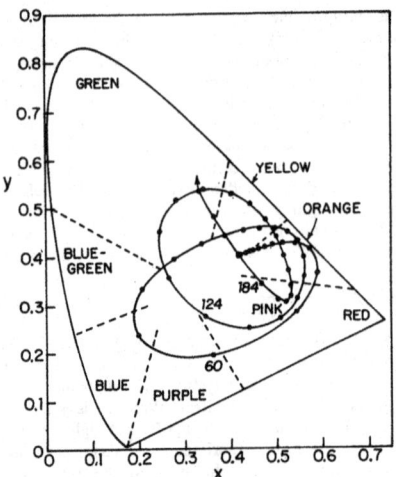

FIG. 2. Calculated chromaticity coordinate (x,y) vs film thickness for 70.7 films with $n_0 = 1.60$ and $d = 29.8$ Å assuming a 2655-K light source incident at $\theta = 25°$. The solid line indicates the locus of the chromaticity changes, the dots on the line indicate every fourth layer starting with $N = 4$.

$(N \simeq 136)$, yellow $(N \simeq 153)$, pink $(N \simeq 178)$, pale green, etc. Although the observed colors repeat, the first two loops are sufficiently different that with experience, the orders (for example, the $N = 62$ blue and the $N = 127$ blue) can be distinguished. Between $N = 200$ and 500 layers the color cycles between pale pink, green, and white, becoming paler until the thickest films appear a milky white. This method allowed the thickness to be determined to within ~ 2 layers for blue films and to within ~ 5 layers for yellow films. This technique is very sensitive to the uniformity of the film; in favorable cases, such as with blue or gray films, film thickness nonuniformities of a single layer can be easily detected. We found it important to monitor the film uniformity since the film occasionally thinned or became nonuniform at several of the phase transitions we studied.

In this study, the reported film thicknesses for films thinner than 15 layers are exact. For films between 16 and 25 layers the uncertainty is ±1 layer and for films between 25 and 200 layers the uncertainty is ±5%.

B. Optical observations

Many of the phase transitions and the macroscopic symmetry of the different phases could be observed qualitatively using polarized optical microscopy.[22] We used a simple $25\times$ microscope and crossed polarizers in the specular direction for films illuminated with white light. Since the molecules are very anisotropic, all of the tilted phases (smectic-C, -F, -I and CrG) are strongly

birefringent. The phase transitions between nontilted phases or between the different tilted phases are very easy to observe even in very thin films. By correlating the observed textures with the x-ray observations for a few thicknesses, the textures can be identified with specific phases and the thickness-dependent phase diagram can be completed optically. Because of the extreme sensitivity of the optical observations, and because of the marked difference in the textures of the two phases, it is easy to observe the difference between a homogeneous fluid (smectic-C) film and a film with a single hexatic (smectic-I) overlayer on the two exterior surfaces. Of course x-ray measurements are necessary to firmly identify the surface layers as smectic-I.[4]

C. X-ray measurements

The low-resolution measurements were performed using the Harvard Materials Research Laboratory 12 kW Rigaku RU-200 rotating-anode x-ray generator operating with an effective source size of 0.3×0.3 mm^2 and Cu K_α radiation ($\lambda = 1.542$ Å). The generator was operated at 4.5 kW (50 kV, 90 mA). The scattering angle (2θ)

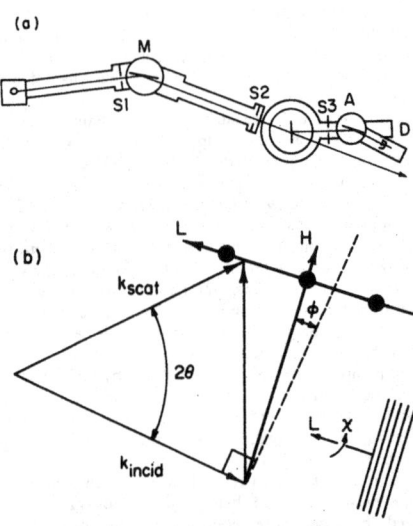

FIG. 3. (a) Low-resolution spectrometer configuration consisting of a bent graphite monochromator (M), a flat graphite analyzer (A), three antiscatter slits (S1,S2,S3), and a NaI scintillation detector (D). (b) Schematic illustration of the experimental kinematics: $\mathbf{k}_{incid}$ and $\mathbf{k}_{scat}$ are the incident and scattered wave vectors, respectively. The reciprocal-lattice vectors H and L are parallel and normal to the lattice planes, respectively. When the angle $\phi = 0$, the incident wave vector is normal to the smectic layers. In order to observe scattering from a particular point in the H-L plane, the film must be rotated by an angle ϕ. The angle χ defines rotations around the layer normal, L.

and the oven orientation (θ) was controlled with a Huber 420 two-circle goniometer. The film orientation (χ) with respect to the scattering plane was controlled with a rotation stage inside the oven. The measurements were made using the geometry shown in Fig. 3. The monochromator was a vertically focusing pyrolytic graphite (002) crystal, 2 in. high with a 115-mm radius of curvature. The source-to-monochromator and monochromator-to-sample distances were equal, producing horizontal Bragg-Brentano parafocusing. The hot spot of the beam at the sample position was 1 mm wide and 3 mm high. Because the focusing was imperfect, there were many stray photons which could not be eliminated by slits S1 and S2. A brass aperture inside the oven located 1 mm from the sample on the source side with a diameter 1 mm less than the 7 mm diameter of the film prevented the incident photons from hitting the bulk material at the edge of the sample aperture. This was very important for the thin-film studies where the bulk edge scattering would be a serious problem.

The analyzer was a flat (1×1)-in.^{2} pyrolytic graphite (002) crystal. To maximize the signal intensity, the detector was placed as close as possible to the sample, ~ 350 mm. This produced a measured χ full width at half maximum (FWHM) of 12° for $2\theta = 20°$. The measured zero-arm 2θ FWHM of the spectrometer was 0.65° and the measured resolutions for $2\theta = 20°$ were $\Delta L = 9.2 \times 10^{-3}$ Å^{-1}, $\Delta H = 4.7 \times 10^{-2}$ Å^{-1}. The intensity of the direct beam was $\sim 2 \times 10^8$ photons/sec.

The high-resolution studies were done using beam line VII-2 at the Stanford Synchrotron Radiation Laboratory (SSRL).[2,3,23] The oven was oriented with the four-circle Huber 5020 goniometer. The monochromator contained a pair of asymmetrically cut Ge(111) crystals and the analyzer was a LiF(200) crystal. The longitudinal resolution was $\Delta H = 3.8 \times 10^{-3}$ Å^{-1} and the transverse resolution was limited by the sample mosaic. The FWHM in χ was about 2°. For the synchroton studies, the brass aperture inside the oven was unnecessary since the beam was well collimated.

D. Samples

Two different commercial sources of 7O.7 were used for these studies.[24] Although the transition temperatures drifted down by as much as 1 K for some of the material, neither the phase sequence nor any of the observed phase transitions were noticibly affected by the decreased transition temperatures. Recrystallization of the material in petroleum ether restored the original transition temperature. When the sample was maintained in the oven for long periods, the transition temperatures also drifted down. To ensure that the observed thickness dependence of the smectic-(SmF)-to-CrG transition was not due to transition temperature drifts, we alternated measurements of thick, thin, and then thick films without reloading the sample; films of the same thickness were measured many times and the measured correlation lengths versus temperature matched when the transition temperatures were shifted to agree. All of the data presented in this study have

been corrected (when necessary) to correspond to the unshifted nondegraded transition temperatures.

III. EXPERIMENTAL RESULTS

A. Observed phases

The thickness-dependent phase diagram for 7O.7 is surprisingly rich. Nine of the phases in this system are shown in the phase diagram in Fig. 4. Above 69 °C bulk 7O.7 has smectic-C, smectic-A, nematic, and isotropic phases. The crystalline-G phase persists down to about 33 °C.

The crystalline-B (CrB) phases are three-dimensional (3D) crystals[7,11,25] with the director oriented normal to the layers. The layer spacing c is about 30.6 Å with a small temperature dependence ($\sim 2 \times 10^{-4}$ Å/°C). No jumps in the layer spacing were observed at the restacking transitions.[16] The in-plane nearest-neighbor distance of the hexagonal lattice a is about 5.053 Å. In the remainder of this paper we will express the components of the in-plane reciprocal lattice H in units of $4\pi/\sqrt{3}a = 1.4357$ Å^{-1} and the normal component L in units of $2\pi/c = 0.2053$ Å^{-1}. These structures all have very similar in-plane lattice-producing sharp crystalline peaks at $H = 1.0$ in these units.

The small temperature- and phase-dependent distortions have been described in detail previously.[14-16] There are six different CrB structures observed in this

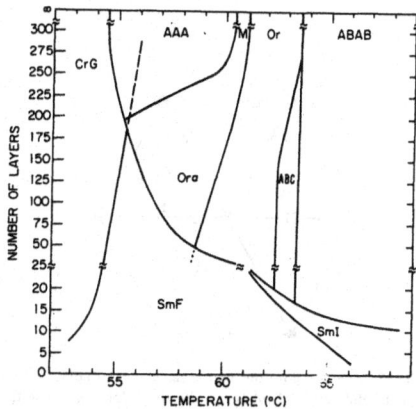

FIG. 4. Thickness vs temperature phase diagram for 7O.7. The solid lines indicate the reversible phase boundaries. The dashed (dotted) lines indicate the phase boundaries of the metastable phases observed kinetically on heating (cooling). The phases are crystalline-G (CrG), smectic-F (SmF), smectic-I (SmI), and the crystalline-B (CrB) modifications: hexagonal-close-packed-$ABAB$ ($ABAB$), hexagonal ABC (ABC), orthorhombic (Or), orthorhombic-a (Ora), monoclinic (M), and Hexagonal-AAA (AAA). The smectic-C phase (SmC) occurs above about 69 °C.

system. These structures all have very similar in-plane lattices producing sharp crystalline peaks at $H = 1.0$. The different interlayer stackings produce peaks at different characteristic L values. The stacking of the thick-film phases has been discussed in detail previously.[14,15] We report here a new hexagonal-ABC (ABC) and a new orthorhombic (Ora), which only appear in thinner films, and which have not been reported previously. The modulations of the thick-film orthorhombic, monoclinic, and hexagonal-AAA phases are discussed in detail in a companion paper.[16] Figure 5 shows the locations of the molecules for the different stacking structures, and Fig. 6 shows representative L scans for these phases.

A single-domain $ABAB$ sample has peaks at $L = 0$ and $\frac{1}{2}$ [Fig. 6(a)] for all six values of χ ($0°, \pm 60°, \pm 120°, 180°$). The observed peaks at $L = \frac{1}{2}$ are approximately three times the intensity of those at $L = 0$ as predicted from the structure factor for this stacking. The signature for monodomain ABC samples is $L = \frac{1}{3}$ peaks for $\chi = 0°, \pm 120°$ and $L = \frac{2}{3}$ peaks for $\chi = \pm 60°, 180°$ [Fig. 6(b)]. Samples with an equal mixture of ABC and ACB stacking will have peaks at $L = \frac{1}{3}$ and $\frac{2}{3}$ for all six χ positions.[20] The orthorhombic phase has peaks at $L = 0$ for $\chi = 0°, 180°$ and peaks at $L = \frac{1}{2}$ for the other four values of χ [Fig. 6(c)]. The monoclinic phase has peaks at $L = 0$ for $\chi = 0°, 180°$ and at $L = \pm s$ for $\chi = \pm 60°, \pm 120°$, where s is a temperature-dependent shear parameter with $0 < s < 0.5$ [Fig. 6(e)].[14] The structure and signature of the new Ora phase is discussed below. The AAA phase has peaks at $L = 0$ for all six χ

values [Fig. 6(d)]. In addition to the peaks listed above, all the phases also have peaks at $L \pm 1$ (actually $L \pm m$ with m an integer), generated by the layer periodicity. The molecular form factor and the Debye-Waller factor produce an overall envelope peaked at $L = 0$. At each of the six values of χ, there is a ridge of strong diffuse scattering (with $H = 1$) due to vibrational modes associated with the layers sliding over each other.[15,25,26]

Similar scans for the crystalline-G phase are shown in Fig. 10.[3,10,12,27] Peaks are observed at $H = 0.925$ for $L = 2.6$, $\chi = 0$ and $L = -2.6$, $\chi = 180$ and at $H = 0.980$ for $L = 1.3$, $\chi = \pm 60$ and $L = -1.3$, $\chi = \pm 120$. In addition to these peaks there are also peaks at those L values

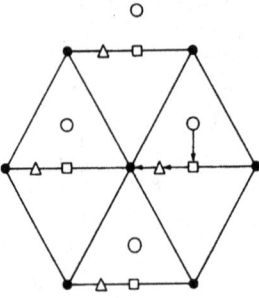

FIG. 5. Relative positions of the molecules in adjacent layers in the different CrB phases. The solid dots represent the reference layer positions. The open symbols represent the layer positions in the next layer above for lattices with a single-layer unit cell. For lattice with a two-layer unit cell, the position of the layer below is the same as that of the layer above. The open circles represent both the hexagonal-ABC structure (one-layer unit cell) and the hexagonal-$ABAB$ structure (two-layer unit cell). The squares indicate the orthorhombic structure positions. The triangles indicate both the monoclinic (one-layer unit cell) and the Ora (two layers per unit cell) structure. In the hexagonal-AAA phase the molecular positions in the adjacent layers are the same as those in the reference layer.

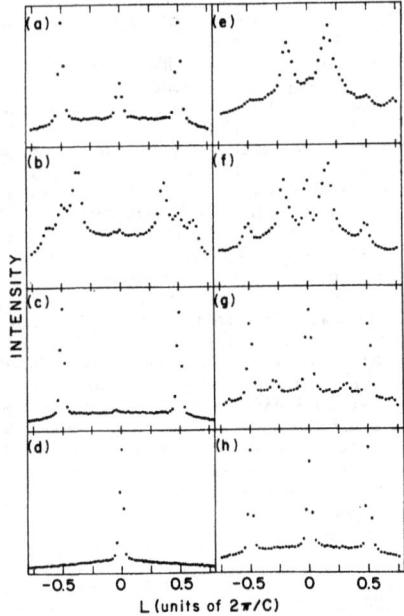

INTENSITY

L (units of $2\pi/C$)

FIG. 6. Typical low-resolution L scans in the different crystalline-B phases. (a) The hexagonal-$ABAB$ phase. The peaks at $L = \pm \frac{1}{2}$ are about three times as strong as the peaks at $L = 0$. (b) The hexagonal-ABC phase coexisting with the orthorhombic or the hexagonal-$ABAB$ phase. There are peaks at multiples of $L = \frac{1}{3}$. (c) The orthorhombic phase. There are peaks at $L = \pm \frac{1}{2}$ at four χ positions ($\chi = \pm 60°, \pm 120°$) as shown. There are also peaks at $L = 0$ at the other two χ positions ($\chi = 0°, 180°$), as shown in (d). (d) The hexagonal-AAA phase. There are only peaks at $L = 0$. (e) The monoclinic phase in a very thick film. There are peaks at $L = s$ and $L = 1 - s$. (f) The monoclinic phase coexisting with the Ora phase in a 400-layer film. (g) The orthorhombic phase coexisting with the monoclinic phase in a 200-layer film. (h) The orthorhombic phase in a 100-layer film. There are peaks at $L = \pm \frac{1}{2}$ and $L = 0$.

plus integral multiples of ± 1.08. Like the crystalline-B phases, there are also ridges of strong diffuse scattering. The peaks of the form factor and Debye-Waller factor are at the $HL\chi$ Bragg peak positions, indicating that the molecules are tilted by $\sim 22°$. (The tilt varies from $22°$–$24°$ with temperature.) The crystalline-G phase is a 3D crystal with the molecules stacked end to end; the tilt of the primitive translation vectors coincides with the molecular tilt. The molecular tilt direction is towards the face of the hexagon formed by its nearest in-plane neighbors. The lattice parameters[12] are $a = 9.45$ Å, $b = 5.15$ Å, $c = 30.6$ Å, and $\beta = 111.9°$.

The diffuse peaks in the smectic-F phase[3,27–31] occur at the same positions as the crystalline peaks of the crystalline-G phase. The interlayer correlation lengths determine the width along L and the in-plane correlation lengths determine the width along the H direction. The peaks are located in the same positions in these two phases because they have the same local packing.[32] The SmF phase is a stacked tilted hexatic phase[33] with short-range in-plane positional order; the in-plane positional correlations decay exponentially with a decay length of about 50 molecular diameters.

The smectic-I phase[3,30,29,34–36] is a stacked tilted hexatic phase, with the molecular tilt direction toward the corner of the hexagon formed by its nearest neighbors. The diffuse peaks appear at $H = 0.988$, $L = 0$, $\chi = 0, 180$; $H = 0.955$, $L = 2$, $\chi = 60, 120$; and $H = 0.955$, $L = -2$, $\chi = 240, 300$. The crystalline analogue of the smectic-I phase is the crystalline-J phase, which does not occur in 7O.7. The lattice parameters for the smectic-I structure are $a = 5.34$ Å, $b = 8.86$ Å, $c = 30.6$ Å, and $\beta = 109°$.

B. Restacking transitions

Below about 200 layers, the hexagonal-ABC phase appeared between the hexagonal-$ABAB$ and the orthorhombic phases. This was a very difficult region in the phase diagram to study because the films tended to rupture in the ABC phase. Films about 60 layers thick ruptured soon after the ABC phase formed (either by cooling or heating). Films over 100 layers thick were stable longer; one 200-layer film was stable for days. However, in the stable films, the ABC structure coexisted with either the orthorhombic or the $ABAB$ structure. The films did not break if they were cooled or heated quickly through the ABC temperature range. A typical L scan for the coexisting ABC structure is shown in Fig. 6(b). The insertion of the ABC phase between the $ABAB$ and the orthorhombic phase, produces the same phase sequence as that found in bulk n-[4-(n-pentyl)oxybenzylidene]-4'-(n-hexyl)aniline (5O.6).[37]

In films thicker than ~ 300 layers, the monoclinic phase[15] was always observed between the orthorhombic and the AAA phases [Fig. 6(e)]. However, weak peaks at $L = \pm \frac{1}{2}$ were always observed in the monoclinic phase.[37] These peaks became weaker and disappeared as the temperature was lowered towards the AAA phase. Below about 150 layers, the transition between the orthorhombic and the monoclinic phase gradually changed, until it occurred without any evidence for

peaks at the L values associated with the monoclinic phase [Fig. 6(h)]. The x-ray signature for this transition was that the orthorhombic domains with peaks at only $L = \pm \frac{1}{2}$ developed peaks at $L = 0$. As the temperature was lowered, the $L = \pm \frac{1}{2}$ peaks continuously weakened and the $L = 0$ peak continuously strengthened, until only the $L = 0$ peak remained. The orthorhombic domains with peaks at $L = 0$ remained unchanged. Between about 150 and 300 layers, this continuous orthorhombic–to–hexagonal-AAA transition coexisted with small monoclinic peaks [Fig. 6(g)].

This transition can be interpreted either as a new intermediate orthorhombic-a phase (Ora) between the orthorhombic and the hexagonal-AAA phases or as a coexistence between the hexagonal-AAA and the orthorhombic phases. In the monoclinic phase the displacement of each layer with respect to the one below it occurs in the same direction. This results in a tilted monoclinic structure with one molecule per unit cell. The observed x-ray scattering from the new Ora phase is consistent with a structure produced by displacing the layers in alternating directions so that the macroscopic symmetry is orthorhombic. The resolution-limited peaks at both $L = 0$ and $\pm \frac{1}{2}$ for the same χ position require an Ora unit cell with at least two molecules per unit cell. The real-space structures for the observed crystalline-B phases are shown schematically in Fig. 7; the monoclinic and the Ora structures have the same local packing between adjacent layers, but differ in their second-neighbor packing. The weak $L = \frac{1}{2}$ peaks in the thick films may be due to Ora surface layers on the exterior surfaces of the monoclinic interior. The observed x-ray intensities are consistent with about 75 Ora layers on each surface.

The second interpretation of the data is that there is a coexistence of the orthorhombic and the AAA phases. High-resolution x-ray measurements of an 80-layer film showed that the L widths of the Bragg peaks were limited only by the 80-layer finite thickness, and did not broaden as the peak intensities diminished. Thus if there is coexistence, it must occur in the direction parallel to the film surface and the lateral domain size of the coexisting domains must be smaller than the (1×2)-mm^2 size of the beam. Since this effect was observed in different batches of 7O.7 we tend to rule out impurities as the cause of the coexistence. While we cannot absolutely rule out coexistence, we have no explanation for its appearance.

The optical observations (with crossed polarizers) showed that the crystalline-B phase, produced by cooling from the smectic-C phase, initially formed without any visible observable texture. However, as the temperature was lowered, a weak birefringent texture (much weaker in contrast than the SmC or the lower-temperature tilted phases) was observed. This texture became progressively stronger as the temperature was lowered. This texture is apparently due to the formation of tilted surface layers on the two exterior surfaces, as originally noted by Farber[38]—on close inspection the observed tilt domains clearly overlap, indicating two distinct sets. The actual phase of the surface layers has not been determined; the optical texture could be caused by

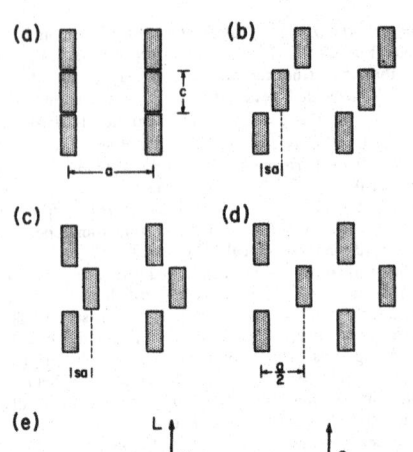

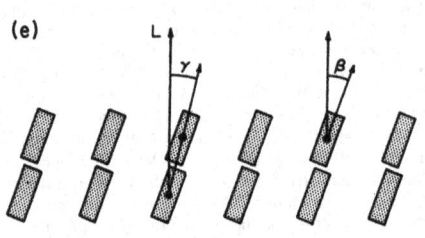

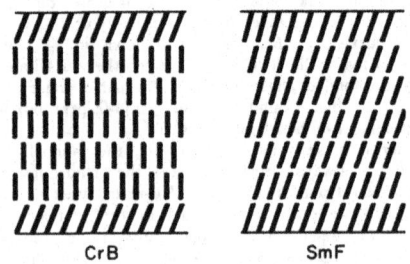

FIG. 7. Schematic view of the projection of the molecular axes onto the (100) plane showing the in-plane projections for the (a) hexagonal-AAA phase, (b) monoclinic phase, (c) orthorhombic phase, (d) Ora phase, and (e) the crystalline-G phase. The angle between the layer normal and the primitive translation vector is γ and is determined from the position of the Bragg peak. The angle β between the long molecular axis and the layer normal is somewhat model dependent, however, as discussed in the text, measurements indicate that the two are indistinguishable.

smectic-F, smectic-I, crystalline-G, or crystalline-J surface layers. An x-ray study to determine the surface phase is planned. We recently used x-ray methods to demonstrate the formation of surface smectic-I layers on the two exterior surfaces of 7O.7 smectic-C films.[4]

C. Tilted phases

Using the low-resolution spectrometer at Harvard the thickness dependence of the smectic-F-to-crystalline-G transition was studied versus film thickness by measuring the x-ray scattering profiles along the L direction.[3] This spectrometer allowed measurements of the interlayer correlations for films thicker than about eight layers. The longitudinal resolution of this spectrometer was too broad to allow measurements of the in-plane positional evolution. The difference in optical textures between the smectic-F and crystalline-B phases made it quite easy to visually determine the phase boundary between the two phases for films thicker than about 14 lay-

ers. In thinner films, there was less contrast and the change at the transition was quite subtle, making it difficult to determine the transition optically. Consequently, for thinner films the x-ray signature (the diffuse peaks in the smectic-F phase becoming resolution- and finite-size-limited Bragg peaks in the crystalline-G phase) was used to establish the phase boundary. The composite transition temperature versus film thickness diagram is shown in Fig. 4. At first there is a gradual decrease in the transition temperature as the thickness is reduced, then below about 25 layers there is a substantial decrease.

The interlayer correlations were studied using L scans through the peaks at $L = \pm 2.6$ and $L = \pm 1.3$; the peaks at $L = \pm 1.3$ were more than twice as wide as those at $L = \pm 2.6$, indicating an anisotropy in the interlayer disorder correlated with the molecular tilt direction. Typical L scans at these positions are shown in Fig. 9. The anisotropy is presumably produced by the anisotropic steric interactions produced by the molecular tilt. A remnant of the anisotropy is also visible in the diffuse CrG scattering as shown in Fig. 10. The diffuse scattering ridge produced by the soft modes corresponding to the layers sliding over each other is approximately twice as strong (relative to the Bragg peaks) near the $L = \pm 1.3$ peaks as it is near the $L = \pm 2.6$ peaks.

The temperature dependence of the interlayer correlations was also studied. With our low-resolution spectrometer, it was much easier to study the $L = \pm 2.6$ peaks than the $L = \pm 1.3$ peaks. The low-resolution measurements presented here are for the $L = \pm 2.6$ peaks. Because the samples were always unoriented χ powders, the appropriate line shape is a square-root Lorentzian (SRL) profile.[39,40] The data were fit to independent SRL's at $L = 2.6$ 3.68, and 1.52 multiplied by the Gaussian envelope produced by the molecular form factor. A background term was also included. The measured smectic-F interlayer correlation length versus temperature and film thickness is shown in Fig. 11. The observed correlation lengths are independent of the film thickness and exhibit identical temperature dependence. However, in thinner films the SmF-to-CrG transition occurs at a lower temperature and the SmF correlations develop further before the transition occurs. The ob-

FIG. 8. Proposed structures of the smectic-F (SmF) and of the crystalline-B (CrB) with tilted surface layers.

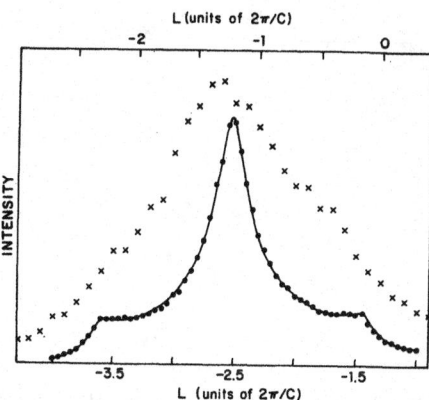

L (units of $2\pi/C$)

FIG. 9. Typical L scans in the SmF phase for a 125-layer film at $T = 55\,°C$. Scans through the $L = -2.6$ peak are shown by dots, scans through the $L = -1.3$ peaks are shown by crosses. The solid line is the fit to square-root Lorentzian peaks centered at $L = -2.6$, -3.68, and -1.52 multiplied by the Gaussian envelope produced by the molecular form factor. These data were collected with the low-resolution spectrometer. The peak counting rates were 300 counts/sec for $L = -2.6$ and 100 counts/sec for $L = -1.3$.

served temperature dependence appears to follow a universal curve; extrapolation of this curve to thinner films than we have measured indicates that the SmF-to-CrG transition may change from first order to second order or be continuous for films thinner than about four layers. Unfortunately, this could not be studied with the low-resolution spectrometer, since, as shown in Fig. 11, the widths along the L direction eventually became dominated by the finite film thickness in thin films. The in-plane correlations can be measured using the high-resolution synchrotron based spectrometer, and we plan to study this transition in thin films in the future. (See Fig. 13.)

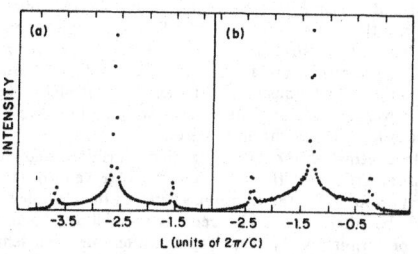

FIG. 10. Low-resolution L scans through the (a) $L = -2.6$ and (b) $L = -1.3$ peaks in the crystalline-G phase. The weaker secondary peaks offset by ± 1.08 in L are visible. The relative intensity of the diffuse ridge is about twice as strong for the $L = -1.3$ peak as it is for the $L = -2.6$ peak.

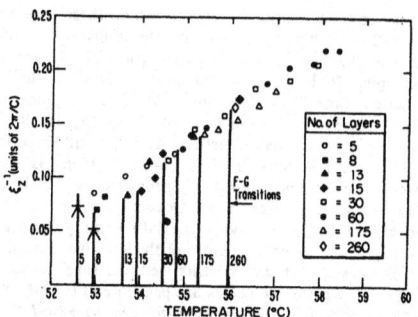

FIG. 11. Observed inverse correlation lengths from the low-resolution measurements of the temperature dependence of the L linewidths for the $L = 2.6$ peak of the smectic-F phase of 7O.7 vs film thickness N. For each thickness, the widths follow a universal curve until they undergo the first-order jump (indicated by the solid bar) at their respective smectic-F-to-crystalline-G transitions. The two arrows indicate the finite-size limits for five- and eight-layer films, respectively.

In thicker films, our previous high-resolution measurements have shown that there is a large pretransitional evolution of the in-plane and the interlayer correlations followed by a small first-order jump at the SmF-to-CrG transition. The observed linewidths for 13-layer films are shown in Fig. 12. In addition to the interesting pre-

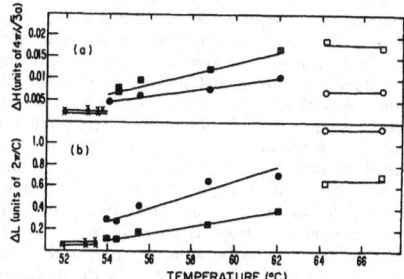

FIG. 12. Temperature dependence of the measured linewidths along the H direction (a) and along the L direction (b) in the crystalline-G, smectic-F, and smectic-I phases of 7O.7, taken with the high-resolution spectrometer at SSRL for a 13-layer film. The solid lines are guides for the eye. SmI phase: Open circles indicate $H = 0.988$, $L = 0$; open squares indicate $H = 0.955$, $L = 2$. SmF phase: Closed circles indicate $H = 0.975$, $L = 1.3$; closed squares indicate $H = 0.925$, $L = 2.6$. (a) Anisotropic in-plane linewidths [half widths at half maximum (HWHM)] from the H scans. The crosses indicate the instrumental resolution measured in the CrG phase. The widths are presented in H units. (b) Linewidths (HWHM) illustrating the anisotropic interlayer correlations from the L scans. The crosses indicate the combined mosaic and finite-size limit, measured in the CrG phase. The widths are presented in L units.

transitional correlation length evolution, these data also show that both the in-plane and the interlayer correlations of the smectic-F and the smectic-I phases are anisotropic. In both cases, the peaks closest to the tilt direction (SmF $L=\pm2.6$, SmI $L=\pm2$) are broader in the in-plane direction and narrower in the interlayer direction than the peaks farthest from the tilt direction (SmF $L=\pm1.3$, SmI $L=0$). This anisotropy is consistent with the steric anisotropy produced by the tilt.

Previous measurements by Doucet and Levelut[12] of the crystal structure (giving the lattice tilt γ) and the layer spacing (giving a measure of the effective molecular tilt β) showed that the two tilts were the same to about 1°; when $\gamma=\beta$ the molecules are packed end to end. We have investigated the packing in the smectic-F and smectic-I phases of 7O.7. The angle between the reciprocal-lattice-vector direction (with orientation ϕ and 2θ) and the smectic-layer direction is $\phi-\gamma=(2\theta)/2$. The molecular tilt direction β can be determined quite accurately by measuring the difference in the L values between primary and secondary L peaks since $\beta=\cos^{-1}(2/\Delta L)$. Typical data for the $L-2.6$ CrG primary–to–secondary-peak separation are $\Delta L=2.146\pm0.005$, yielding $\beta=21.3\pm0.3$ and a corresponding γ value of $\gamma=\phi-(2\theta)/2=21.6\pm0.1°$. The molecular tilt and the layer tilt are coincident to $\pm0.4°$ in the CrG phase are presumably in both the smectic-F and the smectic-I phases.

We also investigated the molecular packing of the crystalline-G phase. High-resolution measurements of the Bragg peaks showed that magnitude of the reciprocal-lattice vector of the $L=2.6$ peak was greater than that of the $L=1.3$ peak by $0.75\%\pm0.02\%$. Assuming that the molecules are packed end to end (with $\gamma=\beta$), this requires a closer packing, in the plane normal to the long molecular axes, of the molecules along the tilt direction than in that perpendicular to the tilt direction, by 0.75%.[41] Similar results have been reported in other studies.[10,12,35,42] If this effect were due to the tilt of the molecules (β), it would require that $3/8[\cos^2(\gamma)/\cos^2(\beta)-1]=0.0075$ or $\beta-\gamma=1.5°$. Our data show that a difference between the two tilt angles cannot be responsible for the measured anisotropy. The observed packing anisotropy is probably caused by the tendency for local herringbone packing produced by the planar geometry of the benzene rings in the molecule. However, there is no long-range herringbone order in the CrG phase. When herringbone order is observed, the resulting distortion is very pronounced.[42]

The evolution of the tilt in the SmI, SmF, and CrG phases was determined from the inclination of the peaks relative to the plane of the layers. With decreasing temperature the tilt increased continuously from its value of ~19° in the SmI phase at a rate of about 0.5 degrees/K. For films thicker than about 10 layers there was no observed thickness dependence of the tilt. The tilt evolution at the SmF-to-SmI transition appeared to be continuous, consistent with previous work.[34]

Optically, the SmF and SmI phases have the same texture. The SmI-to-SmF transition is visible because the domains change their relative brightness, corresponding

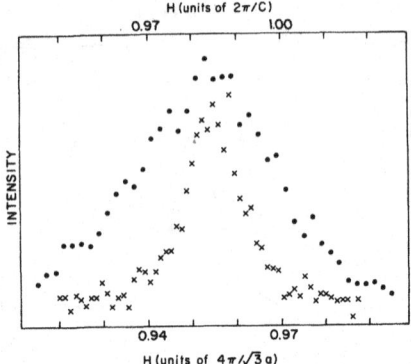

FIG. 13. H scans in the SmI phase of a 13-layer film at $T=62.3$°C. The $L=2$ data are shown by dots and the $L=0$ data are indicated by crosses. These data were collected at SSRL. The peak counting rate was about 200 counts per minute.

to the change in the tilt direction at the transition.[22] Although these two phases cannot be distinguished optically, the combination of a few x-ray measurements with an optical determination of the phase boundary provides a complete determination. The SmF-to-SmI phase boundary shown in Fig. 4 was determined with combined optical and x-ray measurements.

D. Nontilted-to-tilted phase transitions

Optical and x-ray observations of the transitions from the nontilted crystalline-B phase to the tilted SmI, SmF, and CrG phases were made. In bulk 7O.7, there is a direct, CrB-to-CrG transition at about 54.5°C. In films thinner than about 180 layers, there is a SmF phase between the CrB and the CrG phases. As shown in Fig. 4, the SmF-to-CrB transition temperature increases with decreasing thickness. Below about 20 layers there is also a SmI phase. The SmI phase first appears between the CrB and the SmF phases in films about 20 layers thick. As the thickness is reduced, the SmI range grows and the CrB range shrinks until about 10 layers where the CrB phase disappears and there is a direct SmC-to-SmI transition. The transitions between the CrB phase and the lower-temperature tilted phases were all found to be first order with significant hysteresis.

In contrast to the CrB restacking transitions and the SmF-to-CrG transition which were observed to take minutes to occur, the transitions between the CrB phases and the tilted phases all occurred very slowly (hours). By performing x-ray scans rapidly, after the oven temperature equilibrated but before the sample reached its equilibrium structure, it was possible to map out the reversible CrB restacking transitions at temperatures where the equilibrium structure was SmF. These transient phase boundaries are indicated on the phase diagram (Fig. 4) by dotted lines. Similarly, by performing rapid scans while heating out of the CrG phase it was possible to observe a reversible CrG-to-SmF transition in the temperature range in which the equilibrium structure was CrB. This is indicated by the dashed line in Fig. 4.

It is possible to construct a simple model which correctly describes the observed thickness dependence of the CrB-to-SmF transition temperature. Because the optical observations show that the tilted molecules on the two free surfaces of the CrB films (see Fig. 8), we assume that the surface field produces SmF surfaces. Let N denote the total number of layers in the film and let n denote the number of surface layers ($2 \leq n \leq N$). The excess free energy per surface layer is $F(T) = F_F(T) - F_B(T)$ where $F_F(T)$ and $F_B(T)$ are the free energies per layer of the SmF and CrB phases. We assume the simplest form for the temperature dependence, of the excess free energy, $F(T) = a(T - T_c) = at$, where t is the reduced temperature and T_c is the temperature where a bulk CrB-to-SmF transition would occur if the CrG phase did not intervene. We also assume that the two surfaces are always forced to be tilted by the surface field (consistent with the optical observations). The total free-energy cost for the tilted surface layers is then $nat + F_w$, where F_w is the energy of the wall between the tilted surface phases and the nontilted interior. We assume a constant wall cost of $c/2$ per wall, so that

$F_w = c$, and note that the minimum surface cost occurs for two surface layers and is equal to $2at + c$. Comparing this cost to the cost of converting the entire film to the tilted form (Nat) shows that the relative energy cost is $\Delta = -(N - 2)at + c$. This model predicts a transition when the relative cost vanishes, $\Delta = 0$, or $t = c/a(N - 2)$. The agreement of this prediction with the observed CrB-SmF phase boundary is shown in Fig. 14 for $c/a = 180$ K and $T_c = 54.5$ °C.

More detailed models might explain the observed increasing birefringence in the CrB phase as the temperature is lowered. We have not pursued this further since we do not have sufficient data to distinguish between increased tilt of single surface layers and an increased number of tilted surface layers. Further x-ray and optical studies are planned.

IV. DISCUSSION

The observed anisotropy of the in-plane and interlayer correlation lengths in the tilted hexatic phases must be produced by a combination of the intrinsic molecular anisotropy and the tilted hexatic structure. The peaks closest to the tilt direction have smaller in-plane correlation lengths than the peaks that are closer to the perpendicular direction. This is consistent with simple steric considerations which imply that the intralayer lattice spacing in the direction closest to the tilt will be more strongly affected by tilt-orientation fluctuations producing in-plane decorrelation. In addition, the homogeneous relative motions between adjacent layers along the direction of the tilt are suppressed, producing larger relative layer displacements perpendicular to the tilt direction than parallel to it. This produces the decreased L widths for the nearly perpendicular peaks.

As noted in the Introduction, some of the observed phase boundaries move to lower temperatures with decreasing film thickness and some of the phase boundaries move to higher temperature. The crystalline-B-to-smectic-F transition moves to higher temperature as the film thickness is decreased. This is an unusual transition because the low-temperature phase, smectic-F which is a tilted hexatic, is less positionally ordered than the high-temperature phases, crystalline-B which is a nontilted crystal. As the simple model in Sec. III demonstrates, this is exactly the kind of behavior expected for a surface field which favors tilted molecules at the film-vacuum interface. Very similar behavior has been observed for the smectic-A–to–smectic-C transition in thin films[43] where the surface field is also believed to favor tilt at the surface. There is a number of liquid crystal systems where the surface field produces a more ordered surface phase, including the formation of nontilted crystalline-B surface layers on smectic-A films,[39,44,45] the formation of surface tilted hexatic smectic-I layers on smectic-C films,[4] and the production of surface-induced smectic-A order at the surface-vacuum interface of bulk nematic or isotropic samples.[46,47]

The observed decrease in the transition temperature with decreasing film thickness observed for the smectic-

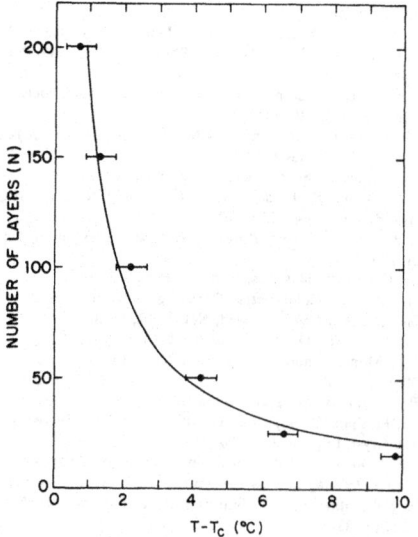

FIG. 14. Predicted thickness dependence (solid line) of the smectic-F-to-crystalline-B phase boundary compared to the observed transition temperatures (data with error bars). The model parameters are $c/a = 180$ K and $T_c = 54.5$ °C.

SIROTA, PERSHAN, SORENSEN, AND COLLETT

F-to-crystalline-G transition is consistent with the predicted suppression of the ordered phase in two dimensions due to the increased importance of the thermal fluctuations. To our knowledge this is the first observation of a suppressed transition in a liquid-crystal system. Since the local order in the smectic-F and crystalline-G phases are the same, this transition is allowed by symmetry to be second order. Because the films are freely suspended there is no periodic substrate to influence the transition, and this transition should be a superb model for the two-dimensional tilted hexatic to tilted crystal transition.[48,49] The observed decrease in the size of the first-order jump with film thickness suggests that this transition might become second order for thicknesses below about four layers, with a universal correlation length evolution versus temperature in the hexatic phase.

It seems very likely that there will be similar restacking sequences and thin-film hexatic phases in many liquid-crystal materials. We have made measurements for the two 7O.7 homologos,n-[4-(n-pentyl)oxyben-zylidene]-4'-(n-hexyl)aniline (5O.6) and n-[4-(n-nonyl)oxybenzylidene]-4'-(n-butyl)aniline (9O.4), which exhibit bulk hexatic (smectic-F) phases. Both of these materials have a correlation-length anisotropy similar to that discussed above for 7O.7. In addition, the smectic-F-to-crystalline-G phase boundary of 9O.4 shows qualitatively the same suppression with decreasing thickness as 7O.7. The crystalline-B-to-smectic-F phase boundary of 5O.6 also has qualitatively the same thickness

dependence as 7O.7. The observed thin-film restacking sequence in 7O.7 ($ABAC \rightarrow ABC \rightarrow$ orthorhombic $\rightarrow$ monoclinic $\rightarrow AAA$) is also observed in bulk 5O.6 films,[20] and except for the final AAA phase, in another homolog (4O.8) which has crystalline-B phases (9O.4 does not).

In conclusion, the striking changes in the phase sequence and phase boundaries of 7O.7 (and similar) films versus film thickness provide a wealth of information about the influence of reduced dimensionality and the strong surface fields on these systems. The smectic-F-to-crystalline-G transition, which exhibits a suppressed transition temperature with decreasing thickness, appears very promising for future two-dimensional melting studies in a substrate free system. The crystalline-B-to-smectic-F transition is well described by a simple surface field model.

ACKNOWLEDGMENTS

We wish to acknowledge helpful conversations with Allan Farber, B. I. Halperin, D. R. Nelson, and D. E. Moncton. We also wish to acknowledge the help of Kelby Chan, Frank Molea, Alan Braslau, George Pisiello, and the staff at SSRL. This work was supported by the National Science Foundation under Grant No. DMR-85-13523 and DMR-80-20247. Part of this work was done at the Stanford Synchroton Radiation Laboratory, which is supported by the U.S. Department of Energy, Office of Basic Sciences.

*Present address: Corporate Research Science Laboratories, Exxon Research and Engineering Company, Route 22 East, Annandale, NJ 08801.

†Present address: IBM, Department 242/040-2, Rochester, MN 55901.

[1]R. J. Birgeneau and J. D. Litster, J. Phys. (Paris) Lett. 39, L399 (1978).

[2]J. Collett, P. S. Pershan, E. B. Sirota, and L. B. Sorensen, Phys. Rev. Lett. 52, 356 (1984).

[3]E. B. Sirota, P. S. Pershan, L. B. Sorensen, and J. Collett, Phys. Rev. Lett. 55, 2039 (1985).

[4]E. B. Sirota, P. S. Pershan, S. Amador, and L. B. Sorensen, Phys. Rev. A 35, 2283 (1987).

[5]G. W. Smith, Z. G. Gardlund, and R. J. Curtis, Mol. Cryst. Liq. Cryst. 19, 327 (1973).

[6]G. W. Smith and Z. G. Gardlund, J. Chem. Phys. 59, 3214 (1973).

[7]D. E. Moncton and R. Pindak, Phys. Rev. Lett. 43, 701 (1979).

[8]A. J. Leadbetter, J. C. Frost, and M. A. Mazid, J. Phys. (Paris) Lett. 40, 325 (1979).

[9]A. J. Leadbetter, M. A. Mazid, B. A. Kelly, J. W. Goodby, and G. W. Gray, Phys. Rev. Lett. 43, 630 (1979).

[10]A. J. Leadbetter, M. A. Mazid, and R. M. Richardson, in Proceedings of the Bangalore Conference on Liquid Crystals, 1979, edited by S. Chandreasekhar (Hayden and Sons, London, 1980).

[11]P. A. C. Gane, A. J. Leadbetter, and P. G. Wrighton, Mol. Cryst. Liq. Cryst. 66, 245 (1981).

[12]J. Doucet and A. M. Levelut, J. Phys. (Paris) 38, 1163 (1977).

[13]A. M. Levelut, J. Doucet, and M. Lambert, J. Phys. (Paris)

35, 773 (1974).

[14]J. Collett, L. B. Sorensen, P. S. Pershan, R. J. Birgeneau, J. D. Litster, and J. Als-Nielsen, Phys. Rev. Lett. 49, 553 (1982).

[15]J. Collett, L. B. Sorensen, P. S. Pershan, and J. Als-Nielsen, Phys. Rev. A 32, 1036 (1985).

[16]E. B. Sirota, P. S. Pershan, and M. Deutsch, following paper Phys. Rev. A 36, 2902 (1987).

[17]K. Chan, Ph.D. thesis, Harvard University, 1984.

[18]C. Y. Young, R. Pindak, N. A. Clark, and R. B. Meyer, Phys. Rev. Lett. 40, 773 (1978).

[19]M. Born and E. Wolf, Principles of Optics (Pergamon, Oxford, 1975).

[20]J. A. Collett, Ph.D. thesis, Harvard University, 1983.

[21]Committee on Colorimetry, Optical Society of America, in The Science of Color (Crowell, New York, 1953).

[22]A. S. Farber, Ph.D. thesis, Brandeis University, 1985.

[23]D. E. Moncton and G. S. Brown, Nucl. Instrum. Methods 208, 576 (1983).

[24]CPAC Organix, Inc. and Frinton Laboratories, Inc.

[25]P. S. Pershan, G. Aeppli, J. D. Litster, and R. J. Birgeneau, Mol. Cryst. Liq. Cryst. 67, 205 (1981).

[26]D. E. Moncton and R. Pindak, in Proceedings of the Conference on Ordering in Two Dimensions, Lake Geneva, Wisconsin, 1980, edited by S. K. Sinha (North-Holland, Amsterdam, 1980), pp. 83-90.

[27]A. J. Leadbetter, J. P. Gaughan, B. Kelly, G. W. Gray, and J. Goodby, J. Phys. (Paris) Colloq. 40, C3-178 (1979).

[28]J. J. Benattar, J. Doucet, M. Lambert, and A. M. Levelut, Phys. Rev. A 20, 2505 (1979).

[29]P. A. C. Gane, A. J. Leadbetter, J. J. Benattar, F. Moussa,

and M. Lambert, Phys. Rev. A **24**, 2694 (1981).

[30]J. J. Benattar, F. Moussa, M. Lambert, and C. Germain, J. Phys. (Paris) Lett. **42**, 67 (1981).

[31]F. Moussa, J. J. Benattar, and C. Williams, Mol. Cryst. Liq. Cryst. **99**, 145 (1983).

[32]D. Guillon, A. Skoulios, and J. J. Benattar, J. Phys. (Paris) **47**, 133 (1986).

[33]D. R. Nelson and B. I. Halperin, Phys. Rev. B **21**, 5312 (1980).

[34]J. J. Benattar, F. Moussa, and M. Lambert, J. Chim. Phys. **80**, 99 (1983).

[35]J. Budai, R. Pindak, S. C. Davey, and J. W. Goodby, J. Phys. (Paris) Lett. **45**, 1053 (1984).

[36]J. Doucet, P. Keller, A. M. Levelut, and P. Porquet, J. Phys. (Paris) **39**, 548 (1978).

[37]E. B. Sirota, J. Collett, P. S. Pershan, and L. B. Sorensen (unpublished).

[38]A. Farber (private communication).

[39]S. C. Davey, J. Budai, J. W. Goodby, R. Pindak, and D. E. Moncton, Phys. Rev. Lett. **53**, 212 (1984).

[40]B. M. Ocko, A. R. Kortan, R. J. Birgeneau, and J. W. Goodby, J. Phys. (Paris) **45**, 113 (1984).

[41]E. B. Sirota and P. S. Pershan (unpublished). This distortion has also been measured in the CrG phases of 9O.4, 5O.6, and 4-(2'-methylbutyl)phenyl 4'-n-octylbiphenyl-4-carboxylate

(8SI) and values of 0.72, 0.81, and 0.38 %, respectively, have been obtained for the relative contraction in the direction of tilt. The distortion in the CrJ phase (crystal analogue of SmI) of p-hexyloxybenzylidene-p'-amino-2-chloro-α-propyle-cinnamate, (HOBACPC), 8SI, and 4-(2'-methylbutyl)phenyl 4'-n-octylbiphenyl-4-carboxylate, (8OSI) were found to be 0.30, 0.30, and 0.32 %, respectively, with the lattice being contracted perpendicular to the direction of tilt.

[42]J. Doucet, A. M. Levelut, and M. Lambert, Phys. Rev. Lett. **32**, 301 (1974).

[43]S. Heinekamp, R. A. Pelcovits, E. Fontes, E. Yi Chen, R. Pindak, and R. B. Meyer, Phys. Rev. Lett. **52**, 1017 (1984).

[44]D. E. Moncton, R. Pindak, S. C. Davey, and G. S. Brown, Phys. Rev. Lett. **49**, 1865 (1982).

[45]D. J. Bishop, W. O. Sprenger, R. Pindak, and M. E. Neubert, Phys. Rev. Lett. **49**, 1861 (1982).

[46]P. S. Pershan and J. Als-Nielsen, Phys. Rev. Lett. **52**, 759 (1984).

[47]B. M. Ocko, A. Braslau, P. S. Pershan, J. Als-Nielsen, and M. Deutsch, Phys. Rev. Lett. **57**, 94 (1986).

[48]B. I. Halperin and D. R. Nelson, Phys. Rev. Lett. **41**, 121 (1978).

[49]D. R. Nelson and B. I. Halperin, Phys. Rev. B **19**, 2456 (1979).

7. TILTED HEXATIC PHASES

7.1. Benattar, J. J., Doucet, J., Lambert, M. and Levelut, A. M. (1979). Nature of the smectic F phase. *Phys Rev.* **A20**, 2505–2509   333

7.2. Collett, J., Pershan, P. S., Sirota, E. B. & Sorensen, L. B. (1984). Synchrotron x-ray study of the thickness dependence of the phase diagram of thin liquid-crystal films. *Phys. Rev. Lett.* **52**, 356–359   338

7.3. Sirota, E. B., Pershan, P. S., Sorensen, L. B. and Collett, J. (1985). X-ray studies of tilted hexatic phases in thin liquid-crystal films. *Phys. Rev. Lett.* **55**, 2039–2042   342

7.4. Aharony, A., Birgeneau, R. J., Brock, J. D. and Litster, J. D. (1986). Multicriticality in hexatic liquid crystals. *Phys. Rev. Lett.* **57**, 1012–1015   346

7.5. Brock, J. D., Aharony, A., Birgeneau, R. J., Evans-Lutterodt, K. W., Litster, J. D., Horn, P. M., Stephenson, G. B. and Tajbakhsh, A. R. (1986). Orientational and positional order in a tilted hexatic liquid crystal phase. *Phys. Rev. Lett.* **57**, 98–101   350

PHYSICAL REVIEW A   VOLUME 20, NUMBER 6   DECEMBER 1979

Nature of the smectic F phase

J. J. Benattar, J. Doucet, M. Lambert, and A. M. Levelut

Laboratoire de Physique des Solides associé au Centre National de la Recherche Scientifique,
Université Paris-Sud—Bâtiment 510, 91405 Orsay, France

(Received 12 February 1979; revised manuscript received 24 May 1979)

X-ray diffraction patterns of single-domain samples of the smectic F phase of terephthalydene-bis-4-n-pentylaniline, obtained by melting single crystals of the crystalline phase, have enabled us to collect important information concerning the nature of this phase. The smectic layers are almost uncorrelated, and within the layers the order is characterized by a pseudohexagonal order analogous to that observed in the smectic B phase with tilted molecules, but less well defined: the smectic F phase is intermediate between the smectic B and smectic C phases. An intensity profile analysis performed on single-domain patterns is consistent with the two-dimensional structure of the smectic F phase, and the investigation of the powder pattern indicates in addition the existence of another disorder type which remains to be determined.

I. INTRODUCTION

Among the various mesogenic compounds only a few exhibit a smectic F phase. This phase, since its identification from miscibility criteria by Demus *et al.*,[1] has remained somewhat elusive.

Recently, Goodby and Gray synthesized several compounds in which a smectic F phase appears[2,3] and revealed the presence of such a phase in the compound TBPA (terephthalydene-bis-4-n-pentylaniline).[4] We thought it would be of special interest to study this compound since it is an homolog of the well-known TBBA (terephthalydene-bis-4-n-butylaniline) (Ref. 5) from which it only differs by the aliphatic terminal chains (butyl in TBBA, pentyl in TBPA). Therefore, it is not surprising that the polymorphism of TBPA is similar to that of TBBA, but it also shows an extra feature, the S_F phase:

$$\text{Cr} \xrightarrow{76.5°C} S_H \xrightarrow{144°C} S_F \xrightarrow{153.5°C} S_C \xrightarrow{182.5°C} S_A \xrightarrow{213.5°C} N \xrightarrow{232.5°C} I.$$
$$\searrow {57°C}\; S_G$$

[The smectic H phase (S_H in our notation) is named SmH by Goodby *et al*; but its structure is actually the same as the structure of the tilted S_B phase of TBBA which is sometimes named SmB$_C$. The same remark also applies to the so-called SmG phase of TBPA, which presents the same structure as that of the tilted S_B phase of TBBA (sometimes called SmE$_C$).]

Leadbetter *et al.*[6] have recently reported some results about the structural aspect of the S_F of this compound; from x-ray patterns obtained with samples oriented by a magnetic field, they suggested the existence of hexagonal packing within the layers. In order to study this phase in a more straightforward way, we successfully grew good single crystals of TBPA which give single domains

by melting. These samples enabled us to complete Leadbetter's study; in addition we present here an analysis of the intensity profiles, which provides valuable information as regards the nature of the S_F phase.

II. EXPERIMENTAL PROCESSES

A. Preparation of samples

The single crystals of TBPA were grown in the dark in a solution of ethanol and chloroform by evaporation of the solvents over a period of about 20 days; the single crystals grown by this method are transparent and yellow, and have a platelike morphology and a parallelepiped shape with typical dimensions of the order of 10×0.5 mm.[3]

B. Experiments

The x-ray patterns of single domains were taken by means of a transmission device with a monochromatic convergent beam (Cu$K\alpha$, 1.54Å) perpendicular to a flat film. The sample is fixed, and held at temperatures within $\pm 0.5°C$. From various S_H phase patterns, we have deduced the orientation of the long molecular axes $\vec{c}$ with respect to the external faces of the samples. We observe that the orientation of TBPA molecules differs from one sample to another: the layer planes are not always parallel to the larger external faces. We have represented in Fig. 1(A) the two geometries A and B used in x-ray experiments; they allow, respectively, examination of the ($\vec{a}^*, \vec{c}^*$) and ($\vec{a}^*, \vec{b}^*$) reciprocal planes. [The $\vec{a}$ and $\vec{b}$ axes are chosen in the smectic layer plane as shown in Fig. 1(B)].

Powder patterns were recorded with a Guinier camera equipped with an electric stage and using a focused monochromatic beam (Co$K\alpha_1$, 1.789Å).

334

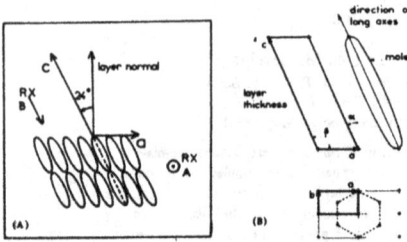

FIG. 1. (A) The two orientations A and B of the x-ray beam with respect to the long molecular axis (c) and the smectic layers. (B) S_H lattice orientation with respect to the smectic layers [the plane (a, b) is parallel to the smectic layers].

This device permits us to obtain lattice constants in the ordered smectic phases and the layer thickness in all the smectic phases.

III. EXPERIMENTAL RESULTS

A. S_H phase

The two x-ray patterns of the S_H phase of TBPA [Figs. 2(A), 2(B)] are typical of a S_H phase, with molecules tilted by an angle α relative to the layer normal; they are quite similar to the homologous patterns of TBBA.[1-9]

i. When the direct beam is almost parallel to the smectic layers (geometry A) the corresponding pattern [Fig. 2(A)] exhibits different Bragg reflections: the very sharp $00l$ spots [Fig. 2(A),(a)], the $hk0$ spots —characteristic of the order within the layers —and hkl spots ($l = \pm 1, \pm 2$) [Fig. 2(A),(b)]. The $hk0$ and hkl reflections are in fact located along arcs of a circle, which in that kind of sample is usual; this phenomenon is indicative of a disorientation of the normal direction to the smectic layers up to 20°, whereas it is less than 5° in TBBA. The most significant point is that these reflexions are as sharp as those usually observed with crystalline phases. This is quite clear on

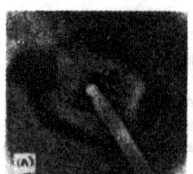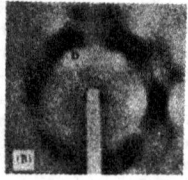

FIG. 2. X-ray patterns of the S_H phase. (a) in geometry A: (a) designates $00l$ Bragg spots, (b) hkl Bragg spots, and (c) the diffuse sheets. (b) in geometry B: we can clearly see the pseudohexagonal packing (d) designates the diffuse spots.

the powder patterns of the S_H phase which exhibit very sharp rings as in the TBBA case.[7] The existence of sharp hkl ($l = \pm 1, \pm 2$) reflections is indicative of a three-dimensional order.

The pattern also shows diffuse equidistant parallel lines [Fig. 2(A), (c)], which are the traces on the film of diffuse sheets parallel to the $(\vec{a}^*, \vec{b}^*)$ reciprocal plane. This effect results from the x-ray diffraction by periodic chains of molecules oriented along $\vec{c}$; the molecular longitudinal displacements are correlated along one chain but not from one chain to another. It may be noticed that such a description in terms of collective longitudinal displacements along chains is completely consistent with the three-dimensional order of the structure; another equivalent description of the same effect would be to introduce soft phonons polarized along the molecular direction and having their wave vectors parallel to the $(\vec{a}^*, \vec{b}^*)$ reciprocal plane; the resulting molecular displacements would be correlated only in the chain direction. These "undulation modes"[8] are compatible with the concept of layers correlated in a three-dimensional lattice. Such an effect is related to the large anisotropy of the thermal motion of the molecules (~2-Å amplitude in the $\vec{c}$ direction), the mean positions of these molecules being distributed on the three-dimensional lattice which is defined below.

ii. When the direct beam is parallel to the long molecular axes (geometry B), the pattern [Fig. 2(B)] exhibits six $hk0$ Bragg spots forming a pseudohexagon. We noticed again the existence of some hkl ($l = \pm 1, \pm 2$) Bragg spots close to each $hk0$ Bragg spot, the existence of which has been discussed above. Besides, the hkl spots are distributed on an arc of a circle, bearing evidence that the layers are disoriented by up to 25° (5° in TBBA); it is not possible to specify whether this disorientation occurs between neighboring layers or between stacks of layers with parallel axes.

The three-dimensional structure can be described by a C-face-centered monoclinic cell whose parameters are accurately determined from powder patterns. At (132 ± 2)°C, we find

$$a = 9.62 \pm 0.02 \text{Å}, \quad b = 5.14 \pm 0.01 \text{Å},$$

$$c = 31.2 \pm 0.2 \text{Å}, \quad \beta = 113.6 \pm 0.5°,$$

which gives a tilt angle of 23°6′ [Fig. 1(B)].

Finally, the presence of twelve diffuse spots located outside the $hk0$ Bragg spots [Fig. 2(B),(d)] is indicative, as for the S_B phase of TBBA, of a local "herringbone packing" of the molecular sections within the layers, inside small domains whose size is here or the order[9] of 20–30 Å.

B. S_F phase

(i) When the direct beam is parallel to the smectic layers (geometry A), the x-ray pattern of this phase [Fig. 3(A)] appears to be different from the S_H one. Indeed, the diffuse sheets observed in the S_H phase have, in this case, almost vanished, the $hk0$ Bragg spots are now transformed into diffuse peaks, and the hkl Bragg spots have then disappeared. These phenomena are indicative of the loss of three-dimensional order: the smectic layers are now very weakly coupled.

(ii) We can observe in Fig. 3(B), where the direct beam is parallel to the long molecular axes (geometry B), that the x-ray pattern, characteristic of the order within the layers, presents a diffuse ring whose intensity is modulated with maxima arranged in six diffuse peaks at the same place as the $hk0$ spots of the S_H phase. The intensity distribution evokes a hexagonal packing similar to that of the S_H phase, but far less well defined. Nevertheless, this intensity distribution is clearly indicative of the persistence of *some orientational correlation* between layers.

Another interesting feature can be seen on the original pattern corresponding to Fig. 3(B)(g): the intensity of the twelve previous diffuse spots of the S_H phase is uniformly distributed around the diffuse ring, thus indicating the persistence of a local herringbone packing.

The Debye-Scherrer patterns yield less information about the S_F phase than the S_H phase since they only exhibit a diffuse ring instead of sharp rings at large diffraction angles. Consequently, no accurate cell dimensions can be given; from the analysis of single domain patterns [Figs. 3(A) and 3(B)], we can only state that the local cell parameters are roughly the same as in the S_H phase. And on that point, we disagree with the results of Leadbetter et al.,[6] which report accurate cell parameter values for the S_F phase.

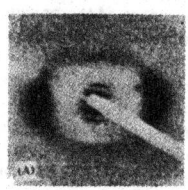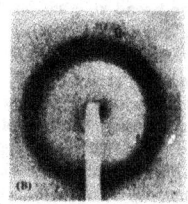

FIG. 3. X-ray patterns of the S_F phase. (A) in geometry A, the diffuse sheets almost vanished and the $hk0$ Bragg spots are transformed into diffuse peaks (e). (B) in geometry B, the diffuse ring is modulated by intensity maxima (f), and the diffuse spots (d) give the intensity (g) distributed around the ring.

Indeed, such values cannot be calculated with precision from the x-ray patterns, which only exhibit broad reflections, excepting the $00l$ reflections.

The only accurate deduction we can make from the powder patterns of the S_F phase is the layer thickness d. Assuming that the molecular length l remains equal to the length of the c parameter measured in the S_H phase (31.2 Å at 132 ℃), we can infer the tilt angle α, which is given by the formula $d = l \cos\alpha$. We obtain

$$d = 28.6 \pm 0.2 \text{Å}, \quad \alpha = 23.6 \pm 1° \ (S_H, 132°C),$$

$$d = 29.4 \pm 0.2 \text{Å}, \quad \alpha = 20 \pm 1° \ (S_F, 150°C),$$

$$d = 30.9 \pm 0.2 \text{Å}, \quad \alpha = 8 \pm 3° \ (S_C, 155°C).$$

These values show that the S_F is also an intermediate state between the S_H and the S_C phase as regards the tilt angle.

IV. INTENSITY PROFILES

A. Single-domain patterns

It proved to be impossible to analyze in the S_H phase the diffuse scattering around the $hk0$ reflections because, on account of the disorientations within the sample (cf. Sec. II A i), hkl reflections appear around the $hk0$ reflections; nevertheless, it is very probable that the diffuse scattering intensity varies as in the TBBA case, where longitudinal phonons can propagate[10]; i.e., this intensity $I(\vec{q})$ is such that $I(\vec{q}) \propto |q|^{-2}$, with $\vec{q} = \vec{S} - \vec{\tau}$ (where $\vec{S}$ is the scattering vector parallel to $\vec{\tau}$ and $\vec{\tau}$ a reciprocal lattice position).

In contrast, such an analysis *is* possible in the S_F phase. We tried to fit its intensity profile with a function such as $|q|^{-x}$ (here $\vec{q}$ is parallel to $\vec{\tau} = 2\vec{a}^*$), and we obtained $x = 1.50 \pm 0.05$. This result only indicates that the S_F phase behaves quite differently from that of $S_H(S_{B_C})$. The value $x = 1.50$ is only suggestive, since no deconvolution has been done from the intrinsic instrumental line shape and the sample mosaics. The instrumental resolution is sufficiently good to have no measurable effect on the linewidth for a three-dimensional ordered crystal. If we assume a two-dimensional order (uncorrelated layers), the sample mosaics combined with the two-dimensional features of the scattering, which consists of rods perpendicular to the $(\vec{a}, \vec{b})$ plane, will contribute a large part of the measured width and is difficult to take into account quantitatively. Since we did not succeed in preparing a better-oriented sample, powder experiments were then performed in order to simplify the analysis of the intensity profile.

In the above discussion, we did not consider that optical modes can contribute to the diffuse scattering. We have assumed this contribution as

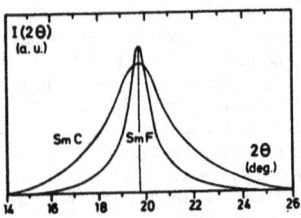

FIG. 4. Comparison of the intensity profiles between the S_F and S_C phases ($\lambda = $ Cu $K\alpha$).

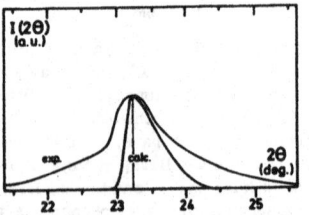

FIG. 5. The experimental and calculated intensity profiles of the diffuse ring on powder diagrams in the S_F phase ($\lambda = $ Co $K\alpha$).

a negligible one since we do not think that low-frequency optical modes exist in the S_F phase; this had been checked through neutron experiments in the S_H phase of TBBA,[10] and we assume that this is also the case in the S_F phase of TBPA.

It is also interesting to compare the intensity profiles of the S_F and S_C phases (Fig. 4). The intensity profile of the S_F phase is less broad and less homogeneous than in the S_C phase [the angular widths at midheight are, respectively, $\Delta(2\theta)_F = 1°$ and $\Delta(2\theta)_C = 3°$]. This finding corroborates the preceding ones concerning the intermediate nature of the S_F phase between the S_H and S_C phases.

B. Powder patterns

Powder patterns of the S_F phase exhibit only two sharp reflections (001 and 002) at small angles and a broad diffuse ring at large angles. We have calculated its theoretical intensity profile, assuming a complete disorder between parallel layers and a perfect long-range order within the layers in order to see if the lack of correlation between layers is the only phenomenon which is responsible for the broadening of the ring.

Under such assumptions, the intensity is localized in reciprocal space along bars, parallel to $\vec{c}^*$ whose length[11] is about $2/C$ Å^{-1}. The electronic density is chosen such that

$$\begin{vmatrix} \rho(x) = 1 & \text{for } -C/2 \lesssim x \leqslant C/2 \text{,} \\ \rho(x) = 0 & \text{elsewhere.} \end{vmatrix}$$

The intensity along the bars is therefore given by the square of the Fourier transform:

$$I(q) \propto c^2 (\sin\pi qc / \pi qc)^2 \text{;}$$

the intensity of the diffuse ring $I(2\theta)$ is then obtained by means of a graphic method, summing the intensity projections of the 110 and 200 bars on the 2θ axis.[12] This theoretical profile is compared with the experimental one in Fig. 5. We can see that its width at midheight is about two times smaller than the experimental one, which implies

that the disorder responsible for the broadening of the ring at large diffraction angles cannot be described roughly merely by the lack of correlation between the layers; it involves extra disorder phenomena probably corresponding to a loss of the long-range order within the layers or to lattice-dynamic effects. From our x-ray experiments we cannot distinguish between the two possibilities; further experiments, such as neutron experiments, are required to determine the true nature of this extra disorder.

It is worthwhile to compare the powder diagrams of the S_F phase of TBPA with that of the SmIII phase of HOBACPC, which is a ferroelectric compound exhibiting a similar stacking of uncorrelated layers[13]; the two patterns show a diffuse ring at large diffraction angles, but the SmIII diffuse ring is sharper than that of S_F, which indicates that the lateral hexagonal order within the layers extends to longer distances and is better defined in the SmIII phase than in the S_F phase.

V. CONCLUSION

As do all the other smectic phases (except the S_D phase), the S_F phase of TBPA presents a clearly defined layered structure indicated by the sharpness of the 00l Bragg spots. At large diffraction angles, the patterns only exhibit one diffuse broad ring whose intensity is modulated into six maxima, which are reminiscent of the pseudo-hexagonal structure of the ordered smectic phases. Thus the structure of the S_F phase appears to be intermediate between the structure of the S_C phase (liquid order) and the structure of the S_B phase. We found that one of the relevant features of the S_F phase is the nearly complete absence of positional correlation between successive layers, while some orientational correlation is retained. We have to deal with a two-dimensional structure with regard to the positional order of molecules, whereas the other ordered smectic phases are characterized by three-dimensional order. In such conditions, the de Gennes–Sarma model[14], which had been formerly proposed for the structural description of the S_B phase, seems to be more appropriate for the S_F phase. From our experiments, it is difficult to ascertain whether the loss of the three-dimensional order is accompanied by a loss of the long-range order within the layers or by some kind of molecular motion. It is worth noting that the $S_H \rightarrow S_F$ transition which corresponds to the transformation of the three-dimensional order into two-dimensional order is discontinuous, since this transition is a first-order one.

While this work has clarified some of the essential properties of the S_F phase, it leaves unanswered the question as to what factors influence its formation. Further investigations on the nature and behavior of the S_F phase are needed before these factors can be ascertained.

ACKNOWLEDGMENTS

We are very grateful to Dr. C. Germain for synthesizing the compound TBPA and to Mrs. M. C. Comes for revising the English manuscript.

[1]D. Demus, S. Diele, M. Klapperstück, V. Link, and H. Zaschke, Mol. Cryst. Liq. Cryst. 15, 161 (1971).

[2]J. W. Goodby and G. W. Gray, Mol. Cryst. Liq. Cryst. 41, 145 (1978).

[3]J. W. Goodby, G. W. Gray, and A. Mosley, Mol. Cryst. Liq. Cryst. 41, 183 (1978).

[4]S. Sakagami, A. Takase, and M. Nakamizo, Mol. Cryst. Liq. Cryst. 36, 261 (1976).

[5]J. Doucet, J. P. Mornon, R. Chevalier, and A. Lifchitz, Acta Crystallogr. B 33, 1701 (1976); J. Doucet, A. M. Levelut, and M. Lambert, ibid. 33, 1710 (1976).

[6]A. J. Leadbetter, J. P. Gaughan, B. Kelly, G. W. Gray, and J. Goodby, in Proceedings of the International Conference on Liquid Crystals, Bordeaux, 1978, J. Phys. (Paris) 40, C3, 178 (1978).

[7]J. Doucet, A. M. Levelut, and M. Lambert, Phys. Rev. Lett. 32, 301 (1974).

[8]W. Helfrich, in Ref. 6; J. Phys. (Paris) 40, C3, 105 (1979).

[9]A. M. Levelut, J. Phys. (Paris) Colloq. C3-37, C-51, (1976).

[10]J. Doucet, M. Lambert, A. M. Levelut, P. Porquet, and B. Dorner, J. Phys. (Paris) 39, 173 (1978); J. J. Benattar, A. M. Levelut, L. Liebert, and F. Moussa, J. Phys. (Paris) 40, C3, 115 (1978).

[11]A. Guinier, X-Ray Diffraction (Freeman, San Francisco, 1963).

[12]A. Tardieu, V. Luzzati, and F. C. Reman, J. Mol. Biol. 75, 711 (1973).

[13]J. Doucet, P. Keller, A. M. Levelut, and P. Porquet, J. Phys. (Paris) 39, 548 (1978).

[14]P. G. de Gennes and G. Sarma, Phys. Lett. A 38, 219 (1972).

Synchrotron X-Ray Study of the Thickness Dependence of the Phase Diagram of Thin Liquid-Crystal Films

Jeffrey Collett,[a] P. S. Pershan, Eric B. Sirota, and L. B. Sorensen[b]

Division of Applied Sciences, Harvard University, Cambridge, Massachusetts 02138

(Received 7 October 1983)

The phase diagram of freely suspended thin films of heptyloxybenzylidene–heptylaniline shows dramatic changes for thicknesses below 22 layers. The most surprising feature of the phase diagram is the inclusion of two phases lacking long-range crystalline order (smectic-F and hexatic-B phases) between two crystalline phases (crystalline smectic B and smectic G). Neither the smectic F nor the hexatic B occurs in bulk samples. Between sixteen and ten layers the width, in temperature, of the hexatic-B phase increases.

PACS numbers: 64.70.-p, 61.30.-v

Freely suspended liquid-crystal films provide one system for studying the crossover from three to two dimensions.[1] Interest in this problem has been stimulated by suggestions that transitions in three-dimensional (3D) smectic liquid crystals may be related to transitions within individual 2D layers[2,3] and by the development of a detailed theory of 2D melting.[4] The most experimentally accessible prediction of the theory of 2D melting is the existence of an intermediate hexatic phase with short-range positional order but with quasi-long-range bond-angle orientational order. A recent study of melting in variable-thickness films of $\overline{1}4S5$ has shown the existence of an intermediate hexatic phase in a three-layer film which is not observed in bulk.[5] In this study we report the existence of a smectic-F phase (hexatic layers with a molecular tilt) in films as thick as 22 layers and the appearance of a hexatic-B phase at sixteen layers in a system which has no such phases in bulk samples.

We have examined freely suspended films of 4-n-heptyloxybenzylidene 4-n-heptylaniline (7O.7) varying in thickness from 5 to 22 layers. Previous work[6] has shown that thick films (> 500 layers) of 7O.7 exhibit rich structure with a series of three transitions involving restacking of the layers and subtle changes in the intralayer packing. In order of decreasing temperature, the observed crystal structures are hexagonal-close-packed (hcp), orthorhombic face-centered (ortho-F), monoclinic C-centered (mono-C), and simple hexagonal (hex-AA) structures. The last three of these have a one-dimensional modulation of the smectic layers which has a wavelength of 95 Å, a wave vector in the plane of the molecular layers directed along a reciprocal-lattice direction, and a polarization normal to the plane of the smectic layers. Immediately below the hex-AA structure is a smectic-G phase, a monoclinic structure with the long axes of the molecules

tilted by 24 deg with respect to the layer normal. The projection of the long axes of the molecules onto the 2D triangular in-plane lattice is directed halfway between nearest neighbors.[7] This complex behavior is undoubtedly due to a combination of intralayer and interlayer effects. The present study was undertaken to clarify the origin of the restacking transitions and to look for the effects of reduced dimensionality.

X-ray diffraction studies were carried out on the triple-axis spectrometer installed on beam line VII-2 at the Stanford Synchrotron Radiation Laboratory.[8] A pair of asymmetrically cut Ge[111] crystals were used in the double-crystal monochromator. Low-resolution measurements were made with a pyrolytic graphite [002] analyzer which resulted in a longitudinal resolution of $\Delta Q = 22 \times 10^{-3}$ Å^{-1}. Higher-resolution measurements were done with a LiF[200] analyzer which provided a longitudinal resolution of $\Delta Q = 3.8 \times 10^{-3}$ Å^{-1}. In both cases a 5-mrad horizontal divergence of the input beam was accepted by the focusing mirror. The films were oriented with the normal to the smectic layers always in the scattering plane. Scans were done along Q_{xy}, parallel to the planes of the smectic layers, or along Q_z, normal to the layers. A rotation angle χ, about an axis normal to the smectic layers, specified the direction of Q_{xy} within the smectic layers.

We find that the phase sequence of 7O.7 undergoes dramatic changes as a function of film thickness. Part of the thickness-temperature plane of the phase diagram is shown in Fig. 1. The measured points are shown together with speculative phase boundaries. The most striking feature of this phase diagram is the inclusion of phases lacking long-range crystalline order (smectic-F and hexatic-B phases) between the crystalline-B and smectic-G phases, both of which are crystalline. For thicknesses below 22 layers, we find

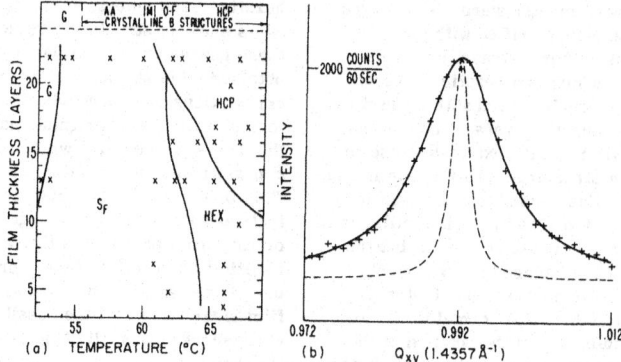

FIG. 1. (a) Thickness vs temperature phase diagram for 7O.7. The phases are denoted by G, crystalline G; AA, hex-AA crystalline B; M, mono-C crystal; O-F, ortho-F; HCP, hcp; S_F, smectic F; and HEX, hexatic B. Measured points are marked by crosses with speculative phase boundaries shown. (b) Q_{xy} scan in hexatic-B phase of thirteen-layer film at $T = 65.07\,°C$. The solid line is a fit by the sum of a Lorentzian and a constant background. The dashed line is the spectrometer resolution displayed on the same background.

that a smectic-F phase occupies the temperature region corresponding to the hex-AA and mono-C phases in bulk samples. As the sample thickness is reduced to sixteen layers, a hexatic-B phase appears between the smectic-F and the crystalline-B phases. Between sixteen and ten layers the hexatic-B phase gradually replaces the crystalline-B phase. Below ten layers the crystalline-B phase seems to disappear.

All of the four phases (crystalline B, hexatic B, smectic F, and smectic G) observed in thin films are observed in thirteen-layer films. Figures 1(b) and 2 illustrate the interlayer and intralayer correlations for a thirteen-layer film. Figure 2(a) shows a Q_z scan in the crystalline-B phase at $T = 67.4\,°C$. There are definite peaks at $Q_z = \pm\frac{1}{2}$ but no peak at $Q_z = 0$. The $Q_z = 0$ peak is more difficult to observe because the theoretical intensity is only $\frac{1}{3}$ that of the $Q_z = \pm\frac{1}{2}$ peaks and stacking faults reduce its relative intensity even further.[9]

At $65.5\,°C$ there is a transition to a hexatic-B phase. This has the properties theoretically predicted[9] and observed in other smectic materials.[10] These properties include a finite intralayer correlation length, a lack of interlayer correlations, and the presence of bond-angle orientation order. Figure 1(b) shows a Q_{xy} scan at $Q_z = 0$ which has a width corresponding to a correlation length of 130 Å. This is similar to other hexatic-B materials and is much longer than the correlation lengths observed in the smectic-

A and -C phases of 7O.7. Figure 2(b) shows a Q_z scan through the peak of the scan shown in Fig. 1(b). This scan shows a single broad peak with full width at half maximum of $1.7(2\pi/d)$. This width is similar to the width expected from the molecular form factor and indicates that there are no significant interlayer correlations. Finally, χ scans in this phase show considerable struc-

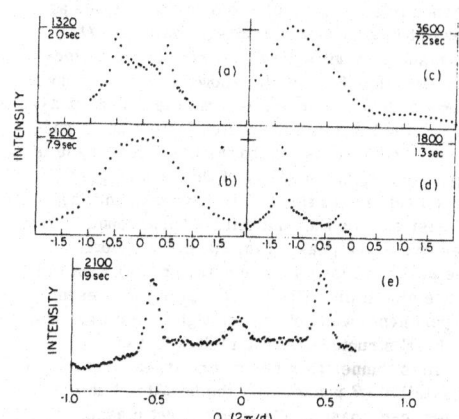

FIG. 2. Q_z scans: (a) thirteen-layer crystalline B at $T = 67.4\,°C$, (b) thirteen-layer hexatic B at $T = 62.8\,°C$, (c) thirteen-layer smectic F at $T = 60.8\,°C$, (d) thirteen-layer crystalline G at $T = 51.2\,°C$, and (e) seventeen-layer crystalline B at $T = 67.6\,°C$.

ture although single domains were not observed and the domain structure drifted with time.

At 61 °C there is an abrupt transition to a smectic-F phase, which is a hexatic phase with long axes of the molecules tilted with respect to the normal to the smectic layers. The correlation length is similar to the hexatic-B phase and χ scans also show structure. The Q_z scan in Fig. 2(c) differentiates this phase from the hexatic-B phase; the width of the broad peak is similar to the scan in the hexatic phase [Fig. 2(b)] but the center of the peak is displaced to nonzero Q_z. This is because the molecular form factor has its maximum in the direction normal to the long axis of the molecule. From the position of the peak the molecular tilt is estimated to be 20 deg in the smectic-F phase.

At $T = 51.6$ °C the thirteen-layer film crystallized into a smectic-G phase. Figure 2(d) shows a Q_z scan in this phase which shows sharp peaks characteristic of the monoclinic structure. The Q_{xy} scans in this phase show no excess width with graphite resolution, implying that the intralayer ordering is long ragne.

In films with thicknesses greater than ten molecular layers, the first phase which appears at temperatures below the smectic-C phase is the crystalline-B hcp structure. This phase is characterized by resolution-limited peaks in the $Q_z = 0$ plane. When Q_{xy} is scanned through one of these in-plane peaks, results similar to the dashed line in Fig. 1(b) are obtained. Peaks appear at integer and half-integer values of Q_z (measured in units $2\pi/d$, where d is the molecular length). Figure 2(e) shows a scan in a seventeen-layer film; here the intensities of the peaks are in the 3:1:3 ratio expected for the hcp structure. As the transition to the hexatic-B phase is approached, the intensity of the peak at $Q_z = 0$ drops relative to the half-order peaks and the intensity of the thermal diffuse background grows relative to the peaks [Fig. 2(a)]. The widths of the peaks in these Q_z scans are greater than the finite size limit. The resolution of the present experiments was not high enough to distinguish an ortho-F structure from a faulted hcp structure.

Films thinner than ten layers appear to have no crystalline-B phase, only the hexatic-B and smectic-F phases. We have not yet measured very thin films in the region of the smectic-F to smectic-G transition, but we expect the G phase to occur below the F. There is one difference between the data taken on five-layer films and those obtained from thirteen-layer films. The

hexatic-B and smectic-F phases in the five-layer films show no structure in χ scans. We identify these phases by the intralayer correlation lengths obtained from Q_{xy} scans and from the interlayer correlations measured with the Q_z scans. The loss of measurable orientational order in very thin films of materials which have bulk hexatic-B phases has been noted in other experiments.[11] This is assumed to be related to both static defects and the quasi-long-range nature of the orientational ordering in the two-dimensional hexatic.[11,12] The interlayer correlations measured in the smectic-F phase for the five-layer film clearly rule out the possibility of a smectic-C phase with unusually long intralayer correlation lengths.

The nature of this transformation of the system from a crystalline solid to a hexatic with short-range positional order is not understood. At present we do not know at what thickness the crystalline phases characteristic of bulk 7O.7 are replaced by the smectic-F phase. The fact that a tilted hexatic phase replaces the crystalline structures supports the suggestion that molecular tilt is involved in the restacking transitions observed in the bulk.[6] The crossover from the crystalline phases to the smectic-F phase is probably due to a competition between bulk and surface energies. As the thickness of the sample is reduced, the contribution of the surface term to the total free energy becomes more important and ultimately results in the instability of the crystalline structures. The free-energy difference between the crystalline structures and the bulk smectic-F structure may be very small, since materials such as 5O.6 have a phase sequence in bulk (crystalline B–smectic F–smectic G) which is identical to the one observed in 22-layer films of 7O.7.[7]

This study reveals an amazing richness of structure which poses a new set of questions to be answered. A systematic study of crystalline-B line shapes as a function of thickness may give further insight into the details of the crystal to hexatic transition. Preliminary data suggest that the smectic-F to smectic-G transition in the thin films may be a second-order transition. An understanding of this system may lead to the understanding of other bulk liquid-crystal systems such as 5O.6 which have a liquidlike phase sandwiched between two crystalline phases.[7] Further studies of these systems promise to enhance our understanding of both three-dimensional layered systems and the role of reduced dimensionality.

VOLUME 52, NUMBER 5   PHYSICAL REVIEW LETTERS   30 JANUARY 1984

This research was supported by the National Science Foundation under Grant No. NSF-DMR82-12189. The work reported herein was performed at the Stanford Synchrotron Radiation Laboratory, which is supported by the U. S. Department of Energy, Office of Basic Energy Sciences; the National Science Foundation, Division of Materials Research; and the National Institutes of Health, Biotechnology Resource Program, Division of Research Resources. We wish to acknowledge the assistance of the staff at the Stanford Synchrotron Radiation Laboratory, especially T. Porter, in carrying out these measurements. We would also like to thank R. Pindak and D. E. Moncton for helpful discussions of the experiments.

(a)Present address: IBM Research Laboratory, Yorktown Heights, N.Y. 10598.

(b)Present address: Department of Physics FM-15, University of Washington, Seattle, Wash. 98195.

[1]D. E. Moncton and R. Pindak, Phys. Rev. Lett. 43, 701 (1979).

[2]P. G. de Gennes and G. Sarma, Phys. Lett. 38A, 219 (1972); B. A. Huberman, D. M. Lublin, and S. Doniach, Solid State Commun. 17, 485 (1975).

[3]R. J. Birgeneau and J. D. Litster, J. Phys. (Paris), Lett. 39, L399 (1978).

[4]D. R. Nelson and B. I. Halperin, Phys. Rev. B 19, 2457 (1979).

[5]D. E. Moncton, R. Pindak, S. C. Davey, and G. S. Brown, Phys. Rev. Lett. 49, 1865 (1982).

[6]J. Collett, L. B. Sorensen, P. S. Pershan, J. D. Litster, R. J. Birgeneau, and J. Als-Nielsen, Phys. Rev. Lett. 49, 553 (1982).

[7]A. J. Leadbetter, M. A. Mazid, and R. M. Richardson, in Liquid Crystals, edited by S. Chandrasekhar (Cambridge Univ. Press, London, 1980), pp. 65–79; J. W. Goodby, G. W. Gray, A. L. Leadbetter, and M. A. Mazid, J. Phys. (Paris) 41, 591 (1980).

[8]D. E. Moncton and G. S. Brown, Nucl. Instrum. Methods 208, 579 (1983).

[9]A. Gunier, X-Ray Diffraction (Freeman, San Francisco, 1963).

[10]R. Pindak, D. E. Moncton, J. W. Goodby, and S. C. Davey, Phys. Rev. Lett. 46, 1135 (1981).

[11]J. Budai, S. C. Davey, R. Pindak, and D. E. Moncton, Stanford Synchrotron Radiation Laboratory Report No. 83101, 1983 (unpublished), p. VII-102.

[12]T. F. Rosenbaum, S. E. Nagler, P. M. Horn, and R. Clark, Phys. Rev. Lett. 50, 1791 (1983).

VOLUME 55, NUMBER 19   PHYSICAL REVIEW LETTERS   4 NOVEMBER 1985

X-Ray Studies of Tilted Hexatic Phases in Thin Liquid-Crystal Films

E. B. Sirota and P. S. Pershan

Division of Applied Sciences, Harvard University, Cambridge, Massachusetts 02138

L. B. Sorensen

Department of Physics, University of Washington, Seattle, Washington 98195

and

J. Collett[a]

*Division of Applied Sciences, Harvard University, Cambridge, Massachusetts 02138,
and IBM Thomas J. Watson Research Center, Yorktown Heights, New York 10598*

(Received 22 January 1985)

X-ray-diffraction studies of the structures and phase transitions of the tilted hexatic phases (smectic F and smectic I) in thin liquid-crystal films of 4-n-heptyloxybenzylidene-4-n-heptylaniline (7O.7) are reported. The measured correlation lengths were strongly anisotropic in both phases. The smectic-I to smectic-F transition is first order as expected from the symmetry change. The smectic-F to smectic-G transition is first order with strong pretransition effects and becomes nearly second order as the film thickness is decreased.

PACS numbers: 64.70.−p, 61.30.−v

There is currently great interest in the hexatic phases of matter. The hexatic phases have order intermediate between that of liquids and solids, with liquidlike short-range order of the in-plane positional correlations and solidlike quasi long-range order [in two dimensions (2D)] or long-range order [in three dimensions (3D)] of the orientation of the geometric bonds connecting the in-plane neighboring molecules.[1] Liquid-crystal systems provide ideal samples to study these remarkable phases because they have an extremely weak interlayer coupling which allows these materials to exhibit 3D stacked hexatic phases in addition to the standard 2D phases.[2] There are three different 3D smectic hexatic phases[3]: one untilted hexatic (hexatic B) and two tilted hexatics (smectic F and smectic I). Although most of the experimental studies of the liquid-crystal hexatic phases have concentrated on the structure of the hexatic-B phase and on the hexatic-to-liquid phase transition (hexatic B to smectic A)[4] there have been a few basic studies of similar phenomena in tilted systems.[5-10] In this Letter we report the first systematic x-ray-scattering study of the temperature dependence and anisotropy of the correlation lengths in the two tilted hexatic phases.

In freely suspended films of 4-n-heptyloxy-benzylidene-4-n-heptylaniline (7O.7) the smectic-F (SmF) and the smectic-I (SmI) phases only occur for films thinner than approximately 280 and 25 layers, respectively.[11] We also report similar anisotropies in the SmF phase of thick films of two homologous compounds, 5O.6 and 9O.4, suggesting that the anisotropy is a universal feature of the tilted hexatics. Since the observed positions of the diffuse peaks in the SmF phase and the Bragg peaks in the crystalline-G (CrG)

phase are essentially coincident, this is an explicit demonstration that the short-range structure is identical in these two phases and thereby establishes that the hexatic SmF phase is a faulted version of the CrG phase.[6]

The high-resolution x-ray-diffraction studies were carried out on the triple-axis spectrometer installed on beam line VII-2 at the Stanford Synchrotron Radiation Laboratory (SSRL).[12] A pair of asymmetrically cut Ge[111] crystals were used in the double-crystal monochromator and a single LiF[200] crystal was used as an analyzer; this resulted in a longitudinal resolution of $\Delta Q = 3.8 \times 10^{-3}$ Å^{-1}. The low-resolution measurements were made with a rotating-anode source of Cu $K\alpha$ radiation and a triple-axis spectrometer at Harvard University. The monochromator was a vertically focusing pyrolytic graphite [002] crystal and the analyzer was a flat pyrolytic graphite [002] crystal; this resulted in a longitudinal resolution of $\Delta Q = 5 \times 10^{-2}$ Å^{-1}.[13] The freely suspended films were drawn across a 7-mm-diam aperture in an oven which has been described in detail elsewhere[14]; the pressure inside the oven was maintained at 1 Torr and the temperature was regulated to ±5 mK. The films were oriented with the normal to the smectic layers always in the scattering plane. Scans were done along Q_{xy}, parallel to the planes of the smectic layers, or along Q_z, normal to the layers. A rotation stage inside the oven allowed rotation about an axis normal to the smectic layers thereby varying the direction of Q_{xy} within the smectic layers.

The phase diagram for 7O.7 is shown in Fig. 1 as a function of both temperature and film thickness. For thicknesses greater than ~280 layers, the structures

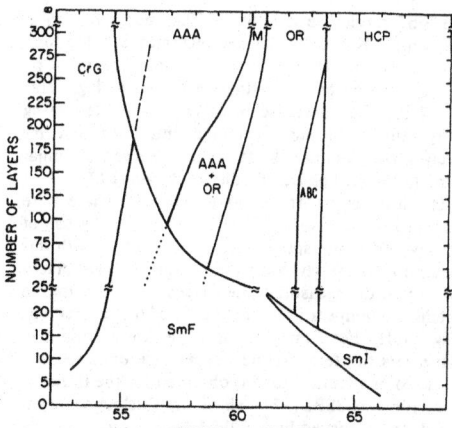

FIG. 1. Thickness vs temperature phase diagram for 7O.7. The solid lines indicate the reversible phase boundaries; the dashed (dotted) lines indicate boundaries of unstable phases observed kinetically on heating (cooling). The phases are crystalline G (CrG), smectic F (SmF), smectic I (SmI), hexagonal AAA (AAA), orthorhombic (OR), hexagonal ABC (ABC), hexagonal close packed (hcp), and monoclinic (M).

of the various crystalline phases and their respective phase transitions have been described elsewhere.[11,14-16] The crystalline-B (CrB) phase between ~ 69 and $63.5\,°C$ is a hexagonal close-packed lattice with $ABAB$ stacking and two molecules per primitive unit cell. At $63.5\,°C$ there is a first-order dislocation-mediated transition to a lattice in which the local stacking is ABC with one molecule per primitive unit cell.[16] For thick samples, the dislocations interact to form regular arrays of low-angle grain boundaries that result in a modulated structure with orthorhombic macroscopic symmetry. At slightly lower temperatures there are further transitions to phases that are still locally ABC but with macroscopic symmetries that are first monoclinic ($61\,°C$) and then hexagonal with AAA stacking ($60.5\,°C$). At $\sim 55\,°C$ the system undergoes a first-order transition to a CrG phase in which the molecules are tilted by about $22\,°C$ to the layer normal. With decreasing film thickness, starting at ~ 280 layers, the monoclinic phase is gradually replaced by a coexistence between the orthorhombic and the AAA phases. In addition, the direct first-order transition from the $ABAB$ to the orthorhombic phase is replaced by a region in which the orthorhombic phase coexists with a hexagonal phase with macroscopic ABC stacking.

The most surprising change in the phase diagram[11] is the inclusion of two tilted hexatic phases (SmF and

SmI) which do not appear in bulk 7O.7. The SmF phase starts appearing as a reversible intervening phase between the AAA and the CrG phases when the film thickness is reduced below about 180 layers. However, when we heat (films with thickness between 180 and 280 layers) from the CrG phase to the AAA phase, a kinetic CrG-AAA coexistence gives way to a kinetic SmF-AAA coexistence along the dashed line in Fig. 1. The SmI phase first appears at a film thickness of about 25 layers. When the SmF and the SmI phases first appear they exist only over narrow temperature ranges, but as the thickness is reduced further, these ranges increase substantially. Because the phase diagram is so rich, a wide variety of phase transitions can be studied by variation of the temperature at a fixed film thickness. These transitions include several solid-to-hexatic transitions (AAA to SmF, orthorhombic to SmF, ABC to SmI, and hcp to SmI), a hexatic-to-hexatic transition (SmI to SmF), and a hexatic-to-solid transition (SmF to CrG). All of these transitions, except for the SmF-to-CrG transition, are expected to be first order because of the change in symmetry across the transition and they were all observed to be first order. The SmF-to-CrG transition is allowed by symmetry to be a second-order transition and, as discussed below, appeared to become more nearly second order as the film thickness was decreased.

High-resolution measurements of the in-plane and the interlayer correlations were made at SSRL for a series of thirteen-layer films [see Figs. 2(a) and 2(b)]. The in-plane correlation lengths are proportional to the reciprocals of the half widths at half maxima (HWHM) of the appropriate longitudinal (Q_{xy}) scans, and the interlayer lengths are similarly related to the HWHM of the transverse (Q_z) scans. In view of the fact that the films were composed of many domains, providing only a partial χ average, we did not attempt detailed fits to theoretical line shapes. The molecular tilt distorts the in-plane hexagonal structure sufficiently that the instrumental resolution in the in-plane direction allowed us to characterize the anisotropic correlations in the SmF phase. In the CrG phase each layer consists of a hexagonal 2D lattice which is slightly distorted by the molecular tilt, which results in a centered rectangular lattice. This produces two inequivalent in-plane reciprocal-lattice vectors $(Q_{xy})_1 = 0.925(4\pi/\sqrt{3}a)$ and $(Q_{xy})_2 = 0.98(4\pi/\sqrt{3}a)$, where a is the hexagonal lattice parameter in the CrB phases; $a = 5.05$ Å. Since the molecular tilt of $22\,°$ from the layer normal is accompanied by lateral displacements of adjacent layers the Bragg peaks are also displaced along Q_z. Together these effects result in Bragg peaks at $(Q_z)_1 = 2.6(2\pi/l)$ and $(Q_z)_2 = 1.3(2\pi/l)$, where the molecular length is $l = 30.6$ Å. The corresponding Q_{xy} and Q_z scans are resolution limited in the CrG phase. On heating into

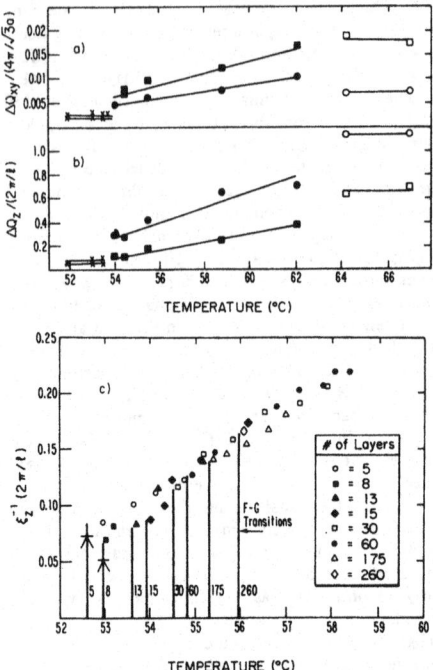

FIG. 2. The temperature dependence of the measured linewidths in the crystalline-G (CrG), smectic-F (SmF), and smectic-I (SmI) phases of 70.7. (a) High-resolution measurements illustrating the in-plane anisotropic linewidths (HWHM) of Q_{xy} scans for a thirteen-layer film at SSRL. SmI phase: open circles, $(Q_z)_2/(2\pi/l) = 0$; open squares, $(Q_z)_1/(2\pi/l) = 2.0$. SmF phase: solid circles, $(Q_z)_2/(2\pi/l) = 1.3$; solid squares, $(Q_z)_1/(2\pi/l) = 2.6$. The crosses indicate the instrumental resolution measured in the SmG phase. The solid lines are guides for the eye. The units are $4\pi/\sqrt{3}a = 1.436$ Å^{-1} and $2\pi/l = 0.205$ Å^{-1}. (b) High-resolution measured linewidths (HWHM) illustrating the interlayer anisotropy observed in the Q_z scans for a thirteen-layer film at SSRL. SmI phase: open circles, $(Q_{xy})_2 = 0.988(4\pi/\sqrt{3}a)$; open squares, $(Q_{xy})_1 = 0.955(4\pi/\sqrt{3}a)$. Sm$F$ phase: solid circles, $(Q_{xy})_2 = 0.975(4\pi/\sqrt{3}a)$; solid squares, $(Q_{xy})_1 = 0.927(4\pi/\sqrt{3}a)$. The crosses indicate the mosaic limit measured in the SmG phase. The solid lines are guides for the eye. The units are as above. (c) Fitted reciprocal correlation lengths from low-resolution measurements of the temperature dependence of the observed Q_z linewidths for the SmF phase of 70.7 as a function of the film thickness, n. For each thickness, the measurements followed a single universal curve until they underwent the first-order jump (indicated by the solid bar) at their respective SmF-to-CrG transitions. The arrows indicate the finite-size limits for the 5 to 8 layer films.

the SmF phase the resulting diffuse peaks remain at the same positions which demonstrates that the SmF phase has the same short-range order as the CrG phase. The observed widths are shown in Figs. 2(a) and 2(b). The decrease in HWHM with decreasing temperature indicates increased interlayer and intralayer correlations. Budai et al. observed a similar trend in the SmI phase of a different material.[9]

The interlayer correlations in the SmF phase were studied as a function of the film thickness by use of the low-resolution spectrometer at Harvard. With this spectrometer, nearly complete χ averages were practical so that the measured line shapes could be fitted to obtain the temperature dependence of the correlations [Fig. 2(c)]. For films ranging in thickness from 8 to 260 layers the temperature dependence observed was found to be identical to that observed for the thirteen-layer films at SSRL. For all of the thicknesses measured, the observed linewidths fall on a single universal curve with the observed first-order jump at the SmF-to-CrG transition becoming smaller as the film thickness (and the SmF-to-CrG transition temperature) decreased. The observed correlation lengths appear to be independent of both the film thickness and the temperature range of the SmF phase and to depend only on the absolute temperature. This suggests that although the SmF phase in 70.7 is only observed in thin films, it is actually a stable phase that would exist in bulk were it not preempted by another phase with lower free energy. According to this model the chemical potentials of the SmF and CrB (AAA) phases in 70.7 must be nearly equal, with the CrB phase having a slightly lower value in bulk. Presumably surface-induced tilt causes the SmF to have a lower chemical potential for films thinner than ~180 layers. This observation is clear evidence for the idea that the near equality of the chemical potentials for various ordered phases is responsible for the variety of phase sequences for closely related chemical homologs. We plan to study the behavior of two-layer films to see if the trend toward a second-order transition continues.

Anisotropic correlations were also observed in the SmF phases of the homologous compounds, 50.6 and 90.4, with use of a low-resolution spectrometer.[13] The observed HWHM for 90.4 were $(\Delta Q_z)_1 = 0.13$ Å^{-1} and $(\Delta Q_z)_2 = 0.17$ Å^{-1}, and $(\Delta Q_{xy})_1 = 0.014$ Å^{-1} and $(\Delta Q_{xy})_2 = 0.011$ Å^{-1} at 68.5 °C. Only the interlayer correlations were studied for 50.6 with the results $(\Delta Q_z)_1 = 0.12$ Å^{-1} and $(\Delta Q_z)_2 = 0.18$ Å^{-1} at 41.4 °C. The temperature dependence for these materials is currently under study.[17] We were able to align the tilt field in the SmF phase of 90.4 producing well-aligned single-domain samples by slowly cooling from the SmA into the SmF phase. The sixfold hexatic order in the SmF phase was measured directly by the observation of the scattering intensity as a function of

χ around the contour of maximum molecular form factor. The sixfold order was clearly visible and explicitly demonstrated the long-range bond orientational order in the tilted hexatic phase. We were not able to produce monodomain samples of 5O.6, possibly because the phase sequence of 5O.6 is CrB to SmF.

The SmI phase differs from the SmF in the orientation of the molecular tilt relative to the near-neighbor positions.[3,7,10] In the SmI (SmF) phase the molecules tilt towards the corner (face) of the real-space hexagon formed by the local in-plane packing. The transition from the SmF to the SmI phase is clearly first order [see Figs. 2(a) and 2(b)], as expected from the symmetry change between the phases. The principal peak positions in the SmI phase are $(Q_{xy})_1 = 0.955(4\pi/\sqrt{3}a)$, $(Q_z)_1 = 2.0(2\pi/l)$, $(Q_{xy})_2 = 0.988(4\pi/\sqrt{3}a)$, and $(Q_z)_2 = 0$ and the correlation lengths do not appear to be temperature dependent.

This study has revealed the underlying anisotropy of the tilted hexatic phases and the amazing richness of the thickness-dependent phase diagram. The spontaneous appearance of the tilted hexatic phases in thin 7O.7 films and the trend towards a second-order transition for very thin films is strongly reminiscent of the behavior of the hexatic phases predicted by the theory of 2D melting.[18,19] In particular, the decreasing SmF-to-CrG transition temperature with decreasing thickness appears to be an example of the suppression of the melting temperature expected to be associated with reduced dimensionality.[1] Further studies of the 3D to 2D crossover in these systems are in progress and promise to enhance greatly our understanding of these remarkable phases.

This work was supported by the National Science Foundation under Grants No. DMR 82-12189 and No. DMR 80-20247. Part of this work was done at the Stanford Synchrotron Radiation Laboratory, which is supported by the U.S. Department of Energy, Office of Basic Energy Sciences. We would like to thank the staff at SSRL, especially T. Troxel, for their assistance. Conversations with D. R. Nelson are acknowledged.

(a)Present address: Department 242/040-2, IBM, Rochester, Minn. 55901.

[1]For a review see D. R. Nelson, in *Phase Transitions and Critical Phenomena*, edited by C. Domb and M. S. Green (Academic, New York, 1983), Vol. 7, p. 1.

[2]R. J. Birgeneau and J. D. Litster, J. Phys. (Paris), Lett. 9, L399 (1978).

[3]G. W. Gray and J. W. Goodby, *Smectic Liquid Crystals — Textures and Structures* (Heyden, Philadelphia, 1984), pp. 153–154.

[4]S. C. Davey, J. Budai, J. W. Goodby, R. Pindak, and D. E. Moncton, Phys. Rev. Lett. 53, 2129 (1984); R. Pindak, D. E. Moncton, S. C. Davey, and J. W. Goodby, Phys. Rev. Lett. 46, 1135 (1981); C. C. Huang, J. M. Viner, R. Pindak, and J. W. Goodby, Phys. Rev. Lett. 46, 1289 (1981).

[5]F. Moussa, J. J. Benattar, and C. William, Mol. Cryst. Liq. Cryst. 99, 145 (1983).

[6]J. J. Benattar, J. Doucet, M. Lambert, and A. M. Levelut, Phys. Rev. A 20, 2505 (1979).

[7]J. J. Benattar, F. Moussa, M. Lambert, and C. Germain, J. Phys. (Paris), Lett. 42, L67 (1981).

[8]A. J. Leadbetter, J. P. Gaushan, B. Kelly, G. W. Gray, and J. W. Goodby, J. Phys. (Paris), Colloq. 40, C3-178 (1979).

[9]J. Budai, R. Pindak, S. C. Davey, and J. W. Goodby, J. Phys. (Paris), Lett. 45, L1053 (1984).

[10]J. Doucet, P. Keller, A. M. Levelut, and P. Porquet, J. Phys. (Paris) 39, 548 (1978).

[11]J. Collett, P. S. Pershan, E. B. Sirota, and L. B. Sorensen, Phys. Rev. Lett. 52, 356 (1984), and 52, 2190(E) (1984).

[12]D. E. Moncton and G. S. Brown, Nucl. Instrum. Methods 208, 579 (1983).

[13]The measurements on 5O.6 and 9O.4 were done with a position-sensitive detector rather than a crystal analyzer.

[14]J. Collett, L. B. Sorensen, P. S. Pershan, and J. Als-Nielsen, Phys. Rev. A 32, 1036 (1985).

[15]J. Collett, L. B. Sorensen, P. S. Pershan, J. D. Litster, R. J. Birgeneau, and J. Als-Nielsen, Phys. Rev. Lett. 49, 553 (1982).

[16]J. P. Hirth, P. S. Pershan, J. Collett, E. Sirota, and L. B. Sorensen, Phys. Rev. Lett. 53, 473 (1984).

[17]E. B. Sirota and P. S. Pershan, to be published. Recent measurements show the temperature and thickness dependence of the SmF-to-CrG transition in 9O.4 to be qualitatively the same as for 7O.7.

[18]D. R. Nelson and B. I. Halperin, Phys. Rev. B 21, 5312 (1980).

[19]R. Bruinsma and D. R. Nelson, Phys. Rev. B 23, 402 (1981).

346

Multicriticality in Hexatic Liquid Crystals

Amnon Aharony,[(a)] R. J. Birgeneau, J. D. Brock, and J. D. Litster

Center for Materials Science and Engineering and Department of Physics, Massachusetts Institute of Technology,
Cambridge, Massachusetts 02139

(Received 17 June 1986)

A theory is presented for the successive sixfold Fourier components, C_{6n}, in the bond-orientational order in the neighborhood of a smectic-A–hexatic-B phase transition. Near the transition we predict that $C_{6n} \sim C_6^{\sigma_n}$ with $\sigma_n = n + x_n n(n-1)$ where x_n depends weakly on n. General arguments are presented for the topology of the phase diagram in the vicinity of the smectic "liquid-hexatic-crystal" triple point which lead to the existence of a tricritical point on the smectic-A–hexatic-B line.

PACS numbers: 64.70.Md, 61.30.Gd, 64.60.My

In recent years it has been realized that there are bulk phases of matter with bond-orientational long-range order as in a solid but positional short-range order as in a fluid. Examples include the hexatic-B phase of liquid crystals and icosahedral metallic glasses.[1] In the liquid-crystal hexatic phases there is long-range sixfold-symmetric orientational alignment of the lines connecting neighboring molecules in the smectic planes. This order is characterized by a local order parameter $\psi(\mathbf{r}) = e^{6i\theta(\mathbf{r})}$, where θ is the angle between the "bonds" and some reference axis.[2] In-plane positional order, characterized by the density Fourier components $\rho_{\mathbf{q}}$, is achieved at a lower temperature. Recent research has concentrated on two main phase sequences. In some materials, such as n-hexyl-4'-n-pentyloxybiphenyl-4-carboxylate (65OBC) and mixtures containing it, one observes[3] the sequence smectic $A \rightarrow$ hexatic $B \rightarrow$ crystal E ($S_A \rightarrow S_{BH} \rightarrow S_E$). The temperature range of the hexatic phase is usually narrow. In other liquid crystals, such as 4O.8, a direct transition from S_A to a crystal, S_{BC}, is observed.[4] Mixtures of the two types of materials exhibit a triple point, at which the S_A, S_{BH}, and S_{BC} phases coexist.[5] In many of these systems, the transition from S_A to S_{BH} is continuous, but has a tricritical specific-heat exponent, $\alpha \approx 0.5$.[6] In others, the transition appears to be weakly first order. For cases in which the crystal phase also has long-range herringbone order such as S_E, the existence of such a tricritical point has been attributed to the coupling of the bond-orientational order with herringbone order.[7] An alternative phase sequence involves smectic $C \rightarrow$ tilted hexatic $I \rightarrow$ crystal J ($S_C \rightarrow S_I \rightarrow S_J$), as in racemic 4-(2-methylbutyl)phenyl 4'-(octyloxy)-(1,1')-biphenyl-4-carboxylate (8OSI).[8] In this case the coupling to molecular tilt induces long-range hexatic order even in the S_C phase.[2,8] This coupling allows the growth of single-domain samples, which in turn has recently made possible a direct measurement of many of the $6n$-fold order parameters[8] $C_{6n} = \mathrm{Re}\langle \psi^n \rangle = \mathrm{Re}\langle e^{6in\theta} \rangle$.

In the present Letter we discuss in detail the theoretical origins of the power laws $C_{6n} \sim C_6^{\sigma_n}$ observed in the synchrotron x-ray studies of the $S_C \rightarrow S_I$

transition in 8OSI. We note that the exponents σ_n are related to a sequence of crossover exponents,[9] associated with symmetry-breaking terms which describe multicritical crossover from the XY-model critical behavior into that of uniaxial ($n = 20$),[10] three-state Potts model ($n = 3$),[11] cubic ($n = 4$), hexagonal ($n = 6$),[12] etc., symmetry. Although the crossover exponents for $n = 2$ and $n = 3$ were measured separately before,[10,11] the present method allows a simultaneous measurement of many of these exponents, including those with $n \geq 4$ which usually only represent corrections to scaling, and which are accordingly very difficult to measure. The sequence σ_n represents an infinite set of independent critical exponents, all of which characterize the critical behavior. As we discuss below, a detailed analysis of the n dependence of σ_n is quite illuminating. Since averages like $\langle \psi^n \rangle$ always appear as Fourier coefficients in phase transitions characterized by a complex order parameter,[9] similar phenomena should occur near a variety of incommensurate phase transitions,[9] and in other systems such as certain graphite intercalates.[13]

As we show below, the data on 8OSI are fully consistent with the theoretical predictions on C_{6n}. In addition, we find some indications that the transition may be nearly tricritical, similar to those in many $S_A \rightarrow S_{BH}$ systems.[6] Since the coupling to the herringbone order is unimportant in 8OSI, we present an alternative thermodynamic argument, which suggests that the $S_A \rightarrow S_{BH} \rightarrow S_{BC}$ phase diagram should always have the structure shown in Fig. 1, with a tricritical

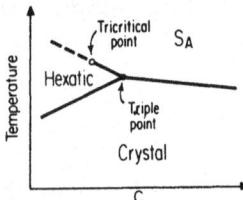

FIG. 1. Generic temperature-concentration phase diagram near the S_A-S_{BH}-S_{BC} coexistence triple point. The broken (full) lines indicate second- (first-) order transitions.

point and a triple point. The same structure should apply to the $S_C \to S_I \to S_J$ sequence although the second-order $S_C \to S_I$ transition will be rounded by the small hexatic field created by the molecular tilt. We hope that the present paper will stimulate much more detailed experiments in the vicinity of the triple point as well as detailed measurements of the successive Fourier components in other systems mentioned above.

Our theoretical analysis of the S_A-S_{BH} and S_C-S_I transitions starts from the Ginzburg-Landau Hamiltonian,[2]

$$\bar{H} = \int d^d r \{ \tfrac{1}{2} |\nabla \psi|^2 + \tfrac{1}{2} r |\psi|^2 + u_4 |\psi|^4 + u_6 |\psi|^6 + h \,\mathrm{Re}\psi \}. \tag{1}$$

Here the ordering field h is determined[2] by the average tilt order parameter, ϕ ($h \sim \phi^6$); it is zero for S_A phases while the experimental results indicate that h is quite small in S_C phases. For $h = 0$, this model exhibits XY-model critical behavior, provided that u_4 is larger than a tricritical value, u_{4t}. At $u_4 = u_{4t}$, Eq. (1) has a tricritical point, characterized for $d \geq 3$ dimensions by the Gaussian fixed point, with mean-field exponents and logarithmic corrections.

To study $C_{6n} = \mathrm{Re}\langle \psi^n \rangle$, we add to Eq. (1) a field term $\bar{H}_n = g_n \int d^d r \,\mathrm{Re}(\psi^n)$. The essential observation is that if one writes $\psi = x + iy$, one sees that successive terms $n = 2, 3, 4$, etc., scale like the uniaxial ($x^2 - y^2$), Potts ($x^3 - 3xy^2$), cubic $4(x^4 + y^4) - 3|\psi|^4$, etc., anisotropies.[9-11] Asymptotically close to the XY-model fixed point, the free energy should scale as $F(t, g_n) = |t|^{2-\alpha} f(g_n/|t|^{\phi_n})$, where $t = (T - T_c)/T_c$, α is the XY specific-heat exponent, and ϕ_n the appropriate crossover exponent.[9,10] Thus,

$$C_{6n} = (\partial F/\partial g_n)_{g_n = 0} \sim |t|^{2-\alpha-\phi_n} \sim C_y^{\sigma_n}, \tag{2}$$

with

$$\sigma_n = \frac{2-\alpha-\phi_n}{2-\alpha-\phi_1} = \frac{2(d-\lambda_n)}{d-2+\eta}, \tag{3}$$

where $\lambda_n = \phi_n/\nu$. This behavior holds only asymptotically close to the XY transition. As we show below, in that limit $\sigma_n = n + x_n n(n-1)$ with x_n weakly dependent on n.

Generally, one is not asymptotically close to the XY transition so that C_6 may not be small and there is typically a nearby tricritical point. In mean-field theory $\sigma_n = n$, so that at a tricritical point one should have simply $C_{6n} \sim C_6^n$ plus logarithmic corrections. Therefore, it is necessary to consider carefully the crossover from Gaussian to XY behavior. To leading order in $\bar{u}_4 - (u_4 - u_{4t})$, the renormalization-group recursion relation for g_n is $dg_n/dl = y_n^0 g_n - 4K_d \bar{u}_4 n(n-1) g_n$, where $y_n^0 = d - n(d-2+\eta)/2$. The factor $n(n-1)$ comes from the combinatorics of picking two out of the n factors in ψ^n, and appears in all higher-order terms as well.[10,14] Using the solutions[15] $\bar{u}_4(l) = \bar{u}_4 e^{\epsilon l}/Q(l)$, $t(l) = te^{2l}/Q(l)^{2/5}$, we find that

$$g_n(l) = g_n \exp(y_n^0 l)/Q(l)^{n(n-1)/10},$$

with $Q(l) = 1 + (\bar{u}_4/u_4^*)(e^{\epsilon l} - 1)$, $4K_d u_4^* = \epsilon/10$, and $\epsilon = 4 - d$. Substituting in $F(t, g_n) = e^{-dl} F(t(l), g_n(l))$, we find that

$$C_{6n}(t, h) = \exp[-(d - y_n^0)l] Q(l)^{-n(n-1)/10} C_{6n}(t(l), \exp(y_1^0 l) h).$$

We now iterate our recursion relations until $C_6(t(l^*), \exp(y_1^0 l^*) h) = 1$. Since at that point C_{6n} will also be of order unity, we conclude that

$$C_{6n} \simeq C_6^n [1 + (\bar{u}_4/u_4^*)(C_6^{-2\epsilon/(d-2+\eta)} - 1)]^{-n(n-1)/10}. \tag{4}$$

Note that the dependence on t and h has dropped out.

Equation (4) is correct for all dimensionalities $d \geq 3$ near the tricritical point where $\bar{u}_4$ is small. It should also work well whenever C_6 is not too small, so that l^* is not too large. When the ratio $\rho = C_6^2/(\bar{u}_4/u_4^*)$ is much less than 1, one is in the asymptotic XY regime and (4) reduces to (2) with $\sigma_n = C_6^{\sigma_n}$, with $\sigma_n = n + x_n n(n-1)/(d-2+\eta)$ and $x_n = \epsilon/5 + O(\epsilon^2)$. When ρ is not small, the C_6^{-2} term must remain inside the brackets of (4), and can represent an important correction because of the $n(n-1)$ exponent. In the 8OSI experiments, the C_{6n} are most easily measured when C_6 is large, so that ρ may not be small. Even under these circumstances, both (2) and (4) predict that the correction to the mean-field result $C_{6n} \sim C_6^n$ scales like a temperature-dependent constant raised to the power $n(n-1)$. In fact, if we take $x_n = \lambda(T)$ to be independent of n, the two forms are identical with $(\bar{u}_4/u_4^*) = (C_6^{-10\lambda(T)} - 1)/(C_6^{-2} - 1)$ in $d = 3$.

We next consider the asymptotic XY regime $\rho \ll 1$. Using the ϵ-expansion results[14] to order ϵ^3, one has

$$x_n \simeq \tfrac{1}{5}\epsilon \{ 1 + \tfrac{1}{5}\epsilon(3 - n) + \tfrac{1}{25}\epsilon^2(n^2 + 3.366\,455 n - 22.322\,657) \}.$$

Clearly, extrapolation to $\epsilon = 1$ is not trivial. Simple substitution of $\epsilon = 1$ in this equation shows that x_n increases from ~ 0.15 to ~ 0.4 for $n = 2$ to 7. An alternative estimate of x_n may be obtained by use of the values of u_4^* and the diagrammatic integrals in $d = 3$ directly rather than in $d = 4 - \epsilon$. Taking the numbers of Jug[16] to order $(u_4^*)^2$

348

we find $x_n \simeq 0.3 - 0.008n$ or $x_n \simeq 0.3/(1+0.027n)$. This form yields a much weaker dependence on n, with x_n varying from 0.3 to 0.25 for $n = 2$ to 7. To summarize, at $d = 3$ Eq. (4) holds for $C_6 > (\bar{u}_4/u_4^*)^{1/2}$; for smaller C_6 it simplifies to $C_{6n} \sim C_6^{\sigma_n}$ with $\sigma_n = n + x_n n(n-1)/(d-2+\eta)$.

Given the above theoretical predictions, we have reanalyzed the data on 8OSI.[8] We first fitted the measured angular structure factor at each temperature by the form

$$S(\chi)$$
$$= I_0 \left\{ \frac{1}{2} + \sum_{n=1}^{\infty} C_{6n} \cos[6n(90-\chi)] \right\} + I_{BG}, \quad (5)$$

with $C_{6n} = C_6^{\sigma_n}$. The resulting average effective exponents σ_n, already reported in Ref. 8, are shown in Fig. 2. As emphasized in Ref. 8, Eq. (5) is chosen such that as $\psi \to 1$ each $C_{6n} \to 1$ so that the C_{6n} are properly normalized. Thus it is plausible that the proportionality sign in Eq. (2) would become an equality and this has been assumed in the analysis. As seen in the figure, the results fit well with $\sigma_n = n + 0.295n(n-1)$ consistent with the theory. From Eq. (3), with the XY values $\nu \simeq 0.67$, $\eta \sim 0.03$, our results give $\phi_2 = 1.16 \pm 0.07$ and $\phi_3 = 0.4 \pm 0.17$, consistent with earlier measurements.[10,11] Our results also yield $\beta_6 = (d - \lambda_6)\nu = 5.1 \pm 0.4$, which is higher than the ϵ-expansion estimate 3.59 in Ref. 2.

We next fitted Eq. (5) to the data of Ref. 8 assuming $C_{6n} = C_6(T)^{n+\lambda(T)n(n-1)}$, and the resulting $C_6(T)$ and $\lambda(T)$ are shown in Fig. 3. For $T < 77.6\,°C$, that is $C_6 > 0.47$, $\lambda(T)$ is practically a constant, $\lambda(T) \simeq 0.295 \pm 0.02$. This may be compared with the theoretical XY value $x_n = 0.3 - 0.008n$. Fixing $\lambda(T)$ at 0.295 for all T indeed gives comparable fits, with the goodness-of-fit parameter χ^2 typically between 1 and 2. Identical fits are obtained with Eq.

(4) with $\bar{u}_4/u_4^* \simeq 1.7$ so that $\rho \sim 0.4$. In the transition region $\lambda(T)$ appears to decrease towards zero suggesting a crossover to mean-field (tricritical?) behavior; however, much more precise data are required to establish this definitively.

Finally, in Ref. 8 fits were performed with no restrictions whatsoever on the C_{6n}. If we write $C_{6n} = a_n C_6^{\sigma_n}$ and fit all of the data for $T < 76.8\,°C$ for each C_{6n} we find excellent fits with a_n varying from 0.99 to 0.73 as n varies from 2 to 6 and the exponents follow the law $\sigma_n = n + 0.22n(n-1)$. These results are consistent with the constrained fits at individual temperatures discussed above.

As noted above, the essential feature of both Eqs. (2) and (4) is that the correction to the mean-field result $C_{6n} \sim C_6^n$ scales like some temperature-dependent constant raised to the power $n(n-1)$. Since Eq. (4) applied away from the critical point, choosing the amplitude factor to be 1 is correct to leading order; however, writing $C_{6n} = C_6^{\sigma_n}$ in the critical regime is a much stronger assumption. The 8OSI data, nevertheless, seem to support this assumption; presumably the fact that Eqs. (2) and (4) must connect continuously means that the proportionality factor in Eq. (2) cannot deviate significantly from 1. The actual value for $\lambda(T)$ agrees remarkably well with the theoretical estimate, $x_n = 0.3 - 0.008n$ for the asymptotic XY regime in spite of the fact that C_6 is as large as 0.9. Much better data than those currently available, especially in the regime where C_6 is small, will be required to differentiate between Eqs. (2) and (4) and to determine whether or not the crossover to mean-field behavior [$\lambda(T) \to 0$] near T_c suggested by Fig. 3 is real.

The beauty of the above analysis is that the finite ordering field h in the tilted hexatic has not hidden in-

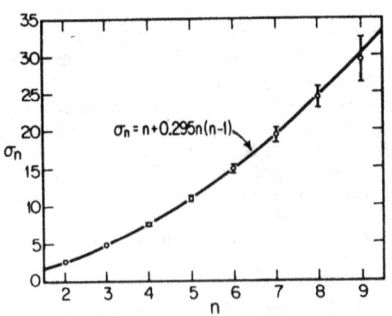

FIG. 2. Measured exponents σ_n from Ref. 8. The line is $\sigma_n = n + 0.295n(n-1)$.

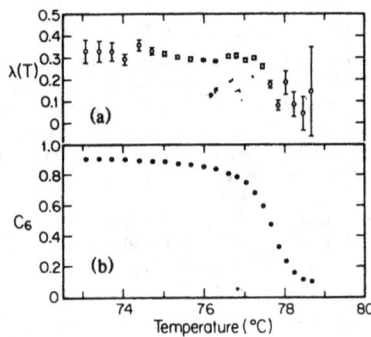

FIG. 3. (a) $\lambda(T)$ and (b) $C_6(T)$, from fits of the 8OSI data of Ref. 8 to Eq. (5), with $C_{6n} = C_6(T)^{\sigma_n}$, $\sigma_n = n + \lambda(T)n(n-1)$.

teresting behavior but rather has made possible a direct measurement of C_{6n}. The situation is more difficult if we wish to identify other critical exponents, such as β or δ. We fitted C_6 to a parametric equation of state,[17] $h = \theta (1 - \theta^2) r^{\beta + \gamma}$, $C_6 = K \theta r^\beta$, $t = (1 - b^2 \theta^2) r$, in which β, γ, T_c, K, and h were parameters. Data from 73 to 81 °C could be fitted, but gave a small value of β suggesting the possibility of a weakly first-order transition. When only $C_6 < 0.8$ were fitted, we obtained $\gamma \simeq 1$ and $\beta \simeq 0.25$. Fits of the same range of data to Eq. (1) without the gradient term gave small negative values of u_4. These results are also suggestive of a tricritical point, as found in Refs. 3–6. The sharp changes in the order parameter C_6 near 77 °C are reminiscent of the behavior near a wing tricritical line close to a tricritical point.[18]

We now present arguments explaining why such a tricritical point must occur, due to the nearby coexistence point of S_A, S_{BH}, and S_{BC} or, ignoring the rounding effect of the induced h, S_C, S_I, and S_J. The transition $S_A \rightarrow S_{BC}$ or $S_C \rightarrow S_J$ is always first order, due to the cubic term in the crystalline order parameter,[19] $\rho_{\mathbf{k}_1} \rho_{\mathbf{k}_2} \rho_{\mathbf{k}_3}$ with $\mathbf{k}_1 + \mathbf{k}_2 + \mathbf{k}_3 = 0$. The same cubic term survives in the S_{BH} (S_I) phase, and turns the $S_{BH} \rightarrow S_{BC}$ ($S_I \rightarrow S_J$) transition first order as indeed observed in 8OSI. These two first-order lines meet at the triple point with different slopes (see Fig. 1), since the coupling between ψ and ρ (or order $|\psi|\rho^2$) shifts the transition $S_{BH} \rightarrow S_{BC}$ ($S_I \rightarrow S_J$) relative to the continuation of the $S_A \rightarrow S_{BC}$ ($S_C \rightarrow S_J$) line. This discontinuity in slope at the triple point now implies a discontinuity across the $S_A \rightarrow S_{BH}$ ($S_C \rightarrow S_I$) line in that vicinity, turning it first order.[20] As one moves away from the triple point the effects of ρ on fluctuations in ψ decrease, and the effective coefficient $\bar{u}_4$, obtained after elimination of ρ from the partition function, may change sign.[21]

Since the $S_I \rightarrow S_J$ transition in 8OSI occurs only about 4 °C in temperature below the rounded $S_C \rightarrow S_I$ transition, we expect a triple point to exist nearby in some extended parameter space such as concentration or pressure. Further experiments are needed to confirm Fig. 1 for 8OSI and for the many cases discussed in Refs. 3–6.

We acknowledge gratefully illuminating discussions with G. Aeppli, P. Bak, A. N. Berker, M. E. Fisher, P. M. Horn, C. C. Huang, D. L. Johnson, T. C. Lubensky, D. R. Nelson, and R. Pindak. This work was supported by the National Science Foundation, Materials Research Laboratory, under Grant No. DMR84-18718.

(a)Permanent address: Tel Aviv University, Tel Aviv, Israel.

[1]R. J. Birgeneau and J. D. Litster, J. Phys. (Paris), Lett. 39, 1399 (1978); B. I. Halperin and D. R. Nelson, Phys. Rev. Lett. 41, 121 (1978); P. W. Stephens and A. Goldman, Phys. Rev. Lett. 56, 1168 (1986).

[2]D. R. Nelson and B. I. Halperin, Phys. Rev. B 21, 5312 (1980); R. Bruinsma and D. R. Nelson, Phys. Rev. B 23, 402 (1981).

[3]R. Pindak, D. E. Moncton, S. C. Davey, and J. W. Goodby, Phys. Rev. Lett. 46, 1135 (1981); C. Rosenblatt and J. T. Ho, Phys. Rev. A 26, 2293 (1982); R. Mahmood, M. Lewis, R. Biggers, V. Surendranath, D. L. Johnson, and M. E. Neubert, Phys. Rev. A 33, 519 (1986); J. M. Viner, D. Lamey, C. C. Huang, R. Pindak, and J. W. Goodby, Phys. Rev. A 28, 2433 (1983).

[4]P. S. Pershan, G. Aeppli, J. D. Litster, and R. J. Birgeneau, Mol. Cryst. Liq. Cryst. 67, 205 (1981); D. E. Moncton and R. Pindak, Phys. Rev. Lett. 43, 701 (1979).

[5]J. W. Goodby, Mol. Cryst. Liq. Cryst., Lett. Sect. 72, 95 (1981); J. W. Goodby and R. Pindak, Mol. Cryst. Liq. Cryst. 75, 233 (1981).

[6]T. Pitchford, G. Nounesis, S. Dumrongrattana, J. M. Viner, C. C. Huang, and J. W. Goodby, Phys. Rev. A 32, 1938 (1985); C. C. Huang, private communication.

[7]R. Bruinsma and G. Aeppli, Phys. Rev. Lett. 48, 1625 (1982).

[8]J. D. Brock, A. Aharony, R. J. Birgeneau, K. W. Evans-Lutterodt, J. D. Litster, P. M. Horn, G. B. Stephenson, and A. R. Tajbakhsh, Phys. Rev. Lett. 57, 98 (1986).

[9]R. A. Cowley and A. D. Bruce, J. Phys. C 11, 3577 (1978).

[10]M. E. Fisher and P. Pfeuty, Phys. Rev. B 6, 1889 (1972); F. J. Wegner, Phys. Rev. B 6, 1891 (1972); for reviews see (theory) A. Aharony, in Phase Transitions and Critical Phenomena, edited by C. Domb and M. S. Green (Academic, New York, 1976), p. 357; (experiment) Y. Shapira, in Multicritical Phenomena, edited by R. Pynn and A. Skjeltrup (Plenum, New York, 1984), p. 35.

[11]A. Aharony, K. A. Müller, and W. Berlinger, Phys. Rev. Lett. 38, 33 (1977), and references therein.

[12]D. R. Nelson, Phys. Rev. B 13, 2222 (1976).

[13]See, for example, P. Bak, Phys. Rev. Lett. 44, 889 (1980).

[14]F. J. Wegner and A. Houghton, Phys. Rev. A 10, 435 (1974); D. J. Wallace, in Phase Transitions and Critical Phenomena, edited by C. Domb and M. S. Green (Academic, New York, 1976), p. 293.

[15]D. R. Nelson and J. Rudnick, Phys. Rev. B 13, 2208 (1976).

[16]G. Jug, Phys. Rev. B 27, 609 (1983).

[17]J. T. Ho and J. D. Litster, Phys. Rev. B 2, 4523 (1970).

[18]See, e.g., D. Mukamel and M. Blume, Phys. Rev. A 10, 610 (1974).

[19]S. Alexander and J. P. McTague, Phys. Rev. Lett. 40, 702 (1978).

[20]J. Wheeler, J. Chem. Phys. 61, 4474 (1974).

[21]A similar phenomenon was discussed by E. Domany, D. Mukamel, and M. E. Fisher, Phys. Rev. B 15, 5432 (1977).

Orientational and Positional Order in a Tilted Hexatic Liquid-Crystal Phase

J. D. Brock, A. Aharony,[a] R. J. Birgeneau, K. W. Evans-Lutterodt, and J. D. Litster

Department of Physics, Massachusetts Institute of Technology, Cambridge, Massachusetts 02139

P. M. Horn and G. B. Stephenson

I.B.M. Thomas J. Watson Research Center, Yorktown Heights, New York 10598

and

A. R. Tajbakhsh

Department of Chemistry, The University, Hull HU6 7RX, England
(Received 17 April 1986)

We present a synchrotron-x-ray study of smectic-C (S_C) and smectic I (S_I) phases in a *single-domain* freely suspended film. Weak bond-orientational order in the S_C phase evolves continuously into the S_I phase so that the two phases are not thermodynamically distinct. From a Fourier analysis we obtain the hexatic order parameter C_6, the coefficient of $\cos 6\chi$. The higher harmonics behave as $C_{6n} = C_6^{\sigma_n}$, with $\sigma_n \simeq n + 0.3n(n-1)$, in agreement with a scaling theory.

PACS numbers: 64.70.Md, 61.30.Gd

It has been known for many decades that condensed-matter systems may be described by independent positional and bond-orientational (BO) order parameters.[1] Further, there is no symmetry reason why the positional and BO order must vanish at the same temperature. Thus it is possible to have a phase of matter in which the positional order is short range as in a liquid or glass but the BO order is long range as in a crystalline solid. Indeed, by generalizing concepts developed in the context of theories of two-dimensional dislocation-mediated melting it has been suggested that such an unusual phase could occur in certain smectic-liquid-crystal materials.[2] This phase has been labeled "stacked hexatic" for hexagonal-symmetry systems. Pindak and co-workers[3] first observed this novel phase in freely suspended liquid-crystal films. However, in the materials studied by that group it was not possible to produce single-domain samples so that the BO order could not be studied quantitatively.

Very interesting theoretical predictions have been made for the smectic-C (S_C), smectic-F (S_F), and smectic-I (S_I) phases, in which the molecules are tilted with respect to the layer normal.[4] The S_F and S_I are hexatic phases with molecules tilted between and towards near neighbors, respectively. However, coupling to the molecular tilt should induce long-range hexatic order even in the fluid S_C phase. This prediction by itself presents an interesting problem for study. An equally attractive feature of the tilted hexatics is that the coupling between the tilt and BO order provides a possible avenue for the growth of *single-domain* samples. This would, in turn, make possible direct measurements of the BO order, an essential aspect of BO-ordered phases which has not yet been studied in

substrate-free systems.

In this paper we report a study of the $S_C \rightarrow S_I$ transition in the liquid-crystal material racemic 4-(2-methylbutyl)phenyl 4'-(octyloxy)-(1,1')-biphenyl-4-carboxylate (8OSI). 8OSI has a rich phase diagram[5]:

$$I \xrightarrow{174.5°} N \xrightarrow{170.0°} S_A \xrightarrow{133.4°} S_C \xrightarrow{79.9°} S_I \xrightarrow{75.1°} S_J \xrightarrow{61.9°} S_K.$$

The phase sequence $S_A \rightarrow S_C \rightarrow S_I$ on cooling was an essential factor in choosing this material. S_J and S_K are crystalline phases without and with herringbone order, respectively. The smectic layers were aligned by pulling a freely suspended S_A film across an 8-mm hole in a 0.64-mm-thick piece of polished stainless steel. This yielded a film of approximate thickness 10 μm. The sample was maintained at 1 Torr of nitrogen in a two-stage oven which controlled the temperature to better than 5 mK. The tilt field, $\phi(\mathbf{r})$, of the sample was aligned as the film cooled from the S_A phase into the S_C phase by two SmCo$_5$ magnets ($H \sim 0.1$ T) which held the tilt field in place thereafter. Samples consisting of a single tilt domain could be routinely produced and maintained for weeks. Optical measurements show that the tilt mosaic is $< 1.5°$. Over the course of the experiment (~ 3 weeks) the $S_C \rightarrow S_I$ transition temperature drifted $< 1°$C. These measurements were the first done on beam line X-20C at the National Synchrotron Light Source (NSLS). Running in an unfocused configuration, the monochromator consisted of a pair of Si(111) crystals set to Cu-K edge ($\lambda = 1.3796$ Å). The analyzer was a Ge(111) single crystal. [$\Delta Q_{\parallel} = 2 \times 10^{-4}$ Å^{-1}, $\Delta \chi < 0.05°$ (the χ axis is normal to the smectic plane)]. The illuminated spot size was set by a 1.6-mm-diam hole in a tantalum plate directly in front of the sample. A Huber six-circle diffractometer provided the neces-

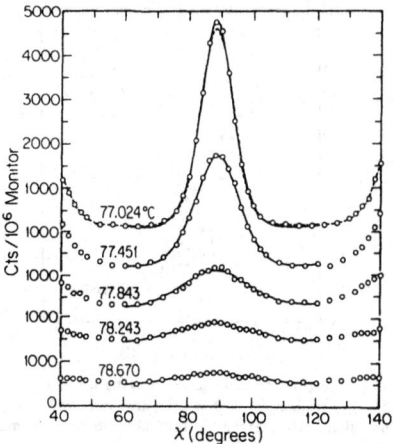

FIG. 1. Angular scans in χ along the peak of the form factor and structure factor as a function of temperature in 8OSI. The solid lines are the results of fits to Eq. (1) with $C_{6n} = C_6^{2.6(n-1)}$ while the dashed line for $T \sim 77.024\,°C$ is the result of a fit to Eq. (2).

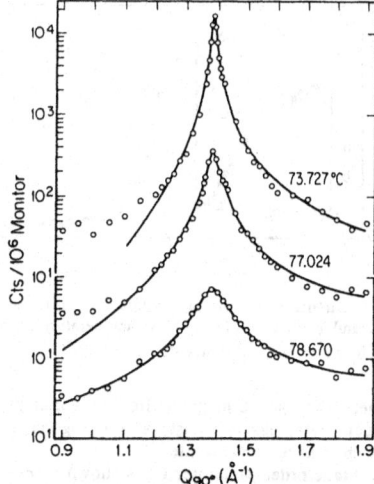

FIG. 2. Longitudinal scans through the $\chi = 90°$ peak. The solid lines are the results of fits to Eq. (2).

sary sample and analyzer motions.

The in-plane x-ray-scattering profile of the S_A phase is a diffuse ring characteristic of the fluid order in the smectic layers. As hexatic ordering develops in a single-domain sample, one expects the ring to develop a sixfold modulation, eventually breaking up into six symmetrically placed diffuse spots. In the case of the S_C phase, the shape of the diffuse ring is more complex. As a result of the tilt of the molecules, the molecular form factor is tilted with respect to the plane of the smectic layers, effectively tilting the diffuse ring. Accordingly, it is necessary that χ scans be done in the tilted plane which contains the maximum of the molecular form factor.[3] Further, because of the tilt, the BO order will have overall twofold symmetry although the sixfold aspect should dominate.

Figure 1 shows χ scans along the peak of the molecular form factor at several temperatures near the previously reported S_C to S_I phase transition. At high temperatures the ring is uniform to within our counting statistics. As the sample is cooled, a measurable sinusoidal modulation of the ring develops. At $T \sim 77.5\,°C$ the χ scan shows definite peaks every 60° indicating substantial amounts of hexatic BO ordering. The magnetic field direction is $\chi = 0°$ so that the BO-order peaks indeed come at the angles expected for a S_I phase.[3] Figure 2 shows longitudinal scans through the same peak. At high temperatures we see a broad diffuse scattering profile indicative of short-range positional order. As the sample is cooled, the width of the

peak narrows simultaneously with the development of the BO order, suggesting that the enhanced positional correlations are due to a coupling to the BO order. At no point do we approach the resolution of our spectrometer; the positional order is always of short range.

In order to characterize the BO order quantitatively, we performed a nonlinear least-squares fit of the χ scans between 60° and 120° to the Fourier cosine series

$$S(\chi) = I_0 \left[\frac{1}{2} + \sum_{n=1}^{\infty} C_{6n} \cos 6n(90° - \chi) \right] + I_{BG}, \quad (1)$$

where χ is the angle between the in-plane component of $\mathbf{Q}$ and $\mathbf{H}$. The coefficients C_{6n} measure the amount of $6n$-fold ordering in the sample. These fits neglect the residual twofold effects arising from the molecular tilt which breaks the hexagonal symmetry; this approximation does not affect our principal conclusions. With the constant term chosen as $\frac{1}{2}$ in Eq. (1), the C_{6n} approach 1 for perfect BO and positional order. From the scans deep in the S_I phase we find a background, I_{BG}, of 20 counts per 10^6 monitor counts. The fits show explicitly that the system smoothly develops first sixfold order, then twelvefold order, then eighteenfold order, etc. There is no abrupt transition; rather the evolution is smooth and continuous. This behavior confirms the prediction that coupling between the tilt and hexatic fields induces hexatic ordering in the S_C phase, destroying the $S_C \rightarrow S_I$ phase boundary.[4] The situation is similar to that of a fer-

VOLUME 57, NUMBER 1   PHYSICAL REVIEW LETTERS   7 JULY 1986

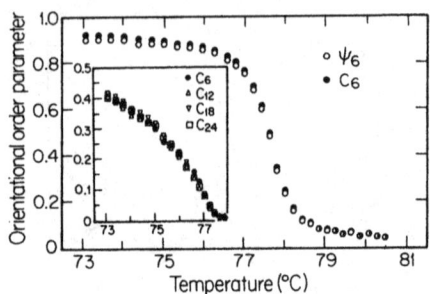

FIG. 3. Orientational order parameters C_6 [Eq. (1) unrestricted] and ψ_6 [$\sigma_n = 2.6(n-1)$] vs temperature. Inset: the first four Fourier coefficients plotted as $C_{6n}^{\sigma_n/\sigma_n}$.

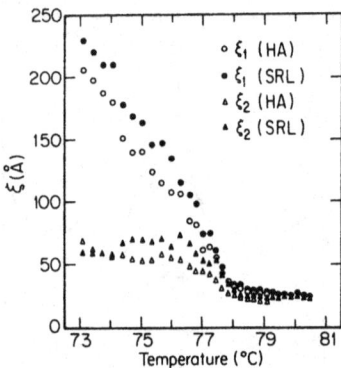

FIG. 4. Longitudinal correlation lengths for the 90° (ξ_1) and 30° (ξ_2) peaks. ξ(SRL) is the value obtained from the SRL line shape while ξ(HA) is the value obtained from the harmonic approximation Eq. (2).

romagnet in an applied magnetic field or a triangular, incommensurate, hexatic phase on a hexagonal substrate such as xenon on graphite.[6]

The hexatic order parameter C_6 is shown in Fig. 3; as noted above, C_6 evolves continuously. In analyzing the results of the fits we discovered a simple scaling relation $C_{6n} \sim C_6^{\sigma_n}$. As average values we find $\sigma_2 = 2.57 \pm 0.05$, $\sigma_3 = 4.80 \pm 0.14$, $\sigma_4 = 7.59 \pm 0.27$, $\sigma_5 = 10.96 \pm 0.44$, $\sigma_6 = 14.85 \pm 0.70$, and $\sigma_7 = 19.2 \pm 1.4$. With this scaling, all C_{6n} out to $n = 7$ fall on the same curve to within the fitting error over the complete temperature range with no adjustment in the amplitude. This is illustrated for the first four harmonics in the inset of Fig. 3. We note that the successive σ_n are increasingly determined by the large-C_6 data.

In an attempt to understand this novel scaling behavior, we first noted empirically that to a good approximation $\sigma_n = 2.6(n-1)$ for $1 < n < 6$; this form allows one to sum Eq. (1) exactly. The solid lines in Fig. 1 are, in fact, calculated with this approximation and the order parameter so obtained is plotted as ψ_6 in Fig. 3. Clearly ψ_6 and C_6 obtained from the unrestricted fits agree well. More recently, stimulated by these experiments, a multicritical scaling theory for this transition has been developed.[7] The theory predicts that the effective exponents should have the form $\sigma_n = n + \lambda(T)n(n-1)$. Indeed, the values quoted above fit this prediction excellently with $\lambda(T) = 0.295$. Plots using this form are indistinguishable from those in Fig. 1 although the goodness-of-fit parameter χ^2 improves by a factor ~ 4 deep in the S_I phase. The order parameter obtained in these latter fits agrees within the errors with C_6 and ψ_6 in Fig. 3.

To characterize quantitatively the positional order, it is necessary to use a simple model. For a hexatic with perfect BO order but only short-range positional order, a reasonable *Ansatz* is a sixfold symmetric pattern of circular Lorentzian spots.[6] Inclusion of BO fluctua-

tions produces an effective mosaic averaging of these circular Lorentzian spots. In the harmonic approximation the consequent line shape is[8]

$$S(\mathbf{Q}) \propto \int_{-\pi/6}^{\pi/6} d\psi \, \frac{\exp[-\frac{1}{2}\psi^2/\langle|\delta\psi|^2\rangle]}{\kappa^2 + Q^2 + G^2 - 2QG\cos(\theta - \psi)},$$

(2)

where θ is the angle between the in-plane component of $\mathbf{Q}$ and $\mathbf{G}$.

For $|\mathbf{Q} - \mathbf{G}| < G|\delta\psi|$, the BO fluctuations distort the longitudinal line shape so that they are well approximated by the square root of a Lorentzian. In previous studies it has been found that the square root of a Lorentzian (SRL) line shape fits smectic-phase in-plane longitudinal scans quite well.[3,9] The ratio of the longitudinal to the transverse linewidths is 1:5 at $\chi = 90°$. Interestingly, this same aspect ratio was found for monolayer xenon on graphite.[6] Therefore, we fit the 90° longitudinal scans in the region $G \pm 5\kappa$ to a SRL line shape. The SRL line shape works well for this range of Q over the entire temperature range. The out-of-plane component of the line shape is due to the molecular form factor and is well described by a Lorentzian of width κ_z. Including this form factor in Eq. (2), we can simultaneously fit the longitudinal and χ scans. This fit should hold over a larger range of Q. The solid lines in Fig. 4 are results of fits to Eq. (2); the SRL results are indistinguishable. As expected, Eq. (2) holds over a much larger range of Q. It should be noted that Eq. (2) has been multiplied by a linear scale factor $1 + B(Q - G)$ to account for the overall asymmetry[9] of the profile about G.

As noted previously, the molecular tilt breaks the

hexagonal symmetry so that the peaks $\pm 90°$ differ from those at $\pm 30°$ and $\pm 150°$; specifically, the in-plane lattice constants differ by about 10%. We have fitted the two types of peaks separately to Eq. (2); the results for the 30° and 150° peaks are less reliable because the beam was blocked by the $SmCo_5$ magnets for $\chi < 32°$ and $\chi > 155°$. With this caveat, the transverse angular spread $\langle |\delta\psi^2| \rangle^{1/2}$ is identical for the two peaks while as shown in Fig. 4, $\xi(30°) = \xi(90°) = \xi_1$ in the S_I phase.[10] In general, the simplified SRL line shape gives results quite similar to those obtained from the full theory, Eq. (2).

These experiments have enabled us to obtain rather complete information about the BO and positional correlations in the S_C and S_I phases of 8OSI. The BO order is locked to the tilt and grows continuously as one cools from the S_C to S_I phase so that the two phases are not, in fact, thermodynamically distinct. Successive Fourier components of the hexatic order grow progressively with the simple law, $C_{6n} = C_6^{\sigma_n}$. The positional correlations grow anisotropically in the S_I phase with $\xi(90°)$ reaching ~ 250 Å before the transition into the crystalline S_I phase.

We are especially grateful to the members of both the Massachusetts Institute of Technology and IBM groups, who have contributed greatly to the development of the IBM-MIT beam lines at NSLS. We thank R. Pindak for advice on the oven design. We have enjoyed stimulating discussions of this work with G. Aeppli, S. G. J. Mochrie, and P. S. Pershan. Two of us (J.D.B. and J.D.L.) acknowledge support from the National Science Foundation under Grant No. DMR83-19985. The MIT contributions to the operation and construction of the beam lines at the NSLS are supported by the National Science Foundation, Materials Research Laboratory program under Grant No. DMR84-18718 and by the Joint Services Electronics Program under Contract No. DAAL03-86-K-0002. Two of us (R.J.B. and K.W.E.-L.) also acknowledge the support of this latter grant. The National Synchrotron Light Source, Brookhaven National Laboratory is supported by the U. S. Department of Energy, Division of Materials Sciences and Division of Chemical Sciences.

Note added.— An elegant experiment demonstrating BO order in a S_I phase via the defect structures was recently reported by Dierker, Pindak, and Meyer.[11]

[a]On leave from Tel Aviv University, Tel Aviv, Israel.

[1]R. E. Peierls, private communication; L. D. Landau and E. M. Lifshitz, *Statistical Physics* (Addison-Wesley, Reading, MA, 1969), p. 466.

[2]R. J. Birgeneau and J. D. Litster, J. Phys. (Paris), Lett. 39, 1399 (1978); B. I. Halperin and D. R. Nelson, Phys. Rev. Lett. 41, 121 (1978).

[3]R. Pindak, D. E. Moncton, S. C. Davey, and J. W. Goodby, Phys. Rev. Lett. 46, 1135 (1981); J. Budai, R. Pindak, S. C. Davey, and J. W. Goodby, J. Phys. (Paris), Lett. 45, L1053 (1984).

[4]D. R. Nelson and B. I. Halperin, Phys. Rev. B 21, 5312 (1980); R. Bruinsma and D. R. Nelson, Phys. Rev. B 23, 402 (1981).

[5]P. A. C. Gane, A. J. Leadbetter, and P. G. Wrighton, Mol. Cryst. Liq. Cryst. 66, 247 (1981).

[6]See, for example, S. E. Nagler, P. M. Horn, T. F. Rosenbaum, R. J. Birgeneau, M. Sutton, S. G. J. Mochrie, D. E. Moncton, and R. Clarke, Phys. Rev. B 32, 7373 (1985).

[7]A. Aharony, R. J. Birgeneau, J. D. Brock, and J. D. Litster, to be published.

[8]G. Aeppli and R. Bruinsma, Phys. Rev. Lett. 53, 2133 (1984).

[9]B. M. Ocko, A. R. Kortan, R. J. Birgeneau, and J. W. Goodby, J. Phys. (Paris) 45, 113 (1984).

[10]Similar behavior has been seen previously in thin film S_F and S_I phases; E. B. Sirota, P. S. Pershan, L. B. Sorensen, and J. Collett, Phys. Rev. Lett. 55, 2039 (1985).

[11]S. B. Dierker, R. Pindak, and R. B. Meyer, Phys. Rev. Lett. 56, 1819 (1986).

8. SMECTIC-D

8.1. Diele, S., Brand, P. and Sackmann, H. (1972). X-ray diffraction and polymorphism of smectic liquid crystals. II. D and E modifications. *Mol. Cryst. Liq. Cryst.* **17**, 163–169   357

8.2. Tardieu, A. and Billard, J. (1976). On the structure of the "smectic D modification". *J. Phys. (Paris) Colloq.* **37**, C3-79–C3-81   364

8.3. Etherington, G., Leadbetter, A. J., Wang, X. J., Gray, G. W. and Tajbakhsh, A. (1986). Structure of the smectic D phase. *Liquid Crystals,* **1**, 209–214   367

Molecular Crystals and Liquid Crystals. 1972. Vol. 17, 163-169
© Gordon and Breach Science Publishers S.A.
Reset with permission

X-RAY DIFFRACTION AND POLYMORPHISM OF SMECTIC LIQUID CRYSTALS. II. D AND E MODIFICATIONS[†]

S. Diele, P. Brand and H. Sackmann
Martin-Luther-Universität
402 Halle-Saale, Sektion Chemie
German Democratic Republic

Received June 2, 1971; in revised form September 7, 1971

Abstract The X-ray patterns of two new smectic modifications (smectic D and smectic E) are investigated. The D-modification is optically isotropic. A cubic structure is presumed. In the E-modification the pattern exhibits three sharp outer rings besides the inner ring, which points to a higher order compared with the A, B and C-Phases.

1. Introduction

The great majority of smectic liquid-crystalline forms are modifications of type A, B and C (see Table 1 of Part I[1]). Some forms are found which could not be associated with these modifications by their miscibility. They are designated as D and E. Now we have looked into the question whether these modifications exhibit characteristic features in their X-ray diffraction patterns.

2. The Smectic D-Modification

SUBSTANCES

No. 1: 4′-*n*-Hexadecyloxy-3′-nitrodiphenyl-4-carboxylic acid

$$\text{solid} \xleftrightarrow{126.8} s_C \xleftrightarrow{171.0} s_D \xleftrightarrow{197.2} s_A \xleftrightarrow{201.9} \text{isotropic}$$

[†]Presented at the Third International Liquid Crystal Conference in Berlin, August 1970.

Molecular Crystals and Liquid Crystals

No. 2: 4'-*n*-Octadecyloxy-3'-nitrodiphenyl-4-carboxylic acid

$$\text{solid II} \overset{85}{\longleftrightarrow} \text{solid I} \overset{124.6}{\longleftrightarrow} s_C \overset{158.9}{\longleftrightarrow} s_D \overset{195.0.}{\longleftrightarrow} \text{isotropic}$$

The two homologues of the nitrodiphenyl-4-carboxylic acid show a new liquid crystalline modification D. This modification occurs in substance No. 1 between the smectic A and C types and in substance No. 2 as a high-temperature form in addition to the C-modification. Some details of the description of the modifications are found in Ref. 2. Frequently, modification D originates as an isotropic, polygonal form in microscopic observations. In liquid-crystalline modifications "pseudo-isotropy" caused by a spontaneous orientation of the sample is often found.[3] Also, the D-modification was considered as pseudo-isotropic at first. Investigations for the double refraction have established the optical isotropy of modification D. This modification is not double-refracting.[4]

The X-ray patterns are recorded by the flat-film method (see Part I).[1] The sample placed into a heated glass capillary tube was irradiated perpendicular to the cylinder axis with Ni-filtered Cu K_α radiation. The electric furnace does not permit rotation of the sample about the cylinder axis.

The patterns of the C- and A-modifications show the usual picture:[1] an inner sharp ring and outer ring. The outer ring is weak and blurred.

In the D-modification a very weak and blurred ring is also found in a Bragg-angle range of about 10°. The inner ring degenerates into spot-like interferences (appearing at a Bragg-angle of about 1°). Their number and position may be different (Fig. 1, the arrow indicates the direction of the cylinder axis of the sample). However in substance No. 1 it was possible to get diagrams with a clearly hexagonal arrangement of the reflections (Fig. 2). Figure 3 gives a schematic graph. The open circles represent the hexagonally arranged interference spots at 1°. The filled circles (dots) indicate very weak reflections visible on the original. The very weak and blurred ring at 10° is sketched on the figure.

The interpretation of the results as a fibre-texture with a hexagonal symmetry must be eliminated because the formation of such a texture, vertical to the cylinder axis, is hardly conceivable.

The existence of the spot-like reflexes with a hexagonal arrangement, in addition to the liquid-like outer interference, requires a model which contains well-ordered parts and statistically distributed ones. Layer-structures are incompatible with the optical isotropy.

D and E Modifications-II

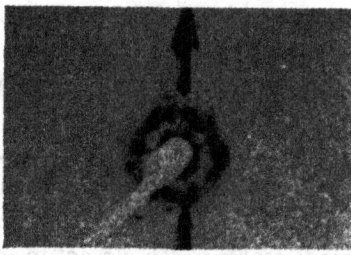

Figure 1. Substance No. 1, $T = 175^{\circ}$C, smectic D, the arrow indicates the direction of the sample.

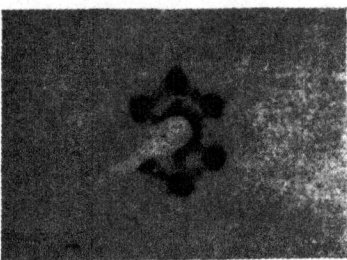

Figure 2. Substance No. 1. $T = 182^{\circ}$C, smectic D.

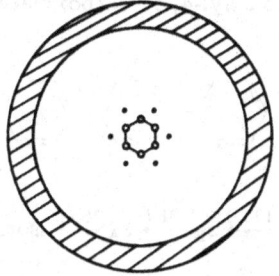

Figure 3. Schematic graph of the reflections from Fig. 2.

Thus an attempt has been made to describe the patterns on the basis of a cubic model. On the formation of the modification, small irregularly distributed parts with a defined arrangement of the molecules occur, giving rise to a pattern like Fig. 1. Some of these parts grow to a liquid crystalline "single crystal" in the viscous melt. In Fig. 2 such a "single crystal" was examined, whose size is equal to or greater than the irradiated volume, and whose body-diagonal was in this case nearly parallel to the incident beam. An exact evaluation of the Bragg angle from Fig. 2 is not possible because the interference spots which have different shapes and a noticeable width, do not form a regular hexagon. For the calculation we have used a Bragg-angle $\theta = 1.03°$, but we must take into account an error of $\pm 0.08°$. If we assume these are $(\bar{1}10)$-reflections, we obtain for the lattice distance $a = 61 Å$. So far we do not know the density of this modification. For the present we have used the following model: the lattice points are formed by sphere (micelles) existing of the aromatic parts of several molecules. But the hydrocarbon chains should be distributed irregularly to cause the weak ring at large angles. But more detailed investigations are necessary to confirm these speculations.

3. The Smectic-E-Modification

SUBSTANCES

No. 1: Diethyl-*p*-terphenyl-4, 4''-carboxylate[5]

$$C_2H_5OOC-\!\!\!\bigcirc\!\!\!-\!\!\!\bigcirc\!\!\!-\!\!\!\bigcirc\!\!\!-COOC_2H_5$$

$$\text{solid} \xleftrightarrow{173.0} s_E \xleftrightarrow{188.5} s_A \xleftrightarrow{259.0} \text{isotropic}$$

No. 2: Dipropyl-*p*-terphenyl-4, 4''-carboxylate[5]

$$\text{solid} \xleftrightarrow{122.0} s_E \xleftrightarrow{137.1} s_A \xleftrightarrow{293.2} \text{isotropic}$$

No. 3: 2-(4-*n*-decyloxydiphenyl)-quinoxalline[6]

$$C_{10}H_{21}-O-\text{[structure]}$$

$$\text{solid} \overset{97/98}{\longleftrightarrow} s_E \overset{120}{\longleftrightarrow} s_A \overset{194}{\longleftrightarrow} \text{isotropic}$$

No. 4: 1-(4-*n*-pentyloxydiphenyl-(4')-glyoxal-2-phenylhydrazine[6]

$$C_5H_{11}-O-\text{[structure]}-C-CH=N-NH-\text{[phenyl]}$$
$$\overset{|}{O}$$

$$\text{solid} \overset{130}{\longleftrightarrow} S_E \overset{165}{\longleftrightarrow} S_A \overset{253}{\longleftrightarrow} \text{isotropic}$$

In the substances named, a low-temperature form besides the smectic high-temperature for A is observed. The investigations of miscibility carried out so far[5] show the existence of a new smectic modification, which has been given the designation E.

The X-ray patterns obtained in non-oriented samples of the four substances in modification E show the characteristics in Fig. 4. The diagrams were obtained with a Gunier camera using Cu K_α radiation (monochromatized with a cut and curved quarts crystal). Besides a sharp interference at small angles (and its second order) three sharp interferences are found at large angles. The inner ring indicates the existence of the smectic layers.

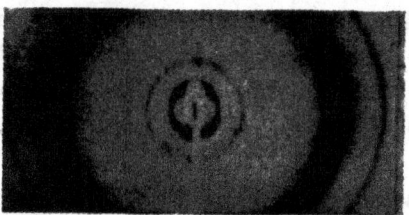

Figure 4. Substance No. 2. $T = 129°$ C, smectic E.

An X-ray pattern from substance No. 1 in the crystalline phase at room temperature is shown in Fig. 5. A band structure of the reflections is visible. Such a band structure is found in molecular lattices, unit cells of which exhibit in one direction a significantly larger dimension than in the other directions (for example in the case of paraffins). This band structure is diminished on the pattern of modification E. This observation suggests that the three-dimensional order is largely removed.

A unique interpretation of the three reflections based upon a two-dimensional lattice with a quadratic or a hexagonal net was not possible. For the determination of an oblique cell further reflexes are necessary.

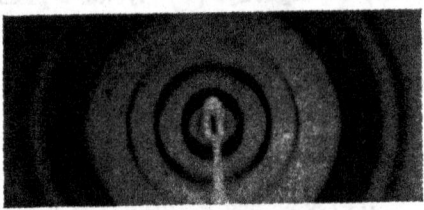

Figure 5. Substance No. 1. Room temperature, crystalline phase.

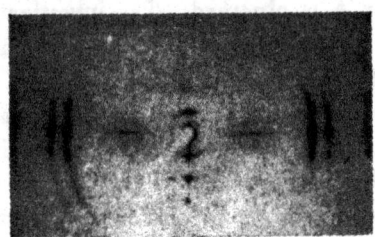

Figure 6. Oriented sample of substance No. 2. $T = 124°C$, smectic E.

The X-ray pattern of a sample oriented in a magnetic field[1] and recorded the flat-film method (see Sec. 2) is shown in Fig. 6. Layer line reflections on the meridien are found up to the third order. The reflections at large angles are only visible up to the first layer-line. This could be a hint that the order in the third dimension is not well marked.

4. X-ray Diffraction and Polymorphism

The investigations of X-ray diffraction in the smectic modifications A, B, C, D and E presented in a previous paper (Part I) and in this paper confirm the characterization obtained by investigations of the miscibility in binary systems. The modifications designated in this way exhibit X-ray patterns in each case which tally in their essential characteristics. The new results found for F and G modifications[7] seem to prove this statement. It follows that the described modifications in each case also exhibit the same structure.

The well known concept of the basic structure of smectic modifications formed by layers of molecules is true of the A, B and C-modifications. The results show that the A modification is distinguished by a largely parallel arrangement of molecules standing vertically in the layers. The distribution of the molecular axes within the layers is largely statistical in the A modification.

In the B modification a rigid arrangement should be assumed. At present it is not possible to discern special characteristics of the structure of the B modification from our experiments. In the C modifications some investigations of oriented samples point to an inclination of the molecular axes to the layer. But the difference in the X-ray patterns of several investigators should be pointed out.

In the D and E modifications indications of a more highly ordered structure exist. For the D modification a cubic structure must be considered. This raises the question whether the designation "smectic" for the D modification is allowed. In the E modification a higher order of the molecules within the layers must be assumed. In this modification signs are present that the order goes beyond that within single layers.

References

1. Diele, S., Brand, P. and Sackmann, H., *Mol. Cryst. and Liq. Cryst.,* Part I, in print.
2. Demus, D., Kunicke, G., Neelsen, J. and Sackmann, H., *Z. Naturforschung,* **23a**, 84 (1968).
3. Sackmann, H. and Demus, D., *Mol. Cryst.,* **2**, 81 (1966).
4a. Pelzl, G., Dissertation Halle (S) 1969.
4b. Pelzl, G. and Sackmann, H., *Trans. Faraday Soc.,* in preparation.
5. Kölz, K.-H., Diplomarbeit Halle (S) 1966.
6. Herrmann, I., Dissertation Halle (S) 1970.
7. Demus, D., Diele, S., Klapperstück, M., Link, V. and Zaschke, H., *Mol. Cryst. and Liq. Cryst.,* **15**, 161 (1971).

JOURNAL DE PHYSIQUE

Colloque C3, *supplément au n⁰ 6, Tome 37, Juin 1976, page* C3-79

ON THE STRUCTURE OF THE « SMECTIC D MODIFICATION »

A. TARDIEU

Centre de Génétique Moléculaire du C. N. R. S., 91190 Gif-sur-Yvette, France

J. BILLARD

Laboratoire de Physique de la Matière Condensée, Collège de France, 75231 Paris Cedex 05, France

Résumé. — La structure de la phase *smectique de type D* a été réétudiée par diffraction des rayons X. Six ordres de réflexions ont été observés, qui correspondent à un réseau cubique centré, de groupe de symétrie Ia3d. Nous pensons qu'il s'agit de la même structure que celle d'une phase cubique observée avec les lipides.

Abstract. — The structure of the *smectic D modification* is reinvestigated by X-ray diffraction techniques. Six reflexions orders are observed which are in agreement with a body centered cubic lattice, space group Ia3d. We propose that the structure is similar to that of one cubic phase observed with lipids.

1. Introduction.

The *smectic D modification* was first observed and described by the Halle group in 1968 [1]. It has the particularity of being optically isotropic. This modification was latter on studied, also by the Halle group by X-Ray diffraction techniques, using a powder sample and a monodomain [2]. The powder diagram contained too many spots to be interpretable. However, on the basis of six reflexions organized in a pseudo hexagonal array observed with the monodomain, a cubic structure was proposed [2]. We report here a new series of X-Ray experiments, in which it has been possible to increase the number to visible reflexions. The cubic structure is confirmed and the symmetry group is determined.

2. Experimental part.

2.1 SUBSTANCES.

Only two compounds are actually known to present the « smectic D modification » :

No. 1 : 4'-n-Hexadecyloxy-3'-nitrodiphenyl-4-carboxylic acid

$$C_{16}H_{33}-O- \text{(ring)} -COOH \quad (NO_2)$$

$$\text{solid} \overset{126.8}{\longleftrightarrow} S_C \overset{171.0}{\longleftrightarrow} S_D \overset{197.2}{\longleftrightarrow} S_A \overset{201.9}{\longleftrightarrow} \text{isotropic} \quad (2)$$

No. 2 : 4'-n-Octadecyloxy-3'-nitrodiphenyl-4-carboxylic acid

$$\text{solid}_{II} \overset{85}{\longleftrightarrow} \text{solid}_I \overset{124.6}{\longleftrightarrow} S_C \overset{158.9}{\longleftrightarrow} S_D \overset{195.0}{\longleftrightarrow} \text{isotropic} . \quad (2)$$

The sample used, (hexadecyloxy-) was a gift from Pr. Sackmann.

2.2 X-RAY.

The pictures are Debye Scherrer diagrams, obtained with a Guinier type camera, with bent quartz monochromator, operating in vacuum. The radiation if Cu Kα ($\lambda = 1.54$ Å). The sample, about 0.7 mm thick, was held in a closed sample holder with mica windows. In order to avoid a spotted appearance for the X-Ray diagrams, the sample is rotated around its axis during the experiment. The sample seems damaged after repeated heating ; because of the small amount of sample available (a few mgs) only two pictures were taken, with and without rotation of the sample.

3. Results.

As observed by Diele *et al.* [2], the X-Ray diagrams show a diffuse reflexion, at high angles around $s \simeq (4.5 \text{ Å})^{-1}$ indicating that the hydrocarbon chains are in a *liquid* state. At low angles and without rotation, the diagram consists in a large series of spots, again similar to the results of Diele *et al.* ([2], Fig. 1).

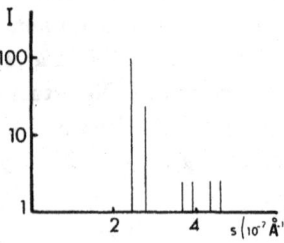

FIG. 1. — Schematic representation of the low angle part of the X-Ray diagram. The ordinate represents the intensity of each reflexion (estimated by eye), and the abscissa gives the diffraction angle, in the form of $s = 2 \sin \theta/\lambda$).

A. TARDIEU AND J. BILLARD

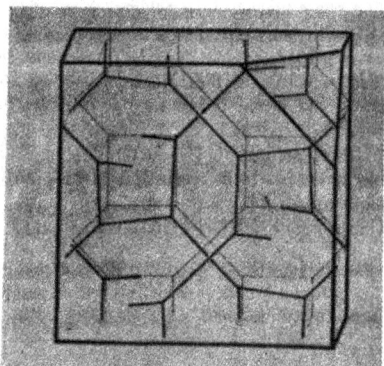 

FIG. 2. — The cell is centred cubic space group Ia3d. The left frame represents a perspective view of one unit cell. The axis of the rods that sit on crystallographic two fold axis are represented by thin lines. The heavy lines show the limits of one unit cell. The right frame shows the distribution of water and lipids (the water regions are hatched, the polar groups of the lipids are represented by black dots and the hydrocarbon chains by curly lines, for a structure of type II (see text) as observed with egg lecithin.

When the sample is rotated, a series of six reflexion orders may be observed. The two first are very strong and the four others rather faint. Because of the difficulty of reproducing correctly the picture, it is just schematically drawn in figure 1. The spacing ratios of the six reflexions are : $\sqrt{3}$, $\sqrt{4}$, $\sqrt{7}$, $\sqrt{8}$, $\sqrt{10}$, $\sqrt{11}$. Because of the presence of $\sqrt{7}$, the reflexions are only in agreement with a body centered cubic lattice, and have to be indexed (211), (220), (321), (400), (420) and (332). Thus the parameter of the unit cell is $a = 102$ Å. Assuming a density of 0.89 g/cm³, this corresponds to about 1 150 molecules within the unit cell. Diele et al. [2] obtained a parameter $a = 61$ Å in indexing the first reflexion as $\overline{1}10$; if it is now indexed 211, a becomes equal to 105 Å, a value in very good agreement with our measurements when considering the size of the spots in figure 2 of reference [2].

If we assume that the high number of absent reflexions has to be explained by the symmetry group, the choice of the space group Ia3d (No. 230 of the International Tables [3]) is quite unambiguous in this case. This space group is the same as that of the cubic phase of lipid-water systems which is commonly observed between a lamellar phase and an hexagonal one. Although the number of observed reflexions is smaller here than is usually observed with lipids, the analogy of the intensity distribution [4] further suggests that in all the cases the structure is similar.

The structure of the cubic phase observed with lipids was solved by Luzzati and Spegt [5]. The structure consists in rods of finite length all identical and crystallographically equivalent, which lie along two-fold axes. These rods are linked three by three and define two infinite three dimensional networks, mutually interwoven and unconnected. The structure

is depicted in figure 2. For this class of structures in which a topological distinction can be made between the inside and the outside of the structure elements, it is proper to distinguish two types of distributions : type II if the head groups (with water if present) are localized along the rods, the paraffin chains filling the rest of the space, and type I if the paraffin chains are inside the rods and the water outside. Various examples are given in reference [4]. The choice between type I and type II in lipid-water systems can be guided by several considerations (see ref. [4]) such as by example the succession of the phases with increasing water content (type I if the succession is lamellar, cubic, hexagonal, type II if the succession is hexagonal, cubic, lamellar). An additional verification is provided by the comparison of observed and calculated intensities. Unfortunately, with the smectic D modification the small number of observed reflexions does not allow a precise crystallographic verification. Thus other experiments, using these and other compounds presenting this phase, would be needed in order to apply the criteria used with lipid systems to the smectic D system. One argument in favour of a type II structure would be that the smectic D modification is observed dry, condition in which with lipids type I structures are never observed ; also a type II structure would allow the formation of dimers or polymers by the means of hydrogen bonding between the head groups.

Miscibility experiments with dry lipids (strontium soaps and lecithin) presenting this cubic phase have been attempted without success.

4. Discussion. — In this study we have been able to establish that the smectic D modification is a cubic phase, space group Ia3d. Because of the similarity

with a cubic phase observed with lipids, we have proposed the same structure with the ambiguity that it cannot be established if it is of type I or of type II. Other experiments, with powders or with mono-crystals, using pure compounds or mixtures, will be needed to clarify this point. A precise study of the adjacent phases could also be helpful.

The *smectic D modification* is interesting in several respects. It is the first observation in Thermotropic Liquid Crystals of a phase presenting a well developed long range order in three dimensions, while remaining highly disordered at the atomic level (diffuse band at $(4.5 \text{ Å})^{-1}$) ; in this respect it is completely different from a plastic crystal. It is also the first observation in a compound which otherwise displays the classical

polymorphism of Thermotropic Liquid Crystals (S_C, S_A) of a phase typical of the polymorphism of Lyotropic Liquid Crystals. The reason for this peculiar behaviour can perhaps be related to the rough similarity of chemical structure of the compound studied and of lipids (a head group linked to a long hydrocarbon chain).

As already pointed out [2], the terminology adopted for this phase has to be reviewed as it does not fulfill the requirements [6] of a smectic phase.

Acknowledgments. — We thank Pr. Sackmann and Dr. Demus who kindly provided us with a sample of this precious compound.

References

[1] DEMUS, D., KUNICKE, G., NEELSEN, J., SACKMANN, H., Z. *Naturforsch.* **23a** (1968) 84.

[2] DIELE, S., BRAND, P., SACKMANN, H., *Mol. Cryst. Liq. Cryst.* **17** (1972) 163.

[3] *International Tables for X-Ray Crystallography*, 1 (The Kynoch Press, Birmingham) 1952.

[4] LUZZATI, V., TARDIEU, A., GULIK-KRZYWICKI, T., RIVAS, E., REISS-HUSSON, F., *Nature* **220** (1968) 485.

[5] LUZZATI, V., SPEGT, P. A., *Nature* **215** (1967) 701.

[6] FRIEDEL, G., *Ann. Phys. Chem.* **18** (1922) 273.

LIQUID CRYSTALS, 1986, VOL. 1, No. 3, 209–214

Structure of the smectic D phase

by G. ETHERINGTON, A. J. LEADBETTER and X. J. WANG†

Rutherford Appleton Laboratory, Chilton, Didcot, Oxon OX11 OXQ, England

G. W. GRAY and A. TAJBAKHSH

Department of Chemistry, Hull University, Cottingham Road, Hull HU6 7RX, England

(Received 7 January 1986; accepted 7 March 1986)

The so-called 'smectic' D phase of 4'-n-octadecyloxy-3'-cyanobiphenyl-4-carboxylic acid (one of only four materials known to exhibit this phase) has been shown unambiguously by X-ray diffraction to be characterized by a primitive cubic space group. The space group is either $P23$ or $Pm3$, and the lattice parameter a_0 is 86 Å. It is shown that data from previous studies of this phase may be re-intepreted to be consistent with these findings. In view of these conclusions it is clearly inappropriate to refer to the phase as smectic.

1. Introduction

At the present time, four materials are known to exhibit the 'smectic' D phase. These are the biphenyl carboxylic acids with the general formula

$$C_nH_{2n+1}O{-}\!\!\bigcirc\!\!{-}\!\!\bigcirc\!\!{-}COOH$$

with substituent R

where $n = 16$ or 18 and R is either NO_2 or CN. For brevity, these materials will be referred to in the remainder of this paper simply by specifying n and R, and in the light of our conclusions we shall refer to the phase simply as the D phase. This phase was first observed in $16:NO_2$ and $18:NO_2$, by Gray and co-workers in 1957, although the phase was not defined as 'smectic D' until these acids were investigated by Demus *et al.* [1]. Subsequently, Pelzl and Sackmann [2] showed that the D phase is optically isotropic, and suggested that this indicated a structure possessing cubic symmetry.

Attempts to confirm this point using X-ray diffraction have been hampered because, in the relatively high temperature range where the D phase occurs, the $16:NO_2$ and $18:NO_2$ acids deteriorate quite rapidly. The D phases of $16:CN$ and $18:CN$ first observed by Goodby and Gray [3] have now been found to be appreciably more stable, and this paper reports the first X-ray diffraction measurements on a new preparation of one of these materials ($18:CN$). This has the following phase behaviour:

$$K \underset{131°C}{\rightleftharpoons} S_C \underset{138°C}{\overset{156°C}{\rightleftharpoons}} D \underset{200°C}{\overset{201°C}{\rightleftharpoons}} I.$$

†Present address: Physics Department, Tsinghua University, Beijing, Peoples Republic of China.

The CN and NO_2 compounds are completely miscible in the D phase so that it is clear that this phase is the same for all four compounds.

Two X-ray diffraction studies of the D phase of the NO_2 acids have been reported in the literature. In each of these studies, a weak and diffuse ring corresponding to a d spacing of $\sim 4\cdot5$ Å was always observed, which presumably arises from liquid-like intermolecular ordering.

Diele *et al.* [4] examined the $16:NO_2$ and $18:NO_2$ using a flat film technique and, for the former, reported a non-regular hexagonal set of Bragg reflections at small scattering angles. They assumed a cubic unit cell and indexed these as $\{110\}$ reflections, which would correspond to the [111] direction lying parallel to the incident beam. This yielded a lattice parameter of 61 Å. A structure was postulated in which spherical aggregations of the aromatic parts of molecules are packed in a cubic lattice, while disorder between hydrocarbon chains gives rise to the diffuse scattering.

Tardieu and Billard [5], on the other hand, made powder diffraction measurements of a $16:NO_2$ sample using the Debye–Scherrer method with a rotating sample. Six reflections were observed and were indexed assuming a cubic unit cell according to the a_0/d ratios $\sqrt{3}$, $\sqrt{4}$, $\sqrt{7}$, $\sqrt{8}$, $\sqrt{10}$, and $\sqrt{11}$. Because of the presence in this series of $\sqrt{7}$, a body-centred cubic lattice was proposed (space group $Ia3d$) with the first reflection indexed as (211), giving a lattice parameter of 102 Å. By analogy with lipid–water systems (Luzzati and Spegt [6]), a model was proposed consisting of two interpenetrating infinite three-dimensional networks of interconnected rods, where the rods may represent either the positions of the aromatic groups or of the hydrocarbon chains. Tardieu and Billard appeared to favour the former option, although they did not discuss space-filling or X-ray intensities for either possibility.

2. Experiment

The sample of the $18:CN$ acid was tightly sealed in a cylindrical beryllium cell of $1\cdot0$ mm i.d. and $3\cdot0$ mm o.d. which was mounted on a modified and automated Hilger–Watts four-circle diffractometer. Diffraction patterns were measured using a multi-wire position-sensitive area detector. Data were acquired in the form of a 400×400 array with a position resolution of $\sim 0\cdot5$ mm. A similar detector system has been described in detail by Bateman *et al.* [7, 8]. With a sample to detector distance of about $1\cdot1$ m, a scattering angle range of $0\cdot5°$ to $5°$ was covered, giving a minimum d-spacing of approximately 17 Å, using $\lambda = 1\cdot54$ Å.

Initial measurements on the D phase showed a large number of reflections at all sample orientations, characteristic of randomly oriented domains but not of a good powder sample. The sample was held at a temperature of 175°C in the D phase for 3 weeks, during which time the number of domains decreased and the domain size increased. By suitably orienting the sample, it was then possible to identify particular reflections from a major domain, although the diffraction patterns still contained reflections from neighbouring domains.

3. Results

The figure shows diffraction patterns measured at two sample orientations. The diffuse ring at $d \sim 4.5$ Å is outside the range of the detector for these measurements, although it was observed in measurements made with a smaller sample to detector distance. The peaks shown can be indexed unambiguously as the (200), (210), (220),

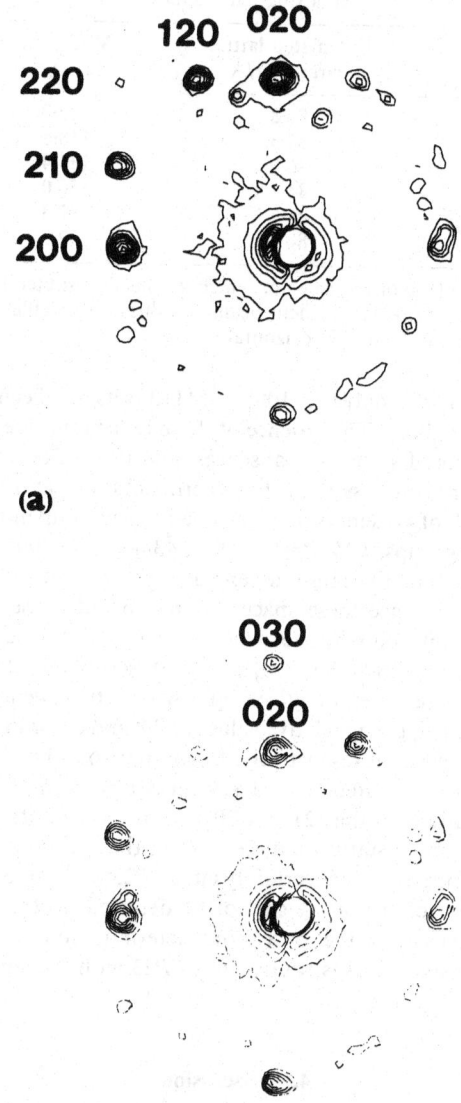

Contour map of the diffraction pattern of the D phase of the 18:CN acid, (*a*) showing (200) and (210) reflections, plus a (220) spot. (*b*) showing a (030) reflection. Contours at the centre of the pattern result from scattering by the beam stop.

(120), (020) and (030) reflections of a cubic unit cell, from a single domain. Other measurements which are not presented here show the {111} and {211} reflections from the same domain. Note that the figure does not give a true representation of relative intensities because it was taken with a fixed sample orientation chosen simply to show the maximum number of reflections. The lattice parameter is $86 \pm 2 \text{ Å}$. The calculated and observed parameters for these peaks are compared in the table.

G. Etherington *et al.*

Indexing parameters.

hkl	Measured d spacing/Å	Calculated lattice parameter/Å	Measured χ/deg	Calculated χ/deg
020	42·9	85·8	92·9	92·0
120	37·8	84·5	118·8	118·6
220	30·4	86·0	136·4	137·0
210	37·9	84·7	156·0	155·4
200	43·4	86·8	182·3	182·0
030	28·1	84·3	92·9	92·0

χ is the angle in the plane of the detector which a reflection subtends with the horizontal line intersecting the position of the incident beam. χ values are calculated assuming that the [200] direction is inclined at 2° to the horizontal.

For $(h^2 + k^2 + l^2) < 9$, only the $\{100\}$ and $\{110\}$ sets of reflections have intensities below the measurement limit. The absence of these reflections does not correspond to any of the possible sets of systematic absences, and it must be concluded that these intensities are non-zero, but less than the experimental detection limit.

Because of the lack of systematic absences, the space group must be one of the five primitive cubic space groups, $P23$, $Pm3$, $P432$, $P\bar{4}3m$ and $Pm3m$. These space groups can be further differentiated because in general the (hkl) and (khl) intensities differ only for $P23$ and $Pm3$, since these space groups do not possess four-fold axes of symmetry. The inequality only applies when $h \neq k \neq l$, so that, of the reflections measured, only the intensities of the $\{210\}$ and $\{120\}$ sets of reflections might be found to be unequal. The intensities of all except two of the twenty-four $\{210\}/\{120\}$ reflections were measured, giving relative values of 7·8 and 5·2 with standard deviations of 19 per cent and 15 per cent respectively. In comparison, the intensities of all eight $\{111\}$ reflections were measured with a standard deviation of 12 per cent. Thus although the distributions of the $\{210\}/\{120\}$ intensities are broader than would be expected from the counting statistics, it seems clear that these intensities are different and that the space group is almost certainly either $P23$ or $Pm3$. Because the presence or absence of a centre of symmetry cannot be detected on the basis of diffraction measurements (Friedel's Law), the space group determination cannot be taken further using X-ray methods since $Pm3$ is equivalent to $P23$ with the addition of a centre of symmetry.

4. Discussion

4.1. *Structural considerations*

An estimate of the density of 18 : CN indicates that a cubic unit cell with a lattice parameter of 86 Å must contain approximately 700 molecules. Since intensity measurements have been made for only nine independent reflections (including (100) and (110)), it is obvious that a complete and unique structural determination cannot be made. The number of intensity measurements is limited by the rapid decrease in intensity with increasing scattering angle, so that the amount of structural information derivable from diffraction studies is inherently limited. Therefore, in order to attempt a structural interpretation of the diffraction data, it is necessary to make some simplifying assumptions.

The simplest starting structural model is based on micelles and was first proposed in general terms by Diele *et al.* [4]. The simplest approach in the light of the findings presented here is to site micelles, made up of the aromatic regions of the 18 : CN molecules, on the equivalent positions of *P*23 or *Pm*3. Furthermore, in order for the model to predict very low relative intensities for the {100} and {110} reflections, the micelles must be sited on the special positions for which the fractional co-ordinates are given by $x = y = z = 1/4$. It is implicit in this siting that the micelles can no longer be spherical but must have three-fold point symmetry, since otherwise a space group of higher symmetry would result.

In the *P*23 space group these special positions lie at the corners of a regular tetrahedron within the unit cell, while for *Pm*3, a cubic arrangement results. For this type of model, space filling considerations indicate that *P*23 is more likely for the following reasons. The maximum (fully extended) molecular length for 18 : CN is about 35 Å, and it is highly likely that these molecules are dimerized across the carboxylic acid groups, giving a maximum dimer length of 70 Å. The latter dimension may be taken as a rough estimate of the size of a micelle. In the micellar model based on *P*23, the distance between micelle centres for a unit cell of 86 Å is 60 Å which is reasonable for interpenetrating and partially melted tails. For *Pm*3, however, the inter-micelle distance would be 43 Å, which seems unreasonably short. Detailed modelling calculations based on these considerations are in progress.

4.2. *Comparison with previous diffraction studies*

While this study has shown that the D phase of 18 : CN is almost certainly characterized by the space group *P*23 (or *Pm*3) with a lattice parameter of 86 Å, the two diffraction studies of the D phase of the 16 : NO_2 acid discussed earlier came to different conclusions. Although it may be conceivable that the space group of the 'D' phase may be different for the different acids, it is interesting to see how the data of these earlier studies can be interpreted in the light of the present findings.

4.2.1. *Diele* et al. *[4]*

Figure 2 of [4] shows six reflections, which were indexed as (110) reflections. However, an alternative indexing scheme is possible. If the [110] direction is assumed to lie in the direction of the incident beam, then two reflections may be re-indexed as (200), while the remaining four may be re-indexed as (111). This indexing has the advantage that it explains the non-regularity of the hexagonal arrangement of Bragg reflections which is apparent in the figure. As a check, *d* spacing ratios were measured from an enlargement of figure 2 of [4], and were found to be within 5 per cent of those expected from this indexing, while the angles between the (200) and (111) reflections are within 4° of the values expected. This re-indexing gives a new lattice parameter of 74·7 Å. Direct comparison with the lattice parameter determined in this study for 18 : CN is not valid, but the value is not unreasonable given the 3 Å difference in molecular lengths which corresponds to a difference in lattice parameter of about 9 Å for the simple micelle model which we have discussed.

4.2.2. *Tardieu and Billard [5]*

Tardieu and Billard indexed their power diffraction data according to the $\sqrt{(h^2 + k^2 + l^2)}$ sequence: $\sqrt{3}, \sqrt{4}, \sqrt{7}, \sqrt{8}, \sqrt{10}$ and $\sqrt{11}$. In comparison, the data presented here are indexed in the sequence: $\sqrt{3}, \sqrt{4}, \sqrt{5}, \sqrt{6}, \sqrt{8}$ and $\sqrt{9}$. Clearly, it is possible to re-index the first two reflections quoted by Tardieu and Billard as (111) and (200) but the remainder are not in agreement with our data. However, since they gave no actual data but only a schematic representation of their powder diffraction results, with no values of *d* spacings, it is not possible to resolve this point. Indexing the first and second peaks as (111) and (200), a lattice parameter of 72·1 Å is obtained, which is again a reasonable value.

Thus, when allowance is made for the difficulty in indexing poor powder diffraction data, it can be seen that the data from the two previous diffraction studies can be reinterpreted to be consistent with the results presented here.

5. Conclusions

The structure of the D phase of 4'-*n*-octadecyloxy-3'-cyanobiphenyl-4-carboxylic acid is based on a primitive cubic space group which is almost certainly either *P*23 or *Pm*3. If a model of the micelle type is assumed, then the available evidence suggests that the space group is *P*23. While diffraction techniques are not capable of differentiating between the two space groups, this can be done in principle through measurements of optical second harmonic generation on which work is now in progress, together with detailed calculations on structural models.

The D phase is now clearly established as having three-dimensional periodicity and is definitely cubic. It should not therefore be called a smectic phase but it will be convenient for identification purposes to continue to refer to it as the D phase.

References

[1] DEMUS, D., KUNICKE, G., NEELSEN, J., and SACKMANN, H., 1968, *Z. Naturf. (a)*, **23**, 84.
[2] PELZL, G., and SACKMANN, H., 1971, *Symp. chem. Soc. Faraday Div.*, **5**, 68.
[3] GOODBY, J. W., and GRAY, G. W., 1984, *Smectic Liquid Crystals* (Leonard Hill), pp. 72 and 73.
[4] DIELE, S., BRAND, P., and SACKMANN, H., 1972, *Molec. Crystals liq. Crystals*, **17**, 163.
[5] TARDIEU, A., and BILLARD, J., 1976, *J. Phys. Paris, Coll.* C3, **37**, C3–79.
[6] LUZZATI, V., and SPEGT, P. A., 1967, *Nature, Lond.*, **215**, 701.
[7] BATEMAN, J. E., CONNOLLY, J. F., STEPHENSON, R., TAPPERN, G. J. R., and FLESHER, A. C., 1983, *Nucl. Instrum. Meth.*, **217**, 77.
[8] BATEMAN, J. E., 1984, *Nucl. Instrum. Meth.*, **221**, 131.

9. STRUCTURE OF SURFACES

9.1. Pershan, P. S. and Als-Nielsen, J. (1984). X-ray reflectivity from the surface of a liquid crystal: surface structure and absolute value of critical fluctuations. *Phys. Rev. Lett.* **52**, 759–762   375

9.2. Ocko, B. M., Braslau, A., Pershan, P. S., Als-Nielsen, J. and Deutsch, M. (1986). Quantized layer growth at liquid crystal surfaces. *Phys. Rev. Lett.* **57**, 94–97   379

9.3. Pershan, P. S., Braslau, A., Weiss, A. H. and Als-Nielsen, J. (1987). Smectic layering at the free surface of liquid crystals in the nematic phase: X-ray reflectivity. *Phys. Rev.* **A35**, 4800–4813   383

X-Ray Reflectivity from the Surface of a Liquid Crystal: Surface Structure and Absolute Value of Critical Fluctuations

P. S. Pershan

Gordon McKay Laboratory, Harvard University, Cambridge, Massachusetts 02138

and

J. Als-Nielsen

Risø National Laboratory, DK-4000 Roskilde, Denmark
(Received 1 November 1983)

X-ray reflectivity from the surface of a nematic liquid crystal is interpreted as the coherent superposition of Fresnel reflection from the surface and Bragg reflection from smectic order induced by the surface. Angular dependence of the Fresnel effect yields information on surface structure. Measurement of the intensity of diffuse critical scattering relative to the Fresnel reflection yields the absolute value of the critical part of the density-density correlation function.

PACS numbers: 61.30.-v, 61.10.Fr

Although there has been considerable recent interest in both structure and critical phenomena at all types of interfaces there have been very few experimental techniques available for probing the microscopic properties of liquid-vapor interfaces.[1-10] One purpose of this manuscript is to demonstrate that, with use of a synchrotron source, x-ray reflectivity from liquid surfaces is a practical technique for obtaining structural information at the molecular level. In the particular case of the nematic liquid crystal octyloxycyanobiphenyl (8OCB) using $\lambda = 1.54$ Å, the observed reflection follows the Fresnel reflection formulas[11,12] for incident angles between grazing and $1.3°$, independent of temperature. For larger angles surface structure and Bragg reflection from surface-induced smectic order (i.e., layering) result in temperature-dependent deviations.

A second purpose of the manuscript is to demonstrate that since the x-ray reflectivity can be calculated from the Fresnel formulas,[11] the ratio of the scattering from critical fluctuations in the nematic phase of bulk 8OCB to the signal reflected from the surface yields the absolute value of the critical part of the density-density correlation function. Similarly the absolute value of the surface-induced smectic order in the nematic phase is also measured.

The experiments have been carried out at HASYLAB in Hamburg, Germany, and details of the spectrometer have been reported previously.[12,13] The geometry is shown schematically in Fig. 1. The horizontal synchrotron beam is deflected downward by an angle θ, with a spread in the vertical direction $\Delta\theta_1$, to produce the incident wave vector $\vec{k}_1$. The height of the horizontal liquid crystal surface is controlled such that the

beam strikes a fixed position on the sample. The detector assembly can be either a tilted crystal or a simple slit arrangement. In either case, if the spectrometer is tuned for specular reflection the wave vector of the detected beam, $\vec{k}_2$, is inclined upward by the same average angle θ. The detector accepts an angular spread $\Delta\theta_2$ in the vertical direction and $\Delta\psi_2$ in the horizontal out-of-plane direction. If the detector is centered on the plane of the figure the average scattering vector $\vec{Q} = \vec{k}_2 - \vec{k}_1 = \hat{z}2k\sin\theta$. Moving the detector out of the plane scans Q_x. The spectrometer records all scattered x rays for which $\vec{Q}$ is contained in the resolution volume Δ^3Q indicated by the parallelogram.

As discussed previously,[13] a consequence of

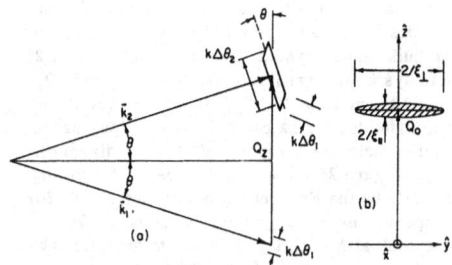

FIG. 1. (a) Schematic of the spectrometer geometry. Typically $\lambda = 1.54$ Å, $\theta \approx 1.5°$, $\Delta\theta_1 \approx 0.003°$, and $\Delta\theta_2 \approx 0.01°$ and $0.73°$ for high and low resolution, respectively. The out-of-plane angular resolution $\Delta\psi_2$ is typically $0.003°$ and $0.4°$ for high and low resolution, respectively. (b) Schematic shape of the full width at half maximum for critical scattering from the bulk nematic phase.

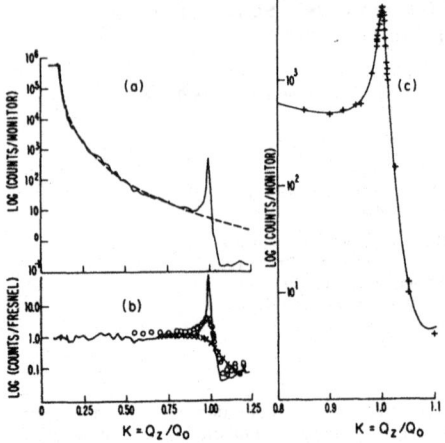

FIG. 2. (a) High-resolution scan of the reflected intensity from the surface of 8OCB; $T - T_{NA} = 0.05$ °C. The peak at $K = 1$ corresponds to $Q_z = Q_0 = 0.1989$ Å^{-1}. The dashed line is the calculated Fresnel reflection law. (b) Reflected intensity vs K, at high resolution divided by the Fresnel law. The solid line is for $T - T_{NA} = 0.05$ °C; open circles and crosses are for $T - T_{NA} = 2.8$ and 11.06 °C, respectively. (c) High-resolution scan, $T - T_{NA} = 0.40$ °C. The solid line is the fit by the model.

translational invariance parallel to the surface, i.e., in the x-y plane, is that the scattering cross section for x rays that are either specularly reflected from the flat surface, or Bragg reflected from smectic layers parallel to it, is proportional to $\delta(Q_x)\delta(Q_y)$. A spectrometer with small $\Delta^3 Q$ discriminates against diffuse scattering from the bulk liquid crystal. The solid line in Fig. 2(a) displays a high-resolution scan from $K = Q_z/Q_0 \simeq 0.07$ to $\simeq 1.25$ at $T - T_{NA} = 0.050$ °C. The dashed line indicates the calculated Fresnel law for radiation incident on a planar dielectric discontinuity. Figure 2(b) displays the result of dividing the data by the Fresnel reflection law $R_F(K)$ for temperatures $T - T_{NA} = 0.05$, 2.8, and 11.06 °C. The peak at $K = Q_z/Q_0 = 1$ is due to smectic layers localized to a depth of $\xi_\parallel(T)$ below the surface.[13] The full width at half maximum $\Delta Q_z/Q_0$ is equal to $2/Q_0\xi_\parallel(T)$, where $\xi_\parallel(T)$ is the longitudinal correlation length of the bulk nematic[14]; see Fig. 1(b). For $K \lesssim 0.9$ the reflected intensity is essentially described by the Fresnel law independent of temperature. For $K \gtrsim 1.05$ it falls below the Fresnel value. For the scan displayed in

Fig. 2(c), $T - T_{NA} = 0.40$ °C, the condition $Q_x = Q_y = 0$ was continuously monitored for each Q_z and scattering that did not obey the selection rule was subtracted. The solid line in Fig. 2(c) is the result of fitting the data by the convolution of the experimental resolution with the cross section of the model to be described below.

Standard treatments for the reflection coefficient from a surface with structure yield $R(Q_z) = R_F(Q_z)|\Phi(Q_z)|^2$, where $\Phi(Q_z) = (1/\bar{\rho})\int_{-\infty}^{\infty}(d\rho/dz)\times\exp[-i(Q_z z)]dz$.[15,16] The density is taken as the sum of two terms, $\rho(z) = \rho_0(z) + \rho_1(z)$. The sharp peak at $K \simeq 1$ is due to smectic order induced by the surface[13]; $(\bar{\rho})^{-1}d\rho_0/dz = B_s Q_0 \sin(Q_0 z)\exp(-z/\xi_\parallel)$[13] gives a Lorentzian-type term, $\Phi_0 \simeq +(B_s Q_0/2)\{+(Q_z + Q_0)^{-1} - \xi_\parallel[(Q_z - Q_0)\xi_\parallel - i]^{-1}\}$. Data of the type shown in Fig. 2(b) demand that the Fourier transform of the second term be flat for $K \lesssim 0.8$ and fall off reasonably rapidly in the vicinity of $K \simeq 1$. One form that has this property is $\Phi_1(Q_z) = 0.5\{\text{erf}[(Q_z + Q_1)/\sqrt{2}\sigma] - \text{erf}[(Q_z - Q_1)/\sqrt{2}\sigma]\}$, with $Q_1/Q_0 \approx 1$ and $\sigma/Q_0 \ll 1$. The Fourier transform of this is $(\bar{\rho})^{-1}d\rho_1/dz = (\pi z)^{-1}\exp[-0.5(\sigma z)^2]\times\sin Q_1 z$. It describes a surface of width $\sim \pi/Q_1$, followed by oscillations that decay as z increases.

The nonlinear least-squares fit shown in Fig. 2(c) yields $Q_1/Q_0 = 1.04 \pm 0.01$, $\sigma/Q_0 = 0.062 \pm 0.015$, $B_s = 0.067 \pm 0.003$, and $\xi_\parallel Q_0 = 145 \pm 3$ with a $\chi^2 = 6$. The value of $\xi_\parallel Q_0$ agrees with published values for 8OCB.[13,14] There is a previous measurement of the temperature dependence of B_s; however, this is the first measure of its absolute value.[13] Although the specific function chosen for Φ_1 is ad hoc the general form dictated by the data results in a $\rho_1(z)$ with smectic density oscillations near the surface. The sum of $\rho_0 + \rho_1$ thus gives a nonexponential decay of the smectic order. Since $B_s \xi_\parallel Q_0$ diverges as $T - T_{NA}$, the peak intensity is essentially given by $|\Phi_0|^2$, and accurate knowledge of Φ_1 is not necessary for determination of the critical properties of B_s. Asymmetry between the intensities in the wings for $K > 1$ and $K < 1$ is due to interference effects between Φ_0 and Φ_1.

A second method of taking data, in which spectrometer alignment is less critical, is to lower the resolution by increasing the height and width of the detector slit. In Fig. 1 this means increasing $\Delta\theta_z$ and also the slit width along the $\hat{x}$ direction. As the resolution volume is scanned in the vertical direction it sweeps out a strip in reciprocal space of width $\theta \times (\Delta\theta_1 + \Delta\theta_z)k$ thereby integrating over a typical value of $\Delta Q_y \cong 0.003 Q_0$. In addition to integrating over Q_x the wider slit increases the intensity of diffuse critical scattering relative to the sharp surface scattering. With neglect of $Q_\perp^4$ corrections, the cross section for diffuse critical scattering is proportional to $S_0(\vec{Q})$ $= \sigma_0 / [1 + \xi_\parallel (Q_z - Q_0)^2 + \xi_\perp^2 Q_\perp^2]$, where $S_0(\vec{Q})$ is defined to have the units of volume. The ratio of critical scattering to Fresnel reflection is obtained by integrating

$$I(\vec{Q}) = (Q^4 \beta / 32\pi^2) S_0(\vec{Q}) / (Qk^2 \cos\theta \Delta\theta_1) \qquad (1)$$

over the resolution volume $\Delta^3 Q$ defined by the slits, where $\beta = 0.2$ cm is the absorption length in 8OCB for 8-keV x rays. In general, $\xi_\parallel / \xi_\perp \sim 10$ and the half-intensity contour of $I(\vec{Q})$ has the qualitative form sketched in Fig. 1(b). As $Q_z - Q_0$ is scanned through zero with $2/\xi_\perp \gg \Delta Q_y$, there is a range for which the convolution of the resolution volume and $I(\vec{Q})$ is approximately given by $\int\int_{-\infty}^{\infty} I(\vec{Q}) \times dQ_x\, dQ_z$, essentially independent of 2θ. The spectral shape of the signal from the bulk has a flat top, falling off sharply for $|Q_z - Q_0| \gtrsim \Delta\theta_2 k / 2$. Figure 3(a) displays data observed with the low-resolution geometry for three temperatures. The solid line is a theoretical curve that results from the sum of the surface signal $|\Phi(Q)|^2$ plus the result of integrating Eq. (1) over the measured resolution volume. This interpretation is confirmed by comparison of scans in which a narrow detector slit ($\Delta\psi_2 \simeq 0.07°$) is moved out of the scattering plane, i.e., in the Q_x direction. Figure 3(b) displays a transverse scan at $T - T_{NA} = 0.796$ °C, with $K_z = 1.0$. The narrow peak at $Q_x = 0$ is due to the surface and the broader peak to the bulk. The solid line in Fig. 3(b) is obtained with use of previously measured values of $\xi_\parallel(T)$ and $\xi_\perp(T)$ [14] and integration of $S_0(\vec{Q})$ (including $Q_\perp^4$ corrections) over the resolution volume.

From the data in Fig. 3(a) and other similar data, nonlinear least-squares fits with only two adjustable parameters yield $B_s \xi_\parallel Q_0 = (0.45 \pm 0.02) t^{-(0.43 \pm 0.04)}$ and $Q_0^3 \sigma_0 = (0.039 \pm 0.002) \times t^{-1.26 \pm 0.01}$; $t = (T/T_{NA}) - 1$.[17] The temperature dependence of σ_0 is consistent with previous measurements of bulk 8OCB [14]; however, the absolute value of σ_0 has not previously been measured.

The physical significance of σ_0 is better appreciated upon taking the Fourier transform of $S(Q)$. Defining $R^2 = [x^2 + y^2 + (\xi_\perp z / \xi_\parallel)^2] (Q_0 / 2\pi)^2$ one obtains $(\bar{\rho})^{-2} \langle \delta\rho(\vec{r}) \delta\rho \rangle \simeq [\sigma_0 Q_0^3 / 4\pi^2 (Q_0 \xi_\parallel)(Q_0 \xi_\perp)] R^{-1} \times \exp(-2\pi R / Q_0 \xi_\perp) \cos(Q_0 z)$. Since $\xi_\parallel \sim t^{-0.71 \pm 0.04}$ and $\xi_\perp \sim t^{-0.58 \pm 0.04}$, the ratio $\sigma_0 / \xi_\parallel \xi_\perp \sim t^{-0.03 \pm 0.08}$ is essentially temperature independent.[14] The same is true for the five other materials for which published data exist.[18] Thus the absolute value of $\sigma_0 Q_0 / 4\pi^2 \xi_\parallel \xi_\perp = (0.26 \pm 0.01) \times 10^{-2} t^{0.03 \pm 0.08}$ is a useful measure of the absolute value of the critical part of the density-density correlation function.

To summarize, we have demonstrated the utility of x-ray reflectivity to obtain information on fluid surfaces. In particular, the $Q_z = Q_y = 0$ selection rule facilitates empirical separation of surface reflectivity from other scattering processes. Further, we have demonstrated that since the Fresnel reflection laws can be applied to x-ray wavelengths, the absolute intensity of other scattering processes can be obtained from their intensity relative to Fresnel reflection. For the liquid crystal 8OCB we have demonstrated that the interference between density oscillations localized at the surface and surface-induced smectic order that decays into the bulk as $\exp(-z/\xi_\parallel)$ is responsible for the shape of the peak at $K \simeq 1$. It is tempting to try to explain the contraintuitive temperature dependence of B_s[13] with Landau-type mean-field theory containing a term $\delta F \simeq a(t) \rho_1$ capable of inducing coherent smectic order. A naive guess might take $a \sim t$ since the surface would quench phase fluctuations that are responsible for renormalization effects in the bulk. One might then expect $\rho_0 \sim (t \xi_\parallel) \exp(-z/\xi_\parallel) \times \cos(Q_0 z)$ or $B_s \xi_\parallel Q_0 \simeq t^{-0.42 \pm 0.08}$. This is in fact, the observed temperature dependence. Finally, we have obtained a value for the critical part of $\langle \delta\rho(r) \delta\rho \rangle$.

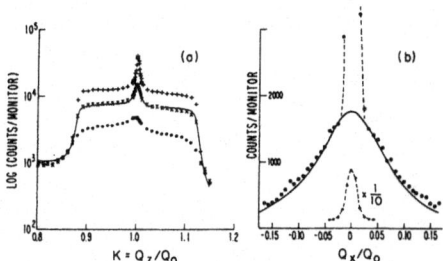

FIG. 3. (a) Low-resolution signals vs K for $T - T_{NA}$ $= 0.011$, 0.164, and 1.105 °C, represented by plusses, crosses, and circles, respectively. The solid line through the 0.164 °C data is the theoretical model discussed in the text. (b) Transverse out-of-plane scans for the same height as in (a) but with narrow width, $\Delta\psi_2 = 0.07°$.

The excellent research conditions provided by HASYLAB and the competent assistance of Risø technical staff E. Dahl Petersen, S. Bang, J. Linderholm, and J. Munck are gratefully acknowledged. Alan Braslau, Kelby Chan, and Alex Weiss assisted in some of the measurements and data analysis. This work was supported in part by grants from the Danish National Science Foundation, by the Risø National Laboratory, by the National Science Foundation through Grant No. DMR 82-12189, and by the Joint Services Electronics Program (U. S. Army, Navy, and Air Force) through Grant No. N00014-75-C-0648.

[1]E. Brezin, B. I. Halperin, and S. Leibler, Phys. Rev. Lett. 50, 1387 (1983).

[2]J. Rudnick and D. Jasnow, Phys. Rev. Lett. 48, 1059 (1982).

[3]M. Robert and C. Stuart, Phys. Rev. Lett. 49, 1434 (1982).

[4]S. Alvarado, M. Campagna, and H. Hopster, Phys. Rev. Lett. 48, 51 (1982).

[5]R. Lipowsky, Phys. Rev. Lett. 49, 1575 (1982).

[6]H. Nakanishi and M. E. Fisher, Phys. Rev. Lett. 49, 1565 (1982).

[7]C. Frank and S. E. Schnatterly, Phys. Rev. Lett. 48, 763 (1982).

[8]D. Beysens and S. Leibler, J. Phys. (Paris), Lett. 43, L133 (1982).

[9]M. P. D'Evelyn and S. A. Rice, Phys. Rev. Lett. 47, 1844 (1981); D. S. Sluis, M. P. D'Evelyn, and S. A. Rice, J. Chem. Phys. 78, 1611 (1983).

[10]J. Meunier and D. Langevin, J. Phys. (Paris), Lett. 43, L185 (1982).

[11]See, for example, G. H. Vineyard, Phys. Rev. B 26, 4146 (1982), or D. Oxtoby, F. A. Novak, and S. A. Rice, J. Chem. Phys. 76, 5278 (1982).

[12]J. Als-Nielsen and P. S. Pershan, Nucl. Instrum. Methods 208, 545 (1983).

[13]J. Als-Nielsen, F. Christensen, and P. S. Pershan, Phys. Rev. Lett. 48, 1107 (1982).

[14]J. D. Litster, J. Als-Nielsen, R. J. Birgeneau, S. S. Dana, D. Daridov, F. Garcia-Golding, M. Kaplan, C. R. Safinya, and R. Schaetzing, J. Phys. (Paris), Colloq. 40, C3-339 (1979).

[15]P. Beckmann and A. Spizzichino, The Scattering of Electromagnetic Waves From Rough Surfaces (MacMillan, New York, 1963).

[16]E. S. Wu and W. W. Webb, Phys. Rev. A 8, 2065 (1973).

[17]Quoted error estimates do not include the systematic error due to ±0.005 °C uncertainty in what we estimate to be the variation in sample temperature normal to the surface. This is most serious for large $\xi_\parallel$ as $T \rightarrow T_{NA}$.

[18]C. W. Garland, M. Meichle, B. M. Ocko, A. R. Kortan, L. J. Yu, D. Litster, and R. J. Birgeneau, Phys. Rev. A 27, 3234 (1983).

Quantized Layer Growth at Liquid-Crystal Surfaces

B. M. Ocko, A. Braslau, and P. S. Pershan

Division of Applied Sciences, Harvard University, Cambridge, Massachusetts 02138

J. Als-Nielsen

Risø National Laboratory, DK-4000 Roskilde, Denmark

and

M. Deutsch

Bar Ilan University, Ramat Gan, Israel
(Received 14 March 1986)

We report x-ray reflectivity measurements on the free surface of dodecylcyanobiphenyl (12CB) at the isotropic to smectic-A phase transition. At about 10 °C above T_{IA}, smectic-A-like ordering develops at the surface while the bulk phase remains isotropic. The angular dependence of the specular reflectivity is consistent with a sinusoidal density modulation, starting at the surface and terminating abruptly, after an integral number of bilayers. As the transition is approached the number of layers increases in quantized steps from zero to five before the bulk undergoes a first-order transition to the smectic-A phase.

PACS numbers: 64.70.Md, 68.10.−m

The phase behavior of liquid-crystal systems is strongly influenced by the proximity of interfaces. Such effects can alter the orientational ordering and phase-transition temperatures, and, in some cases, new surface phases are introduced. In the isotropic phase of a nematogen, nematic ordering can be induced by a wall or free interface. For example, on cooling of the isotropic phase of pentylcyanobiphenyl (5CB) towards the first-order isotropic to nematic phase transition temperature (T_{IN}), the penetration of the homeotropic nematic alignment induced by the surface diverges to produce an example of first-order orientational wetting.[1–3] Below T_{IN} the nematic phase completely wets the free surface. In the nematic phase, above the nematic to smectic-A transition temperature, T_{NA}, the surface induces further positional (smectic) order that penetrates exponentially into the bulk nematic.[4–5] Since the measured surface penetration length is exactly the same as the bulk correlation length and since the latter diverges as $T \to T_{NA}$, this is an example of critical wetting. In this paper we report measurements of the smectic ordering induced by the free surface in the isotropic phase of dodecylcyanobiphenyl (12CB), which does not have a nematic phase but undergoes a first-order transition from the isotropic to the smectic-A phase at $T_{IA} = 57.7$ °C. As the temperature approaches T_{IA} we observe five discrete transitions corresponding to the formation of single additional layers.

The layering transitions at the surface of a liquid-crystal phase closely resemble multilayer-physiadsorption phenomena on attractive substrates.[6–10] For these phenomena the relative strengths of the ad-atom–adatom versus the adatom-substrate energies determine whether the condensed liquid forms droplets or completely wets the interface. For the wetting case, the layering may proceed discretely via the deposition of monolayers (krypton[11] and ethylene on graphite[12, 13]) or continuously if the interface to the vapor is rough (krypton on gold[14]). The thickness of the wetting layer may diverge at coexistence (complete wetting)[8, 9, 15] or remain finite (incomplete wetting). All of the adsorbed-gas problems suffer from the complications of a periodic substrate that interacts strongly with both the liquid and solid adsorbate phases. In the latter case, strains induced by differences in the two lattice parameters exclude the possibility of complete wetting. Since there is no substrate at a free surface of a liquid, the liquid-crystal problem does not suffer from this complication. One major difference between the liquid-crystal and the adsorbed-gas problems is that in the latter the physics can generally be studied as a function of both temperature and partial pressure, i.e., chemical potential, whereas for single-component liquid-crystal samples the temperature is the only free variable. Fortuitously, in mixtures, variations in the concentrations of homologs act as an additional field variable.

These experiments have been carried out by use of synchrotron radiation at HASYLAB at DESY in Hamburg, Germany. We present the essential technical details; however, a more comprehensive review can be found elsewhere.[16] A germanium (111) monochromator crystal is set to Bragg reflect the bending magnet radiation at $\lambda = 1.54$ Å. The incident angle is adjusted by our tilting the monochromator down. This requires simultaneously adjusting the sample height. The spectrometer resolution, along the normal to the surface $\hat{z}$, as determined by the divergence of the incoming beam, $\Delta q_z < 10^{-3}$ Å^{-1}, is over an order of magnitude narrower than the width of the narrowest line shape. To account for the effects of bulk scattering and other nonspecular events, we have measured the intensity with the spectrometer misset from the specular condition. The background is structureless over the region of interest and is of the order of 2×10^{-7} of the direct beam. The liquid-crystal sample was prepared by placement of 1 g of 12CB, as received from BDH, Ltd., on a silane-treated glass disk which is enclosed in a two-stage oven. This temperature was controlled to ± 0.002 °C over several hours. The drift in T_{IA} was less than 0.001 °C/h if the temperature was kept within 1° of T_{IA}.

The experiment is carried out by measurement of the specular reflectivity as a function of the wave-vector transfer normal to the surface, $q_z' = (4\pi/\lambda) \times \sin(\alpha)$, at various temperatures in the isotropic phase of 12CB, where α is the incident angle relative to the surface. We note that the q vector in the material is $q_z = (q_z'^2 - q_c^2)^{1/2}$, where for 12CB $q_c = 0.0215$ Å^{-1} is the critical q vector for total external reflection, below which the reflectivity is close to 100%. In Fig. 1, the reflectivity at $t = (T - T_{IA})/T_{IA} = 6.3 \times 10^{-4}$ is plotted versus q_z'/q_0, where $q_0 = 2\pi/D = 0.1605$ Å^{-1} is the smectic layering q vector for 12CB. The solid line, $R_F(q_z')$, which corresponds to the theoretical reflectivity predicted from the Fresnel law of optics for a simple step-function interface, agrees very well with the displayed data for all $q_z'/q_0 < 0.50$. In Fig. 2 the reflectivity normalized to R/R_F is plotted versus q_z/q_0 at representative temperatures. At the highest temperature (Fig. 2, curve f) the data are structureless, but as the transition is approached the profile becomes more complex corresponding to an increasing number of layers. At the lowest-temperature point (Fig. 2, curve a) the temperature is only 0.010 °C above the smectic-A phase and the profile may be smeared out by a small inhomogeneity in T_{IA} or the temperature over the illuminated surface area. An essential feature of the data, over the temperature range $3 \times 10^{-4} < t < 4 \times 10^{-3}$, is the nearly complete destructive interference at q_z/q_0 slightly greater than 1.

The theoretical reflectivity for an arbitrary density

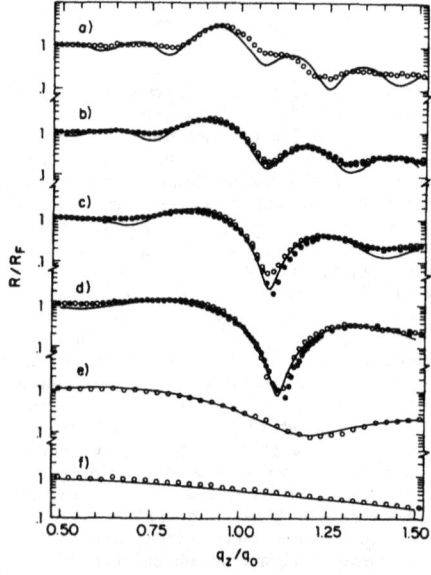

FIG. 2. The measured x-ray reflectivity in the isotropic phase, divided by the theoretical Fresnel reflectivity for a step interface, is plotted vs q_z/q_0 at curve a, $t = 3 \times 10^{-5}$; curve b, $t = 8 \times 10^{-5}$ (open circles) and $t = 1.4 \times 10^{-4}$ (filled circles); curve c, $t = 3.0 \times 10^{-4}$ (open circles) and $t = 8.3 \times 10^{-4}$ (filled circles); curve d, $t = 1.1 \times 10^{-3}$ (open circles) and $t = 3.0 \times 10^{-3}$ (filled circles); curve e, $t = 1.9 \times 10^{-2}$; and curve f, $t = 6.1 \times 10^{-2}$. The solid lines are for a density model with a sinusoidal modulation terminated after an integral number of periods: five for curve a through zero for curve f.

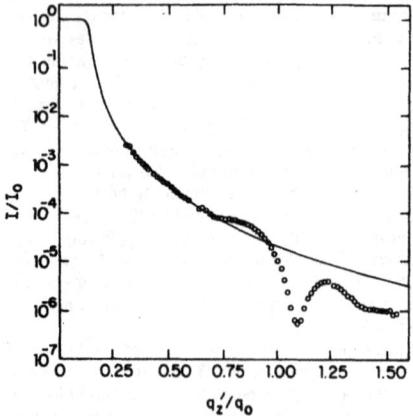

FIG. 1. The measured x-ray reflectivity from the surface of 12CB at $t = (T - T_{IA})/T_{IA} = 6.3 \times 10^{-4}$ is plotted vs q_z'/q_0 as circles. The solid line is the theoretical Fresnel reflectivity for a step-function interface.

distribution can be expressed in terms of $R_F(q_z)$:

$$\frac{R(q_z)}{R_F(q_z)} = \left| \frac{1}{\langle \rho \rangle} \int_{-\infty}^{\infty} \left\langle \frac{\partial \rho}{\partial z} \right\rangle e^{iq_z z} dz \right|^2, \quad (1)$$

in the limit of the Born approximation,[4] i.e., $q_z \gg q_c$. The data can be modeled by a density function $\rho(z)$ that is obtained by a smoothed version of the function

$$\rho(z) = \Theta(z - z_0) + H_n(z) B_s \sin(2\pi z/D), \quad (2)$$

where $\Theta(z - z_0)$ is a step function and $H_n(z) = 1$ for $0 < z < nD$ and zero elsewhere. The smoothing, due to thermal motion and finite atomic size, is represented by convolution of the two terms in $\rho(z)$ with Gaussian profiles. In Fig. 2, we show model curves (solid lines) for $n = 0$ (bottom) through $n = 5$ (top) with surface- and layer-smoothing Gaussian widths of 5.5 and 4.5 Å, respectively, a modulation amplitude $B_s = 0.12$, and a phase factor $z_0 = -0.35D$. This phase corresponds to the peak value of the sine wave being displaced from the surface by $0.6D$. The subsidiary maxima and minima above and below q_0 are a direct consequence of the sharp termination of the sine wave. For an exponential decay of the sine-wave amplitude, the reflectivity is Lorentzian in character with no subsidiary structure.[4,5]

At high temperatures $(n = 0)$ the surface is well represented by a step function convoluted by a Gaussian with a width of 5.5 Å. This is the same form which describes the microscopic surface roughness of simple liquids.[2] For finite numbers of layers, all of the key features of the data, including the intensities and widths of the primary peak and the positions of the peaks and dips, are well represented by the inclusion of a sine wave terminated after an integral number of layers. There are systematic deviations in the intensities of the subsidiary peaks and dips for $n = 3$, 4, and 5 which are not accounted for by this sharp-interface model. These deviations may result from a rough interface, although such features do not alter the determination of the number of layers.

In the bulk phase it has long been recognized that the key features of the smectic-A phase are well represented by a single sinusoidal modulation. This is a valid approximation when there are a large number of layers, i.e., a small region in reciprocal space, and because of the effects of fluctuations in the molecular centers of mass. For the present case the surface fixes the centers of mass and the relevant region in reciprocal space is large, and it is therefore possible to estimate the electron density near the surface in terms of molecular models.[17] This will be discussed more fully elsewhere[18]; however, these results justify the present model in terms of a sinusoidal modulation.

Figure 3 displays the temperature dependence of the ratio R/R_F at the average peak position $q_z' = 0.15$ Å^{-1},

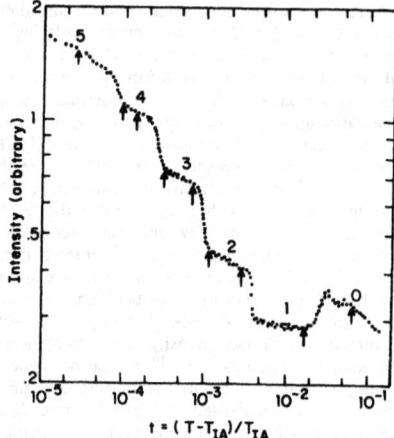

FIG. 3. The measured intensity at $q_z' = 0.15$ Å^{-1} in the isotropic phase upon heating. The quantized nature of the layering is clearly evident from the steplike nature of the intensity. The arrows correspond to the data sets plotted in Fig. 2.

obtained by slewing of the temperature. The slew rate ranged from 0.010 °C/h near the transition to 10.0 °C/h at the highest temperatures. The increase in the intensity at $t \geq 2.5 \times 10^{-2}$ is a consequence of the loss of destructive interference at $q_z' = 0.15$ Å^{-1} when the last layer melts.

The quantized nature of the layering is clearly evident from the steplike nature of the intensity shown in Fig. 3. From the data for the intensity versus t, one might expect that the first step upon cooling from the high-temperature surface phase would correspond to the formation of a single bilayer of smectic A. The analysis of the reflectivity profile supports this notion. Furthermore, each additional step corresponds to the growth of an additional bilayer. As the number of layers increases, the steplike nature of Fig. 3 becomes less distinct, and by five layers the layering transitions are no longer distinguishable. The layering transitions occur at temperatures $t_{0 \to 1} = (2.5 \pm 0.1) \times 10^{-2}$, $t_{1 \to 2} = (3.8 \pm 0.1) \times 10^{-3}$, $t_{2 \to 3} = (9.6 \pm 0.3) \times 10^{-4}$, $t_{3 \to 4} = (2.5 \pm 0.2) \times 10^{-4}$, $t_{4 \to 5} = (9.0 \pm 2.0) \times 10^{-5}$. The reduced range of each step, given by $\log(t_{n-1 \to n}/t_{n \to n+1})$, increases as the number of layers decreases.

In the de Gennes model the coupling between the nematic and smectic order parameters greatly affects the behavior of the smectic correlation function for the nematic to smectic-A transition.[19] However, for the isotropic to smectic-A transition there are no models for the smectic correlation function and there is no experimental evidence for smectic ordering in the bulk. In order to determine the effect of orientational ordering on the isotropic to smectic-A transition we have also measured the surface ordering close to the nematic–smectic-A–isotropic triple point which occurs for nearly equal mixtures of 9CB and 10CB. For 9.55CB, which corresponds to 55% by weight 10CB and 45% 9CB, the reflected intensity near q_0 decreases monotonically with temperature with no apparent signs of quantized layering. The reflectivity profiles do not exhibit subsidiary peaks centered at q_0. Therefore, we believe that the nearby nematic phase is responsible for a rougher interface between the smectic layers and the isotropic bulk. For triple-point wetting, the excess surface density, i.e., the smectic length, should diverge as $t \to 0$.[10] We do not believe that this can occur in the liquid-crystal problem since the smectic-A correlation length is always finite along the nematic to smectic-A line between the tricritical point (9CB) and the triple point (9.45CB).[20] Along this line, as the triple point is approached, the transition becomes more first order[21] and the correlation length decreases;[20] therefore, complete wetting cannot occur at the isotropic to smectic-A transition.

In conclusion, we note that although the free surface of the isotropic phase of 12CB is incompletely wetted by the smectic-A phase, the partial wetting that does occur is discrete. The density profile normal to the surface is well described by a step function plus a surface sine wave terminated after an integer number of periods, convoluted by Gaussian smoothing functions. As the smectic-A transition is approached, discrete layering transitions, up to five layers, are observed. There are no appropriate predictions which incorporate the effects of the nematic order parameter at the surface.

We would like to thank Carl Garland, Mehran Kardar, Michael Schick, and Jan Thoen for enlightening discussions; Robert Fiedenhansl, Francois Grey, and Mourits Nielsen for technical advice at DESY; and Dr. M. G. Pellatt (BDH, Ltd.) and Cyrus Safinya for providing the samples. This work was supported in part by grants from the Danish National Science Foundation, by the Risø National Laboratory, by the National Science Foundation through Grants No. DMR82-12189, No. DMR 83-16979, and No. INT-83-11841, by the Joint Services Electronic Program through Grant No. N00014-84K-0465, and by the United States–Israel Binational Science Foundation.

[1]K. Miyano, Phys. Rev. Lett. **43**, 51 (1979).

[2]D. Beaglehole, A. Braslau, M. Deutsch, P. S. Pershan, A. H. Weiss, J. Als-Nielsen, and J. Bohr, Phys. Rev. Lett. **54**, 114 (1985).

[3]Ping Sheng, in *Liquid Crystals and Ordered Fluids*, edited by Julian F. Johnson (Plenum, New York, 1984), Vol. 4, p. 889.

[4]J. Als-Nielsen, F. Christensen, and P. S. Pershan, Phys. Rev. Lett. **48**, 1107 (1982).

[5]P. S. Pershan and J. Als-Nielsen, Phys. Rev. Lett. **52**, 759 (1984).

[6]J. G. Dash, Phys. Today **38**, No. 12, 26–32 (1985).

[7]Rahual Pandit, M. Schick, and Michael Wortis, Phys. Rev. B **26**, 5112 (1982).

[8]D. A. Huse, Phys. Rev. B **30**, 1371 (1984).

[9]A. R. Lipowski, Phys. Rev. B **32**, 1731 (1985).

[10]R. Pandit and M. E. Fisher, Phys. Rev. Lett. **51**, 1772 (1983).

[11]A. Thomy and X. Duval, Appl. Opt. **21**, 1894–1895 (1982).

[12]M. Sutton, S. G. J. Mochrie, and R. J. Birgeneau, Phys. Rev. Lett. **51**, 407 (1983).

[13]M. Drir, H. S. Nham, and G. B. Hess, Phys. Rev. B **33**, 5145 (1986).

[14]J. Krim, J. G. Dash, and J. Suzanne, Phys. Rev. Lett. **52**, 640 (1984).

[15]John D. Weeks, Phys. Rev. B **26**, 3998 (1982).

[16]J. Als-Nielsen and P. S. Pershan, Nucl. Instrum. Methods **208**, 545 (1983).

[17]E. F. Gramsbergen, W. H. de Jeu, and J. Als-Nielsen, J. Phys. (Paris), to be published.

[18]J. Als-Nielsen, A. Braslau, B. M. Ocko, and P. S. Pershan, to be published.

[19]P. G. de Gennes, Solid State Commun. **10**, 753 (1972).

[20]B. M. Ocko, R. J. Birgeneau, J. D. Litster, and M. E. Neubert, Phys. Rev. Lett. **52**, 208 (1984).

[21]J. Thoen, H. Marynissen, and W. Van Dael, Phys. Rev. Lett. **52**, 204 (1984).

PHYSICAL REVIEW A   VOLUME 35, NUMBER 11   JUNE 1, 1987

Smectic layering at the free surface of liquid crystals in the nematic phase: X-ray reflectivity

P. S. Pershan, A. Braslau, and A. H. Weiss[*]

Division of Applied Sciences, Harvard University, Cambridge, Massachusetts 02138

J. Als-Nielsen

Risø National Laboratory, Postbus 49, DK-4000 Roskilde, Denmark

(Received 19 September 1986)

Smectic layering at the surface and smectic fluctuations in the bulk can be studied simultaneously by x-ray specular reflectivity and scattering measurements. There is a peak in the angular dependence of the specular reflectivity due to surface-induced smectic layering that penetrates into the bulk as $\exp(-z/\xi_{\parallel})$ where $\xi_{\parallel}$ is one of the nematic critical correlation lengths. This is illustrated by measurements of the liquid-crystal materials octyloxycyanobiphenyl (8OCB) and butyloxybenzylidene octylaniline (4O.8). In both cases the critical divergence of the peak in the specular reflectivity is weaker than $\xi_{\parallel}^2$, suggesting unexpected surface physics. The absolute value of the critical part of the density pair correlation function is determined by comparison to the Fresnel reflected intensity. Different geometries of the x-ray spectrometer are also discussed.

I. INTRODUCTION

Although x-ray reflection from material surfaces is well known, there have been relatively few attempts to use the phenomena as a general technique for characterizing surfaces.[1-8] In fact, classical techniques that have been employed at optical wavelengths are generally applicable in the x-ray range, with the advantage that the surface structure is probed on the scale of angstroms rather than microns. Current developments in the field of x-ray optics testify to the accuracy of this assertion.[9-11] In this paper we describe a series of experiments carried out at the synchrotron facility HASYLAB (Hamburger Synchrotronstrahlungslabor, Deutsches Elektronen-Synchrotron) in Hamburg, Germany, on the x-ray reflectivity from the surface of the nematic phase of liquid crystals.[12-14] The techniques developed for these experiments are generally applicable to any liquid surface.[15,16] The principle physics addressed in this paper concerns the smectic-like order induced by a free surface of a nematic liquid crystal. The important feature of the nematic to smectic-A phase transition is the critical divergence of the smectic susceptibility within the nematic phase.[17] By simultaneously measuring the surface density profile and the critical scattering from the bulk, one can extract the absolute value of the critical fluctuations.

The two materials studied, octyloxycyanobiphenyl (8OCB) and butyloxybenzylidene (4O.8), have the phase sequence isotropic to nematic to smectic A.[18,19] The transition temperatures (T_{IN}, T_{NA}) are approximately (80 °C, 67.1 °C) for 8OCB and (79 °C, 63.5 °C) for 4O.8. In both cases the isotropic-to-nematic transition is first order and the nematic-to-smectic-A is second order; however, they differ in that for 4O.8 the smectic layer spacing d is essentially equal to the length of the molecule, while for 8OCB, the molecules overlap each other to form bilayers. The layer spacing is approximately 1.6 times the length of the individual molecules.[20]

Classical treatment of the interaction between bound electrons and electromagnetic radiation at frequencies large compared to the electron binding energies obtains an x-ray dielectric constant $\epsilon = 1 - \rho r_0 \lambda^2 / \pi$, where ρ is the electron density, r_0 the classical electron radius e^2/mc^2, and λ the x-ray wavelength.[21]

Because $\epsilon < 1$, x rays incident on the surface at glancing angle ϕ_i will be refracted *toward* the surface at angle ϕ_i'. By Snell's law, $\cos\phi_i = \sqrt{\epsilon}\cos\phi_i'$. Defining a critical angle ϕ_c by $\cos^2\phi_c \equiv \epsilon$ and utilizing that the angles are small, i.e., $\phi_c^2 = \lambda^2 \rho r_0 / \pi$, an approximate alternative form of Snell's law is $\phi_i^2 = \phi_i'^2 + \phi_c^2$.

If ϕ_i is less than the critical angle ϕ_c, ϕ_i' is imaginary and, if one neglects scattering and absorption, the incident beam will be 100% reflected. If the discontinuity in the dielectric constant at the surface is sharp, the falloff in reflectivity with increasing ϕ_i can be calculated from the Fresnel formulas $R_F(\phi_i)$ of classical electromagnetic theory. Neglecting absorption, and for small ϕ_i,

$$R_F(\phi_i) = \left[\frac{\phi_i - (\phi_i^2 - \phi_c^2)^{1/2}}{\phi_i + (\phi_i^2 - \phi_c^2)^{1/2}} \right]^2$$

$$= \left[\frac{\phi_c}{\phi_i + (\phi_i^2 - \phi_c^2)^{1/2}} \right]^4 . \tag{1}$$

The effects of either a rough surface or a gradual surface profile will be discussed below. Nevertheless, whether the surface profile is sharp or not, an elementary property of classical optics is that for any flat surface the incident and reflected angles are equal to one another, and the reflected wave is in the plane defined by the incident wave vector and the surface normal. This is equivalent to the statement that if $\mathbf{k}_i$ and $\mathbf{k}_s$ are the incident and reflected wave vectors, respectively, then $\mathbf{Q} = \mathbf{k}_s - \mathbf{k}_i$ is normal to the surface. Alternatively, the scattering cross section for specular reflection from a flat surface (i.e., the x-y plane) can be described as being proportional to a two-dimensional δ function $\delta(Q_x)\delta(Q_y)$. This property of specular reflection allows it to be unambiguously separated from all other scattering processes.

The Fresnel reflectivity can be expressed in terms of the wave-vector transfer $Q \equiv 2k \sin\phi_i$ and its critical value $Q_c \equiv 2k \sin\phi_c$. For $Q \gg Q_c$, Eq. (1) becomes

$$R_F(Q) = \left| \frac{Q_c^2}{4Q^2} \right|^2 . \tag{1'}$$

For incident angles ϕ_i that are large compared to the critical angle ϕ_c, the effects of a gradual surface profile can be understood in terms of the reflection coefficient for an electromagnetic wave incident on an infinitesimally thin slab of dielectric material, as given, for example, by Warren[21]

$$\frac{\delta E}{E_0} = 4\pi i r_0 \left[\frac{\rho(z)dz}{Q} \right] \exp(iQz) , \tag{2}$$

where $\rho(z)dz$ is the number of electrons per unit area in the slab of thickness dz at height z. Since the refraction of x rays is negligible for $\phi_i \gg \phi_c$ and since the absorption is also not important for the materials and wavelengths of interest, the amplitude of the wave reflected from the graded surface is obtained by simply summing the amplitudes from successive thin layers. For a semi-infinite sample it is convenient to do one integration by parts to obtain

$$\frac{E}{E_0} = - \left[\frac{4\pi r_0 \rho(\infty)}{Q^2} \right] \Phi(Q) = - \left[\frac{Q_c^2}{4Q^2} \right] \Phi(Q) , \tag{3}$$

where $\rho(\infty)$ is the density far from interface and

$$\Phi(Q) = \rho(\infty)^{-1} \int \left[\frac{d\rho}{dz} \right] \exp(iQZ) dz . \tag{4}$$

For an infinitely sharp surface, $\Phi(Q)$ is unity and Eqs. (3) and (1') are consistent. The reflected intensity $R(\phi_i)$ is thus related to the reflectivity predicted by the Fresnel theory $R_F(Q)$ by

$$\frac{R(Q)}{R_F(Q)} = |\Phi(Q)|^2 , \quad Q \gg Q_c . \tag{5}$$

In the special case that the derivative of the surface density is a Gaussian, e.g.,

$$\rho(\infty)^{-1} \frac{d\rho}{dz} = (\sigma^2/2\pi)^{1/2} \exp(-\sigma^2 z^2/2) ,$$

$$|\Phi(Q)|^2 = \exp\left[\frac{-Q^2}{\sigma^2} \right] . \tag{6}$$

If $Q_c/\sigma \ll 1$, Eq. (6) is applicable for all Q. For $Q < Q_c$, and in the absence of either scattering or absorption, the energy cannot propagate into the bulk of the material regardless of the value of σ. In this case, the reflectivity is essentially 100%, independent of the density profile. For larger values of Q, structure, such as oscillations in the density profile due to smectic-like layers at the surface of

a nematic liquid crystal, will result in a peak in $\Phi(Q)$, and thus a peak in the observed reflectivity, for $Q \approx 2\pi/d$, where d is the smectic layer spacing.

II. SPECTROMETER CONFIGURATIONS

The schematic of an x-ray spectrometer that is capable of scanning the wave-vector transfer component Q_z perpendicular to the liquid surface is shown in Fig. 1(a). Two slits S_1 and S_3 are located at the same height, equal distances L from the sample center. A scan in Q_z is obtained by varying the incident angle ϕ_i and correspondingly the sample height $H_s = L \tan\phi_i$. We shall shortly describe in detail how the horizontal synchrotron beam can be bent downwards at any arbitrary angle ϕ_i. Here it suffices to mention that the effect is obtained by the appropriate tilting of a monochromator crystal, cf. Fig. 2. At the same time, a monochromatic beam is extracted from the "white" synchrotron radiation spectrum. If the slit S_3 is not at the same height as S_1, an in-plane transverse wave-vector component Q_y is selected. It is therefore possible by varying the incident angle, the sample height, and the slit height of S_3 to set the spectrometer for an arbitrary in-plane wave-vector transfer (Q_y, Q_z) and still maintain a horizontal liquid surface as demanded by gravity. S_3 can also be moved horizontally, that is, perpendicular to the drawing plane of Fig. 1(a), and a general three-dimensional wave-vector transfer $Q = (Q_x, Q_y, Q_z)$ is determined.

The resolution implied by finite slit openings is illustrated in Fig. 1(b). The central ray of incident and scattered wave vectors k_i and k_s is shown. The uncertainty $\delta\phi_i$ for the incident angle implies that the end point of possible incident wave vectors is distributed along the vector X_i with a certain characteristic width of $k\delta\phi_i$. Similarly, the end point of possible scattered wave vectors k_s are located along X_s and the actual wave-vector transfers are distributed within the shaded parallelogram spanned by X_i and X_s. To be more precise, if the k_i and k_s distributions are boxlike, that is, constant along X_i and X_s, respectively, and zero outside, then any wave-vector transfer to a point inside the parallelogram will be accepted by the spectrometer, while points that fall outside will be rejected. Alternatively, one might assume Gaussian distributions along X_i and X_s for k_i and k_s. In that case, the spectrometer resolution function that describes the acceptance probability for various wave-vector transfers can be described by a two-dimensional Gaussian characterized by equal probability ellipses, where X_i and X_s are conjugate diameters.[22]

The parallelogram in Fig. 1(b) indicates a long, thin resolution function corresponding to a situation where $\delta\phi_s \gg \delta\phi_i$, obtained by relaxing the slit height of S_3. Our experimental results showing perfect smectic layering at the surface imply that the corresponding cross section is a δ function in the transverse components Q_x and Q_y. The effective Q_z resolution for specular reflection is therefore not the total length of the parallelogram (AA') but rather the section indicated by (δ_z) in Fig. 1(d).

In the bulk nematic phase, critical smectic fluctuations with correlation ranges $\xi_\perp$ and $\xi_\parallel$ occur deep below the

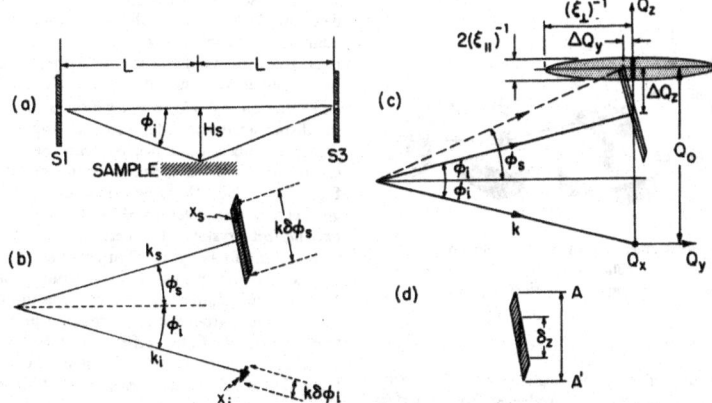

FIG. 1. (a) Schematic illustration of the spectrometer. When the incident angle ϕ_i is varied, the sample height H_s is adjusted so $H_s = L \tan\phi_i$. With S_1 and S_3 at the same height, a ϕ_i scan is equivalent to a longitudinal scan with the wave-vector transfer normal to the liquid surface. (b) The angular resolution $\delta\phi_i$ and $\delta\phi_s$ for the incident and scattered beams $\mathbf{k}_i$ and $\mathbf{k}_s$, respectively, obtains the wave-vector resolution depicted by the two heavy lines $X_i = k\delta\phi_i$ and $X_s = k\delta\phi_s$. The convolution of the two obtains a resolution volume in reciprocal space with the cross-sectional area represented by the shaded parallelogram. (c) The kinematics of the low-resolution spectrometer. The shaded ellipse represents the half height contour for the critical scattering from the bulk nematic liquid crystal. The specular reflection is confined to the vertical Q_z axis and the heavy line that is centered on the ellipse indicates the position in reciprocal space of the peak due to the surface layers. Even though the incident and scattered angles are not equal ($\phi_i \neq \phi_s$), the finite resolution allows specular reflection. The spectrometer settings for this drawing correspond to the smallest value of Q_z for which the critical bulk scattering can be detected. The center of the resolution volume is at $Q_y = 0$, $Q_z = 2k \sin(\phi_i) = Q_0 - \Delta Q_z$ and the end is at $Q_y = -\Delta Q_y$, $Q_z = Q_0$. (d) Detail of the resolution volume. The intersection of the resolution volume with the Q_z axis, i.e., δ_z, is the effective longitudinal resolution for specular reflection.

surface but still within the irradiated volume of the sample. The cross section for this bulk scattering is indicated in Fig. 1(c) by the elliptical, half-height contour ellipse with diameters $\xi_{\parallel}^{-1}$ and $\xi_{\perp}^{-1}$ along the Q_z axis and Q_y axis, respectively. This illustration is for the case when the spectrometer is set for $Q_y = 0$. The intensity of bulk

scattering is then proportional to the overlap of the resolution function and the bulk cross-section ellipse.

The third wave-vector transfer dimension, perpendicular to the plane of the figures is readily included in the discussion. From this the following occurs.

(i) Relatively fine Q_z resolution for specular reflection may be obtained even if the slit height of S_3 is relaxed so long as the illuminated portion of the surface is sufficiently flat.

(ii) The ratio of specular reflection to bulk scattering depends on the resolution and improves with narrow resolution.

(iii) The bulk scattering intensity may be determined separately by setting the spectrometer at a finite value of Q_y (or Q_x) such that the signal for specular reflection at $Q_x = Q_y = 0$ is outside of the resolution volume. By scanning Q_y (or Q_x) at fixed Q_z, one may by extrapolation obtain the bulk intensity contribution at zero transverse wave-vector transfer and thereby separate surface from bulk scattering. An example is shown below in Fig. 14(b). Since the Q_y resolution is always very high, one can also subtract off nonspecular reflection by scanning along Q_z at a very small offset Q_y.

The bending of the "white" incident beam downward to any arbitrary glancing angle ϕ_i is illustrated in Fig. 2. The white, horizontal synchrotron beam is incident from the left on the slit S_1. The box behind S_1 indicates an

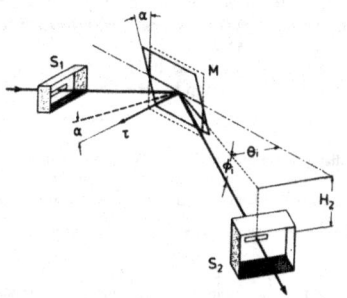

FIG. 2. The incident beam is bent downwards by the angle ϕ_i by tilting the monochromator M by the amount α. The "boxes" behind S_1 and S_2 indicate the ionization chambers monitoring the beam intensity.

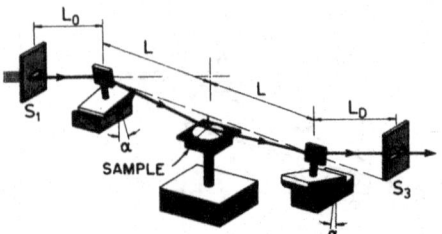

FIG. 3. A spectrometer configuration having a tilted analyzer crystal is used to obtain a resolution function in which the contributions from input and output sides are equal.

ionization chamber for monitoring the beam intensity after S_1. The beam is then Bragg reflected from a perfect Si or Ge crystal in the (111) or (220) orientation. The reciprocal-lattice vector τ, normal to the reflecting planes and of magnitude $2\pi/D$, D being the lattice spacing, is at angle α with horizontal. The Bragg scattered beam is consequently bent downwards by an amount $\phi_i \simeq (\tau/k)\alpha$, where $k = 2\pi/\lambda$ is the magnitude of the wave vector. The Bragg angle is composed of the horizontal part θ_i and the vertical part ϕ_i. In principle, this Bragg angle, and thereby the wavelength, varies if θ_i is kept fixed and ϕ_i is varied. Since $\theta_i \gg \phi_i$ in practice, the wavelength shift is very small and since the synchrotron spectrum is smooth, the corresponding intensity variation is negligible. This is monitored by the ionization chamber after a second slit S_2. In principle, S_2 can be wide open since the angle ϕ_i is completely determined by the tilt angle α and $\delta\phi_i$ is determined by the heights of S_1 and the electron beam 20 meters upstream. However, it is convenient to have a narrow slit S_2 just in front of the sample to provide a well-defined beam "footprint" on the sample surface and also because, in practice, the height setting of H_2 gives a more accurate determination of ϕ_i than the goniometer setting α. A slight error in α does not matter in that case because the counting time is determined by the S_2-monitor rate. In the measurements reported here, S_2 was left wide open. The transverse, out-of-plane (i.e., horizontal) resolution is determined by the smaller of the two slit *widths* of S_1 and S_2, by the electron beam width (which are all of the order of 1 mm), and by the large distance (20 m) from the

slits to the source. If the direction of the detected beam is determined solely by the slit S_3 shown in Fig. 1, then the dominant contribution to the Q_x resolution will be derived from the widths of the footprint on the sample surface together with the slit width of S_3 relative to the distance L from sample to S_3, which is much shorter than the distance from sample to source. However, if a narrow Q_x resolution is desirable, one can use a tilted analyzer crystal similar to the monochromator crystal, as shown in Fig. 3.[13] The Q_x resolution in this nondispersive geometry is then determined by the Darwin widths of the two perfect crystals. Furthermore, if the Q_x resolution should not only be narrow but also have a very steep falloff (which may be useful in separating bulk scattering from surface scattering), one can use multibounce crystals.[23] Here a groove is cut through a perfect crystal and three successive Bragg reflections take place, two from one face at the entrance and exit of the groove and one from the other face in the middle of the length of the groove.

Typical dimensions for both high resolution using an analyzer crystal and low resolution using a slit to define the scattered beam direction are given in Tables I and II.

III. EXPERIMENTAL RESULTS

A. 8OCB high-resolution experimental results

The surface reflectivity of octyloxycyanobiphenyl (8OCB) was measured using the high-resolution spectrometer described above. Figure 4 is a schematic drawing of the liquid-crystal oven. The samples were deposited on the surface of a glass flat, approximately 57 mm in diameter, that had been treated with a surfactant, N,N-dimethyl-N-octadecyl-3-aminopropyl-trimethoxysiyl chloride (DMOAP), in order to insure alignment of the long molecular axis normal to the glass surface.[24] The amount of material (1 g) was chosen such that when heated into the isotropic phase it would form a broad flat drop, approximately 50 mm in diameter and 0.2 mm deep. By repeated slow cooling and heating through the nematic-to-smectic-A transition it was possible to obtain a surface that is flat over most of the sample. The x-ray reflectivity measurements to be described below are the best demonstration of this result.

The glass flat is mounted inside a sealed copper can with a glass window on top to view the sample and beryl-

TABLE I. Spectrometer parameters (high-resolution parameters). The distance, defined in Fig. 3, between the slit S_1 and the monochromator $L_0 = 140$ mm for all configurations. In addition, for configurations 12.83 there was a slit 0.3 mm high by 20 mm wide, 3500 mm before slit S_1. ^{3}Si refers to a channel-cut crystal with three successive reflections. All slit dimensions are expressed as height times width.

	S_1 (mm × mm)	Monochromator and analyzer crystals	θ_0 (degrees)	λ (Å)	L (mm)	L_D (mm)	S_3 (mm × mm)
6.82	0.02 × 2	^{1}Ge(111)	13.642	1.540	575	600	1.0 × 10
7.83	0.1 × 0.7	^{3}Si(111)	14.221	1.537	545	545	0.1 × 3.6^a

^{a}Effective width due to width of channel in analyzer crystal.

TABLE II. Spectrometer parameters (low- and intermediate-resolution parameters). See Table I for definitions.

Measurement	S_1 (mm×mm)	Monochromator	θ_0 (degrees)	λ (Å)	L (mm)	S_3 (mm×mm)
8OCB 12.82	0.1×0.7	[1]Si(111)	14.211	1.536	575	7 ×4
8OCB 3.83	0.1×0.2	[3]Si(440)	23.651	0.768	575	4 ×0.2
	0.1×0.7	[3]Si(220)	14.257	0.943	575	4 ×0.2
	0.1×0.7	[3]Si(220)	23.651	1.536	575	4 ×0.2
4O.8 12.83	0.1×0.7	[3]Si(111)	14.221	1.537	620	0.2×0.7

lium or Kapton x-ray windows on the oven sides. Visible inspection of the isotropic-to-nematic phase transition, together with x-ray measurements of critical phenomena, indicate that the sample temperature was homogeneous to better than a few millikelvin. The temperature of the oven was controlled by an electronic feedback circuit and varied by less than 10 mK over periods of the order of hours.

The solid line in Fig. 5(a) is the x-ray reflectivity of an 8OCB sample at 50 mK above the nematic-to-smectic-A transition temperature T_{NA} as a function of normalized wave vector $K = Q_z/Q_0$ measured with the high-resolution spectrometer listed as "6.82" in Table II. The quantity $Q_z = (4\pi/\lambda)\sin\phi_i$ and $Q_0 = 2\pi/d$, where $d = 31.6$ Å, is the layer spacing in the smectic-A phase of 8OCB. As mentioned above, with high resolution the only significant intensity to be detected is when the spectrometer is aligned exactly for specular reflection. Since the precision of the various goniometers is such that tilt of either the monochromator or analyzer was invariably accompanied by slight rotations compared to Darwin widths, each of these scans was taken by repeated optimization of the analyzer rotation and sample height for varying values of Q_z.

The dashed line in Fig. 5(a) is the Fresnel reflectivity $R_F(Q)$ for a flat surface with an abrupt discontinuity in the dielectric constant. The absolute value of the Fresnel reflectivity was measured to be greater than 90% below the critical angle ϕ_c. In calculating $R_F(Q)$, absorption was included by considering a complex dielectric constant: The real part is given by

$$\epsilon' = 1 - \left[\frac{\rho\lambda^2}{\pi}\right] \cdot \left[\frac{e^2}{mc^2}\right] \tag{7}$$

and the imaginary part ϵ'' can be expressed in terms of the adsorption length l

$$\epsilon'' = \frac{\lambda}{2\pi l} . \tag{8}$$

For small $\phi_i = \phi$ the Fresnel reflectivity, Eq. (1), is then given by[25]

$$R_F(\phi) = \frac{(\phi-\beta)^2 + (\epsilon''/2\beta)^2}{(\phi+\beta)^2 + (\epsilon''/2\beta)^2} , \tag{9}$$

where

$$\beta^2 = \frac{1}{2}\{\phi^2 - \phi_c^2 + [(\phi^2-\phi_c^2)^2 + (\epsilon'')^2]^{1/2}\}$$

$$\text{for } \phi > \phi_c \tag{10a}$$

or

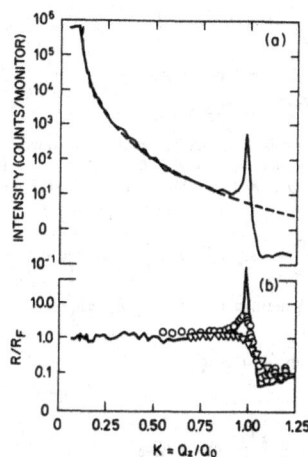

FIG. 5. (a) High-resolution longitudinal scan of the reflected intensity from the surface 8OCB at $T - T_{NA} = 0.05\,°C$. The Q_z scale is normalized to $Q_0 = 0.199$ Å^{-1}. The dashed line is the calculated Fresnel reflection law. (b) Reflected intensity vs Q_z/Q_0 divided by the Fresnel law. The solid line for $T - T_{NA} = 0.05\,°C$, open circles for $T - T_{NA} = 2.8\,°C$, and triangles for $T - T_{NA} = 11.06\,°C$.

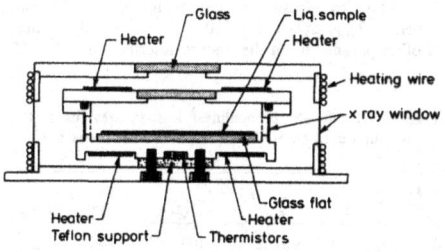

FIG. 4. Side view of the liquid-crystal oven used for reflectivity studies. The walls were made of copper. The sample covers a 2-in-diam area and the thickness is typically 0.2 mm.

$$\beta^2 = \frac{1}{2}\frac{(\epsilon'')^2}{\phi_c^2 - \phi^2 + [(\epsilon'')^2 + (\phi_c^2 - \phi^2)^2]^{1/2}}$$

$$\text{for } \phi < \phi_c . \quad (10b)$$

The effects of absorption are only noticeable very near to the critical angle ($\phi \approx \phi_c$). The solid line in Fig. 5(b) displays the result of dividing the measured reflectivity of Fig. 5(a) by R_F using $\phi_c = 0.152°$ and $\epsilon'' = 1.16 \times 10^{-8}$. The open circles and the triangles display similar results for $T - T_{NA} = 2.8$ and $11.06°$C, respectively.

The half width at half maximum of the temperature-dependent peak at $Q_z = Q_0$ or $K = 1$ is found to be equal to $1/\xi_\parallel(T)$, where $\xi_\parallel(T)$ is the longitudinal correlation length of the critical fluctuations as previously measured in bulk 8OCB.[18] The combination of this, together with the fact that the signal is only observed for the precise specular reflection conditions, shows that the peak is due to smectic layers parallel to the sample surface and penetrating into the bulk of the nematic phase, one bulk correlation length.

It is interesting to note that for $Q_z < 0.8Q_0$ the reflectivity is equal to the Fresnel value independent of temperature. Furthermore, for $Q_z > Q_0$, the reflectivity falls below the Fresnel value rather sharply, with only a slight temperature dependence. The reflectivity line shape, modeled by assuming the average of the electron density over the x-y plane, can be written as the sum of two terms: $\rho(z) = \rho_0(z) + \rho_1(z)$. The intuitively obvious form for the first term is taken to be

$$\frac{1}{\rho(\infty)}\frac{d\rho_0}{dz} = \frac{\partial}{\partial z}\Theta(z - z_0)B_s \exp\left[-\frac{z - z_0}{\xi_\parallel(T)}\right]$$

$$\times \sin[Q_0(z - z_0)] , \quad (11)$$

where

$$\Theta(z - z_0) = \begin{cases} 0, & \text{if } z < z_0 \\ 1, & \text{if } z > z_0 . \end{cases}$$

A second term is necessary to describe the transition from the Fresnel reflectivity for $Q < Q_0$ to the lower reflectivities at $Q > Q_0$. A form which represents the data very well is

$$\frac{1}{\rho(\infty)}\frac{d\rho_1}{dz} = C_1(\pi z)^{-1}\exp\left[-\frac{(\sigma z)^2}{2}\right]\sin Q_1 z , \quad (12)$$

where C_1 is defined to ensure $\int \rho^{-1}d\rho_1/dz = 1$. It follows that

$$\Phi(Q) = \Phi_0(Q) + \Phi_1(Q) , \quad (13)$$

where

$$\Phi_0(Q) = +i\left[\frac{B_s}{2}\right]\exp(-iQz_0)$$

$$\times \left[\frac{\xi_\parallel Q_0 - 1}{(Q_z - Q_0)\xi_\parallel - i} - \frac{\xi_\parallel Q_0 + 1}{(Q_z + Q_0)\xi_\parallel + i}\right]$$

$$(14)$$

and

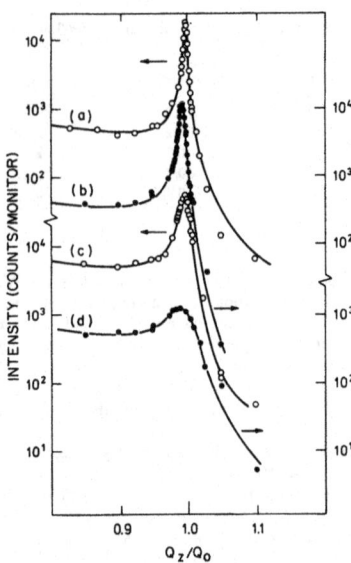

FIG. 6. High-resolution, specular reflectivity data for 8OCB near the peak due to the surface layers. The open circles refer to the left scale and the solid circles to the right scale. The solid lines are calculated from the model described in the text with the parameters in Table III. The temperatures $T - T_{NA}$ are (a) 0.10°C, (b) 0.21°C, (c) 0.40°C, and (d) 1.80°C.

$$\Phi_1(Q) = \left[\frac{C_1}{2}\right]\left[\text{erf}\left[\frac{Q + Q_1}{\sqrt{2}\sigma}\right] - \text{erf}\left[\frac{Q - Q_1}{\sqrt{2}\sigma}\right]\right] ,$$

$$(15)$$

with $C_1^{-1} = \text{erf}(Q_1/\sqrt{2}\sigma)$.

The solid lines in Fig. 6 represent the results of nonlinear least-square fits of the measured results for four different temperatures to the above functional form convoluted with a Gaussian resolution function. The full width at half maximum (ΔQ_z) of the resolution function was obtained by measuring the peak line shape near, but just below, T_{NA}, $\Delta Q_z = (1.7 \times 10^{-3})Q_0$. The temperatures and other parameters in the fits are described in Table III.

TABLE III. Results of nonlinear least-squares fits to the high-resolution data on 8OCB. The parameter z_0 was fixed at $0.25d$.

$T - T_{NA}$ (°C)	B_s	$\xi_\parallel Q_0$	Q_0/σ	χ^2	Q_1/Q_0
0.10	0.043	450 ±30	100 ±80	10	1.001
0.21	0.065	275 ±15	18 ±7	5	1.05
0.40	0.067	145 ±7	16 ±8	7	1.04
1.80	0.099	45.4±7	7.7±1.8	3	1.03

The best fit values of $\xi_\parallel(T)$ agree, within experimental error, with values previously published[18] for the critical correlation length in bulk 8OCB. The value for z_0, the origin of the exponentially damped oscillations, was held fixed at $0.25d$. In real space this value of z_0 locates the maxima in ρ_0 exactly $d/2$ away from the surface as defined by the maxima in $d\rho_1/dz$. Nonlinear least-squares fits in which z_0 is allowed to vary freely converge around $z_0 \approx 0.25d$; however, the fact that the B_S and z_0 are highly correlated, results in random variations that we do not believe to be meaningful. In fact, $z_0 \approx 0.25d$ is necessary to obtain asymmetries in the peak line shapes that are shown in Fig. 6. An example of a best fit keeping z_0 fixed at $0.5d$ is shown in Fig. 7. Clearly $z_0 = 0.25d$ is a better choice.

This particular choice for the representation of ρ_1 emerged as the simplest parametrization of a model which could reproduce both the agreement with the Fresnel law for $Q/Q_0 < 0.8$ and the falloff at larger Q. In particular, the simple Gaussian form that was used to represent either the structureless surface of water[15] or the Φ_1 term used to describe the surface of the isotopic phase of dodecylcyanobiphenyl (12CB) (Ref. 26) does not provide a good fit to the present data. In the absence of a more basic model, the above form has the advantage that it provides a quantitative representation of the fact that the exponentially decaying ρ_0 term described by Eq. (11) does not accurately describe the smectic oscillations near the surface. Furthermore, we believe that the form for ρ_1 given by Eq. (12) does indeed provide an accurate representation of the quantative features of the physical surface.

In view of the fact that the falloff in $\Phi_1(Q)$ occurs in the vicinity of $Q \approx Q_0 \approx Q_1$ its precise shape is partially obscured by the peak in $\Phi_0(Q)$. In particular, without precise measurements in the tails of $\Phi(Q)$, the best that

can be done is to set a lower limit to the Gaussian penetration length, $1/\sigma$. Furthermore, where the peak is relatively intense, the numerical results for B_s and $\xi_\parallel$ are insensitive to the precise value of σ. Figure 8 displays the values of Q_0/σ as obtained from fits to the data sets described in Table III. In addition, we also display results from other data sets that were less complete than the above (and thus not otherwise included here) but which were adequate to set lower limits on Q_0/σ. One can see from Fig. 8, that although the Gaussian penetration length $1/\sigma$ is continually driven to larger and larger values as T approaches T_{NA}, its precise value is rather uncertain. The solid line is the best fit to the exponential correlation length $\xi_\parallel Q_0$ as obtained from critical scattering from the bulk of 8OCB.[18] Since the falloff of ρ_1 in real space is dominated by the Gaussian form, while the falloff in ρ_0 is governed by the longer-range exponential, the density oscillations described by ρ_1 are considerably more localized at the surface than those described by ρ_0. Nevertheless, as T approaches T_{NA}, the ρ_1 oscillations extend deeper and deeper into the sample.

The χ^2 values reported in Table III were based on the assumption that the statistical weight for each point was given by Poisson statistics. The fact that the χ^2 values for these fits are significantly larger than unity is a quantitative demonstration that either the empirical form is not strictly correct, or that there are other statistical uncertainties beyond the counting statistics. Both of these are likely to be true. Nevertheless, in view of the fact that very small systematic errors will make sizable contributions to data sets like this, which encompass large dynamic ranges, these values for χ^2 are not unreasonable.

This particular model, while not necessarily unique, does illustrate our contention that the only way to obtain a functional form that will be flat out to $Q_z \sim Q_0$ and then fall off rather abruptly for $Q_z > Q_0$ is to include spatial

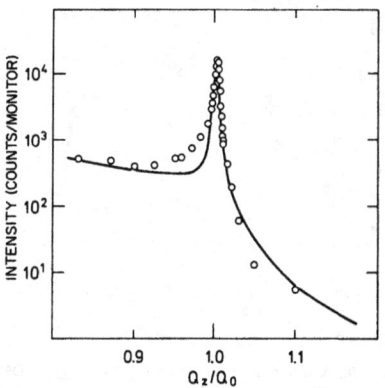

FIG. 7. Effect of varying the phase of the two terms that make up $\Phi(Q)$. The data are the 8OCB data in Fig. 6(a) but the solid line is the best fit, keeping z_0 fixed at a value of $0.5d$.

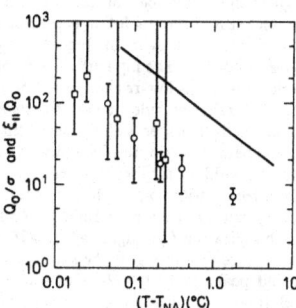

FIG. 8. The Gaussian penetration length Q_0/σ for 8OCB as a function of $T - T_{NA}$. The open circles are the results of the fits to the data shown in Fig. 6. The open squares are from less complete data sets from which it was not possible to obtain upper error limits. The solid line represents the published values for $\xi_\parallel Q_0$ (Ref. 18).

oscillations in the electron density that are localized to the vicinity of the surface. The model-independent physical significance of this is that the exponential decay included in the $\Phi_0(Q)$ term breaks down in the vicinity of the surface. Furthermore, since the surface structure responsible for the principle features of the model for $\Phi_1(Q)$ persist to much higher temperatures than the critical phenomena associated with the bulk, one cannot ignore the possibility of surface phases with separate critical properties from that of the bulk.

B. 8OCB low resolution

We shall now consider the situation when the resolution for the scattered radiation is relaxed by removing the analyzer crystal and using a slit S_3 with a rather open aperture (7 mm high by 4 mm wide, see Table II, entry 12.82). First we shall give a qualitative discussion then present some data, and finally explain how the data are analyzed quantitatively.

Figure 1(c) shows the scattering diagram in reciprocal space. The heavy lines, downwards and upwards at angle ϕ_i represent the incoming and the specular reflected wave vectors, respectively. The high slit height Y_3 of slit S_3 at distance L from the sample implies that the dashed outgoing wave vector at angle $\phi_s = \phi_i + \Delta\phi_s$ will also be detected, $\Delta\phi_s \approx Y_3/2L$. Since the scattering is elastic, this process has a wave-vector transfer component in the Q_y direction, ΔQ_y, as well as the z component ΔQ_z added to the central ray. Inspection of Fig. 1(c) immediately leads to

$$\frac{\Delta Q_z}{Q_0} \approx \left[\frac{k}{Q_0}\right]\Delta\phi_s \approx \frac{\Delta\phi_s}{2\phi_0} , \tag{16}$$

$$\frac{\Delta Q_y}{Q_0} \approx \phi_i \frac{\Delta Q_z}{Q_0} \approx \frac{\Delta\phi_s}{2} , \tag{17}$$

since ϕ_0 is defined such that $2k\phi_0 \approx Q_0$.

The short side of the resolution parallelogram is given by the angular divergence $\delta\phi_i$ of the incident beam as $k\delta\phi_i$. The thickness of the resolution parallelogram along the z axis is thus indicated in Fig. 1(d) by $\delta_z/Q_0 = 2(k/Q_0)\delta\phi_i \approx \delta\phi_i/\phi_0$. The high effective resolution for specular reflection in this low-resolution geometry is quite remarkable. The selection rule that confines specular reflection to the z axis, together with the small tilt ϕ_0 of the long resolution parallelogram, results in a width δ_z/k that is only twice the width of the incident beam $\delta\phi_i$.

There are three contributions to the scattering cross section: simple specular reflection confined to the z axis and decaying with increasing Q_z roughly as $(Q_c/2Q_z)^4$, reflection from the smectic layering at the surface confined to the z axis and peaking at $Q_z = Q_0$ with a half width of $(\xi_{\parallel})^{-1}$, and finally the scattering from critical smectic fluctuations in the bulk peaking at $Q_z = Q_0$, $Q_y = 0$, with half widths $(\xi_{\parallel})^{-1}$ and $(\xi_{\perp})^{-1}$ along the z and x,y directions, respectively. It is evident from Fig. 1(c) that this latter contribution is expected to be roughly constant in a Q_z scan for $|Q_z - Q_0| < \Delta Q_z$, provided that $\Delta Q_y \ll (\xi_{\perp})^{-1}$, and rapidly vanishes outside this interval.

Experimental data obtained with the conditions listed in Table II at entry 12.82 are displayed in Fig. 9. From Table II one obtains the values of $\Delta Q_z = 0.122 Q_0$, $\Delta Q_y = 0.0030 Q_0$, and $\delta_z = 0.004 Q_0$. The data in Fig. 9 were taken at $T - T_{NA} = 0.164\,°C$. From either the published data for $\xi_{\parallel}(T)$ (Ref. 18) or the high-resolution scans like that in Fig. 6, $\xi_{\parallel}Q_0 = 250$. Similarly, the published data obtain $\xi_{\perp}Q_0 = 29$.[18] Thus we have the necessary condition for the critical scattering from the bulk to be constant, i.e., $\Delta Q_y\xi_{\perp} \ll 1$, and we see that the data in Fig. 9 have the expected flat plateau in the interval $|Q_z/Q_0 - 1| < \Delta Q_z/Q_0 = 0.122$. The peak due to the surface layers at $Q_z/Q_0 = 1$ also has the full width at half maximum of $2(\xi_{\parallel}Q_0)^{-1} = 8 \times 10^{-3}$ that is expected when $\delta_z \ll 2/\xi_{\parallel}$. Finally, these two contributions are superimposed on a sloping "Fresnel background" as indicated by the dashed line calculated from Eq. (1).

For this spectrometer, as for the high-resolution spectrometer, the relative intensity between the surface peak at $Q_z/Q_0 = 1$ and the Fresnel background measures the amplitude B_s. However, now the critical scattering from the bulk obscures the part of the specular reflectivity from which it would be possible to determine the Gaussian penetration depth σ. In analyzing data such as those in Fig. 9 for different temperatures, we therefore estimated σ from Fig. 8 and used the fact that the exponential penetration length is equal to $\xi_{\parallel}(T)$, which is available from the published literature. Nonlinear least-squares fits were then carried out for the data in the vicinity of the peak, with the only free parameter being B_s. Best-fit values of B_s versus temperature are listed in Table IV and plotted in Fig. 10 along with the high-resolution results from Table III. The solid line is the best power-law fit to the displayed data: $B_s = 0.81(T - T_{NA})^{(0.28 \pm 0.03)}$.

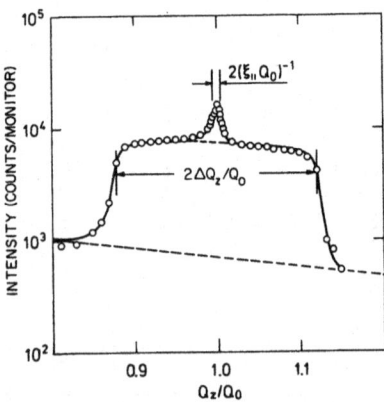

FIG. 9. A longitudinal scan (i.e., a Q_z scan) from 8OCB at $T - T_{NA} = 0.164\,°C$ with the low-resolution spectrometer illustrated in Fig. 1(c). The lower dashed line is the calculated Fresnel reflection law normalized to the intensity of $Q_z/Q_0 = 0.8$. The expected widths $\xi_{\parallel}$ and ΔQ_z are indicated.

PERSHAN, BRASLAU, WEISS, AND ALS-NIELSEN

TABLE IV. The results of nonlinear least-squares fits to low-resolution data for 8OCB. The correlation lengths $\xi_\parallel$ and $\xi_\perp$ were taken from Ref. 18 for bulk 8OCB. Since the critical scattering form the bulk masks the falloff in $\Phi_1(Q)$ near $Q \simeq Q_0$, the fit is insensitive to the value of the Gaussian penetration length. Approximate values for Q_0/σ are taken from Fig. 8.

$T - T_{NA}$	$\xi_\parallel Q_0$	$\xi_\perp Q_0$	$\sigma_0 Q_0^3$	A [a]	Q_0/σ	B_s
0.011	1746	142	1.07×10^4	0.174×10^{-2}	10^3	0.022 ± 0.001
0.017	1279	110	1.01×10^4	0.182×10^{-2}	10^3	0.024 ± 0.001
0.022	1064	94.8	0.75×10^4	0.188×10^{-2}	10^3	0.026 ± 0.002
0.028	895	82.5	0.55×10^4	0.189×10^{-2}	10^3	0.028 ± 0.002
0.034	780	73.7	0.43×10^4	0.189×10^{-2}	10^3	0.031 ± 0.002
0.044	648	64.6	0.31×10^4	0.188×10^{-2}	10^3	0.032 ± 0.001
0.052	575	58.0	0.26×10^4	0.197×10^{-2}	10^3	0.033 ± 0.001
0.088	395	42.5	1.32×10^3	0.199×10^{-2}	110	0.039 ± 0.002
0.164	253	29.6	0.60×10^3	0.203×10^{-2}	50	0.046 ± 0.002
0.284	171	21.6	2.97×10^2	0.204×10^{-2}	20.0	0.054 ± 0.002
0.553	106	14.6	1.24×10^2	0.203×10^{-2}	14.3	0.063 ± 0.003
1.105	64.7	9.8	0.51×10^2	0.204×10^{-2}	7.7	0.073 ± 0.002

[a] $A = \sigma_0 Q_0^3 / [4\pi^2 (Q_0 \xi_\parallel)(Q_0 \xi_\perp)]$.

It is also evident from the data of Fig. 9 that the bulk critical scattering relative to the Fresnel intensity determines the *absolute* value of the amplitude of the critical smectic fluctuations. It is essentially a matter of determining the effective volume V contributing to bulk critical scattering. A ray incident at angle ϕ_i and specularly reflected at a depth z below the surface is attenuated due to absorption by the amount $\exp(-2z/l \sin\phi_i)$, where l is the absorption length. By integration over z, the effective depth is therefore $(l/2) \sin\phi_i$. The irradiated area is the cross-sectional beam area A divided by $\sin\phi_i$. The effective scattering volume is therefore simply $V = Al/2$.

The differential cross section $d\sigma/d\Omega$ due to density fluctuations $\delta\rho(\mathbf{r})$ in the volume V of the electron density around its average value ρ_{av} is

$$\frac{d\sigma}{d\Omega} = r_0^2 V \int \langle \delta\rho(\mathbf{r}) \delta\rho(0) \rangle \exp(i\mathbf{Q}\cdot\mathbf{r}) d^3r$$

$$\equiv r_0^2 V \rho_{av}^2 S(\mathbf{Q}), \tag{18}$$

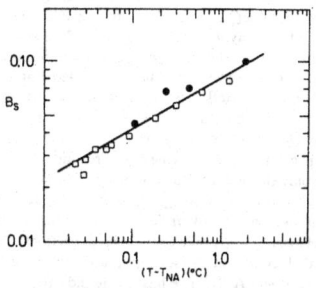

FIG. 10. The results obtained for B_s from fits of the high-resolution data shown in Fig. 6 (solid circles) and low-resolution data like those shown in Fig. 9 (open squares). The solid line represents $B_s = 0.81(T - T_{NA})^{0.28}$.

where the last equation defines the scattering function $S(\mathbf{Q})$.

When the spectrometer is set for the central ray being incident at ϕ_i^0 and scattered to angle ϕ_s^0 in plane and ψ_s^0 out of plane, the recorded intensity I_d is the folding of the cross section with the resolution function $R(\phi_i - \phi_i^0, \phi_s - \phi_s^0, \psi_s - \psi_s^0)$:

$$I_d(\phi_i^0, \phi_s^0, \psi_s^0)$$

$$= I_0 \int \int \int \frac{d\sigma}{d\Omega}(\phi_i, \phi_s, \psi_s) R(\phi_i - \phi_i^0, \phi_s - \phi_s^0, \psi_s - \psi_s^0)$$

$$\times d\phi_i \, d\phi_s \, d\psi_s . \tag{19}$$

Since the cross section is expressed in terms of wave vector $\mathbf{Q}$ rather than (ϕ_i, ϕ_s, ψ_s) it is convenient to carry out the folding integral in $\mathbf{Q}$ space rather than angular space. The relation between the volume element in $\mathbf{Q}$ space and that in angular space can be inferred from the kinematics in Fig. 1,

$$d^3Q = (k^2 Q \cos\phi_i) d\phi_i \, d\phi_s \, d\psi_s . \tag{20}$$

The parallelogram is an area element in Q_y-Q_z space. Denoting the short side by $\mathbf{X}_i$ and the long side by $\mathbf{X}_s$ with lengths $k \, \delta\phi_i$ and $k \, \delta\phi_s$, respectively, we find the area to be

$$k^2 \, \delta\phi_i \, \delta\phi_s \sin(2\phi) = 2(k \sin\phi)(k \cos\phi) \delta\phi_i \, \delta\phi_s$$

$$= Qk (\cos\phi) \delta\phi_i \, \delta\phi_s .$$

The out-of-plane component is of course $k \, \delta\psi_s$ and thereby the volume element is as given by Eq. (20).

Finally, from the definition of $\phi_c^2 = \lambda^2(\rho_{av}) r_0 / \pi$, using Eq. (1a),

$$(\rho_{av} r_0)^2 = \left[\frac{Q^4}{16\pi^2} \right] R_F(Q) , \tag{21}$$

and on substitution into Eq. (18), one obtains the following expression for the bulk intensity at Q^0:

$$I_{\text{bulk}}(\mathbf{Q}^0) = I_{\text{Fresnel}}(Q_z) \left[\frac{Q^4}{16\pi^2} \right] \left[\frac{l}{2} \right] (k^2 Q \cos\phi)^{-1}$$

$$\times \int g(\mathbf{Q}-\mathbf{Q}')S(\mathbf{Q}')d^3Q' , \qquad (22)$$

where $g(\mathbf{Q}-\mathbf{Q}')$ is the resolution function in reciprocal space. For $S(\mathbf{Q})$ we take the form used in Ref. 18,

$$S(\mathbf{Q}) = \sigma_0[1 + \xi_{\parallel}^2(Q_z - Q_0)^2 + \xi_{\perp}^2 Q_{\perp}^2(1 + c\xi_{\perp}^2 Q_{\perp}^2)]^{-1} ,$$

$$(23)$$

with the values of $\xi_{\parallel}$ and $\xi_{\perp}$ and c versus temperature as determined in Ref. 18. The only unknown parameter in Eqs. (22) and (23) is therefore σ_0, which has the dimension of volume. A least-squares fit to the data obtains the values of $\sigma_0 Q_0^3$ given in Table IV and plotted in Fig. 11.

The physical significance of σ_0 is better appreciated upon taking the Fourier transform of $S(\mathbf{Q})$. We define the dimensionless distance s by

$$s^2 \equiv \left[x^2 + y^2 + \left(\frac{\xi_{\perp} z}{\xi_{\parallel}} \right)^2 \right] \left(\frac{Q_0}{2\pi} \right)^2 . \qquad (24)$$

In s space, where the critical fluctuations are isotropic, one obtains the correlation function

$$\rho_{\text{av}}^{-2} \langle \delta\rho(\mathbf{r}) \, \delta\rho(0) \rangle$$

$$\simeq \left[\frac{\sigma_0 Q_0^3}{4\pi^2 (Q_0\xi_{\parallel})(Q_0\xi_{\perp})} \right] s^{-1} \exp\left(\frac{-2\pi s}{Q_0\xi_{\perp}} \right) \cos(Q_0 z) .$$

$$(25)$$

The quantity in square brackets is a measure of the absolute value of the density-density correlation function. It is also listed in Table IV. It is essentially independent of temperature with an average value around 0.0020.

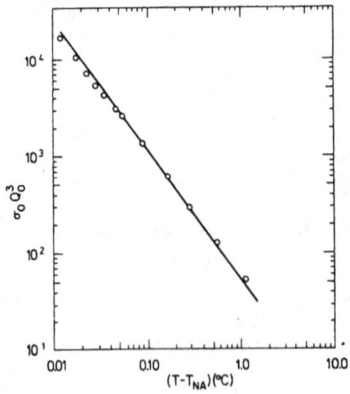

FIG. 11. The amplitude $\sigma_0 Q_0^3$ of the critical scattering from the bulk of 8OCB as a function of $T - T_{NA}$. The solid line has the slope -1.32 that was previously measured in a transmission geometry (Ref. 18).

C. 8OCB, transverse scans, wavelength dependence

So far we have only shown longitudinal scans to support the model of surface smectic layering accompanied by critical scattering from the bulk. Figure 12 displays transverse out-of-plane scans at two different wavelengths. The spectrometer parameters are listed in Table II under entry 3.83. The height of slit S_3 is maintained relatively open at 4 mm; however, its width is reduced to 0.2 mm. As illustrated in Figs. 12(e) and 12(f), the Q_z-Q_y cross section of the resolution volume is still approximated by a long thin parallelogram, as in the case of Fig. 1(c). The scans in Fig. 12 are in the Q_x direction, or normal to the Q_z-Q_y plane, at fixed values of $Q_z/Q_0 = 1.0$ [Figs. 12(a) and 12(c)] and $Q_z/Q_0 = 1.04$ [Figs. 12(b) and 12(d)]. The latter choice is motivated by looking at Fig. 5—here the wave scattered from the surface layering and the ordinary Fresnel wave interfere destructively and the total surface scattering, confined to the Q_z axis, is therefore very

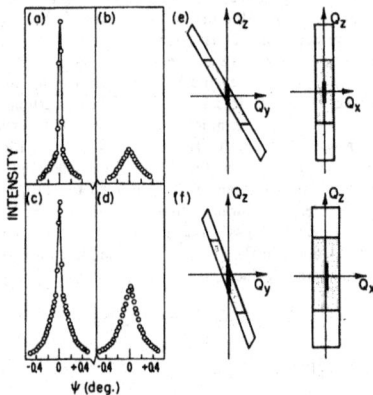

FIG. 12. Transverse out-of-plane scan along Q_x as a function of the angle $\psi \approx Q_x/[k \cos(\phi_i)]$ for 8OCB at $T - T_{NA} = 0.585\,°C$. The x-ray wavelength for (a) and (b) is $\lambda = 1.543$ Å and for (c) and (d) $\lambda = 0.943$ Å. For (a) and (c) the spectrometer is set such that the resolution volume is centered at $Q_y = 0$ and $Q_z/Q_0 = 1$ [see Fig. 1(c)] and the scan shows both the critical scattering from the bulk and the surface peak. For (b) and (d) $Q_y = 0$ and $Q_z/Q_0 = 1.04$. From the data in Fig. 5 one can see that the specular reflection is nearly zero and these scans are essentially only the bulk critical scattering. The signals from the bulk at $Q_z/Q_0 = 1.0$ and at $Q_z/Q_0 = 1.04$ are equal; however, the ratio of specular to bulk varies with wavelength. The intersections between Q_z-Q_y and Q_z-Q_x cross sections of the resolution parallelogram and the scattering are shown for $\lambda = 1.543$ Å (e) and $\lambda = 0.943$ Å (f). The heavy line indicates the specular reflectivity and the crosshatched region indicates the bulk scattering. As discussed in the text the ratio of specular reflection to bulk would be independent of wavelength were it not for irradiated volume effects.

weak and the scattering even at $Q_x \simeq k\psi = 0$ is therefore due to bulk critical fluctuations. Data such as those in Fig. 12(b) are therefore particularly useful to obtain the transverse correlation range $\xi_\perp$. A least-squares fit to the cross section given in Eq. (23), properly folded with the resolution function, gives $\xi_\perp$ versus temperature as shown in Fig. 13 compared to previous bulk data (solid line) from Ref. 18. The agreement is excellent.

We shall now discuss the wavelength dependence of surface scattering relative to bulk scattering. The surface scattering is indicated by the solid line in Figs. 12(a) and 12(c). The data in Figs. 12(b) and 12(d) provide a very accurate determination of the bulk intensity so that the separation into surface and bulk contribution at $Q_z / Q_0 = 1.0$ does not depend on an analysis of two superimposed line shapes. The conditions for the data for $\lambda = 0.943$ Å in Figs. 12(c) and 12(d) are identical to those for $\lambda = 1.543$ Å in Figs. 12(a) and 12(b), except that we here have used another set of monochromator planes, Si(220), and have changed the tilt accordingly.

The bulk scattering relative to the surface scattering is considerably enhanced at the small wavelength, by a factor of about 3.3. The reason is of course that the smaller wavelength radiation penetrates deeper into the bulk and therefore picks up more effective volume.

For a quantitative discussion two effects must be considered. The first is the wavelength dependence of the effective *volume* for bulk scattering, in relation to the effective *area* for specular reflection. We have argued above that the effective volume for bulk scattering is the cross-sectional beam area A times half of the absorption length $l(\lambda)$. The effective area for surface scattering is $A / \phi_i \sim A / \lambda$. Thus the ratio of bulk scattering to surface scattering should vary as $\lambda l(\lambda)$. A second point to consider, but which proves to be unimportant, is the λ dependence of the effective overlap of the resolution function with the specular reflection and with the bulk critical scattering. This is illustrated in Figs. 12(e) and 12(f) for the identical slit configurations used at $\lambda = 1.543$ and

0.943 Å. The shaded areas indicating bulk scattering overlap in the Q_y-Q_z plane are independent of λ, but in the Q_x-Q_z plane, the resolution width varies as k or $1/\lambda$. Altogether the bulk scattering overlap varies as $1/\lambda$. However, the surface scattering overlap, indicated by the heavy bar also varies as $1/\lambda$ because the resolution parallelogram is steeper at shorter wavelengths. We conclude that there is no λ dependence *due to resolution effects* in the *ratio* of bulk-to-surface signal. The measured ratio of bulk-to-surface signal is expected to vary as $\lambda l(\lambda)$ or as λ^{-2} since the measured $l(\lambda)$ varies approximately as λ^{-3}.[27] For the wavelengths in Fig. 12 the ratio is $(0.943/1.543)^{-2} = 2.7$, to be compared with the experimental value of 3.3. In view of the experimental uncertainties, we believe this agreement is reasonable.

In concluding this section we have demonstrated by transverse scans with appropriate slit widths that the bulk scattering is due to critical smectic fluctuations in the bulk with a finite width ($\xi_\perp Q_0$) in contrast to the surface smectic layering which as essentially the infinite correlation of the surface. This interpretation has been confirmed by the wavelength dependence of the bulk signal relative to the surface signal.

D. 4O.8 intermediate resolution

The geometries used for the study of 8OCB were rather extreme: very high resolution obtained by narrow slit heights and a channel-cut, perfect analyzer crystal, or low resolution obtained by a slit aperture in S_3 which was several millimeters either in one or both directions. The entire study of both bulk and surface properties throughout the nematic region can be carried out at an intermediate resolution with no analyzer crystal but with S_3 slit dimensions of a few tenths of millimeters in both directions.

The most important consequence of this is that the Q_y-Q_z projection of the resolution function is changed from the long thin parallelogram of Fig. 1, tilted at angle ϕ_i to the Q_z axis, to a more symmetric figure with the symmetry axis approximately parallel to the Q_z axis.

The material butyloxybenzylidene octylanine (4O.8) was studied using this configuration. The principle difference between 4O.8 and 8OCB is that in 4O.8 the smectic layer spacing d is essentially equal to the length of the molecule, while for 8OCB the molecules overlap each other and the layer spacing is approximately 1.6 times the length of the individual molecules.[20] Data shown in Fig. 14 are for a temperature 0.5 °C above the nematic-to-smectic-A transition temperature. The filled circles in Fig. 14(a) are a longitudinal scan on the Q_z axis with $Q_x = Q_y = 0$. The open circles below are a similar scan but with the out-of-plane wave vector misset, $Q_x / Q_0 = 0.05$. The later data represent the longitudinal bulk scattering. Notice the logarithmic intensity scale. The two peaks appear to have approximately the same widths: a direct demonstration that the penetration depth of surface smectic layering equals the bulk correlation range. The solid lines are least-squares fits of the model described in the preceding sections. The data for $Q_x / Q_0 = 0.05$ determine the bulk correlation range $\xi_\parallel Q_0 = 93$ and the $\sigma_0 Q_0^3 = 256$.

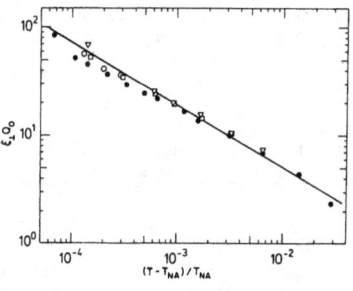

FIG. 13. Results of nonlinear least-squares fits for the transverse bulk correlation length $\xi_\perp$ vs $T - T_{NA}$ for 8OCB from data like those in Fig. 12 for $\lambda = 0.77$ Å (circles), $\lambda = 0.94$ Å (squares), and $\lambda = 1.54$ Å (triangles). The solid line is the previously published result for $\xi_\perp$ (Ref. 18).

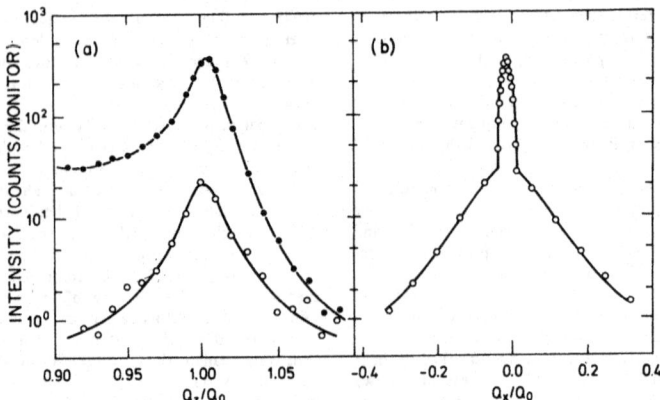

FIG. 14. Intermediate-resolution data for 4O.8 at $T - T_{NA} = 0.5$ °C. (a) Q_z scans at $Q_x = 0$ (filled circles), corresponding to the sum of the specular reflection and the bulk critical scattering and at $Q_x/Q_0 = 0.05$ (open circles), which is only the bulk signal. (b) Q_x scans for $Q_z/Q_0 = 1.0$ in which the specular and bulk are clearly separable.

The data for $Q_x/Q_0 = 0$ determine the penetration depth $\xi_\parallel Q_0 = 99$, the surface amplitude $B_s = 0.095$, and the Gaussian depth $Q_0/\sigma = 7 \pm 3$, assuming the phase $z_0 = 0.25d$ as for 8OCB. The quality of both fits is very good. The data in Fig. 14(b) are a transverse out-of-plane scan similar to those of Fig. 12 for 8OCB. The bulk and surface scattering are clearly separated into a broad and a narrow, resolution-limited peak. These data then determine the transverse bulk correlation range $\xi_\perp Q_0 = 11.2$.

The three types of scans displayed in Fig. 14 contain all information on the free surface of a liquid crystal in the nematic phase. The reflected waves from the surface and the smectic surface layering interfere giving rise to an asymmetric line shape for the total surface scattering. Its width determines the penetration depth of surface layering, its height determines amplitude of the surface-induced smectic order. The bulk scattering is dominated by the critical scattering from smectic-A fluctuations in the nematic phase characterized by the longitudinal and transverse correlation ranges $\xi_\parallel Q_0$ and $\xi_\perp Q_0$, respectively, as well as the absolute magnitude of the density correlation function as expressed by the parameter $\sigma_0 Q_0^3$. Data sets such as those in Fig. 14 were taken at a number of temperatures and the best-fit values of model parameters discussed above are listed in Table V and displayed in Fig. 15. The dashed lines are the results previously obtained for the critical scattering from the bulk of 4O.8 in a transmission geometry,[19] except that in that experiment the absolute value of $\sigma_0 Q_0^3$ was not measured. Best fits to the displayed data obtain $\sigma_0 Q_0^3 = 130(T - T_{NA})^{-1.26 \pm 0.05}$. As with 8OCB, measurements of the bulk properties in the two different geometries agree. The Gaussian penetration length and the amplitude B_s are also similar to those of 8OCB and the best fits obtain $Q_0/\sigma = 5.0(T - T_{NA})^{-0.51 \pm 0.05}$ and $B_s = 0.10(T - T_{NA})^{+0.10 \pm 0.05}$. Although the critical exponents δ and $\nu_\parallel$ and $\nu_\perp$ are essen-

tially equal for 8OCB and 4O.8, the "exponents" for B_s are significantly different.

IV. DISCUSSION

The principle purpose of this manuscript is to present experimental results demonstrating that the x-ray reflectivity of nematic liquid-crystal surfaces display features indicative of surface-induced smectic order. These features are unambiguously distinguished from all other scattering processes by the existence of a two-dimensional selection rule, e.g., $Q_x = Q_y = 0$. The most striking property of this order is the fact that it penetrates into the bulk exponentially with a characteristic length that is equal to the parallel correlation length $\xi_\parallel$ for critical fluctuations. In addition to the sharp peak, the x-ray reflectivity displays characteristic deviations from the Fresnel reflection law. We demonstrated above that the total reflectivity could be modeled by $R(\phi) = R_F(\phi) |\Phi|^2$, where $\Phi = \Phi_0 + \Phi_1$. The first of these Φ_0, the Fourier transform

TABLE V. The results of nonlinear least-squares fits for the intermediate-resolution data from 4O.8.

$T - T_{NA}$ (°C)	$\sigma_0 Q_0^3$	$\xi_\parallel Q_0$	$\xi_\perp Q_0$	Q_0/σ	B_s
0.019	25 870	933	113	a	0.056
0.094	3640	332	46	12.45	0.076
0.134	1554	233	30	8.6	0.082
0.254	682	145	21.1	7.67	0.088
0.510	284	91.8	13.6	6.27	0.095
2.006	45	31.7	5.06	3.56	0.112

[a] Could not be determined because of high uncertainty.

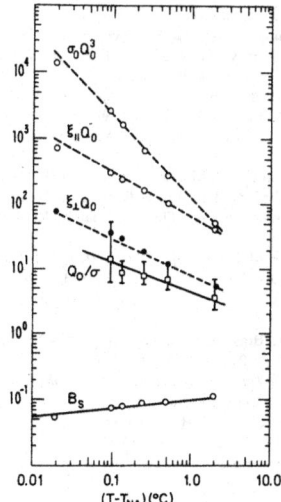

FIG. 15. Temperature dependence of the parameters obtained from nonlinear least-squares fits of the intermediate-resolution data for 4O.8 as listed in Table V. The dashed lines through the data represent previously published results (Ref. 19) for the amplitude and slope, $\xi_\parallel$ and $\xi_\perp$, and for the slope of $\sigma_0 Q_0^3$. The solid lines are fits to the data as discussed in the text.

of the exponentially penetrating surface oscillations, is theoretically motivated and the only adjustable parameters that were not previously known are the amplitude of the oscillations, B_s, and the phase factor, z_0, that defines the phase of the oscillations relative to the surface.

The second, or Φ_1 term was represented by an *ad hoc* form that was simply taken because it was capable of representing the data. On the other hand, the functional form of Φ_1 has features that are essential to any full theoretical representation of the surface layers. For example, from Figs. 8 and 15 one can see that the Gaussian penetration length Q_0/σ of Φ_1 is at least two to three times smaller than the exponential penetration length $\xi_\parallel$ of Φ_0. The implication is that although the simple exponential decay of the Φ_0 term does describe the main peak in the reflectivity it is not adequate to describe the amplitude of the real-space smectic oscillations near the

surface. We believe that the counterintuitive temperature dependence in which B_s approaches zero as T approaches T_{NA} is a reflection of the interplay between these two terms. One consequence of this is that the amplitude of the smectic order parameter near to the surface is not just equal to B_s but is actually the amplitude of the real-space oscillations that are represented here by the sum of the Φ_0 and Φ_1 terms.

An alternative approach that has recently been applied to the interpretation of antiferroelectric layering at the surface of a different type of liquid-crystal material assumes that the surface fixes the position and orientation of the surface molecules and then calculates the electron density profile near the surface from models of the molecular form factor.[28] For both 8OCB (Ref. 29) and 4O.8, such a calculation obtains amplitudes for the Fourier component of the electron density at Q_0 of the order of 0.1. It is interesting to note that the amplitudes for the real-space electron density oscillations associated with the Φ_1 term are of the same order of magnitude. Since in 8OCB, for $T \approx T_{NA}$, B_s is considerably smaller than 0.1, the current model for Φ_1 is in reasonable agreement with the molecular model near to T_{NA}. For 8OCB, at higher temperatures, B_s approaches 0.1 and the predictions from the molecular model must be compared to the density obtained from the sum of the Φ_0 and Φ_1 terms. Since the reasonable agreement that can be obtained is very sensitive to the poorly determined phase z_0, detailed comparisons are not warranted. A similar situation prevails for 4O.8 over the entire measured temperature range. Nevertheless, the chosen phase factor, $z_0 \approx 0.25d$, can be justified by inspection of the molecular models.

Finally, we have made a direct comparison between the diffuse critical scattering from the bulk and the specular reflection from the surface. Since the latter follows the Fresnel reflection law whose value is known absolutely in terms of the Thompson cross section of the electron, this obtains a measure of the absolute value of the critical cross section. We have interpreted this to obtain an absolute value for the critical part of the density-density correlations.

ACKNOWLEDGMENTS

This work was supported in part by grants from the Danish National Science Foundation, by the Risø National Laboratory, by the U.S. National Science Foundation through Grant Nos. DMR-82-12189, DMR-83-16979, and INT-83-11841, and by the U.S. Joint Services Electronics Program (U.S. Army, Navy, and Air Force) through Grant No. N00014-84-K-0465.

*Permanent address: Department of Physics, University of Texas at Arlington, P.O. Box 19059, Arlington, TX 76019.
[1]H. Kiessig, Ann. Phys. 10, 769 (1931).
[2]A. H. Compton and S. K. Allison, *X-Rays in Theory and Experimentation* (Van Nostrand, New York, 1935).
[3]L. G. Parratt, Phys. Rev. 95, 359 (1954).

[4]L. Nevet and P. Croce, Rev. Phys. Appl. 15, 761 (1980).
[5]D. H. Bilderback and S. Hubbard, Nucl. Instrum. Methods 195, 85 (1982); 195, 91 (1982).
[6]J. A. Prins, Z. Phys. 47, 479 (1928).
[7]E. Nahrig, Phys. Z. 31, 401 (1930).
[8]L. A. Smirnov, Opt. Spektrosk. 43, 567 (1977) [Opt. Spectrosc.

(USSR) **43**, 333 (1977)].

[9]*Synchrotron Radiation Research*, edited by H. Winick and S. Doniach (Plenum, New York, 1980).

[10]*Synchrotron Radiation, Techniques and Applications*, edited by C. Kunz (Springer-Verlag, Berlin, 1979).

[11]I. M. Bloch, M. Sansone, F. Rondelez, D. G. Peiffer, P. Pincus, M. W. Kim, and P. M. Eisenberger, Phys. Rev. Lett. **54**, 1039 (1985).

[12]J. Als-Nielsen, F. Christensen, and P. S. Pershan, Phys. Rev. Lett. **48**, 1107 (1982).

[13]J. Als-Nielsen and P. S. Pershan, Nucl. Instrum. Methods **208**, 545 (1983).

[14]P. S. Pershan and J. Als-Nielsen, Phys. Rev. Lett. **52**, 759 (1984).

[15]A. Braslau, M. Deutsch, P. S. Pershan, A. H. Weiss, J. Als-Nielsen, and J. Bohr, Phys. Rev. Lett. **54**, 114 (1985); A. Braslau, B. M. Ocko, P. S. Pershan, J. Als-Nielsen, and M. Deutsch (unpublished).

[16]M. P. D'Evelyn and S. A. Rice, Phys. Rev. Lett. **47**, 1844 (1981); D. S. Sluis, M. P. D'Evelyn, and S. A. Rice, J. Chem. Phys. **78**, 545 (1983).

[17]T. Lubensky, J. Chem. Phys. **80**, 31 (1983).

[18]J. D. Litster, J. Als-Nielsen, R. J. Birgeneau, S. S. Dana, D. Davidor, F. Gareia-Golding, M. Kaplan, C. R. Safinya, and R. Schaetzing, J. Phys. (Paris) Colloq. **40**, C3-339 (1979).

[19]R. J. Birgeneau, C. W. Garland, G. B. Kasting, and B. M. Ocko, Phys. Rev. A **24**, 2624 (1981).

[20]A. J. Leadbetter, C. Frost, J. P. Gaughan, G. W. Gray, and A. Mosley, J. Phys. (Paris) **40**, 375 (1979).

[21]B. B. Warren, *X-Ray Diffraction* (Addison-Wesley, Reading, MA, 1969).

[22]M. Nielsen and H. B. Møller, Acta Cryst. A **25**, 547 (1969); B. Lebech and M. Nielsen, Proceedings of the Neutron Diffraction Conference, Petten, The Netherlands, RCN-234 (1975), p. 466.

[23]V. Bonse and M. Hart, Appl. Phys. Lett. **7**, 238 (1965).

[24]F. J. Kahn, Appl. Phys. Lett. **22**, 386 (1973).

[25]B. K. Agarwal, *X-Ray Spectroscopy* (Springer-Verlag, Berlin, 1979), p. 178.

[26]B. Ocko, A. Braslau, P. S. Pershan, J. Als-Nielsen, and M. Deutsch, Phys. Rev. Lett. **57**, 94 (1986).

[27]*International Tables for X-Ray Crystallography* (Kynoch Press, Birmingham, England, 1962).

[28]E. F. Gramsbergen, W. H. de Jeu, and J. Als-Nielsen, J. Phys. (Paris) **47**, 711 (1986).

[29]B. Ocko (private communication).

10. DISCOTIC PHASES

10.1. Chandrasekhar, S., Sadashiva, B. K. and Suresh, K. A. (1977).
Liquid crystals of disk like molecules. *Pramana* **9**, 471–480   399

10.2. Levelut, A. M. (1983). Structures des phases mesomorphes formées
de molecules discoïdes. *J. Chim. Phys.* **80**, 149–161   409

10.3. Safinya, C. R., Liang, K. S., Varady, W. A., Clark, N. A. and
Andersson, G. (1984). Synchrotron x-ray study of the orientational
ordering D2-D1 structural phase transition of freely suspended
discotic strands in triphenylene hexa-n-dodecanoate. *Phys. Rev.
Lett.* **53**, 1172–1175   422

Pramana, Vol. 9, No. 5, November 1977, pp. 471-480, ©

Liquid crystals of disc-like molecules

S CHANDRASEKHAR, B K SADASHIVA and K A SURESH
Raman Research Institute, Bangalore 560 006

MS received 19 September 1977

Abstract. Thermotropic mesomorphism has been observed in pure compounds consisting of simple disc-like molecules, viz., benzene-hexa-n-alkanoates. Thermodynamic, optical and x-ray studies indicate that the mesophase is a highly ordered lamellar type of liquid crystal. Based on the x-ray data, a structure is proposed in which the discs are stacked one on top of the other in columns that constitute a hexagonal arrangement, but the spacing between the discs in each column is irregular. Thus the structure has translational periodicity in two dimensions and liquid-like disorder in the third.

Keywords. Thermotropic liquid crystals; disc-like molecules; benzene-hexa-n-alkanoates; x-ray diffraction.

1. Introduction

It has long been recognized that the distinctive feature of thermotropic liquid crystals formed by pure compounds is the rod-like or lath-like shape of the molecule. Mesophases composed of large plate-like molecules are known to occur at high temperatures during the carbonization of graphitizable substances, such as petroleum and coal tar pitches (Brooks and Taylor 1965, Zimmer and White 1977), but these are rather complex materials and certainly cannot be regarded as single-component liquid crystalline systems. We report here what is probably the first observation of thermotropic mesomorphism in pure, single-component systems of relatively simple plate-like, or more appropriately disc-like, molecules.

The compounds investigated were benzene-hexa-n-alkanoates (abbreviated, for convenience, as BHn-alkanoates):

$$R = n - C_4H_9 \text{ to } n - C_9H_{19}$$

They were synthesized according to the procedure of Neifert and Bartow (1943) and purified by chromatography and repeated crystallizations from absolute ethyl alcohol.

2. Thermodynamic data

The transitions were studied by differential scanning calorimetry using a Perkin-Elmer Model DSC2.   Four homologues, BHn-hexanoate to nonanoate, show meso-

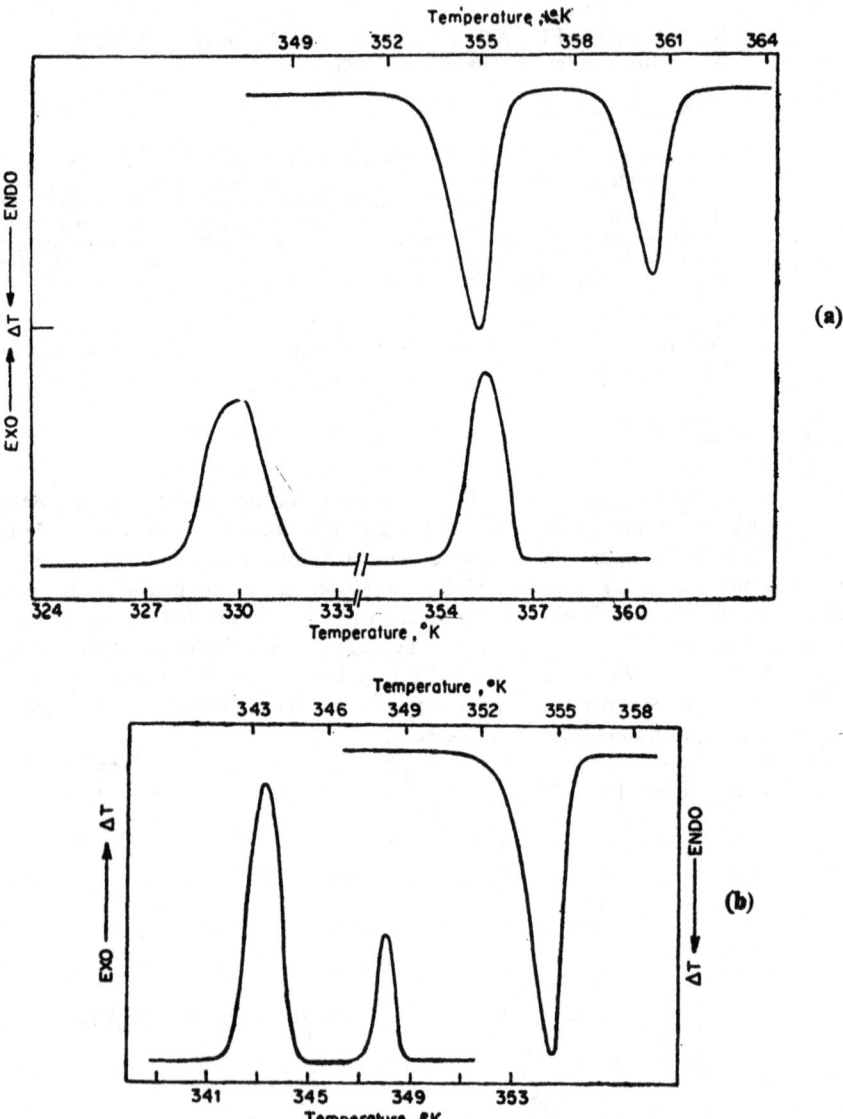

Figure 1. Differential scanning calorimetric curves for (a) BHn-heptanoate, (b) BHn-nonanoate.

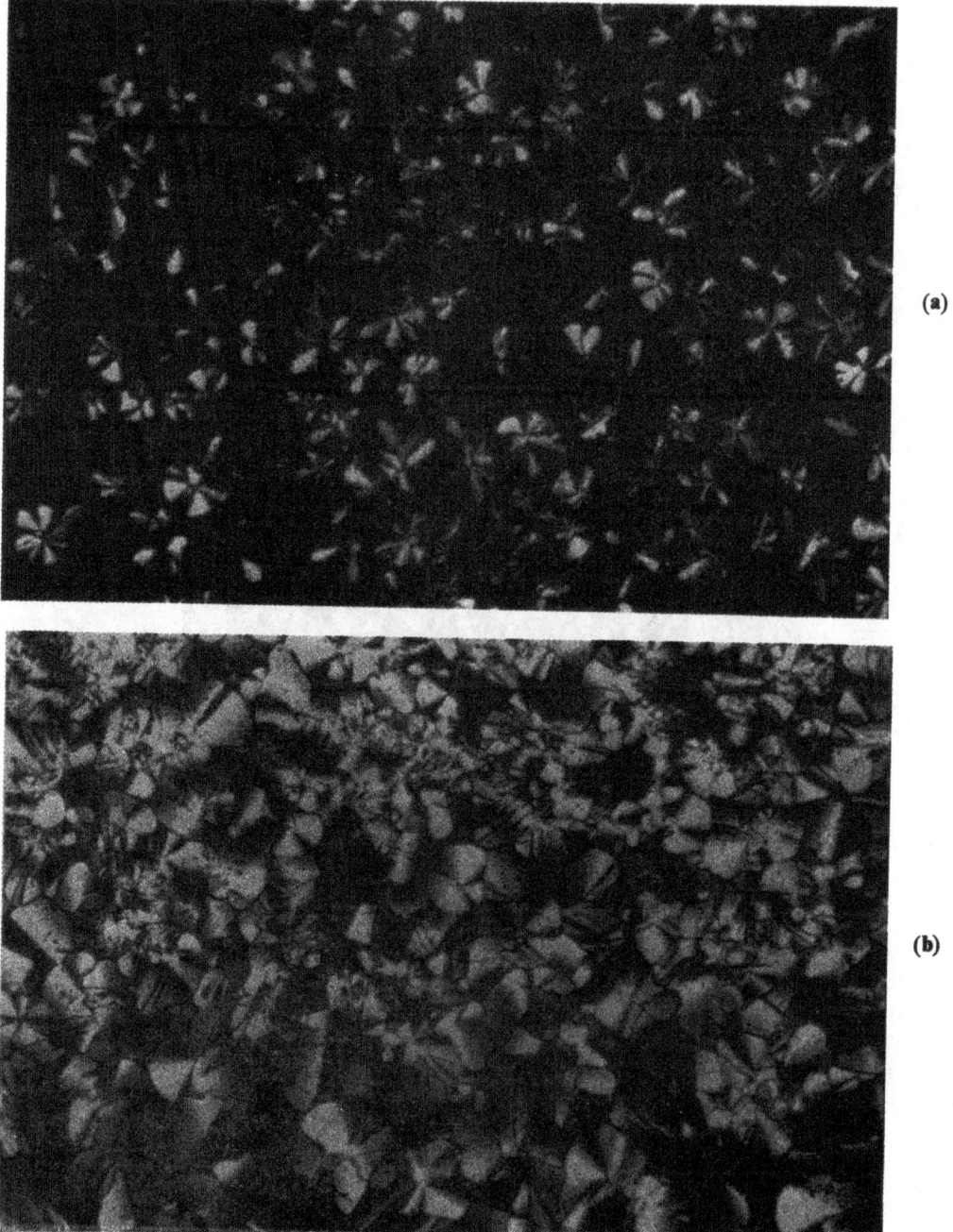

Figure 2. (a) The mesophase appearing as flower-like particles on cooling the isotropic liquid; (b) on further cooling, the particles grow and merge together to fill the entire field of view. Crossed polarizers, × 400.

474   *S Chandrasekhar, B K Sadashiva and K A Suresh*

Figure 3. Broken-fan texture of the mesophase.   Crossed polarizers, × 500.

Table 1. Transition temperatures and heats of transitions

Compound	Transition	Temperature (°C)	Heat of transition kcals/mole
BHn-pentanoate	crystal → isotropic (non-mesomorphic)	106	6·07
BHn-hexanoate	crystal → mesophase	68·3	4·78
	mesophase → isotropic	86·0 ⎫	4·77
	isotropic → mesophase .	83·25 ⎭	
BHn-heptanoate	crystal → mesophase	80·2	6·97
	mesophase → isotropic	86·2 ⎫	4·68
	isotropic → mesophase	83·6 ⎭	
BHn-octanoate	crystal → mesophase	79·4	9·67
	mesophase → isotropic	83·6 ⎫	3·79
	isotropic → mesophase	81·5 ⎭	
BHn-nonanoate	crystal → isotropic	79·6	16·28
	isotropic → mesophase	74·7	3·38
BHn-decanoate	crystal → isotropic (non-mesomorphic)	85·35	18·24

phases. The DSC curves for two of them are reproduced in figure 1. The temperatures and the heats of transitions determined from the curves are given in table 1.

In the three enantiotropic compounds, hexanoate to octanoate, there is supercooling of both the isotropic liquid and the mesophase. Supercooling of the isotropic liquid is not surprising in these cases, since the heat of the isotropic-mesophase transition is quite large—indeed in one compound, BHn-hexanoate, equal to that of the mesophase-solid transition. This suggests a very highly ordered mesophase, probably smectic-like in character.

Though the molecule is by no means globular, we did consider the possibility of the mesophase being a *birefringent* plastic crystal—somewhat analogous to that exhibited by norbornylene and its homologues (Folland *et al* 1975)—with rotational disorder about an axis normal to the disc. However, from the optical and x-ray evidence, which we shall describe presently, we are inclined to think that this is a true liquid crystalline phase in the accepted sense of the term.

3. Optical textures

The transitions could be seen very clearly through the polarizing microscope when the sample was cooled from the isotropic phase. Typical textures are illustrated in figures 2 and 3. Sometimes, the mesophase made its appearance as flower-like particles (figure 2a), which on further cooling grew in size and finally merged together to fill the entire field of view (figure 2b). However, most often the mesophase adopted a ' broken-fan ' texture similar to that of smectic C (figure 3), occasionally with striations running across the fans as in smectics E and F (see Sackmann and Demus 1973). Optical observations with the aid of a phase retardation plate confirmed that the disc-like molecules are oriented with their long molecular axes (i.e., the diameters of the discs) aligned radially in the fan and their short axes (or the normals to the discs) lying in the plane of the sample. Also, like the smectic phases, the material is highly viscous.

The transition to the solid phase on cooling the sample was also quite distinctly observable. However, the change in texture attending the reverse transition, i.e., when the solid transformed to the mesophase on heating, was so slight that it was practically undetectable, but the DSC curves proved that the three compounds hexanoate to octanoate, are in fact enantiotropic.

4. X-ray studies

X-ray diffraction photographs of the three phases of BHn-octanoate using filtered CuK_a radiation are shown in figure 4. Essentially similar patterns were obtained with the hexanoate and heptanoate compounds also. X-ray photographs could not be taken of the mesophase of the monotropic nonanoate compound as it tended to crystallize into the solid phase in a short time.

The crystalline phase gives a large number of sharp maxima as is to be expected of a regular three-dimensional lattice (figure 4a). There is a marked change when the sample goes over to the mesophase (figure 4b) and the pattern now closely resembles that from some of the smectic phases (see de Vries 1973). The outermost ring, which corresponds to a mean intermolecular spacing of about 6 Å, is quite faint and diffuse, and is characteristic of liquid-like disorder. On the other hand, the three inner rings are made up of a large number of fairly sharp spots, which clearly suggests a highly ordered lamellar structure. The Bragg spacings (d) corresponding to these three rings are 20·1, 11·7 and 10·1 Å. (In some photographs, there were one or two weak spots in between the 20·1 and 11·7 Å rings with $d \sim 15$ Å; however, even assuming that these were not due to some spurious effect, it appeared reasonable to ignore them in comparison with the three prominent diffraction rings, at least in a preliminary interpretation of the x-ray data.)

Both the 6 and 20 Å rings are present in the diffraction pattern from the isotropic phase (figure 4c); the intensity and the width of the inner ring shows that smectic-like (cybotactic) ordering persists in this phase also.

From the nature of the diffraction pattern, it seems safe to conclude that the mesophase is not a plastic crystal. For one thing, a large number of fairly clear sharp reflexions is not generally characteristic of the plastic phase (see e.g. Winsor 1974). But more significantly, the presence of the diffuse 6 Å ring in the mesophase pattern and the fact that its appearance hardly changes when the sample goes over to the isotropic phase is a clear indication that long range translational order is absent in the mesophase in at least one dimension. This again favours a liquid crystalline structure for, by definition, a plastic crystal should have an appreciable degree of three-dimensional order.

Now the fact that the thickness and the diameter of the disc-like molecule are respectively of the order of 6 and 20 Å gives a clue to a possible structure of the mesophase. This is illustrated schematically in figure 6. The discs are stacked·one on top of the other in columns that constitute a hexagonal arrangement. The discs in each column are irregularly spaced with a mean separation of 6 Å. Such a structure would explain the diffuse ring as well as the sharp reflexions in the x-ray diffraction pattern. Direct evidence of the hexagonal symmetry of the structure is presented in figure 5. This is an x-ray photograph of a nearly monodomain specimen of the mesophase of BHn-hexanoate, with the incident beam almost exactly along the six-fold axis.

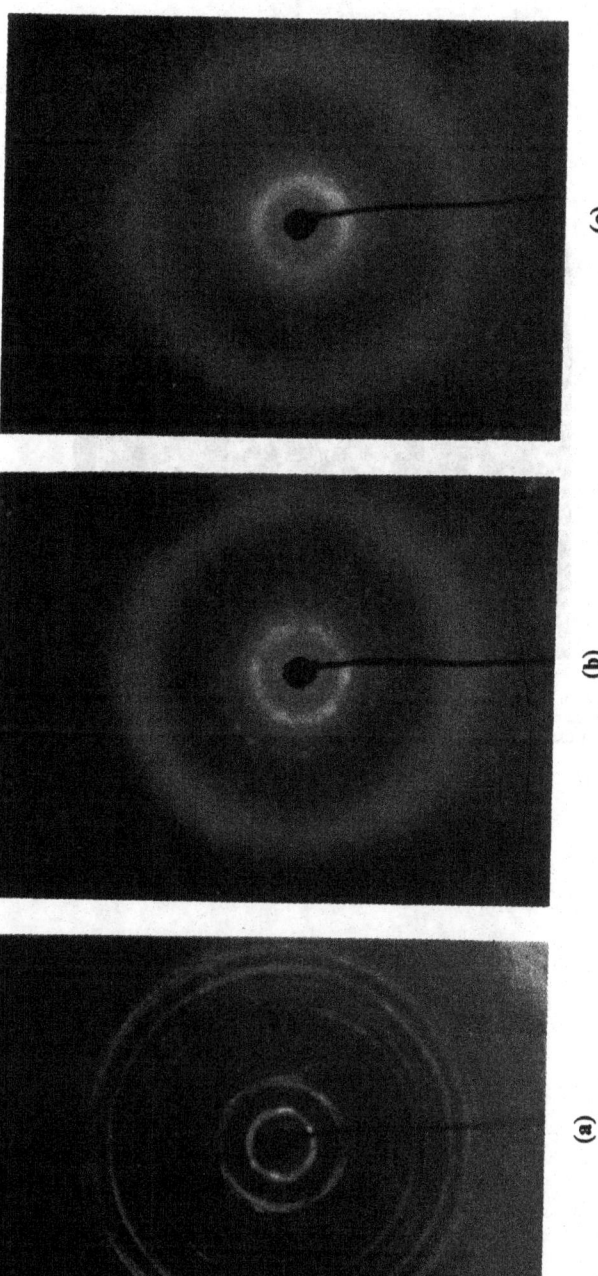

Figure 4. X-ray diffraction photographs of BHn-octanoate. Filtered CuK$_\alpha$ radiation. Specimen to film distance 9·6 cm, (a) crystallinephase, (b) mesophase, (c) isotropic liquid.

Figure 5. X-ray diffraction photograph of a nearly monodomain specimen of the mesophase of BHn-hexanoate. Specimen to film distance 12·5 cm. Incident beam almost exactly along the six-fold axis.

Figure 6. Schematic representation of the structure of the mesophase. The discs are spaced irregularly to form liquid-like columns.

We may now proceed to interpret the x-ray data on the basis of this structure. If a is the lattice constant (which in this case is equal to the diameter of the disc), the Bragg spacings of the first three orders of reflexion from such a lattice should be

$$d_{100} = \frac{\sqrt{3}a}{2}, \quad d_{110} = a/2 \text{ and } d_{200} = \frac{\sqrt{3}a}{4}$$

or

$$d_{100} : d_{110} : d_{200} = 1 : 0.5773 : 0.5$$

The experimental results for the 3 compounds are given in table 2.

Table 2. Molecular spacings for the three compounds

	BHn-hexanoate	BHn-heptanoate	BHn-octanoate
Mean inter-molecular spacing along hexagonal axis Å (expt.)	6	6	6
d_{100} Å (expt.)	18·1	18·9	20·1
d_{110} Å (expt.)	10·6	11·1	11·7
d_{200} Å (expt.)	9·1	9·7	10·1
$d_{100} : d_{110} : d_{200}$ (expt.)	1 : 0·59 : 0·50	1 : 0·59 : 0·51	1 : 0·58 : 0·50
Lattice constant Å (expt.)	21·0	22·2	23·3
Approximate molecular diameter assuming fully extended chains Å	21·5	24·0	26·5

It is seen that the experimental ratios of the Bragg spacings are in reasonable accord with the values for a hexagonal lattice. Also, as is to be expected, the lattice constant a increases systematically from the hexanoate to the octanoate. The experimental value of a is in fair agreement with the calculated diameter of the idealized molecule with *fully* extended end-chains; there is a small discrepancy which increases with increasing chain length, but this is not surprising in view of the fact that the chains become more flexible as they get longer. Thus, on the whole, the proposed structure appears to be essentially correct.

In principle, there is an alternative structure which can give rise to exactly the same type of x-ray diffraction pattern. This structure consists of sheets which are regularly spaced at intervals of 6 Å. Each sheet contains a hexagonal close-packed arrangement of discs, but the different sheets are displaced parallel to themselves in an irregular fashion. A 'planar' disorder of this type should also result in a diffuse 6 Å diffraction maximum in addition to sharp reflexions from the hexagonal lattice (see Guinier 1963). However, such a disorder is probably more easily realized when the intermolecular attraction within each sheet is much greater than that between sheets. With these disc-like molecules, the opposite will be true, and the type of disorder depicted in figure 5 is likely to be more favourable. Moreover, by analogy with smectic A (see Chandrasekhar 1977) one may expect a structure consisting of liquid-like columns to be readily deformed to give the radiating or fan-like arrangement of molecules that is observed in the optical textures.

Further studies are in progress to test the validity of the structure and to investigate other properties of the mesophase.

Acknowledgements

We are much indebted to N V Madhusudana and to G S Ranganath for valuable discussions.

References

Brooks J D and Taylor G H 1965 *Carbon* **3** 185
Chandrasekhar S 1977 *Liquid Crystals* (Cambridge: Cambridge Univ. Press) p. 285
de Vries A 1973 *Proc. Int. Liquid Crystals Conf., Bangalore—Pramana Supplement* **1** 93
Folland R, Jackson D A and Rajagopal S 1975 *Mol. Physics* **30** 1053
Guinier A 1963 *X-ray Diffraction* (London: W H Freeman & Co.) p. 169
Neifert I E and Bartow E 1943 *J. Am. Chem. Soc.* **65** 1770
Sackmann H and Demus D 1973 *Mol. Cryst. Liq. Cryst.* **21** 239
Winsor P A 1974 in *Liquid Crystals and Plastic Crystals* eds. G W Gray and P A Winsor (Chichester: Ellis Horwood Ltd., Vol. 1, p. 51)
Zimmer J E and White J L 1977 *Mol. Cryst. Liq. Cryst.* **38** 177

STRUCTURES DES PHASES MESOMORPHES FORMEES DE MOLECULES DISCOÏDES

par A.M. LEVELUT

Laboratoire de Physique des Solides associé au CNRS, Université Paris-Sud, Bâtiment 510, 91405 Orsay (France)

(Reçu le 9.10.82, accepté le 21.10.82)

RESUME

Les mésophases de molécules discoïdes forment deux classes :
— *Les mésophases en colonnes* sont constituées de colonnes de molécules parallèles entre elles et assemblées dans un réseau bidimensionnel périodique. Le polymorphisme provient essentiellement de la grande variété dans les réseaux rencontrés. Nous décrivons également le mode d'empilement et la conformation des molécules dans chaque colonne ;
— *La phase nématique* est très semblable à la phase nématique ordinaire constituée de molécules allongées bien que le désordre des centres de gravité y apparaisse moins complet.

Il faut noter que cet article a été écrit à un moment où le sujet est encore en évolution rapide et que de nouvelles phases seront probablement découvertes d'ici sa parution.

ABSTRACT

The major part of mesophases formed by rod like molecules belong to one of the two following classes: smectic or lamellar phases and nematic (cholesteric) phase. With disc-like molecules it is also possible to obtain two classes of mesophases:
— *In columnar* (or canonic) mesophases, the molecules are stacked in parallel columns, these columns forming a periodic two-dimensional array. A great variety of lattices have been discovered and are here described with the help of the eighty planar crystallographic space groups. From X-ray diffraction pattern and from the comparison with other physical properties one can give informations about the packing and the conformation of the molecules in each column;
— *The nematic phase of disc-like* molecules is similar to ordinary nematic phase of rod-like molecules. Nevertheless a local order of columns still remains present in this fluid mesophase.

We give the limits for the discotic phases assuming that a disc-like molecule must be (in the mesophase) made of a planar rigid core surrounded by a more or less planar ring of aliphatic chains. Thus we exclude from this frame the systems in which only four chains are linked to the same core and which form lamellar phases.

Our knowledge upon the structures of the mesophases made of disc-like molecules is given at the present state, whereas the subject evolves quickly, new phases appearing quite monthly. Thus, when this issue will be published this paper will certainly not be a complete survey of the polymorphism which seems so attractive.

C'est en 1977 que S. Chandrasekhar et ses collaborateurs[1] ont découvert la première mésophase thermotrope dans un corps pur constitué de molécules en forme de disque (discoïde). Une étude par diffraction des rayons X permettait à ces auteurs de proposer une structure de la mésophase dans le même article. Depuis, de nombreuses molécules discoïdes mésogènes ont été synthétisées et les études de structure ont souvent suivi assez rapidement. C'est ainsi qu'actuellement il est possible d'avoir une première approche de la structure de ces mésophases. Cependant, dans la plupart des cas, il serait nécessaire d'entreprendre un traitement plus complet des données de diffraction des rayons X, afin d'en extraire le maximum d'information ; c'est pourquoi ce qui va suivre ne constitue qu'une mise au point sur nos connaissances actuelles de la structure de ces phases, car ce domaine reste en pleine évolution.

Avant d'aborder la description des diverses structures connues, je vais définir ce que je considère comme étant une molécule discoïde et une mésophase thermotrope.

Les molécules mésogènes sont, dans l'immense majorité des cas, formées d'une partie rigide généralement, mais pas nécessairement, aromatique, allongée, terminée par une chaîne paraffinique souple ; l'association de ces deux parties de caractères différents est nécessaire à l'obtention d'une mésophase thermotrope[2]. De la même manière, les molécules mésogènes discoïdes seront formées d'un cœur rigide plan entouré de chaînes paraffiniques. Il n'est pas nécessaire que le cœur soit de grande symétrie, ni que toutes les chaînes soient de même longueur. Le caractère discoïde

se traduit par le fait qu'une couronne paraffinique est disposée dans le plan du cœur, en l'entourant.

De la mésophase, je soulignerai le caractère fluide associé aux propriétés d'anisotropie ; ceci implique qu'il y a disparition de symétrie de translation dans au moins une direction, mais que les molécules conservent une orientation privilégiée. Ces définitions appellent plusieurs remarques :

— Elles sont assez restrictives, puisqu'elles excluent tous les systèmes pour lesquels on peut définir un réseau cristallin moyen tridimensionnel ; ainsi exclueraije les phases S_B, S_E, S_G, S_H[3] de molécules allongées habituellement classées dans les mésophases ou les phases rotationnelles des dérivés hexasubstitués du benzène[4] tels que les chlorométhyl benzènes ; ce sont des cristaux avec désordre orientationnel ;

— Il y a une autre classe de phases mésomorphes thermotropes qui présentent un ordre cristallin tridimensionnel, mais où la position du centre de gravité de chaque molécule n'est pas mieux définie que dans un liquide classique ; la structure cristalline résulte alors d'un arrangement périodique de lignes singulières. Les phases bleues[5] et la phase S_D[6] de symétrie cubique, appartiendraient à cette classe, mais actuellement il n'a pas été découvert de phase analogue avec des molécules discoïdes ;

— Les conditions de symétrie imposées à la molécule par son caractère discoïde impliquent que les paramètres mesurant l'ordre translationnel, s'ils ne sont pas identiques, restent du moins très comparables pour n'importe quelle direction du plan ; autrement dit, comme dans le cas des molécules allongées, il y a un axe qui joue un rôle particulier, les deux autres étant semblables.

En dernier lieu, il me faut dire un mot des méthodes qui permettent de caractériser la structure. En effet, outre les méthodes de diffraction des rayons X grâce auxquelles on peut donner une description de la structure d'autant plus précise que le système étudié est ordonné et bien orienté, l'examen des propriétés physiques peut aussi apporter des informations utiles. Actuellement, de telles études sur les systèmes discoïdes commencent à se développer, mais les résultats sont encore restreints. Cependant, des informations ont été apportées principalement par l'étude des textures et défauts et par l'étude des conformations moléculaires (résonance magnétique nucléaire, calculs énergétiques conformationnels, etc.).

I. Les deux classes de mésophases — La nomenclature

On peut distinguer deux types de mésophases "discotiques ou discoïdes".

Il existe une phase fluide où l'ordre des centres de gravité des molécules est analogue à celui d'un liquide isotrope. Cependant, les plans des molécules restant plus ou moins parallèles entre eux, cette phase est anisotrope [7]. Elle ne diffère apparemment de la phase nématique constituée de molécules allongées que par la forme des molécules. On l'appelle donc nématique "discotique", l'autre nématique étant "calamitique" [8]. La mésophase carbonée [9], obtenue à partir de substances organiques et étape intermédiaire à la formation de graphite, est constituée d'un mélange de molécules plates et appartiendrait à la même classe de mésophases. Il est possible d'obtenir aussi une phase cholestérique à partir de mésogènes discoïdes chiraux [10].

La première mésophase thermotrope discotique découverte est constituée par des colonnes de molécules [1]. Ces colonnes sont parallèles les unes aux autres et peuvent glisser parallèlement à leur axe. Elles sont disposées aux nœuds d'un réseau plan. Depuis, diverses phases en colonnes ont été observées. Elles diffèrent les unes des autres, soit par le type de réseau que forment les colonnes, soit par le mode d'empilement des molécules dans la colonne. Jusqu'ici, les mésophases discotiques découvertes sont soit nématiques ou cholestériques, soit en colonne avec réseau régulier plan [9]. Les deux grandes classes de mésophases discotiques connues sont donc la classe nématique et celle des phases en colonnes. Ces dernières ont leur équivalent, pour les molécules allongées, dans les phases smectiques, c'est-à-dire lamellaires. Soulignons cependant que les premières phases en colonnes ont été mises en évidence par A. Skoulios et al. [11] dans les savons anhydres, donc pour des molécules allongées, mais nous nous limiterons ici au cas des molécules discoïdes.

Le nombre de phases en colonnes actuellement existantes est suffisamment important pour que se pose le problème de la nomenclature. On peut attribuer à chaque nouvelle phase, dans l'ordre des découvertes, une lettre ou un numéro d'ordre [8] ainsi que cela se pratique pour les smectiques. S'il a l'avantage de la simplicité, ce système présente quelques inconvénients : nécessité de connaître la liste des phases et leurs caractéristiques, risque de confusion à la suite d'erreur d'identification ou de baptême simultané non concerté, changement de phase à l'intérieur d'une phase apparue d'abord comme unique (c'est le cas pour la plus simple des phases smectiques : la phase

smectique A [12]). En fait, le principal intérêt de ce type de nomenclature est de permettre de nommer les phases identifiées, en se basant sur des critères de miscibilité, sans que cela nécessite aucune connaissance de leur structure. En effet, plusieurs mésophases ont été identifiées bien avant que la structure en soit connue. La situation dans le cas des mésophases discotiques est différente, car il a été possible de déterminer au moins le réseau plan pour la plupart des mésophases en colonnes connues. Ces mésophases en colonnes se caractérisent tout d'abord par le réseau formé par les colonnes. Ces réseaux étant périodiques rentrent dans le cadre des quatre-vingt groupes spatiaux plans ; ces groupes sont ceux des deux-cent-trente groupes cristallins pour lesquels il n'y a pas d'éléments de symétrie faisant intervenir des translations parallèlement à l'axe des colonnes. On peut distinguer les réseaux obliques dérivés du triclinique ou du monoclinique, les réseaux rectangles centrés ou non dérivés du monoclinique ou de l'orthorhombique, les réseaux carrés, triangulaires de type trigonal ou hexagonal.

A la connaissance du groupe spatial plan, il faut ajouter des informations sur le mode d'empilement des molécules dans une colonne. Il peut être quasi régulier, ou avec désordre de type liquide. On peut envisager des surstructures, par exemple avec empilement hélicoïdal des cœurs. Enfin, le plan des molécules peut être incliné par rapport à l'axe des colonnes. Cette dernière information peut être incluse dans la connaissance du groupe spatial. Il est évident qu'il n'y a pas a priori de relation biunivoque entre groupe spatial et mésophase définie, et a fortiori entre maille et mésophase, mais cette situation est aussi rencontrée dans la description des phases cristallines. M'intéressant ici aux structures, j'utiliserai la description en groupes spatiaux avec les notations des tables internationales de cristallographie [13], choisissant par convention l'axe c parallèle à l'axe des colonnes. Pour désigner plus rapidement les divers réseaux, j'utiliserai les abréviations h pour hexagonal, r pour rectangle, ob pour oblique en indice d'un D pour phase discotique. De plus, la notion d'ordre dans la colonne sera aussi évoquée en indice. Cette nomenclature n'est certes pas universelle, ni à l'abri d'erreur d'identification, mais me semble appropriée à cet exposé.

II. Structures des phases en colonnes

Pour décrire ces structures, je parlerai successivement du réseau de l'ordre dans la colonne et des conformations des chaînes paraffiniques.

a) Le réseau

Tous les mésogènes connus présentent au moins une phase en colonnes. Ces colonnes forment toujours un réseau régulier plan et ce réseau est donc la caractéristique essentielle de chaque phase. L'existence de ce réseau régulier, ses symétries et ses dimensions sont mises en évidence au moyen de la diffraction des rayons X. La figure 1a donne en exemple des diagrammes de diffractions obtenus avec un échantillon d'hexapentyl-oxytriphénylène (pour plus de détails sur la constitution des molécules citées ici, se reporter à l'article de C. Destrade, H. Gasparoux, P. Foucher, Nguyen Huu Tinh de ce numéro) ; cet échantillon contient essentiellement un gros monocristal, d'autres petits cristaux sont aussi présents dans le volume irradié par le faisceau. Le faisceau de rayons X est monochromatique et convergent, il focalise dans le plan de la photographie. L'échantillon est fixe. Le cliché 1a correspond au cas où les colonnes

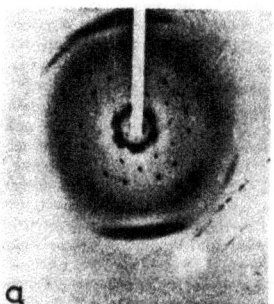

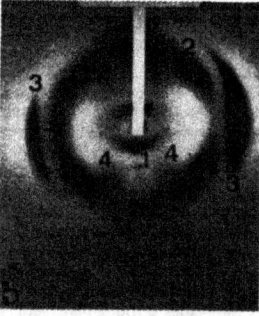

Fig. 1. — Diagrammes de diffraction d'un monocristal d'hexa-pentyloxytriphénylène.

a) Le faisceau monochromatique CuKα est parallèle à l'axe des colonnes. Les taches de diffraction forment un réseau hexagonal et sont reliées par des lignes diffuses. Les traînées diffuses intenses visibles à l'extérieur du cliché proviennent de petits cristaux orientés différemment.

b) Le faisceau fait un angle voisin de 90° avec l'axe ZZ' des colonnes. On distingue quelques taches de diffraction (1), les arcs de cercles provenant des interférences entre chaînes paraffiniques (2) et les lignes diffuses correspondant à l'intersection des plans réciproques perpendiculaires à l'axe ZZ' avec le plan d'observation. En (3) des disques pleins, mais surexposés sur le cliché font la preuve de l'empilement régulier des molécules. En (4) des couronnes larges ont leur origine dans la disposition en hélice des molécules dans chaque colonne.

La distance séparant l'échantillon du film photographique est de 40 mm.

du cristal principal sont sensiblement parallèles au faisceau incident. On observe alors des taches de Bragg aussi fines que celles que l'on obtiendrait avec un monocristal. Ces taches forment un réseau hexagonal plan. Le cliché de la figure 1b est obtenu en tournant le cristal ; les colonnes sont sensiblement perpendiculaires au faisceau, seules 3 taches de Bragg alignées sont visibles. Elles appartiennent au réseau plan hexagonal visible sur la photographie 1a. Puisque les nœuds de Bragg sont tous localisés dans un seul plan réciproque, nous avons donc diffraction par un empilement d'objets ordonnés dans un plan ; ces objets sont les colonnes, elles forment ici un arrangement hexagonal régulier (Fig. 4a). Ces taches de Bragg sont fines dans toutes les directions et quelle que soit la valeur du vecteur de diffusion, ce qui signifie que le réseau est bien défini et que les colonnes sont étendues (leur longueur est grande devant la longueur d'onde des rayons X et elle dépasse plusieurs centaines d'angströms). Remarquons toutefois que de nombreuses taches sont observées simultanément sur le cliché bien que l'échantillon soit fixe ; cela signifie que le monocristal observé n'est pas

parfait ; il y a des désorientations de l'axe des colonnes d'environ ± 6° dans le volume éclairé par le faisceau, soit 0,25 mm² × 1,5 mm. L'intensité diffractée ne semble pas décroître de façon sensible lorsque le vecteur de diffusion augmente, du moins si l'on excepte les six premières taches équivalentes qui ont une intensité d'au moins deux ordres de grandeur au-dessus. Par conséquent, le désordre du réseau, de type agitation thermique, n'est pas très élevé et le déplacement quadratique moyen dans le plan du réseau reste inférieur à l'angström. La forte intensité des 6 premières taches s'explique par le facteur de forme de la colonne ; ce type de contraste est d'ailleurs fréquent dans les cristaux moléculaires. Notons aussi que cette grande intensité les fait apparaître plus larges sur le cliché 1a. En effet, sur ce cliché, nous ne distinguons pas la tache de Bragg, surexposée, de la diffusion d'origine thermique qui l'entoure (un film très peu exposé permettrait de vérifier la finesse de cette tache). En fait, les diagrammes de diffraction de la figure 1 représentent un cas idéal et nous n'avons que rarement obtenu de bons monocristaux des mésophases en colonnes. Cependant, la perfection du réseau bidimensionnel semble générale dans ces phases ; seul le coefficient d'atténuation de l'intensité des raies est souvent plus important, autrement dit le déplacement quadratique moyen des axes des colonnes peut atteindre quelques angströms.

Le réseau le plus simple est le réseau hexagonal. On le rencontre dans la mésophase des hexaethers de triphénylène[14] pour la phase haute température des hexaesters de truxène et de benzo[1,2-b : 3,4-b' 5,6-b''] tris benzo-furanne, c'est-à-dire pour des molécules qui peuvent admettre un axe de symétrie ternaire. Une phase hexagonale similaire s'observe dans les systèmes lyotropes ainsi d'ailleurs que dans les savons secs comme nous l'avons remarqué ci-dessus. En effet, les molécules de savon peuvent s'assembler en cylindres qui s'empilent suivant un réseau hexagonal. On observe alors deux types de phase hexagonale selon que les têtes polaires et leur interface avec l'eau se trouvent à l'intérieur ou à l'extérieur du cylindre. Notre mésophase hexagonale se rapprocherait de la première de ces phases lyotropes.

Compte-tenu de la symétrie des molécules, les colonnes peuvent avoir soit un axe ternaire, soit un axe senaire, soit un axe d'ordre infini. Les groupes spatiaux de plus haute symétrie compatible avec l'existence d'axes ternaires seraient les groupes P $\bar{3}$ 1 2/m ou P $\bar{3}$ 2/m 1. Si nous avons un axe senaire ou infini, le groupe pourrait être P 6/m 2/m 2/m. Le diagramme de rayons X ne permet pas de choisir entre ces groupes. Cependant, dans le cas des hexaethers de triphénylène, des observations de défauts sont en faveur de la symétrie P 6/m 2/m 2/m. En effet, on peut observer deux lignes de disinclinaison S = 1/2 à angle droit[15], une ligne S = 1/2 correspond à une rotation des colonnes d'un demi tour autour d'un axe qui leur est perpendiculaire. Dans le cas du groupe P 6 2/m 2/m 2/m, le défaut se réduit aux deux lignes mises en évidence tandis que le groupe P $\bar{3}$ 1 2/m ou P $\bar{3}$ 2/m 1 impliquerait l'existence d'une paroi parallèle aux colonnes et prolongeant l'une des lignes de disinclinaison (Fig. 2). Le réseau de part et d'autre de cette paroi n'a pas la même orientation, mais les deux orientations sont symétriques par rapport à un plan. Une telle paroi ne peut être décelée optiquement, mais on peut cependant supposer que son existence est énergétiquement peu favorable et rendrait improbable l'observation de deux lignes ainsi associées dans le cas de symétrie $\bar{3}$.

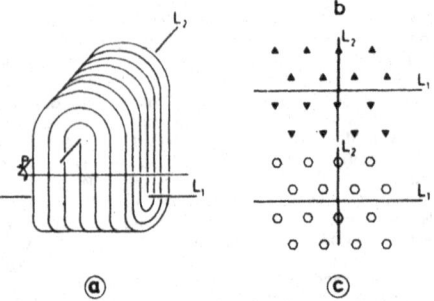

Fig. 2. — Association de deux lignes de disinclinaison L_1 et L_2 à angle droit. La disposition des colonnes dans ce cas est représentée en (a). Une coupe dans le plan P permet de visualiser le réseau plan formé par les colonnes en (b) cas du réseau P $\bar{3}$ 1 2/m. Chaque colonne, de symétrie $\bar{3}$, 2/m, peut être représentée par deux triangles parallèles entre eux, symétriques par rapport à un point situé sur l'axe qui joint leurs centres. Seuls les triangles situés au-dessus du plan P figurent ici. On constate que les deux lignes de disinclinaison s'accompagnent d'une paroi passant par l'une d'elles. Le cas du réseau P $\bar{3}$ 2/m 1 est analogue, mais la paroi est dans une direction perpendiculaire. En (c) cas du réseau P 6 2/m 2/m où chaque colonne est figurée par un hexagone ; aucune paroi n'accompagne les lignes de disinclinaisons.

Un second type de réseau est observé à basse température pour les esters de triphénylène à chaînes longues $n > 10$. La figure 3 donne un exemple de cliché obtenu dans ce cas avec un monocristal. L'apparence hexagonale reste respectée. Cependant, l'apparition de taches supplémentaires indique que la symétrie est celle d'une phase rectangle non centrée à deux colonnes par maille. Les deux colonnes se déduisent l'une de l'autre par une symétrie suivie de translation, miroir avec glissement ou axe hélicoïdal 2_1. Dans un tel réseau, l'axe des colonnes est au plus un axe binaire. On peut donc figurer la section des colonnes par une ellipse. La disposition des colonnes est alors celle représentée figure 4b.

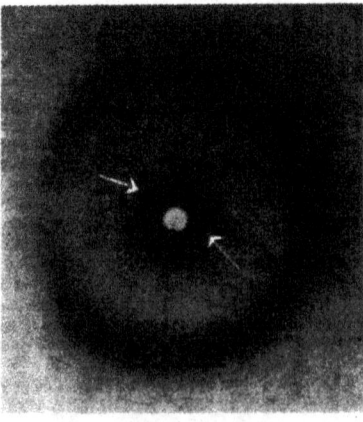

Fig. 3. — Cliché de diffraction d'un monocristal de la phase rectangle de l'hexadodécanoate de triphénylène. Les taches de Bragg supplémentaires qui apparaissent lors de la transition de la phase hexagonale vers la phase rectangle sont repérées par des flèches. Distance film-échantillon : 73 mm.

Cette disposition rappelle l'arrangement en chevrons des cycles phényls des molécules calamitiques en phase S_E ou S_H. Ce réseau est celui que l'on rencontre le plus fréquemment aussi bien pour des molécules peu symétriques comme l'hexaoctanoyloxy-anthraquinone [16] que pour des molécules ayant une symétrie ternaire ou sénaire [17]. Ainsi, contrairement à ce qui avait été annoncé [1], les hexaesters de benzène présentent une phase en colonne ayant cette symétrie ; de même, on la rencontre dans les hexaesters de triphénylène, soit seule, soit succédant à basse température à une phase hexagonale. Ce type de transition est aussi bien observé pour les hexaesters de truxène [18] et de benzotribenzofuranne [19]. Enfin, cette phase est aussi présente dans certains dérivés de l'hexabenzoate de triphénylène [20]. Il faut remarquer que les colonnes sont aux nœuds d'un réseau hexagonal très peu déformé (sauf pour l'anthraquinone qui n'a pas la symétrie ternaire). Cependant, les raies de Bragg 11 et 20 du réseau rectangle qui seraient toutes les deux équivalentes à la raie 10 d'un réseau hexagonal ne sont pas d'intensité équivalente, la raie 20 étant parfois presque éteinte. Ceci suppose donc que les colonnes sont loin d'être circulaires.

Pour comprendre la cause de cette déformation des colonnes, il faut tout d'abord connaître le groupe spatial. Dans le cas des molécules achirales, deux groupes spatiaux $P2_1/a$ et $P2_1/b$, $2_1/a$, 2/m peuvent décrire cette maille. Le second implique que les colonnes possèdent un axe binaire dans leur direction d'allongement, mais nous n'avons pu obtenir de clichés de rayons X permettant de trancher entre les deux (avec des molécules optiquement actives, les groupes spatiaux seraient soit $P2_1$, soit $P2_12_12$). L'observation au microscope permet de trancher. En effet, Franck et Chandrasekhar [21] ont observé les textures de l'hexaoctanoate de benzène. Il est possible de déterminer la direction des colonnes par l'observation de la direction des parois ou lignes de défauts. Ces auteurs ont constaté que, entre polariseurs croisés, il existait deux pinceaux noirs perpendiculaires entre eux, mais ni perpendiculaires, ni parallèles à l'axe des colonnes. Il en résulte que les axes principaux de l'ellipsoïde des indices ne sont ni parallèles, ni perpendiculaires à l'axe des colonnes. Par conséquent, le réseau des colonnes est décrit par le groupe $P2_1/a$; il n'y a donc pas de raison pour que les molécules soient planes et perpendiculaires à l'axe des colonnes. Nous reviendrons plus en détail sur le problème de l'orientation des molécules dans la colonne, lorsque nous décrirons l'ordre dans une colonne. Il faut toutefois remarquer que le réseau $P2_1/a$ implique deux colonnes par maille, ces deux colonnes n'étant pas identiques.

Les deux réseaux que nous venons de décrire sont les réseaux les plus fréquemment rencontrés. Nous avons observé d'autres réseaux, mais ils correspondent à des cas isolés. Ainsi, pour les hexaesters de benzotrisbenzofuranne, à la phase rectangulaire succède une seconde phase de réseau oblique [18] qui correspond à une déformation du réseau rectangle. Les deux vecteurs du réseau sont peu modifiés en module, mais l'angle qu'ils font entre eux passe de 90° dans la phase rectangle à 97°5 (Fig. 4c) dans cette phase oblique. Dans l'hexaoctanoyloxy-anthraquinone, il existe une deuxième mésophase en colonne monotropique. La surface de la maille est doublée. Dans cette phase, il y a donc 4 colonnes par maille [16]. Le groupe d'espace pourrait être P 2/a ou P m a 2 (dans ce dernier cas, les molécules seraient perpendiculaires en moyenne à l'axe des colonnes). Etant donné les symétries observées, on peut proposer une structure dans laquelle

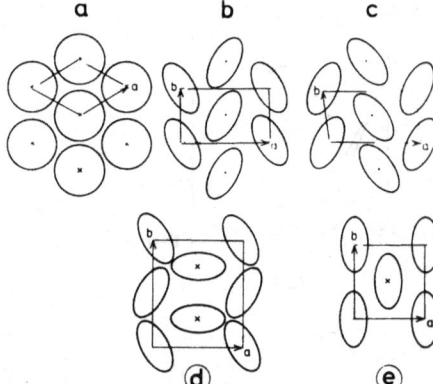

Fig. 4. — Disposition des sections des colonnes (schématisée ici par des cercles ou des ellipses) dans les réseaux rencontrés a) : Réseau hexagonal (P 6 2/m 2/m) ; b) rectangle (P 2_1/a) ; c) oblique (P_1) ; d) réseau rectangle de la phase de basse température de l'hexaoctanoate du ruffigallol (P 2/a) ; e) réseau rectangle centré (C 2/m).

les colonnes de section elliptique prennent trois orientations dans le plan du réseau (Fig. 4d). Enfin, l'hexahexyloxybenzoate de triphénylène ne présente qu'une mésophase en colonnes qui est très différente des autres. Le réseau est rectangle centré et les deux côtés du rectangle sont sensiblement de même longueur (Fig. 4e [20]). Les clichés de rayons X permettent de mettre en évidence la direction des plans des molécules. On constate que ces plans font un angle d'environ 40° avec l'axe des colonnes et que toutes les colonnes sont identiques. Le groupe spatial serait donc C 2/m. Il n'a pas été possible de déterminer la direction de l'axe binaire sur le réseau rectangle, car il n'y a pas de diagramme de monocristaux de qualité suffisante.

Tous les réseaux que nous avons observés correspondent à des déformations plus ou moins importantes du réseau hexagonal. La plus grande déformation est observée pour le réseau rectangle C 2/m et est liée au fait que les molécules sont fortement inclinées sur l'axe des colonnes et ont toutes la même direction. Dans ce cas, la surface moyenne par colonne est faible tandis que pour les autres substances la transition hexagonale → rectangle lorsqu'elle existe ne s'accompagne pas d'un changement notable de la surface par colonne.

Il faut aussi noter que la surface par colonne est toujours inférieure à celle de la molécule prise dans sa configuration la plus étendue. On peut calculer la surface spécifique : rapport entre la surface offerte à une colonne et la masse moléculaire. Cette quantité décroît légèrement lorsque le nombre de carbones des chaînes paraffiniques croît, mais cependant, varie peu d'une molécule à l'autre. On ne note que trois exceptions : l'hexaoctanoyloxy-anthraquinone, l'hexaoctanoyloxy-benzène et l'hexahexyloxybenzoate de triphénylène qui ont une surface spécifique nettement plus petite. Cette distinction est le reflet d'une différence dans l'arrangement des molécules dans les colonnes (distance entre cœurs ou angle d'inclinaison des molécules).

b) L'ordre dans la colonne

Pour avoir des informations complètes sur le mode d'empilement des molécules dans une colonne, il est nécessaire d'obtenir le diagramme de diffraction par des échantillons orientés. Jusqu'à présent, nous ne

disposons de tels diagrammes que dans peu de cas. Pour les autres systèmes, il faut opérer par comparaison, soit entre mêmes phases de divers composés, soit entre les résultats obtenus par diffraction des rayons X et d'autres propriétés physiques. Examinons tout d'abord ce que nous apporte l'analyse d'un cliché sur échantillon orienté tel que nous en avons obtenu figure 1b pour l'hexapentyloxythriphénylène. Dans ce cas, outre les taches de Bragg, nous distinguons d'une part une intensité diffusée localisée sur des arcs de cercle, le rayon du cercle correspondant à un vecteur de diffusion $\dfrac{4\pi \sin \theta}{\lambda} = \dfrac{1}{4,5}$Å^{-1}, d'autre part une intensité diffusée localisée sur des lignes plus ou moins larges de direction perpendiculaire à l'axe des colonnes. Enfin, les taches de Bragg sont entourées de taches diffuses allongées dans la direction de l'axe des colonnes.

L'intensité diffusée localisée sur des arcs de cercle traduit l'organisation des chaînes paraffiniques. Celles-ci sont à l'état fondu et, dans le cas considéré ici, elles ont une orientation privilégiée non perpendiculaire à l'axe des colonnes. La présence de lignes plus ou moins larges sur le cliché provient d'une intensité diffusée dans des plans de l'espace réciproque perpendiculaires à l'axe des colonnes. Ainsi l'espace réciproque est-il constitué d'un plan équatorial dans lequel l'intensité diffractée est localisée aux nœuds d'un réseau plan, et de plans continus tous parallèles au plan équatorial. Sur la figure 1b, on constate qu'il y a deux sortes de plans : les uns sont fins et sont des disques centrés sur l'axe parallèle aux colonnes et passant par l'origine ; les autres environ dix fois plus larges ont une intensité diffusée répartie en couronne. Cette figure de diffraction s'explique entièrement si l'on considère que dans une colonne les cœurs des molécules (triphénylène et oxygènes) sont perpendiculaires à l'axe de la colonne et régulièrement espacés de 3,6 Å. De plus, l'orientation de ces cœurs dans leur plan est définie, puisqu'ils forment une hélice de pas 13,7 Å. La longueur de cohérence pour l'empilement régulier est d'environ 300 Å tandis qu'elle n'est que d'environ 30 Å pour l'enroulement en hélice. Il n'y a pas de corrélations entre les positions et orientations des molécules de deux colonnes voisines ; ceci tient au fait que les chaînes paraffiniques ne s'ordonnent pas comme les cœurs.

Enfin, la diffusion entourant les taches de Bragg est en fait localisée dans des plans parallèles à l'axe des colonnes et passant par des rangées du réseau hexagonal. Cette intensité provient de la diffusion par les modes d'ondulation des colonnes qui sont de basse fréquence.

A partir des conclusions tirées de l'interprétation du cliché de la figure 1b, il nous est possible d'étendre notre analyse à des systèmes peu ou pas orientés.

Notons que nous observons toujours un anneau de diffraction caractéristique de l'état fondu des chaînes paraffiniques. Le plus souvent, l'intensité diffusée par la partie centrale non aliphatique est localisée en dehors de cet anneau. On peut donc distinguer le mode d'empilement des cœurs et le séparer de celui des chaînes. On observe une grande variété dans l'organisation des cœurs. En effet, il peut y avoir un ordre régulier tel que nous l'avons décrit pour l'hexapentyloxytriphénylène ou un ordre de type liquide ; la longueur caractéristique de l'ordre linéaire peut varier, le maximum semblant atteint dans le cas de l'hexapentyloxytriphénylène.

Dans le cas de réseaux non hexagonaux, il n'y a jamais d'ordre régulier des molécules dans une colonne. La largeur de la diffusion correspondante montre que la longueur de cohérence ne dépasse pas quelques

molécules. La distance moyenne entre deux cœurs
est variable. Elle atteint 5,6 Å pour l'hexaoctanoyloxy-
anthraquinone ; elle est d'environ 4 Å dans les hexaesters de triphénylène ou de truxène et la distance de
3,8 Å est observée dans la phase rectangulaire de symétrie C 2/m de l'hexahexyloxybenzoate de triphénylène. Dans le cas des réseaux hexagonaux, on peut
avoir un ordre régulier avec une longueur de corrélation
variable entre 100 et 10 molécules, mais il y a aussi des
cas où l'ordre est de type liquide. Ainsi la transition
de réseau P 6 2/m 2/m → P 2₁/a observée dans les
hexaesters de triphénylène à longue chaîne ne s'accom-
pagne-t-elle d'aucun changement apparent dans la répartition des molécules le long d'une colonne, l'ordre
étant de type liquide dans les deux cas. En revanche,
la même transition s'accompagne d'un changement
de la longueur de corrélation de l'empilement régulier : elle est d'environ 30 Å dans le phase hexagonale alors que l'ordre est de type liquide dans la
phase rectangle. La même transition est également
observée dans des mélanges binaires d'éther et d'ester
de triphénylène. Cette transition n'a lieu que pour un
domaine très étroit en concentration (environ 80 %
d'octanoate pour 20 % d'octyloxytriphénylène). La
longueur de corrélation est là encore d'environ 30 Å
dans la phase hexagonale. La nature des groupes liant
le cœur à la chaîne paraffinique semble influer de
manière prépondérante sur la portée de l'ordre linéaire
dans les colonnes. En effet, tous les éthers de triphénylène ou de truxène ne présentent qu'une phase hexagonale dans laquelle l'ordre linéaire a une portée d'au
moins 100 Å. Au contraire, dans les esters, la phase
hexagonale n'existe pas toujours et n'apparaît ordonnée
que dans le cas de certains esters de truxène. Il apparaît
donc que la présence de groupes carboxyliques amenant
des atomes d'oxygène de part et d'autre du plan formé
par les noyaux aromatiques favorise un empilement
désordonné dans la colonne. Ces observations sont
cohérentes avec les conclusions d'une étude conforma-
tionnelle d'empilement de molécules d'éther et d'ester
de triphénylène [22]. Cependant, l'addition de 50 % de
molécules d'ester isomère dans un éther de triphénylène
ou de truxène stabilise la phase hexagonale (Fig. 5),
la longueur de corrélation de l'ordre linéaire restant élevée.
De plus, la phase hexagonale reste la seule mésophase,
même si la proportion de molécules d'ester est élevée.
Ainsi, dans le mélange d'hexaoctanoyloxy et d'hexaoctyloxytriphénylène, la zone de démixtion sépare-t-elle les
phases hexagonale et rectangle correspond-elle à une
concentration de plus de 75 % de molécules d'esters.
La portée de l'ordre linéaire diminue lorsque la
concentration d'ester passe de 50 à 80 % [18]. Il semble
donc qu'en fait une interaction molécule à molécule
entre alkanoyloxy et alkoxytriphénylène favorise
l'ordre dans la colonne. Le même effet se rencontre
aussi pour un mélange d'éthers et d'esters isomères
de truxène. De plus, l'étude de la série des esters de
truxène permet de constater que, dans une même
série, la phase hexagonale qui, dans ce cas, succède
à une phase rectangle, peut être ordonnée pour les
homologues à chaîne courte ($n < 11$) et désordonnée
pour les homologues supérieurs. Il n'y a pas eu cependant dans ce cas d'expériences permettant de préciser si,
tout en conservant le même type de réseau, il est
possible de passer continûment de colonnes ordonnées
à des colonnes désordonnées. En outre, pour bien définir l'ordre, il faut connaître la figure de diffraction en
détail. Il faut pouvoir comparer les divers ordres de
diffraction correspondant à la même périodicité tant du
point de vue de leur largeur que de leur intensité. En
effet, dans un cristal parfait, même petit, les ordres de

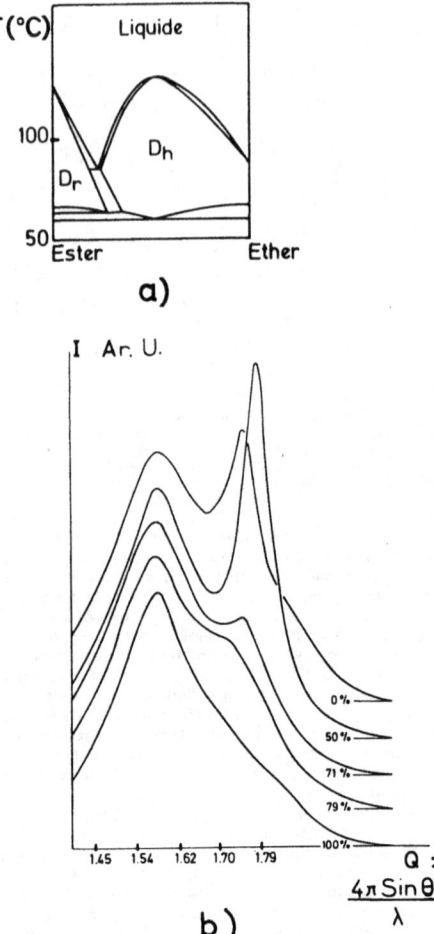

Fig. 5. —
a) Diagramme de phase du mélange binaire hexaoctanoate
de triphénylène (à gauche) et hexaoctyloxytriphénylène (à
droite).
b) Enregistrements microdensitométriques des clichés de
rayons X obtenus à 80°C avec des échantillons partiellement
orientés du mélange à diverses concentrations. L'enregistrement
se fait selon un axe passant par l'origine et parallèle à la direction
d'alignement. On peut ainsi comparer l'intensité diffusée par les
chaînes paraffiniques qui n'évolue pas et a son maximum vers
$q \simeq 1,60 \text{ Å}^{-1}$ à celle provenant de l'interférence entre cœurs
de molécules appartenant à une même colonne, son maximum
situé vers $q \simeq 1,80 \text{ Å}^{-1}$ est le plus pointu pour le mélange
équi-moléculaire.

diffraction successifs ont tous la même largeur angulaire
qui est inversement proportionnelle à la taille du cristal.
La décroissance de l'intensité pour les ordres élevés
est liée alors aux fluctuations de position des atomes
ou molécules autour d'un point moyen parfaitement
défini. Si, au contraire, ce point moyen n'est pas défini
par une relation de périodicité, sauf à courte distance,
les ordres de diffraction sont d'autant plus larges qu'ils
sont plus élevés. Pour des raisons géométriques, nous

n'avons pu observer que deux ordres de diffraction correspondant à l'ordre dans la colonne. Quand il existe, le deuxième est de faible intensité, mais ne semble pas beaucoup plus large que le premier. Ceci prouve que la cohérence est grande, mais que les fluctuations longitudinales des positions des cœurs sont cependant importantes (la valeur quadratique moyenne du déplacement atteint environ 1 Å^2).

Après avoir considéré le mode d'empilement des molécules, il faut aussi définir l'orientation des plans aromatiques par rapport à l'axe de la colonne. En effet, si dans la phase hexagonale ces plans sont en moyenne perpendiculaires à l'axe des colonnes, nous avons vu que l'observation des propriétés optiques prouve qu'il n'en est pas de même dans les autres phases. En fait, nous n'avons pas d'informations claires à ce sujet pour la phase oblique ou pour la phase rectangle basse température du dérivé de l'anthraquinone.

Dans le cas de la phase rectangle centrée observée dans l'hexahexyloxybenzoate de triphénylène, nous avons obtenu des diagrammes de rayons X de gros monocristaux : on voit très nettement une tache diffuse caractéristique de l'arrangement linéaire des cœurs dans la colonne (Fig. 6). Il est ainsi possible de localiser la direction des plans des cœurs et de constater qu'elle fait un angle d'environ 55° avec le plan perpendiculaire aux

colonnes contenant les réflexions par le réseau bidimensionnel. D'après la symétrie du réseau, on peut affirmer que l'orientation des noyaux aromatiques est la même pour toutes les colonnes. La situation est différente dans le cas de la phase rectangle de symétrie P $2_1/a$. En effet, dans ce groupe les deux colonnes de la maille se déduisent l'une de l'autre par la symétrie $2_1/a$. Ainsi a-t-on le choix entre deux configurations représentées figure 7 et proposées par Franck et Chandrasekhar [21]. Il n'est pas possible actuellement de distinguer entre ces deux configurations. En effet, nous n'avons jamais obtenu de monocristal de cette phase permettant de préciser l'orientation des cœurs dans les deux types de colonne et leur orientation par rapport au réseau. Les deux réseaux ne diffèrent que par la longueur de l'axe binaire (petit ou grand côté du rectangle, mais les règles d'extinction sont les mêmes). L'observation optique ne permet pas non plus de choisir : elle donne une idée de l'angle d'inclinaison des cœurs, à supposer que sa conformation soit plane, ce qui n'est sans doute pas le cas à cause des oxygènes des groupes carboxyliques.

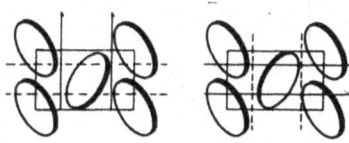

Fig. 7. — Les deux configurations compatibles avec la symétrie P $2_1/a$. Ces deux configurations correspondent au fait que l'axe hélicoïdal 2_1 (→) est parallèle soit au grand côté, soit au petit côté du rectangle, le plan de symétrie avec glissement (---) restant perpendiculaire à l'axe 2_1.

En ce qui concerne la conformation et l'ordre des chaînes paraffiniques, nous savons que c'est un ordre de type liquide, mais leur état est plus ou moins désordonné. En effet, dans l'hexapentoxytriphénylène, la direction des chaînes est bien définie à une vingtaine de degrés au-dessus ou au-dessous du plan des cœurs. Dans l'hexadodécanoate de triphénylène, l'anneau paraffinique est quasiment isotrope pour un échantillon, par ailleurs bien orienté. Dans le cas de l'hexahexyloxybenzoate de triphénylène, nous avons une situation intermédiaire et les chaînes restent préférentiellement dans le plan des cœurs. Nous n'avons eu d'échantillons suffisamment orientés que dans ces trois cas : il est donc difficile d'établir une loi générale.

Ainsi, dans l'ester de triphénylène, le grand désordre des chaînes peut-il résulter de la grande longueur de chaque chaîne et mais aussi du fait que les cœurs de colonnes voisines n'ont pas la même orientation. La connaissance de la surface offerte à chaque colonne nous renseigne également sur la conformation des chaînes. En effet, cette surface est toujours inférieure à la surface d'une molécule isolée avec des chaînes en conformation étirée. Il y a donc déformation ou imbrication des chaînes. Une telle déformation des chaînes est effectivement mise en évidence par l'étude de la résonance magnétique des noyaux de deutérium sur une molécule d'hexahexyloxytriphénylène partiellement deutérié [23]. Il semble aussi d'après ces études que les chaînes ne soient pas situées dans le plan des cœurs. Les études conformationnelles [22] confirment ce résultat au moins pour les premiers chaînons près du noyau aromatique.

Une autre information indirecte sur la conformation des chaînes peut être obtenue par l'étude des distances

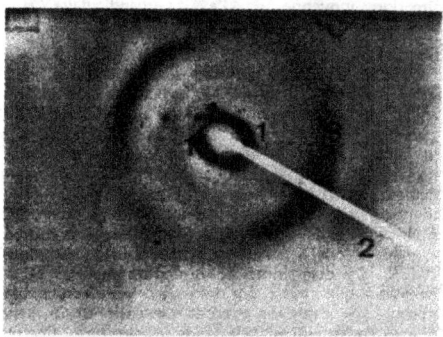

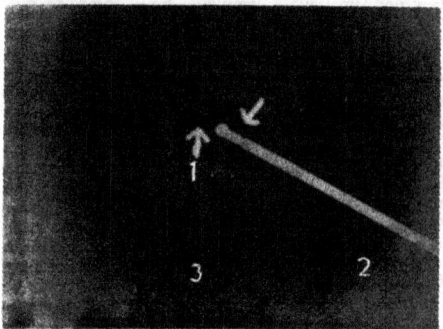

Fig. 6. — Clichés de diffraction de la phase en colonnes de l'hexa-hexyloxybenzoate de triphénylène. Les deux clichés correspondent au même échantillon, la température et les temps de pose étant différents. Plusieurs domaines sont simultanément éclairés par le faisceau de rayons X. Cependant, on voit nettement les réflexions 110 du domaine principal (1) et la tache diffuse provenant des molécules d'une même colonne (2). L'anneau des chaînes paraffiniques (3) est incomplet, car les chaînes sont partiellement orientées par rapport à l'axe des colonnes.

Distance film-échantillon : 59 mm.

entre colonnes dans une série isomorphe. La figure 8 donne la variation du paramètre de réseau dans la phase hexagonale des hexaethers de truxène. La variation de ce paramètre pour des nombres de carbones sur la chaîne variant de 6 à 13 est linéaire. L'accroissement est de 1,04 Å par CH_2 ajouté sur chacune des six chaînes ; ceci correspond à une légère diminution de la densité par unité de surface pour les homologues supérieurs. Toutefois l'augmentation de 1 Å par carbone est inférieure à la variation attendue pour des molécules étendues et non imbriquées. Dans ces composés, l'ordre dans la colonne est bien défini, donc on peut mesurer la distance moyenne entre cœurs, celle-ci varie peu d'un homologue à l'autre et semble passer par un maximum pour les chaînes de longueur intermédiaire. Cependant, la variation reste peu significative.

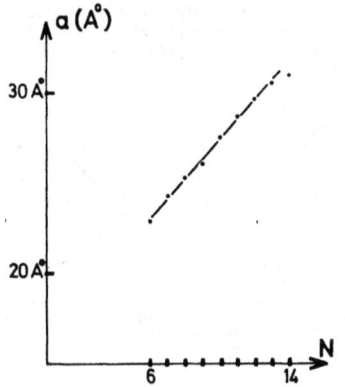

Fig. 8. — Variation du paramètre *a* du réseau de la phase hexagonale des hexaethers de triphénylène en fonction du nombre N de carbones des groupes alcoxy. Mesures effectuées par P. Foucher dans sa thèse de 3e Cycle.

Pour finir, nous donnons un tableau où sont rassemblées les diverses caractéristiques des phases colonnes que nous avons étudiées.

III. La phase nématique

L'existence d'une phase fluide anisotrope constituée d'objets discoïdes tous plus ou moins parallèles entre eux, mais dont les positions des centres de gravité seraient réparties comme dans un liquide isotrope, a été prédite depuis longtemps à partir de considérations théoriques de stabilité [24]. De fait, quelques exemples de telles phases étaient connus avant la découverte des mésogènes thermotropes discoïdes : ainsi, la mésophase carbonée [9], étape intermédiaire dans la graphitation des substances organiques, est-elle une mésophase ayant les caractéristiques de l'ordre nématique et c'est un mélange contenant une proportion importante d'espèces chimiques constituées des noyaux polyaromatiques plans. Cependant, ce mélange est très complexe et nous sommes loin du corps pur bien défini.

Un autre cas se rencontre dans des mélanges ternaires (ou binaires) de savons alcool et eau. Dans certaines conditions bien définies, des micelles ayant la forme d'ellipsoïde de révolution aplaties en suspension dans l'eau, peuvent former une phase nématique [25]. L'existence d'une phase nématique thermotrope de molécules

discoïdes était donc envisageable et cette phase a été en effet mise en évidence pour la première fois par Nguyen Huu Tinh et al. [7] dans les hexa alkoxybenzoates de triphénylène. Cependant, il est vite apparu que l'ordre dans lequel les différentes phases d'un mésogène discoïde se suivent en fonction de la température n'était pas unique. En effet, la phase nématique peut se former à partir de la phase colonne rectangle et à plus haute température, mais l'ordre de ces deux phases peut aussi être inversé, ce qui montre que l'entropie de phase nématique n'est pas nécessairement supérieure à celle de la phase colonne d'un même corps. Il est vrai que certaines molécules allongées peuvent aussi présenter des séquences réentrantes où, en élevant la température, on rencontre une phase nématique, une phase smectique A, une phase nématique [26, 27], avec identité des deux phases nématiques. Cependant, l'inversion de séquence, sinon la rentrance, semble plus fréquente dans le cas des discotiques. Ainsi, si la séquence normale en température croissante D_r (phase colonne rectangle) $\rightarrow N_d$ (nématique discotique) est observée dans les hexalkoxybenzoates de triphénylène, la séquence inverse $N_d \rightarrow D_r$ ou D_{ob} (oblique) est observée dans les dérivés du truxène et du benzotrisbenzofurane, plusieurs phases en colonne pouvant ensuite se succéder. Dans tous les cas, cette phase est relativement fluide et nous pouvons obtenir des échantillons orientés sous l'influence d'un champ magnétique extérieur. Cependant, le directeur (direction moyenne de l'axe perpendiculaire aux disques) se place généralement dans un plan perpendiculaire au champ magnétique. Il est donc nécessaire d'appliquer un champ tournant pour obtenir une orientation unique du directeur perpendiculaire au plan dans lequel tourne ce champ. De fait, au voisinage d'une transition vers une mésophase en colonne, la viscosité du nématique est élevée et l'application d'un champ magnétique successivement dans deux directions perpendiculaires suffit à aligner le nématique, le directeur étant lui-même perpendiculaire aux deux directions du champ.

Nous avons ainsi obtenu des diagrammes de diffraction des rayons X par divers échantillons de nématique (Fig. 9). Ces diagrammes confirment la qualification de nématique que nous avons donné a priori à cette phase, puisqu'il n'y a que des maxima d'intensité diffus, donc absence de périodicité dans toutes les directions. Comme dans un nématique calamitique, nous observons que l'intensité diffractée est localisée dans deux régions distinctes de l'espace réciproque. Aux grands angles, nous observons un maximum localisé au voisinage de l'anneau caractéristique d'une paraffine fondue et cette diffusion est localisée sur deux calottes opposées, elliptiques, ou le plus souvent quasi sphériques. L'effet de l'orientation moyenne des molécules est peu sensible sur cette région du cliché. Il est aussi parfois possible de séparer une diffraction provenant des cœurs mieux ordonnée de celle des paraffines qui sont très désordonnées, mais la séparation reste peu nette. Le caractère d'ordre orientationnel est plus marqué dans la région des petits angles. En effet, on observe de l'intensité diffusée localisée en des points (Fig. 9a) ou sur des lignes parallèles (Fig. 9b) au directeur. Ce dernier cas correspond à ce que l'on attendrait pour le cliché d'une phase nématique et nous pouvons le comparer aux clichés obtenus dans le cas d'un nématique lyotrope discotique [28] (Fig. 10). Les lignes diffuses parallèles au directeur observées dans les deux cas sont en fait l'image de l'intersection d'un cylindre d'axe parallèle au directeur par la sphère d'Ewald ; le diamètre du cylindre est fonction inverse de la distance moyenne entre molécules ou micelles, dans une direction perpendiculaire au directeur. Les calottes sphériques observées aux grands angles sur les clichés de la figure 9 existent aussi sur le

TABLEAU

Caractéristiques du réseau de quelques phases en colonnes.

Nature de la molécule		Réseau des colonnes			Surface spécifique par colonne cm²/g x 10⁷
Coeur	Substituants paraffiniques	Symétrie	Paramètres en Å	Température °C	
Benzène	C_7H_{15}-COO-	P 2₁/a	a = 31,2 b = 18,6	84	1,88
	C_6H_{13}-COO (40)	P 2₁/a	a = 28,5 b = 17,7	86	1,79
Triphénylène	C_5H_{11} 0-	P 6 2/m 2/m	a = 18,95	≃ 80	2,51
	C_7H_{15} 0-	"	a = 22,2	≃ 80	2,82
	C_8H_{17} 0-	"	a = 23,3	≃ 80	2,87
	$C_{11}H_{23}$ COO-	P 2₁/a	a = 44,9 b = 26,4	117	2,52
		P 6 2/m 2/m	a = 26,3	105	2,54
	C_7H_{15} COO-	P 2₁/a	a = 37,8 b = 22,2	100	2,34
	$C_{11}H_{23}$-0-φ-COO-	P 2₁/a	a = 51,8 b = 32,6	165	2,58
	C_6H_{13}-0-φ-COO-	C 2/m	a = 30,7 b = 28,4	185	1,48
Truxène	C_6H_{13} 0-	P 6 2/m 2/m	a = 22,8	80	2,90
	$C_{10}H_{21}$ 0-	"	a = 27,5	80	3,08
	$C_{14}H_{29}$ 0-	"	a = 30,9	80	3,10
	C_9H_{19} COO-	"	a = 26,7	150	2,73
	$C_{13}H_{27}$ COO-	"	a = 30,5	150	2,85
	$C_{11}H_{23}$-0-φ-COO-	C 2/m	a = 44,1 b = 32,7	94	2,08
		C 2/m	a = 50,6 b = 33,8	180	2,47
	$C_{12}H_{25}$-0-φ-COO-	C 2/m	a = 44,6 b = 31,9	84	1,98
		C 2/m	a = 45,6 b = 35,3	180	2,23
Benzo tris benzo furanne	$C_{13}H_{27}$-COO-	P 6 2/m 2/m	a = 30	160	2,75
		P 2₁/a	a = 49,2 b = 29,3	76	2,55
		P1	a = 49,7 b = 28,0 γ = 97,5°	68	2,44
Anthra-quinone	C_7H_{15} COO-	P 2₁/a	a = 34,8 b = 18,1	114	1,78
		P 2/a	a = 34,9 b = 36	95	1,79

cliché de la figure 10, mais à des angles beaucoup plus petits. En effet, parallèlement au directeur, la distance moyenne entre deux micelles de nématique lyotropes est d'environ 40 Å tandis qu'elle est de 4 Å entre deux molécules discoïdes ; la diffraction vient alors se confondre avec l'anneau lié aux chaînes paraffiniques fondues. Notons que ce dernier est présent dans tous les cas, mais l'intensité n'y est pas répartie de la même manière puisque la direction moyenne des chaînes est parallèle au directeur pour le nématique lyotrope et perpendiculaire pour la phase nématique thermotrope discotique. La hauteur du cylindre d'axe parallèle au directeur est du même ordre de grandeur dans les deux cas, mais ceci correspond de fait à des situations fort différentes. Dans le nématique lyotrope, cette hauteur, égale à la distance entre les deux maxima alignés sur le méridien parallèle au directeur, indique que les micelles

ne sont pas empilées les unes au-dessus des autres. Au contraire, dans le nématique thermotrope, les molécules forment encore des colonnes, certes très courtes, puisque leur longueur est inversement proportionnelle à la hauteur du cylindre (soit environ 10 molécules) ; ces fragments de colonnes sont parallèles entre eux et forment des paquets dont la taille est fixée par l'épaisseur du cylindre (environ 5 x 5 colonnes dans un paquet). Nous avons donc formation de groupes cybo-tactiques (29) comme dans le cas des nématiques calami-tiques. Ces groupes sont la manifestation des fluc-tuations critiques existant au voisinage d'une mésophase en colonne. Nous n'avons pas étudié la variation de leur taille avec la température, car le plus souvent la phase nématique n'est stable que sur une dizaine de degrés. Dans l'hexahexylox-benzoate de triphénylène, la phase nématique est stable sur près de 100° C, mais

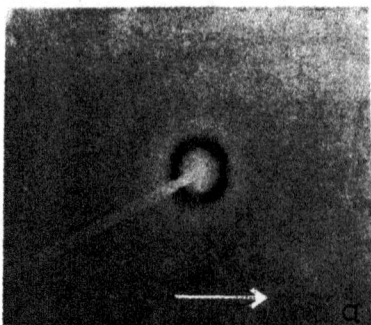

Fig. 9. — Exemples de clichés de la phase nématique de molécules discoïdes alignées dans un champ magnétique. (Les clichés
sont effectués, après avoir tourné l'échantillon de $\pi/2$ autour
d'un axe perpendiculaire au champ, afin d'augmenter le degré
d'orientation). La direction du champ magnétique est repérée
par une flèche.
a) Hexa-hexyloxybenzoate de triphénylène. T = 200°C
   Distance film-échantillon : 67 mm.
b) Hexa-décanoate de truxène. T = 80°C
   Distance film-échantillon : 60 mm.

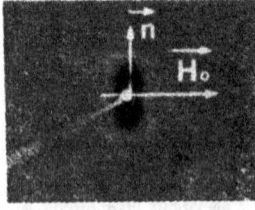

Fig. 10. — Cliché de diffraction d'une phase nématique discotique lyotrope. Le directeur n est perpendiculaire au champ
magnétique d'alignement H_0.
   Distance film-échantillon : 81 mm.
   Cliché transmis grâce à l'obligeance de J. Charvolin.

la gamme de température est trop élevée pour permettre
une telle étude. De plus, le cliché de la figure 9a (qui
correspond à ce composé) diffère de celui décrit ci-
dessus. En effet, le cylindre d'intensité diffractée à
petits angles est renforcé aux deux extrémités (d'où
l'aspect de quatre taches), ce qui signifie que les
colonnes dans un paquet s'empilent dans un plan non
parallèle au directeur, la longueur des colonnes étant
d'ailleurs sensiblement plus grande (environ 30 molé-

cules). Il s'en suit que l'axe des colonnes fait un angle de
40° avec le directeur, les axes des paquets étant répartis
sur un cône. Nous avons donc des groupes cybotactiques
inclinés (Fig. 11). Comme dans les nématiques calamitiques, la présente de groupes cybotactiques n'est pas
liée à l'ordre de succession des phases en fonction de la
température, mais à la symétrie des phases colonnes
voisines, la phase de symétrie C 2/m favorisant la formation de groupes inclinés, la phase de symétrie P 2_1/a
favorisant la formation de groupes droits. Ainsi
l'hexaundecyloxbenzoate de triphénylène, qui possède
une phase nématique entre 176 et 181°C et une phase
D_r de symétrie P 2_1/a entre 160 et 181°, presente-t-il
un diagramme analogue à celui des hexaalkanoates
de truxène dans leur phase nématique, stable au-dessous
de la même phase D_r (P 2_1/a). Le diagramme de la figure
9a a été observé tout récemment[30] dans les deux
régions nématiques des hexaalkoxybenzoyloxytruxène,
la séquence étant :

$$K \text{(cristal)} \leftrightarrow D_1 \text{ (C 2/m)} \leftrightarrow N_d \leftrightarrow D_2 \text{ C 2/m} \leftrightarrow N_d$$

Il semble donc raisonnable de ne parler que d'une
seule phase nématique, puisque l'ordre local seul peut
être modifié d'un système à l'autre cependant que dans
ces mésogènes nous n'avons pas mis en évidence de
"vrai" nématique aussi désordonné que peut l'être le
MBBA (p Methoxybenzylidène p butyl aniline) nématique calamitique. Un tel nématique discotique est mis
en évidence dans le cas des mélanges lyotropes avec des
micelles dont l'anisotropie (rapport du diamètre à l'épaisseur) est d'environ 2 alors qu'elle est d'environ 10 pour
les nématogènes thermotropes discoïdes.

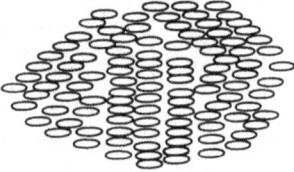

Fig. 11. — Ordre local dans le nématique discotique avec groupes
cybotactiques inclinés.

Nous pouvons même aller plus loin si nous remarquons qu'au voisinage d'une mésophase, le diagramme
de diffraction de la phase liquide isotrope des mésogènes
discoïdes présente deux anneaux, le premier anneau
observé aux petits angles correspond à des distances
moyennes comparables au diamètre de la molécule
(20 à 30 Å). Son intensité et sa largeur ne semblent
pas compatibles avec une image de liquide très désordonné. En effet, on attendrait un maximum large de
faible intensité et correspondant peut-être à des distances plus courtes (< 10 Å). Toutefois, aucune étude
de l'ordre de la phase isotrope n'a été faite. Cependant,
dans des expériences où l'on soumet le liquide isotrope
à des contraintes mécaniques, celui-ci tend à réagir
comme une phase anisotrope[31].
En conclusion à ce paragraphe sur la phase nématique, il faut noter qu'étant moins fréquente que la
phase en colonne, les expériences portant sur cette
phase restent très fragmentaires, en particulier nous
n'avons pas d'informations qui justifieraient de façon
évidente les séquences inversées. Toutefois, remarquons
que les chaînes paraffiniques doivent contribuer de
manière non négligeable à l'entropie et qu'en fait nous
n'avons pratiquement pas d'informations sur les va-

riations de leurs conformations dans les mésophases. Un dernier mot concerne la phase cholestérique [10] analogue de celle des mésogènes calamitiques : elle s'oriente en champ magnétique et le diagramme de diffraction est en tout point semblable à celui d'un nématique (Fig. 12).

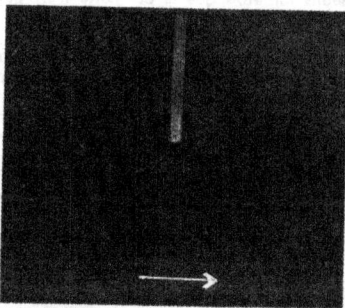

Fig. 12. — Diagramme de diffraction d'un cholestérique aligné par un champ magnétique dont la direction est repérée par une flèche.

IV. Existe-t-il d'autres phases discotiques ?

Nous avons passé en revue les structures des mésophases discotiques et on peut remarquer qu'à une exception près toutes les molécules formant de telles mésophases ont un cœur plat, principalement aromatique et de symétrie ternaire ou sénaire ; de plus, toutes les molécules sont entourées de six chaînes paraffiniques. Ces six chaînes peuvent éventuellement être de longueurs inégales, mais une trop grande différence de longueur entraîne la disparition de la mésophase [32]. D'autres molécules de formes différentes ont été synthétisées ; la formation de mésophases thermotropes a été annoncée pour ces molécules et souvent il a été possible d'étudier la diffraction des rayons X dans ces systèmes. Je vais rapidement donner les caractéristiques structurales que l'on peut extraire de ces études et les conclusions qu'il est possible de donner sur la structure de la mésophase.

Parlons tout d'abord du di-isobutyle silane diol. J.D. Bunning et al. [33] montrent que ce composé connu depuis longtemps présente une mésophase entre 88° et 98° et ils identifient, par miscibilité, cette phase à celle des hexaalkanoates de benzène. Il faut remarquer cependant que les résultats obtenus par diffraction des rayons X par ces mêmes auteurs jettent un doute sur l'identité des deux phases. En effet, deux anneaux de diffraction sont visibles, mais aucun d'entre eux n'est aussi fin que l'anneau aux petits angles d'un smectique. Or, toutes les mésophases en colonne présentent des anneaux fins aux petits angles tout à fait comme les smectiques. Certes, le premier anneau qui correspond à une distance réticulaire de 11 à 12 Å est plus fin que celui d'un liquide isotrope classique, mais semble d'une largeur comparable à celui de la phase nématique classique. On peut, en effet, dans ce cas, estimer à environ 15 % la fluctuation de distance moyenne entre deux molécules dans une direction perpendiculaire au directeur. Le nombre d'anneaux observés est aussi étonnant : en effet, avec un réseau bidimensionnel, l'ordre est en général bien défini et on observe plusieurs ordres

de réflexions, donc plusieurs raies fines aux petits angles. Avec les molécules discogènes habituelles, le premier anneau de la phase hexagonale (ou doublet pour une phase rectangle) est rendu très intense comparé aux suivants à cause du facteur de structure moléculaire. En fait, l'introduction d'atome de silicium dans le cas du silane devrait modifier le rapport des intensités des différents ordres de réflexions et renforcer les ordres supérieurs. De plus, les textures présentées par ce composé sont analogues à celles observées dans les phases hexagonales [34], en particulier, la présence de grandes zones homéotropes [35] laisse supposer que la mésophase est optiquement uniaxe alors que celle des hexaalkanoates de benzène est biaxe. Les éléments apportés jusqu'à présent ne permettent pas de conclure sur la structure de la mésophase du di-isobutyle silane diol qui a priori apparaît moins ordonnée que toutes les autres mésophases en colonnes connues.

Plusieurs autres mésophases ont été annoncées. Citons le cas de molécules substituées par quatre chaînes paraffiniques. Dans ce cas, aucun système ne s'oriente aisément et seuls les diagrammes de diffraction sur poudre ont permis d'obtenir quelques informations sur les structures.

Tout d'abord, j'ai étudié deux dérivés de la série des tetraaryl bipyranyl-4-ylidène [36]. L'un est substitué par quatre chaînes de neuf carbones et l'autre par quatre chaînes de douze carbones. Les diagrammes de poudres de la phase mésomorphe présentent beaucoup de raies fines indiquant une structure très organisée. Ils peuvent s'indexer dans une maille monoclinique, mais, dans les deux cas, l'absence de raies hkl avec h ou k et l non simultanément nuls indique qu'il n'y a pas de corrélations entre les positions de deux plans (ool) successifs. Il s'agit donc de phases lamellaires dans lesquelles les lamelles glissent facilement. Les paramètres dans le plan ab sont similaires pour les deux composés, seule l'épaisseur croît avec la longueur de la chaîne, ce qui conduit à supposer que les chaînes se placent sensiblement perpendiculairement aux lamelles en s'alignant deux à deux de part et d'autre du cœur. Une comparaison avec la structure de la phase cristalline du composé non substitué confirme cette hypothèse [37]. En effet, ce dernier a une structure monoclinique de symétrie P $2_1/c$ avec 2 molécules par maille : $a = 6,046$ Å, $b = 17,07$ Å, $c = 14,139$ Å, $\beta = 123°,44$, $\rho = 1,266$ gcm^{-3} alors que pour les dérivés en C$_9$ et C$_{12}$ on a une maille à 4 molécules. Pour le dérivé en C$_{12}$ (115°C) : $a = 12,17$ Å ; $b = 17,8$ Å ; $c = 32,05$ Å ; $\gamma = 92,5°$; $\rho = 1,25$ gcm^{-3}. Pour le dérivé en C$_9$ (125°) : $a = 12,17$ Å ; $b = 17,11$ Å ; $c = 27,04$ Å ; $\gamma = 91,8°$; $\rho = 1,16$ gcm^{-3}. La symétrie pourrait être P $\overline{1}$, Pm ou P $2/m$.

Mis à part le doublement du paramètre a pour les mésophases, les mailles sont remarquablement semblables dans le plan (a, b). L'augmentation d'épaisseur de la couche est de 0,83 Å par groupement CH$_2$, dans l'hypothèse où les chaînes se placent par paire à chaque interface lamellaire. Cet accroissement est plus faible que celui qui serait obtenu dans le cas de chaînes en conformation trans placées perpendiculairement aux interfaces. On peut donc se demander si chaque couche de molécule n'est pas séparée de la suivante par une région paraffinique partiellement fondue favorisant ainsi la décorrélation entre les couches. Ce système serait donc un empilement de couches cristallines à deux dimensions. Par conséquent, la molécule ne répond pas à la description que j'ai donnée en introduction de la forme discoïde, puisque la partie paraffinique n'est pas disposée en couronne autour du cœur.

Un autre système de forme analogue est obtenu par formation d'un complexe de deux molécules de β dicétone comportant chacune deux chaînes alkyl de 10 carbones avec du cuivre divalent [38]. On obtient donc ici un cœur entouré de quatre chaînes. Le diagramme de poudre de la mésophase n'a pas été dépouillé complètement, car certaines raies de diffraction sont élargies, preuve que l'organisation cristalline est imparfaite. Cependant, il est aisé de voir qu'il s'agit d'une phase lamellaire. L'épaisseur des lamelles étant de 29 Å. Signalons qu'il existe une mésophase dans un dérivé de la porphine substitué par huit chaînes paraffiniques [39]. Toutefois, cette phase est fugitive et il ne semble pas que la structure en soit connue. Enfin, un modèle en colonne a été proposé pour expliquer la formation de mésophase lyotrope dans des solutions basiques de tétraalkanoates de phtalocyanine de cuivre.

Nous pouvons donc dire que les molécules substituées par 6 chaînes alkyles peuvent être mésogènes et que les mésophases obtenues sont, sinon isomorphes, du moins toutes analogues formant ainsi des classes de mésophases analogues aux mésophases calamitiques. Il ne semble pas évident que l'on puisse généraliser et obtenir de telles mésophases avec d'autres molécules.

Conclusion

Nos connaissances actuelles sur les mésophases discotiques sont encore très partielles, mais nous pouvons déjà dégager quelques caractéristiques générales concernant ces phases.

Comme dans les mésogènes calamitiques, on peut trouver une phase nématique ou cholestérique et une classe de phases en colonnes, équivalente à celle des phases smectiques. Cependant, des différences sensibles sont perceptibles, car les mésophases discotiques ressemblent plus aux phases lyotropes. Ces différences tiennent aux structures différentes des molécules mésogènes qui les constituent.

La molécule mésogène en bâtonnet est plus souvent formée par deux ou trois noyaux benzéniques liés entre eux par des groupes déformables. La molécule se termine à ses deux extrémités par des chaînes paraffiniques. Il est généralement admis que la fusion de ces chaînes paraffiniques entraîne la formation de phases smectiques, mais cette fusion s'accompagne de déformations du cœur. La structure moyenne des couches smectiques vue par la diffraction des rayons X prend en compte la molécule dans son ensemble sans que l'on puisse séparer les caractères propres à chaque partie. Dans les mésogènes discotiques au contraire, le cœur reste non déformable et l'on voit très nettement apparaître le rôle des chaînes paraffiniques qui forment une zone tampon désordonnée entre des régions de cœurs régulièrement disposées. Le désordre des chaînes correspond alors à certains caractères spécifiques dans le diagramme des rayons X. On constate que l'organisation des cœurs peut subir de grands changements à l'intérieur de la colonne sans que l'ordre des chaînes paraisse changer. La distance moyenne séparant deux cœurs varie dans de grandes proportions lorsqu'on passe d'une molécule à l'autre ; le diamètre de la colonne varie alors de façon à maintenir une densité constante et par conséquent les chaînes sont plus ou moins ramassées autour du cœur. Enfin, le nombre de chaînes fixées autour d'un seul noyau semble influer sur le type d'organisation observé, puisque toutes les mésogènes discotiques ont six chaînes paraffiniques de longueurs sensiblement équivalentes. Lorsqu'on réduit le nombre de chaînes à quatre, on obtient des phases lamellaires.

Le rapprochement avec les phases lyotropes est aisé : la surface par tête polaire, qui influe beaucoup sur la nature et les paramètres des mésophases lyotropes, trouve son équivalent dans la surface externe du cœur, produit de son épaisseur par le périmètre. Remarquons toutefois que le périmètre n'est pas aisé à définir, car la frontière entre cœur et chaîne n'est pas évidente. Par exemple, dans le cas de l'hexabenzoate de triphénylène, la partie non déformable est restreinte au noyau triphénylène alors que les paraffines sont fixées sur les noyaux phényles externes. En fait, il faut se garder de pousser trop loin l'analogie avec les savons. Par exemple, nous obtenons une grande variété de réseaux périodiques pour l'arrangement des colonnes ; les transitions entre ces réseaux correspondent souvent à des changements très faibles dans les distances entre colonnes. Un autre trait caractéristique de ces phases se trouve dans la possibilité d'avoir des molécules inclinées dans les colonnes et avec des directions d'inclinaisons qui peuvent être différentes dans les deux colonnes d'une même maille.

La phase nématique est à première vue, semblable à la phase nématique des mésogènes calamitiques, mais nous avons vu que des fragments de colonnes subsistent si bien qu'on peut se demander si cette phase ressemble plus à un nématique lyotrope calamitique que discotique. Peut-être rencontrerons-nous des nématiques plus désordonnés, mais il faut souligner que la phase isotrope elle-même semble peu désordonnée.

Je pense que c'est le caractère plat très marqué de ces molécules qui confère aux phases mésomorphes leurs propriétés différentes. En effet, le volume important de la partie paraffinique et la répartition de ces chaînes tout autour du cœur leur confère une grande liberté de conformation, d'où, par exemple, des possibilités de variation importante de la distance entre cœur à l'intérieur d'une colonne, mais peut être aussi une cause des anomalies apparentes dans l'état d'ordre des phases : séquences inversées, apparition d'ordre dans les colonnes à haute température. C'est peut-être la grande anisotropie de ces molécules qui stabilise les fragments de colonnes dans la phase nématique. En effet, le nématique lyotrope discotique est constitué d'objets beaucoup moins plats.

Pour terminer, je rappellerai les informations actuelles sur la structure sont très incomplètes. Par exemple, l'étude de l'intensité des raies de diffraction du réseau bidimensionnel des phases colonnes n'a pas été abordée et devrait renseigner sur la forme des colonnes, en particulier dans les phases rectangles. Si la diffraction des rayons X apporte plus d'information sur la structure des phases en colonnes, à cause de leur caractère de cristal bidimensionnel que sur celle des smectiques, il reste indispensable cependant de confronter nos observations avec d'autres obtenues par des méthodes d'investigation différentes, telles que l'observation sous microscope polarisant. En effet, généralement, nous obtenons des renseignements très incomplets sur l'organisation dans la colonne.

Remerciements

L'ensemble des résultats exposés ici n'a pu être rassemblé que grâce à une collaboration de toute la communauté scientifique française travaillant dans le domaine des phases partiellement ordonnées. Je tiens donc à remercier en premier lieu mes collègues du laboratoire de Physique des Solides de l'université

Paris-Sud pour les nombreuses discussions que nous avons eues ensemble et plus particulièrement les spécialistes de l'étude des défauts et textures de ces mésophases : M. Kleman et P. Oswald ainsi que Y. Bouligand du Centre de Cytologie expérimentale du CNRS à Ivry-sur-Seine, car les moments passés à rapprocher nos visions complémentaires sur la symétrie de ces systèmes m'ont permis d'approfondir la compréhension de leur structure. Il faut aussi souligner que l'effort de synthèse réalisé dans ce domaine en France est à l'origine des études effectuées sur les mésophases discotiques. Que tous les collègues des laboratoires cités ci-dessous soient remerciés pour leur gentillesse à me fournir les échantillons nécessaires aux expériences de diffraction X ainsi que pour les discussions que j'ai pu avoir avec eux :

- Centre de Recherche Paul Pascal (domaine universitaire de Talence),
- Laboratoire de Chimie des Interactions Moléculaires et Laboratoire de Physique de la Matière Condensée du Collège de France (Paris),
- Laboratoire Central de Recherche Thomson-CSF, domaine de Corbeville (Orsay),
- Groupe de Recherches n° 12 du CNRS (Thiais),
- Groupe de Chimie Organique Physique, département de recherches fondamentales du Centre d'Etudes Nucléaires de Grenoble.

Références

(1) S. Chandrasekhar, B.K. Sadashiva, K.A. Suresh. — *Pramana*, 1977, 9, 471-480.
(2) P.G. De Gennes. — *The Physics of Liquid Crystal* (Clarendon Oxford), 1974, chapitre 1.
(3) On trouvera une description succincte des structures des phases smectiques tridimensionnelles dans l'article de Benattar J.J., Moussa F. et Lambert M., dans ce même numéro du *Journal de Chimie Physique*.
(4) C. Brot, I. Darmon. — *J. Chem. Phys.*, 1970, 53, 2271-2280.
(5) S. Meiboom, M. Sammon. — *Phys. Rev. Lett.*, 1980, 44, 882-885.
D.L. Johnson, J.H. Flack, P.P. Crooker. — *Phys. Rev. Lett.*, 1980, 45, 641-644.
(6) D. Demus, G. Kunike, J. Neelsen, H. Sackmann. — *Z. Naturforsch.*, 1968, 23a, 84-90.
A. Tardieu, J. Billard. — *Journal de Physique*, Colloque C3, supplément au n° 6, 1976, 37, C379-C381.
(7) Nguyen Huu Tinh, C. Destrade, H. Gasparoux. — *Phys. Lett.*, 1979, 72A, 251-254.
(8) J. Billard. — *Liquid Crystal of one and two-Dimensional Order*, Ed. W. Helfrich et G. Heppke (Springer-Verlag, Berlin-Heidelberg-New York), 1980, p. 383-394.
(9) H. Gasparoux. — *Liquid Crystals of one and two-Dimensional Order*, Ed. W. Helfrich et G. Heppke (Springer-Verlag Berlin-Heidelberg-New York), 1980, p. 373-382.
(10) C. Destrade, Nguyen Huu Tinh, J. Malthete, J. Jacques. — *Phys. Lett.*, 1980, 79A, 189-192.
(11) A.E. Skoulios, Luzzati. — *Acta Cryst.*, 1961, 14, 278-286.
A. Spegt et A.E. Skoulios. — *Acta Cryst.*, 1962, 16, 301-306.
(12) G. Sigaud, F. Hardouin, M.F. Achard. — *Journal de Physique*, Colloque C3 Supplément au n° 4, 1979, 40, C3356-C3-362.
(13) *International Tables for X-Ray Crystallography*, vol. I, N.F.M. Henry, K. Lonsdale, Ed., Kynoch Press Birmingham, G.B., 1969.
(14) A.M. Levelut. — *Journal de Physique Lettres*, 1979, 40, L81-L84.
(15) P. Oswald. — *C.R. Acad. Sc. Paris*, 1981, 292, 149-152.
(16) J. Billard, J.C. Dubois, C. Vaucher, A.M. Levelut. — *Mol. Cryst. Liq. Cryst.*, 1981, 66, 115-122.
(17) A.M. Levelut. — *Liquid Crystals* (Proceeding of an International Conference held at Bangalore), S. Chandrasekhar, Ed., Heyden, Londres, 1980, 21-27.
(18) C. Destrade, Nguyen Huu Tinh, J. Malthete, A.M. Levelut. — Soumis au *Journal de Physique*.
(19) L. Mamlok, J. Malthete, Nguyen Huu Tinh, C. Destrade, A.M. Levelut. — *Journal de Physique Lettres*, 1982, 43, L641-L647.
(20) A.M. Levelut, F. Hardouin, H. Gasparoux, C. Destrade, Nguyen Huu Tinh. — *Journal de Physique*, 1981, 42, 147-152.
(21) F.C. Franck, S. Chandrasekhar. — *Journal de Physique*, 1980, 41, 1285-1288.
(22) M. Cotrait, P. Marsau, M. Pesquer, V. Volpilhac. — *Journal de Physique*, 1982, 43, 355-359.
(23) D. Goldfarb, Z. Luz, H. Zimmermann. — *Journal de Physique*, 1981, 42, 1303-1311.
(24) R. Alben. — *J. Chem. Phys.*, 1972, 57, 3055-3061.
(25) K.D. Lawson, T.J. Flautt. — *J. Am. Chem. Soc.*, 1967, 89, 54-89.
(26) P.E. Cladis. — *Phys. Rev. Lett.*, 1975, 35, 48-51.
(27) F. Hardouin, G. Sigaud, M.F. Achard, H. Gasparoux. — *Physics Letters*, 1979, 71A, 347-349.
(28) J. Charvolin, E. Samulski, A.M. Levelut. — *Journal de Physique Lettres*, 1979, 40, L587.
(29) A. De Vries. — *Mol. Cryst. Liq. Cryst.*, 1970, 10, 31-37.
(30) Nguyen Huu Tinh, J. Malthete, C. Destrade. — *J. Phys. Lettres*, 1981, 42, L417-L419.
(31) J.F. Palierne et M. Cagnon. — Communication privée.
(32) Nguyen Huu Tinh, M.C. Bernaud, G. Sigaud, C. Destrade. — *Mol. Cryst. Liq. Cryst.*, 1981, 61, 30-316.
(33) J.D. Bunning, J.W. Goodby, G.W. Gray, J.E. Lydon. — *Liquid Crystals on one and two Dimensional Orders*, ed. W. Helfrich, G. Heppke, (Springer Verlag, Berlin-Heidelberg-New York), 1980, 397-402.
(34) J. Billard, J.C. Dubois, Nguyen Huu Tinh, A. Zann. — *Nouv. J. de Chimie*, 1978, 2, 535-540.
(35) J.D. Bunning, J.E. Lydon, C. Eaborn, P.M. Jackson, J.W. Goodby, G.W. Gray. — *J. Chem. Soc. Faraday*, 1982, 72, 713-724.
(36) R. Fugnitto, Strzelecka, A. Zann, J.C. Dubois, J. Billard. — *J.C.S. Chem. Com.*, 1980, 271-272.
(37) D. Chasseau, J. Gaultier, G. Hauw, R. Fugnitto, V. Gianis, H. Strzelecka. — *Acta Cryst.*, 1982, B38, 1629-1631.
(38) A.M. Giroud-Godquin, J. Billard. — *Mol. Cryst. Liq. Cryst.*, 1981, 66, 147-150.
(39) J.W. Goodby, P.S. Robinson, Boom Ten Tao, P.E. Cladis. — *Mol. Cryst. Liq. Cryst. Lett.*, 1980, 56, 303-309.
(40) W.H. de Jeu. — Discussion of paper n° 2. Meeting on Physics Chemistry and Applications of Thermotropic Liquid Crystals. *Phil. Trans. Roy. Soc.*, in press.

VOLUME 53, NUMBER 12   PHYSICAL REVIEW LETTERS   17 SEPTEMBER 1984

Synchrotron X-Ray Study of the Orientational Ordering $D2$-$D1$ Structural Phase Transition of Freely Suspended Discotic Strands in Triphenylene Hexa-n-dodecanoate

C. R. Safinya, K. S. Liang, and William A. Varady

Corporate Research Science Laboratories, Exxon Research & Engineering Company, Annandale, New Jersey 08801

and

Noel A. Clark and G. Andersson

Department of Physics, University of Colorado, Boulder, Colorado 80309
(Received 21 February 1984)

We demonstrate the feasibility of x-ray scattering studies of freely suspended strands of discotic liquid crystals (of triphenylene hexa-n-dodecanoate). We are able to grow strands, which are stable for days, of diameter > 50 μm with a few single-crystal domains. Unexpectedly, we find that the quasi two-dimensional structural phase transition from columnar $D2$ to columnar $D1$ corresponds to the orientational ordering of columns of molecules with the molecules tilted at finite angles to the column axis in both phases.

PACS numbers: 64.70.Ew, 61.30.−v

Recently, disk-shaped molecules have been synthesized that exhibit a new class of columnar liquid-crystalline phases[1,2] in which the molecules segregate into infinite columns close packed in various two-dimensional (2D) lattices. To elucidate their structure and phase transitions, we have initiated synchrotron x-ray scattering studies of freely suspended strands[3] of such columnar phases. Our x-ray studies demonstrate the single-crystal quality of freely suspended strands, which provide the only viable technique for preparation of columnar phase samples suitable for high-resolution x-ray studies. Only high-resolution x-ray studies of such oriented samples allow one to disentangle and understand the intracolumnar 1D interactions and the intercolumnar 2D structure in the columnar phases and at the transitions. We expect these studies will enhance our understanding of collective behavior in lower-dimensional materials, as has been demonstrated for analogously prepared films of rod-shaped molecules.[4,5]

X-ray scattering studies were made on strands of triphenylene hexa-n-dodecanoate (HAT11). Previous x-ray work on HAT11 shows that it exhibits two columnar phases $D2$ and $D1$,[2] but has left unclear the nature of the $D2$-$D1$ structural phase transition from a (2D) hexagonal to a (2D) centered rectangular lattice of columns of molecules. Qualitative optical observations[3] show that the molecules are tilted with respect to the column axis in the $D1$ phase. Our data lead us to propose a model with the molecular tilt remaining finite across the transi-

tion into the $D2$ phase. We find that the structural phase transition is associated with the intercolumn ordering of the molecular tilt orientation about the column axis.

The experiments were carried out at the Stanford Synchrotron Radiation Laboratory (SSRL) with use of the x-ray diffraction spectrometer stationed on beam line VII-2.[6] With use of horizontal and vertical slits, a pair of asymmetrically cut Ge(111) crystals to monochromatize the incident beam (~ 7.1 keV), and LiF(200) crystal to analyze the diffracted beam, we obtain a longitudinal resolution of 1.1×10^{-3} Å^{-1} half-width at half maximum (HWHM). Lower-resolution experiments were carried out on thick strands at Exxon with resolution 5.0×10^{-3} Å^{-1} (Fig. 3) and 1.8×10^{-2} Å^{-1} (Fig. 4) HWHM. In these experiments, we studied ~ 1.5-mm-long strands with diameter between 50 and 200 μm pulled in an oven with two-stage temperature control to ± 0.01 °C.

We find that in the $D2$ (118 °C $> T > 105$ °C) and $D1$ (105 °C $> T > 70$ °C) temperature ranges HAT11 exhibits structures with Bragg peaks *only* in the $(h, k, 0)$ plane and diffuse sheets at $(h, k, \pm 2\pi/d)$ with $d \sim$ molecular spacing within columns, consistent with the previous identification of the structures as two-dimensional lattices of liquid columns [inset, Fig. 1(b)] with no positional correlations between molecules from different columns along the $\hat{z}$ (strand-axis) direction.[2] Across the $D2$-$D1$ transition the system distorts from a hexagonal into a centered rectangular structure with herringbone

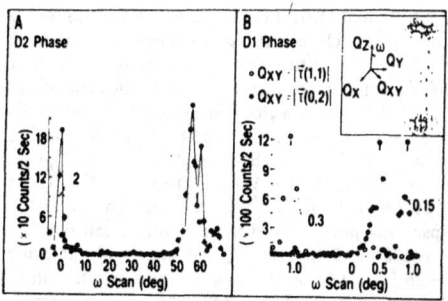

FIG. 1. (a) Wide-range ω scan (rocking curve) through hexagonal (1,0) peaks in the $D2$ phase. (b) Finer ω scan after annealing in the $D1$ phase indicates the single-crystal quality (mosaic $\sim 0.2°$) of discotic strands and as discussed in the text also provides evidence of orientational jumps in strands shown schematically in the inset with the scattering geometry.

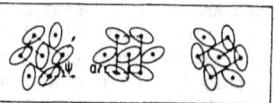

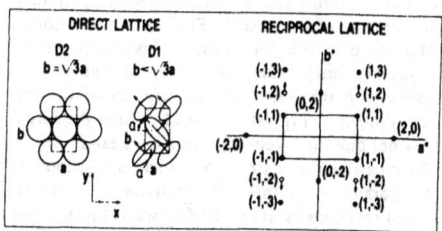

FIG. 2. Top panel: Three orientationally inequivalent (described by ψ) states (each state also doubly degenerate) of herringbone order on a triangular lattice. Bottom panel: Direct and reciprocal lattice of the $D2$ and $D1$ phases indexed on a (2D) rectangular unit cell. The arrows on the ellipses of the $D1$ phase indicate the direction of tilt. The herringbone structure of the $D1$ phase is signaled by the appearance of four new peaks (open diamonds).

symmetry. We show in Fig. 2 the real and reciprocal lattices of the two phases indexed on a two-dimensional rectangular unit cell in the x-y plane defined as $\vec{\tau}(h,k) = ha^*\hat{x} + kb^*\hat{y}$.

We now discuss the structural quality of the freely suspended strands. To determine the degree of the lattice mosaicity, we execute ω (rocking curve) scans in the $D2$ and $D1$ phases with $\vec{Q}$ in the x-y plane. In these scans the amplitude of the momentum transfer is fixed at the peak value of a Q_{xy} reflection [Fig. 1(b), inset]. We show in Fig. 1(a) an ω scan with $Q_{xy} = |\vec{\tau}(0,2)|$ [equivalent to a hexagonal (1,0) peak] over a 70° range in the $D2$ phase. The scan shows two narrow bunches of peaks of 1° to 2° mosaic width spaced by 60° and is typical of what is obtained immediately after pulling a strand. The mosaic structure of a particular bunch evolves continuously in time with significant changes occurring in minutes.[7] Hence, the detailed structure of the two bunches in Fig. 1(a) is not precisely identical because of the structural evolution of the bunches during the scan time of four min. Annealing several hours improves the mosaic quality significantly, a given bunch evolving to a few closely spaced peaks with peaks of width between 0.15° and 0.3° [Fig. 1(b); bunch about $\omega = 0$]. To measure $|\vec{\tau}(1,1)|$ and $|\vec{\tau}(0,2)|$ accurately, at each momentum transfer $\vec{Q}_{xy}$, we rotationally average, with the intensity integrated over ω, segments which contain evolving peaks. In the $D2$ phase at a given temperature, all such rotationally averaged longitudinal scans result in a peak at the same Q_{xy} indicating that $|\vec{\tau}(1,1)| = |\vec{\tau}(0,2)|$; similarly, $|\vec{\tau}(1,3)| = |\vec{\tau}(2,$

0)| and thus the lattice is hexagonal. In the $D1$ phase the system distorts to a centered rectangular lattice with herringbone symmetry, as evidenced by the appearance of the $\vec{\tau}(1,2)$ peaks and a lattice distortion so that $|\vec{\tau}(1,1)| \neq |\vec{\tau}(0,2)|$. Figure 1(b) shows ω scans of the annealed strand in the $D1$ phase in which the mosaic has evolved to two sets of peaks at $|\vec{\tau}(1,1)| = 0.2749$ Å^{-1} and $\vec{\tau}(0,2)| = 0.2783$ Å^{-1}.

The peaks with different wave-vector magnitude Q_{xy} occur closely spaced in ω because the centered rectangular structure of the $D1$ phase can condense on the triangular column lattice in three orientationally distinct directions (Fig. 2). The x-ray beam illuminates sections of the strand with different lattice orientation. Two such distinct regions separated by a soliton-type defect wall[3] are shown schematically in the inset to Fig. 1(b). To further test our observations we have used slits to illuminate ≤ 50 μm sections of the strand to obtain a region with only one lattice orientation. However, the ω scans, while sharp, are usually not single peaked and the in-plane structure contains a few single-column lattice domains separated by a few degrees.

We now discuss the $D2$-$D1$ structural phase transition. For clarity, we describe the hexagonal $D2$ phase by a (2D) centered rectangular structure with $b = \sqrt{3}a$ shown in Fig. 2. The $D2$-$D1$ transition at 105°C, which is first order, is characterized by a lattice distortion and a simultaneous appearance of four twofold symmetric peaks with $Q_{xy} = |\bar{\tau}(1,2)|$ (Fig. 2). The new resolution-limited peaks in the $D1$ phase signify the onset of long-range inter-column molecular tilt orientational correlations about the column axis with herringbone symmetry. The intensity of the peak at $\bar{\tau}(1,2)$ as a function of temperature in the $D1$ phase yields the square of the orientational order parameter. However, because of the time-dependent dynamics of the different domains in the strand, we find that the intensity is not quantitatively reproducible near T_C.

To measure the distortion, which is a secondary order parameter of this $D2$-$D1$ transition, we carried out rotationally averaged $\bar{Q}_{xy}$ scans through the inplane peaks at $\bar{\tau}(1,1)$, $\bar{\tau}(0,2)$, and $\bar{\tau}(1,3)$ in both phases for each temperature. We plot in Fig. 3 $\Delta b/b \equiv (b_2 - b_1)/b_1$ and $\Delta a/a \equiv (a_2 - a_1)/a_1$ as a function of temperature. We find that the lattice contraction is primarily along the b axis with $\Delta b/b \simeq 0.015$, and $(\Delta a/a)/(\Delta b/b) \sim 0.04 \pm 0.02$ across the transition in the $D1$ phase. We show in the lower panel of Fig. 3 the angle between the optic axis tilt and the strand axis as a function of temperature measured by optical microscopy techniques.[3] The optic axis tilt angle is 28° below the $D1$-$D2$ transition and drops abruptly to zero at the transition.

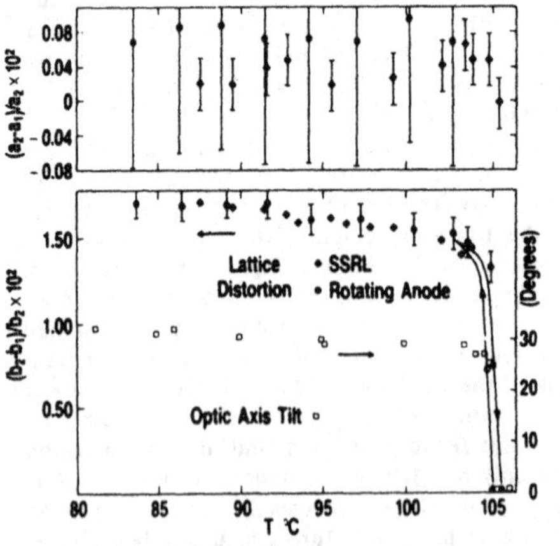

The small distortion $\Delta b/b \sim 0.015$ rules out a herringbone close packing of ellipses resulting from a tilt of the entire molecule; this would yield a significantly larger distortion of the hexagonal lattice.[7] With use of a close-packed hard-ellipse model one has $\Delta b/b \sim 1 - [r(60° + \alpha) + r(60° - \alpha)]/D \simeq 0.092$ with herringbone angle $\alpha \sim 7° \pm 4°$ given by $r(\alpha) = 0.5D(1 - \Delta a/a)$. Here, $r(\theta) = 0.5D/[\cos^2\theta + (1/e)^2\sin^2\theta]^{1/2}$ is the equation for an ellipse in polar coordinates with eccentricity $e = \cos 28° = 0.883$, and $D =$ diameter of molecule ~ 26.5 Å. The optical data measure only the tilt of the central core of the molecules with anisotropic polarizability. The flexible aliphatic tails need not be tilted. The rigid core has a diameter of about 7 Å; hence, a tilt of 28° of only the core of the molecules [inset to Fig. 4(b)] gives a larger eccentricity $e = 1 + (7/26.5)(\cos 28° - 1) = 0.969$ and consequently a smaller distortion $\Delta b/b \sim 0.022$. A close-packed model is clearly not ideal for discotics; thus, the reasonable agreement (0.022 compared to 0.015) with the experimental data is a persuasive argument that only the core is tilted.

We next consider whether the molecular tilt angle is an order parameter of this phase. We show in Fig. 4(a) $(0,0,Q_z)$ scans in both the $D1$ and $D2$ phases along the strand axis through the diffuse sheet which probes intermolecular correlations in columns. The scattering is clearly not single peaked. The solid line through the data is the result of least-squares fits of a sum of two Lorentzians $\sum_{i=1,2} A\xi^2/[1 + \xi^2(Q_z - Q_{iz})^2]$ centered at $Q_{1z} = 1.35$ Å^{-1} and $Q_{2z} = 1.68$ Å^{-1} [indicated by ar-

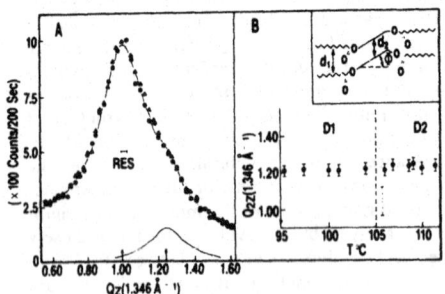

FIG. 3. Top panel: Lattice distortion along the a axis in the $D1$ phase ($a_2 = 26.512$ Å). Bottom panel: Lattice distortion along the b axis and optic axis tilt angle (open square) in the $D1$ phase. The lattice constants were measured to an accuracy of about ± 0.008 Å at SSRL and ± 0.035 Å at Exxon.

FIG. 4. (a) $(0,0,Q_z)$ scans along the strand axis through the double-profile liquid peak (due to the spacings d_1 and d_2 shown in inset) in the $D1$ (closed circle) and $D2$ (open triangle) phases. (b) The high-Q_z peak position vs temperature shows no shift in peak position, which indicates that the molecular tilt remains finite in the $D2$ phase.

rows in Fig. 4(a)] to the data. In these fits, the background is fixed at a value determined from low $(Q_z < Q_{1z})$ and high $(Q_z > Q_{1z})$ scattering points away from the diffuse sheet. As pointed out by Levelut[2] the first peak is due to the liquid structure of the aliphatic tails, while the second peak at $Q_{2z} = 2\pi/d_2 = 1.68$ Å^{-1} (separately shown as solid line), which does not appear in small-core discotics,[7] arises from short-range correlations ($\xi \sim 3$ to 4 molecules) of the large aromatic core with spacing d_2 [inset to Fig. 4(b)]. The tilt behavior across the transition is related to the change of the Q_{2z} peak position. In the simplest model, where the molecular tilt angle $\Phi(\sim 28°)$ drops to zero across the transition, one would expect a significant shift in the peak position $dQ_{2z} \sim Q_{2z}/\cos(28°) - Q_{2z} = 0.13Q_z = 0.18$ Å^{-1} [shown as solid bar in Fig. 4(b)] comparable to the width $\Delta Q_z = 0.20$ Å^{-1} of the scattering. The $(0, 0, Q_z)$ scan in the $D2$ phase is shown as open triangles in Fig. 4. We also show the Q_{2z} peak position versus temperature in Fig. 4(b), which indicates no statistically significant shift in the peak position across the transition. This suggests that while the tilt may change by a few degrees across the transition, it remains finite in the $D2$ phase and consequently rules out the tilt angle as an order parameter of the $D1$ phase.

We thus propose that the $D2-D1$ structural phase change is characterized primarily by the onset of the intercolumn ordering of the molecular tilt orientation about the column axis. In the $D1$ phase, the columns of tilted molecules are close packed with herringbone order. However, there are six ways in which this order can be imposed on a triangular lattice (Fig. 2). In the tilted $D2$ phase, random fluctuations between these states results in an orientationally disordered hexagonal structure with an optic axis along the strand axis.[3]

This research was supported in part by a joint Industry/University National Science Foundation Grant No. DMR-8307157, and in part by the Department of Energy under Contract No. DE-AC02-82RR-1300. We acknowledge the assistance of the staff of the SSRL especially T. Porter. We thank D. E. Moncton, A. M. Levelut, G. Aeppli, and S. Sinha for useful discussions of the experiments, L. Wenzel for technical assistance, and C. Destrade for providing the HAT11 sample.

[1]S. Chandrasekhar, B. K. Sadashiva, and K. A. Suresh, Pramana 9, 471 (1977).

[2]A. M. Levelut, in *Proceedings of the International Liquid Crystal Conference, Bangalore, 1979* (Wiley, New York, 1980); M. Takabatake, J. Appl. Phys. 21, 685 (1982).

[3]David H. Van Winkle and Noel A. Clark, Phys. Rev. Lett. 48, 1407 (1982).

[4]D. E. Moncton and R. Pindak, Phys. Rev. Lett. 43, 701 (1979); R. Pindak, D. E. Moncton, J. W. Goodby, and S. C. Davey, Phys. Rev. Lett. 46, 1135 (1981).

[5]J. Collett, L. B. Sorensen, P. S. Pershan, J. D. Litster, R. J. Birgeneau, and J. Als-Nielsen, Phys. Rev. Lett. 49, 553 (1982).

[6]D. E. Moncton and G. E. Brown, Nucl. Instrum. & Methods 208, 579 (1983).

[7]C. R. Safinya, N. A. Clark, K. S. Liang, W. A. Varady, and G. Andersson, to be published.

Acknowledgement

The publisher would like to thank all authors and the following publishers for permission to reproduce the reprinted papers found in this volume:

Académie des Sciences (*Comptes Rendus de l'Académie des Sciences de Paris*); American Physical Society (*Phys. Rev. A, Phys. Rev. B, Phys. Rev. Lett.*); Gordon and Breach Science Publishers S.A. (*Mol. Cryst. Liq. Cryst.*); Indian Academy of Sciences (*Pranama*); Les Editions de Physique (*J. de Phys., J. de Phys. Lett., J. de Phys. Colloq.*); Pergamon Press (*Solid State Commun.*); Société de Chimie Physique (*J. de Chim. Phys.*); Taylor and Francis Ltd. (*Adv. in Physics, Liquid Crystals*).

To those who have not granted us permission before publication, we have taken the liberty to reproduce their articles without consent. We shall however acknowledge them in future editions of this work.